# Thin Films: Stresses and Mechanical Properties

MATERIALS RESEARCH SOCIETY SYMPOSIUM PROCEEDINGS VOLUME 130

# Thin Films: Stresses and Mechanical Properties

Symposium held November 28-30, 1988, Boston, Massachusetts, U.S.A.

EDITORS:

John C. Bravman
Stanford University, Stanford, California, U.S.A.

William D. Nix
Stanford University, Stanford, California, U.S.A.

David M. Barnett
Stanford University, Stanford, California, U.S.A.

David A. Smith
IBM Research Center, Yorktown Heights, New York, U.S.A.

MRS MATERIALS RESEARCH SOCIETY
Pittsburgh, Pennsylvania

This work was supported by the Air Force Office of Scientific Research, Air Force Systems Command, USAF, under Grant Number AFOSR 89-0142.

CODEN: MRSPDH



Published by:

Materials Research Society
9800 McKnight Road, Suite 327
Pittsburgh, Pennsylvania 15237
Telephone (412) 367-3003

Library of Congress Cataloging in Publication Data

Thin films : stresses and mechanical properties / editors, John C. Bravman, William D. Nix, David M. Barnett and David A. Smith.

p. cm. — (Materials Research Society symposium proceedings : ISSN 0272-9172 ; v. 130)
Bibliography: p.
ISBN 1-55899-003-8
1. Thin films—Mechanical properties. I. Bravman, J.C. (John C.) II. Series: Materials Research Society symposium proceedings ; v. 130.

TA418.9.T45T47 1989 89-3119
620'.44—dc19 CIP

Manufactured in the United States of America

# Contents



*Invited Paper

## PART II: HETEROEPITAXY



*Invited Paper

## PART III: FRACTURE AND RELATED FAILURE MECHANISMS

## PART IV: MECHANICAL PROPERTIES OF THIN FILMS



*Invited Paper

*Invited Paper

# Preface

This volume contains the papers presented at a symposium entitled Thin Films: Stresses and Mechanical Properties, held at the 1988 Fall Meeting of the Materials Research Society. The idea for this symposium originated with the growing awareness of the importance of stresses and mechanical properties in thin film technologies and the recent introduction of new experimental techniques for studying stresses and mechanical properties in thin film media. It was organized to provide the first opportunity for materials scientists and engineers interested in these and related topics to discuss recent progress in the field. Both fundamental and applied problems were discussed, and individuals from a wide spectrum of university, industrial and government laboratories participated.

Materials research is predicated on an interdisciplinary approach to problem solving. In that spirit, four major areas of interest were addressed at the symposium: Testing and Measurement Techniques, Heteroepitaxy, Fracture and Related Failure Mechanisms, and Thin Film Mechanical Properties. In each, both overview and material-specific papers were presented.

The editors believe that the positive response to this symposium, particularly from groups that have not previously been closely associated with MRS, indicates a significant interest in the subject of stresses and mechanical properties of thin films. They hope that this volume furthers the interests of many in this important field of materials research, and that it will help to stimulate new ideas in science and technology.

January, 1989

John C. Bravman
William D. Nix
David M. Barnett
David A. Smith

# Acknowledgments

The editors are pleased to express their appreciation to the Air Force Office of Scientific Research for providing financial support for this Symposium. Their interest and generosity have enhanced the exchange of ideas in this rapidly growing field of materials research.

January, 1989

John C. Bravman
William D. Nix
David M. Barnett
David A. Smith

ISSN 0272 - 9172

Volume 1—Laser and Electron-Beam Solid Interactions and Materials Processing, J. F. Gibbons, L. D. Hess, T. W. Sigmon, 1981, ISBN 0-444-00595-1

Volume 2—Defects in Semiconductors, J. Narayan, T. Y. Tan, 1981, ISBN 0-444-00596-X

Volume 3—Nuclear and Electron Resonance Spectroscopies Applied to Materials Science, E. N. Kaufmann, G. K. Shenoy, 1981, ISBN 0-444-00597-8

Volume 4—Laser and Electron-Beam Interactions with Solids, B. R. Appleton, G. K. Celler, 1982, ISBN 0-444-00693-1

Volume 5—Grain Boundaries in Semiconductors, H. J. Leamy, G. E. Pike, C. H. Seager, 1982, ISBN 0-444-00697-4

Volume 6—Scientific Basis for Nuclear Waste Management IV, S. V. Topp, 1982, ISBN 0-444-00699-0

Volume 7—Metastable Materials Formation by Ion Implantation, S. T. Picraux, W. J. Choyke, 1982, ISBN 0-444-00692-3

Volume 8—Rapidly Solidified Amorphous and Crystalline Alloys, B. H. Kear, B. C. Giessen, M. Cohen, 1982, ISBN 0-444-00698-2

Volume 9—Materials Processing in the Reduced Gravity Environment of Space, G. E. Rindone, 1982, ISBN 0-444-00691-5

Volume 10—Thin Films and Interfaces, P. S. Ho, K.-N. Tu, 1982, ISBN 0-444-00774-1

Volume 11—Scientific Basis for Nuclear Waste Management V, W. Lutze, 1982, ISBN 0-444-00725-3

Volume 12—In Situ Composites IV, F. D. Lemkey, H. E. Cline, M. McLean, 1982, ISBN 0-444-00726-1

Volume 13—Laser-Solid Interactions and Transient Thermal Processing of Materials, J. Narayan, W. L. Brown, R. A. Lemons, 1983, ISBN 0-444-00788-1

Volume 14—Defects in Semiconductors II, S. Mahajan, J. W. Corbett, 1983, ISBN 0-444-00812-8

Volume 15—Scientific Basis for Nuclear Waste Management VI, D. G. Brookins, 1983, ISBN 0-444-00780-6

Volume 16—Nuclear Radiation Detector Materials, E. E. Haller, H. W. Kraner, W. A. Higinbotham, 1983, ISBN 0-444-00787-3

Volume 17—Laser Diagnostics and Photochemical Processing for Semiconductor Devices, R. M. Osgood, S. R. J. Brueck, H. R. Schlossberg, 1983, ISBN 0-444-00782-2

Volume 18—Interfaces and Contacts, R. Ludeke, K. Rose, 1983, ISBN 0-444-00820-9

Volume 19—Alloy Phase Diagrams, L. H. Bennett, T. B. Massalski, B. C. Giessen, 1983, ISBN 0-444-00809-8

Volume 20—Intercalated Graphite, M. S. Dresselhaus, G. Dresselhaus, J. E. Fischer, M. J. Moran, 1983, ISBN 0-444-00781-4

Volume 21—Phase Transformations in Solids, T. Tsakalakos, 1984, ISBN 0-444-00901-9

Volume 22—High Pressure in Science and Technology, C. Homan, R. K. MacCrone, E. Whalley, 1984, ISBN 0-444-00932-9 (3 part set)

Volume 23—Energy Beam-Solid Interactions and Transient Thermal Processing, J. C. C. Fan, N. M. Johnson, 1984, ISBN 0-444-00903-5

Volume 24—Defect Properties and Processing of High-Technology Nonmetallic Materials, J. H. Crawford, Jr., Y. Chen, W. A. Sibley, 1984, ISBN 0-444-00904-3

Volume 25—Thin Films and Interfaces II, J. E. E. Baglin, D. R. Campbell, W. K. Chu, 1984, ISBN 0-444-00905-1

Volume 26—Scientific Basis for Nuclear Waste Management VII, G. L. McVay, 1984, ISBN 0-444-00906-X

Volume 27—Ion Implantation and Ion Beam Processing of Materials, G. K. Hubler, O. W. Holland, C. R. Clayton, C. W. White, 1984, ISBN 0-444-00869-1

Volume 28—Rapidly Solidified Metastable Materials, B. H. Kear, B. C. Giessen, 1984, ISBN 0-444-00935-3

Volume 29—Laser-Controlled Chemical Processing of Surfaces, A. W. Johnson, D. J. Ehrlich, H. R. Schlossberg, 1984, ISBN 0-444-00894-2

Volume 30—Plasma Processing and Synthesis of Materials, J. Szekely, D. Apelian, 1984, ISBN 0-444-00895-0

Volume 31—Electron Microscopy of Materials, W. Krakow, D. A. Smith, L. W. Hobbs, 1984, ISBN 0-444-00898-7

Volume 32—Better Ceramics Through Chemistry, C. J. Brinker, D. E. Clark, D. R. Ulrich, 1984, ISBN 0-444-00898-5

Volume 33—Comparison of Thin Film Transistor and SOI Technologies, H. W. Lam, M. J. Thompson, 1984, ISBN 0-444-00899-3

Volume 34—Physical Metallurgy of Cast Iron, H. Fredriksson, M. Hillerts, 1985, ISBN 0-444-00938-8

Volume 35—Energy Beam-Solid Interactions and Transient Thermal Processing/1984, D. K. Biegelsen, G. A. Rozgonyi, C. V. Shank, 1985, ISBN 0-931837-00-6

Volume 36—Impurity Diffusion and Gettering in Silicon, R. B. Fair, C. W. Pearce, J. Washburn, 1985, ISBN 0-931837-01-4

Volume 37—Layered Structures, Epitaxy, and Interfaces, J. M. Gibson, L. R. Dawson, 1985, ISBN 0-931837-02-2

Volume 38—Plasma Synthesis and Etching of Electronic Materials, R. P. H. Chang, B. Abeles, 1985, ISBN 0-931837-03-0

Volume 39—High-Temperature Ordered Intermetallic Alloys, C. C. Koch, C. T. Liu, N. S. Stoloff, 1985, ISBN 0-931837-04-9

Volume 40—Electronic Packaging Materials Science, E. A. Giess, K.-N. Tu, D. R. Uhlmann, 1985, ISBN 0-931837-05-7

Volume 41—Advanced Photon and Particle Techniques for the Characterization of Defects in Solids, J. B. Roberto, R. W. Carpenter, M. C. Wittels, 1985, ISBN 0-931837-06-5

Volume 42—Very High Strength Cement-Based Materials, J. F. Young, 1985, ISBN 0-931837-07-3

Volume 43—Fly Ash and Coal Conversion By-Products: Characterization, Utilization, and Disposal I, G. J. McCarthy, R. J. Lauf, 1985, ISBN 0-931837-08-1

Volume 44—Scientific Basis for Nuclear Waste Management VIII, C. M. Jantzen, J. A. Stone, R. C. Ewing, 1985, ISBN 0-931837-09-X

Volume 45—Ion Beam Processes in Advanced Electronic Materials and Device Technology, B. R. Appleton, F. H. Eisen, T. W. Sigmon, 1985, ISBN 0-931837-10-3

Volume 46—Microscopic Identification of Electronic Defects in Semiconductors, N. M. Johnson, S. G. Bishop, G. D. Watkins, 1985, ISBN 0-931837-11-1

Volume 47—Thin Films: The Relationship of Structure to Properties, C. R. Aita, K. S. SreeHarsha, 1985, ISBN 0-931837-12-X

Volume 48—Applied Materials Characterization, W. Katz, P. Williams, 1985, ISBN 0-931837-13-8

Volume 49—Materials Issues in Applications of Amorphous Silicon Technology, D. Adler, A. Madan, M. J. Thompson, 1985, ISBN 0-931837-14-6

Volume 50—Scientific Basis for Nuclear Waste Management IX, L. O. Werme, 1986, ISBN 0-931837-15-4

Volume 51—Beam-Solid Interactions and Phase Transformations, H. Kurz, G. L. Olson, J. M. Poate, 1986, ISBN 0-931837-16-2

Volume 52—Rapid Thermal Processing, T. O. Sedgwick, T. E. Seidel, B.-Y. Tsaur, 1986, ISBN 0-931837-17-0

Volume 53—Semiconductor-on-Insulator and Thin Film Transistor Technology, A. Chiang. M. W. Geis, L. Pfeiffer, 1986, ISBN 0-931837-18-9

Volume 54—Thin Films—Interfaces and Phenomena, R. J. Nemanich, P. S. Ho, S. S. Lau, 1986, ISBN 0-931837-19-7

Volume 55—Biomedical Materials, J. M. Williams, M. F. Nichols, W. Zingg, 1986, ISBN 0-931837-20-0

Volume 56—Layered Structures and Epitaxy, J. M. Gibson, G. C. Osbourn, R. M. Tromp, 1986, ISBN 0-931837-21-9

Volume 57—Phase Transitions in Condensed Systems—Experiments and Theory, G. S. Cargill III, F. Spaepen, K.-N. Tu, 1987, ISBN 0-931837-22-7

Volume 58—Rapidly Solidified Alloys and Their Mechanical and Magnetic Properties, B. C. Giessen, D. E. Polk, A. I. Taub, 1986, ISBN 0-931837-23-5

Volume 59—Oxygen, Carbon, Hydrogen, and Nitrogen in Crystalline Silicon, J. C. Mikkelsen, Jr., S. J. Pearton, J. W. Corbett, S. J. Pennycook, 1986, ISBN 0-931837-24-3

Volume 60—Defect Properties and Processing of High-Technology Nonmetallic Materials, Y. Chen, W. D. Kingery, R. J. Stokes, 1986, ISBN 0-931837-25-1

Volume 61—Defects in Glasses, F. L. Galeener, D. L. Griscom, M. J. Weber, 1986, ISBN 0-931837-26-X

Volume 62—Materials Problem Solving with the Transmission Electron Microscope, L. W. Hobbs, K. H. Westmacott, D. B. Williams, 1986, ISBN 0-931837-27-8

Volume 63—Computer-Based Microscopic Description of the Structure and Properties of Materials, J. Broughton, W. Krakow, S. T. Pantelides, 1986, ISBN 0-931837-28-6

Volume 64—Cement-Based Composites: Strain Rate Effects on Fracture, S. Mindess, S. P. Shah, 1986, ISBN 0-931837-29-4

Volume 65—Fly Ash and Coal Conversion By-Products: Characterization, Utilization and Disposal II, G. J. McCarthy, F. P. Glasser, D. M. Roy, 1986, ISBN 0-931837-30-8

Volume 66—Frontiers in Materials Education, L. W. Hobbs, G. L. Liedl, 1986, ISBN 0-931837-31-6

Volume 67—Heteroepitaxy on Silicon, J. C. C. Fan, J. M. Poate, 1986, ISBN 0-931837-33-2

Volume 68—Plasma Processing, J. W. Coburn, R. A. Gottscho, D. W. Hess, 1986, ISBN 0-931837-34-0

Volume 69—Materials Characterization, N. W. Cheung, M.-A. Nicolet, 1986, ISBN 0-931837-35-9

Volume 70—Materials Issues in Amorphous-Semiconductor Technology, D. Adler, Y. Hamakawa, A. Madan, 1986, ISBN 0-931837-36-7

Volume 71—Materials Issues in Silicon Integrated Circuit Processing, M. Wittmer, J. Stimmell, M. Strathman, 1986, ISBN 0-931837-37-5

Volume 72—Electronic Packaging Materials Science II, K. A. Jackson, R. C. Pohanka, D. R. Uhlmann, D. R. Ulrich, 1986, ISBN 0-931837-38-3

Volume 73—Better Ceramics Through Chemistry II, C. J. Brinker, D. E. Clark, D. R. Ulrich, 1986, ISBN 0-931837-39-1

Volume 74—Beam-Solid Interactions and Transient Processes, M. O. Thompson, S. T. Picraux, J. S. Williams, 1987, ISBN 0-931837-40-5

Volume 75—Photon, Beam and Plasma Stimulated Chemical Processes at Surfaces, V. M. Donnelly, I. P. Herman, M. Hirose, 1987, ISBN 0-931837-41-3

Volume 76—Science and Technology of Microfabrication, R. E. Howard, E. L. Hu, S. Namba, S. Pang, 1987, ISBN 0-931837-42-1

Volume 77—Interfaces, Superlattices, and Thin Films, J. D. Dow, I. K. Schuller, 1987, ISBN 0-931837-56-1

Volume 78—Advances in Structural Ceramics, P. F. Becher, M. V. Swain, S. Sōmiya, 1987, ISBN 0-931837-43-X

Volume 79—Scattering, Deformation and Fracture in Polymers, G. D. Wignall, B. Crist, T. P. Russell, E. L. Thomas, 1987, ISBN 0-931837-44-8

Volume 80—Science and Technology of Rapidly Quenched Alloys, M. Tenhover, W. L. Johnson, L. E. Tanner, 1987, ISBN 0-931837-45-6

Volume 81—High-Temperature Ordered Intermetallic Alloys, II, N. S. Stoloff, C. C. Koch, C. T. Liu, O. Izumi, 1987, ISBN 0-931837-46-4

Volume 82—Characterization of Defects in Materials, R. W. Siegel, J. R. Weertman, R. Sinclair, 1987, ISBN 0-931837-47-2

Volume 83—Physical and Chemical Properties of Thin Metal Overlayers and Alloy Surfaces, D. M. Zehner, D. W. Goodman, 1987, ISBN 0-931837-48-0

Volume 84—Scientific Basis for Nuclear Waste Management X, J. K. Bates, W. B. Seefeldt, 1987, ISBN 0-931837-49-9

Volume 85—Microstructural Development During the Hydration of Cement, L. Struble, P. Brown, 1987, ISBN 0-931837-50-2

Volume 86—Fly Ash and Coal Conversion By-Products Characterization, Utilization and Disposal III, G. J. McCarthy, F. P. Glasser, D. M. Roy, S. Diamond, 1987, ISBN 0-931837-51-0

Volume 87—Materials Processing in the Reduced Gravity Environment of Space, R. H. Doremus, P. C. Nordine, 1987, ISBN 0-931837-52-9

Volume 88—Optical Fiber Materials and Properties, S. R. Nagel, J. W. Fleming, G. Sigel, D. A. Thompson, 1987, ISBN 0-931837-53-7

Volume 89—Diluted Magnetic (Semimagnetic) Semiconductors, R. L. Aggarwal, J. K. Furdyna, S. von Molnar, 1987, ISBN 0-931837-54-5

Volume 90—Materials for Infrared Detectors and Sources, R. F. C. Farrow, J. F. Schetzina, J. T. Cheung, 1987, ISBN 0-931837-55-3

Volume 91—Heteroepitaxy on Silicon II, J. C. C. Fan, J. M. Phillips, B.-Y. Tsaur, 1987, ISBN 0-931837-58-8

Volume 92—Rapid Thermal Processing of Electronic Materials, S. R. Wilson, R. A. Powell, D. E. Davies, 1987, ISBN 0-931837-59-6

Volume 93—Materials Modification and Growth Using Ion Beams, U. Gibson, A. E. White, P. P. Pronko, 1987, ISBN 0-931837-60-X

Volume 94—Initial Stages of Epitaxial Growth, R. Hull, J. M. Gibson, David A. Smith, 1987, ISBN 0-931837-61-8

Volume 95—Amorphous Silicon Semiconductors—Pure and Hydrogenated, A. Madan, M. Thompson, D. Adler, Y. Hamakawa, 1987, ISBN 0-931837-62-6

Volume 96—Permanent Magnet Materials, S. G. Sankar, J. F. Herbst, N. C. Koon, 1987, ISBN 0-931837-63-4

Volume 97—Novel Refractory Semiconductors, D. Emin, T. Aselage, C. Wood, 1987, ISBN 0-931837-64-2

Volume 98—Plasma Processing and Synthesis of Materials, D. Apelian, J. Szekely, 1987, ISBN 0-931837-65-0

Volume 99—High-Temperature Superconductors, M. B. Brodsky, R. C. Dynes, K. Kitazawa, H. L. Tuller, 1988, ISBN 0-931837-67-7

Volume 100—Fundamentals of Beam-Solid Interactions and Transient Thermal Processing, M. J. Aziz, L. E. Rehn, B. Stritzker, 1988, ISBN 0-931837-68-5

Volume 101—Laser and Particle-Beam Chemical Processing for Microelectronics, D.J. Ehrlich, G.S. Higashi, M.M. Oprysko, 1988, ISBN 0-931837-69-3

Volume 102—Epitaxy of Semiconductor Layered Structures, R. T. Tung, L. R. Dawson, R. L. Gunshor, 1988, ISBN 0-931837-70-7

Volume 103—Multilayers: Synthesis, Properties, and Nonelectronic Applications, T. W. Barbee Jr., F. Spaepen, L. Greer, 1988, ISBN 0-931837-71-5

Volume 104—Defects in Electronic Materials, M. Stavola, S. J. Pearton, G. Davies, 1988, ISBN 0-931837-72-3

Volume 105—$SiO_2$ and Its Interfaces, G. Lucovsky, S. T. Pantelides, 1988, ISBN 0-931837-73-1

Volume 106—Polysilicon Films and Interfaces, C.Y. Wong, C.V. Thompson, K-N. Tu, 1988, ISBN 0-931837-74-X

Volume 107—Silicon-on-Insulator and Buried Metals in Semiconductors, J. C. Sturm, C. K. Chen, L. Pfeiffer, P. L. F. Hemment, 1988, ISBN 0-931837-75-8

Volume 108—Electronic Packaging Materials Science II, R. C. Sundahl, R. Jaccodine, K. A. Jackson, 1988, ISBN 0-931837-76-6

Volume 109—Nonlinear Optical Properties of Polymers, A. J. Heeger, J. Orenstein, D. R. Ulrich, 1988, ISBN 0-931837-77-4

Volume 110—Biomedical Materials and Devices, J. S. Hanker, B. L. Giammara, 1988, ISBN 0-931837-78-2

Volume 111—Microstructure and Properties of Catalysts, M. M. J. Treacy, J. M. Thomas, J. M. White, 1988, ISBN 0-931837-79-0

Volume 112—Scientific Basis for Nuclear Waste Management XI, M. J. Apted, R. E. Westerman, 1988, ISBN 0-931837-80-4

Volume 113—Fly Ash and Coal Conversion By-Products: Characterization, Utilization, and Disposal IV, G. J. McCarthy, D. M. Roy, F. P. Glasser, R. T. Hemmings, 1988, ISBN 0-931837-81-2

Volume 114—Bonding in Cementitious Composites, S. Mindess, S. P. Shah, 1988, ISBN 0-931837-82-0

Volume 115—Specimen Preparation for Transmission Electron Microscopy of Materials, J. C. Bravman, R. Anderson, M. L. McDonald, 1988, ISBN 0-931837-83-9

Volume 116—Heteroepitaxy on Silicon: Fundamentals, Structures,and Devices, H.K. Choi, H. Ishiwara, R. Hull, R.J. Nemanich, 1988, ISBN: 0-931837-86-3

Volume 117—Process Diagnostics: Materials, Combustion, Fusion, K. Hays, A.C. Eckbreth, G.A. Campbell, 1988, ISBN: 0-931837-87-1

Volume 118—Amorphous Silicon Technology, A. Madan, M.J. Thompson, P.C. Taylor, P.G. LeComber, Y. Hamakawa, 1988, ISBN: 0-931837-88-X

Volume 119—Adhesion in Solids, D.M. Mattox, C. Batich, J.E.E. Baglin, R.J. Gottschall, 1988, ISBN: 0-931837-89-8

Volume 120—High-Temperature/High-Performance Composites, F.D. Lemkey, A.G. Evans, S.G. Fishman, J.R. Strife, 1988, ISBN: 0-931837-90-1

Volume 121—Better Ceramics Through Chemistry III, C.J. Brinker, D.E. Clark, D.R. Ulrich, 1988, ISBN: 0-931837-91-X

Volume 122—Interfacial Structure, Properties, and Design, M.H. Yoo, W.A.T. Clark, C.L. Briant, 1988, ISBN: 0-931837-92-8

Volume 123—Materials Issues in Art and Archaeology, E.V. Sayre, P. Vandiver, J. Druzik, C. Stevenson, 1988, ISBN: 0-931837-93-6

Volume 124—Microwave-Processing of Materials, M.H. Brooks, I.J. Chabinsky, W.H. Sutton, 1988, ISBN: 0-931837-94-4

Volume 125—Materials Stability and Environmental Degradation, A. Barkatt, L.R. Smith, E. Verink, 1988, ISBN: 0-931837-95-2

Volume 126—Advanced Surface Processes for Optoelectronics, S. Bernasek, T. Venkatesan, H. Temkin, 1988, ISBN: 0-931837-96-0

Volume 127—Scientific Basis for Nuclear Waste Management XII, W. Lutze, R.C. Ewing, 1989, ISBN: 0-931837-97-9

Volume 128—Processing and Characterization of Materials Using Ion Beams, L.E. Rehn, J. Greene, F.A. Smidt, 1989, ISBN: 1-55899-001-1

Volume 129—Laser and Particle-Beam Chemical Processes on Surfaces, G.L. Loper, A.W. Johnson, T.W. Sigmon, 1989, ISBN: 1-55899-002-X

Volume 130—Thin Films: Stresses and Mechanical Properties, J.C. Bravman, W.D. Nix, D.M. Barnett, D.A. Smith, 1989, ISBN: 0-55899-003-8

Volume 131—Chemical Perspectives of Microelectronic Materials, M.E. Gross, J. Jasinski, J.T. Yates, Jr., 1989, ISBN: 0-55899-004-6

Volume 132—Multicomponent Ultrafine Microstructures, B.H. Kear, D.E. Polk, L.E. McCandlish, 1989, ISBN: 1-55899-005-4

Volume 133—High Temperature Ordered Intermetallic Alloys III, C.T. Liu, A.I. Taub, N.S. Stoloff, C.C. Koch, 1989, ISBN: 1-55899-006-2

Volume 134—The Materials Science and Engineering of Rigid-Rod Polymers, W.W. Adams, R.K. Eby, D.E. McLemore, 1989, ISBN: 1-55899-007-0

Volume 135—Solid State Ionics, G. Nazri, R.A. Huggins, D.F. Shriver, 1989, ISBN: 1-55899-008-9

Volume 136—Fly Ash and Coal Conversion By-Products: Characterization, Utilization, and Disposal V, R.T. Hemmings, E.E. Berry, G.J. McCarthy, F.P. Glasser, 1989, ISBN: 1-55899-009-7

Volume 137—Pore Structure and Permeability of Cementitious Materials, L.R. Roberts, J.P. Skalny, 1989, ISBN: 1-55899-010-0

Volume 138—Characterization of the Structure and Chemistry of Defects in Materials, B.C. Larson, M. Ruhle, D.N. Seidman, 1989, ISBN: 1-55899-011-9

Volume 139—High Resolution Microscopy of Materials, W. Krakow, F.A. Ponce, D.J. Smith, 1989, ISBN: 1-55899-012-7

Volume 140—New Materials Approaches to Tribology: Theory and Applications, L.E. Pope, L. Fehrenbacher, W.O. Winer, 1989, ISBN: 1-55899-013-5

Volume 141—Atomic Scale Calculations in Materials Science, J. Tersoff, D. Vanderbilt, V. Vitek, 1989, ISBN: 1-55899-014-3

Volume 142—Nondestructive Monitoring of Materials Properties, J. Holbrook, J. Bussiere, 1989, ISBN: 1-55899-015-1

Volume 143—Synchrotron Radiation in Materials Research, R. Clarke, J.H. Weaver, J. Gland, 1989, ISBN: 1-55899-016-X

Volume 144—Advances in Materials, Processing and Devices in III-V Compound Semiconductors, D.K. Sadana, L. Eastman, R. Dupuis, 1989, ISBN: 1-55899-017-8

Tungsten and Other Refractory Metals for VLSI Applications, R. S. Blewer, 1986; ISSN 0886-7860; ISBN 0-931837-32-4

Tungsten and Other Refractory Metals for VLSI Applications II, E.K. Broadbent, 1987; ISSN 0886-7860; ISBN 0-931837-66-9

Ternary and Multinary Compounds, S. Deb, A. Zunger, 1987; ISBN 0-931837-57-x

Tungsten and Other Refractory Metals for VLSI Applications III, Victor A. Wells, 1988; ISSN 0886-7860; ISBN 0-931837-84-7

Atomic and Molecular Processing of Electronic and Ceramic Materials: Preparation, Characterization and Properties, Ilhan A. Aksay, Gary L. McVay, Thomas G. Stoebe, 1988; ISBN 0-931837-85-5

Materials Futures: Strategies and Opportunities, R. Byron Pipes, U.S. Organizing Committee, Rune Lagneborg, Swedish Organizing Committee, 1988; ISBN 0-55899-000-3

Tungsten and Other Refractory Metals for VLSI Applications IV, Robert S. Blewer, Carol M. McConica, 1989; ISSN: 0886-7860; ISBN: 0-931837-98-7

PART I

# Testing and Measurement Techniques

# X-RAY DIFFRACTION DETERMINATION OF STRESSES IN THIN FILMS

T. VREELAND, JR., A. DOMMANN, C.-J. TSAI, AND M.-A. NICOLET
California Institute of Technology, Pasadena, CA 91125

## ABSTRACT

This paper presents the methodology employed in the determination of the stress tensor for thin crystalline films using x-ray rocking curves. Use of the same equipment for the determination of the average stress in poly- or non-crystalline thin films attached to a crystalline substrate is also discussed. In this case the lattice curvature of the substrate is determined by measurement of the shift in the Bragg peak with lateral position in the substrate.

Strains in single crystal layers may be measured using Bragg diffraction from the layers and from the substrate or a reference crystal, with the highest strain sensitivity of any known technique. The difference in Bragg angles for a strained and an unstrained crystal is related to the change in d spacing of the Bragg planes, and the elastic strain is related to this angular difference. The separation of two peaks on an x-ray rocking curve is generally not equal to the difference in Bragg angles of two diffracting crystals, so diffractometer measurements must be carefully interpreted in order to obtain x-ray strains in crystalline films (x-ray strains are strains relative to the reference crystal). The unstrained d spacings of the film and the d spacings of the reference crystal must be known to obtain the elastic strains in the film, from which the stress tensor is determined.

## INTRODUCTION

Measurement of the separation of peaks on x-ray rocking curves (curves of diffracted intensity vs. angle) is commonly employed in the determination of strains in a thin single crystal layer on a crystalline substrate. The very high strain sensitivity of x-ray measurements was demonstrated by Bonse and Hartmann in Si [1]. Using a large Bragg angle and sensitive x-ray intensity measurement, a strain sensitivity of 10E-8 was obtained. A strain sensitivity of ≈ 10E-5 to 10E-4 is typical using x-ray rocking curves with Bragg angles ≈ 45 degrees.

We consider here the determination of the stresses in single, uniformly strained layers. The analysis of multilayer structures is briefly discussed.

The equipment used for obtaining x-ray rocking curves consists of the following components:

1. An x-ray source

Typically, a sealed x-ray tube is used with a Cu target. Other targets are sometimes employed to access different Bragg angles. More intensity is needed for a single, very thin layer (less than about 20 nm), and a rotating anode or synchrotron source is used.

2. An x-ray monochromator

Typically a $K\alpha$ reflection from one single crystal of high quality is used to reduce the wave length spread of the incident beam to the sample. In this case, the $K\alpha_2$ diffraction may be removed by a slit placed between the monochromator crystal and the sample. Sharp diffraction peaks are needed if high angular resolution is to be obtained. The sharpest diffraction peaks from a sample crystal are obtained when the Bragg angle of a single monochromator crystal and the sample are equal. Use of a multiple crystal monochromator, as suggested by Dumond [2], removes the $K\alpha_2$ and does not cause peak broadening when the Bragg angles of the monochromator crystal and sample differ. This increases angular resolution at the expense of some x-ray intensity.

3. Sample goniometer

The goniometer should be capable of angular adjustments of the sample with a step size of ≈ 10E-4 degrees which is useful in some applications. The range of angular adjustments required may exceed 10 degrees for some samples, and a range of ≈ 160 degrees makes the system more versatile.

4. X-ray photon counter

We employ a scintillation crystal (NaI) with a thin Be input window, coupled to a photomultiplier tube, preamplifier, amplifier, single channel analyzer, and pulse counter.

5. Control and recording system

Typical pulse counting times using a sealed x-ray tube source are on the order of 10 s at each angular step, and the diffracted intensity (counts/s) vs. relative goniometer angle is recorded. Digital records of this data permit rapid computer analysis and plotting of rocking curves with the intensities normalized by the incident beam intensity to give the reflecting power, or reflectivity vs. angle.

## ANALYSIS OF THE ROCKING CURVE

### Uncoupled Crystals

In the analysis of x-ray rocking curves of thin films with thick reference crystals (thin and thick compared to the x-ray absorption length), it is often assumed that the diffraction from a layer is not influenced by neighboring layers, so that the kinematic theory of diffraction may be used for the thin layer or layers, and the dynamical theory for the reference [3]. In general, the peak in diffracted amplitude from a highly perfect reference crystal is shifted from the Bragg angle, $\theta_{Br}$ by a small negative increment (about 3.8E-4 degrees with a plane wave CuK$\alpha_1$ and the 440 Si reflection) [4]. Wave length spread in the incident beam from a single monochromating crystal will generally cause the peak intensity to fall within 1E-4 of the Bragg angle of the reference crystal. We will assume the peak intensity from the reference crystal is at the Bragg angle $\theta_{Br}$ in the treatment in this section. We also assume that the kinematic theory is applicable, and we will examine this assumption in a subsequent section.

The difficulty in the determination of absolute angles in a rocking curve is avoided when the Bragg peak from a reference crystal of accurately known d spacing is recorded together with the peak from the film. According to the assumptions in this section, the rocking curve is the sum of the intensities diffracting from the reference crystal and the film. We treat first the case where the diffracting planes of both reference crystal and film are parallel. Two distinct peaks will appear on the rocking curve when $\Delta\theta_B = \theta_{Bf} - \theta_{Br}$ is sufficiently large (the subscripts refer to film and reference). The Bragg peak from a single layer decreases in amplitude and increases in width as the layer thickness decreases, and complicates determination of $\Delta\theta_B$ for thin films.

The difference in Bragg angles is related to the d plane spacings of the film and reference by

$$\Delta\theta_B = \sin^{-1}\lambda/2d_f - \sin^{-1}\lambda/2d_r \qquad (1)$$

where $\lambda$ = x-ray wavelength.

The x-ray strain normal to the parallel Bragg planes of the film and reference crystal is found from Eq. 1 to be

$$\epsilon_{x\text{-ray}} \equiv (d_f - d_r)/d_f = \sin\theta_{Br}/\sin(\theta_{Br} + \Delta\theta_B) - 1 \quad (2)$$

For small x-ray strains Eq. 2 reduces to

$$\epsilon_{x\text{-ray}} \cong -\Delta\theta_B/\tan\theta_r \quad (3)$$

Heteroepitaxial films with relatively large x-ray strains are often analysed. The error in $\epsilon_{x\text{-ray}}$, introduced by use of Eq. 3 instead of Eq. 2 for stress-free films with a Si reference and $FeK\alpha_1$, is shown in Fig. 1. The error is not negligible in the systems with large lattice mismatch such as GaAs on Si.

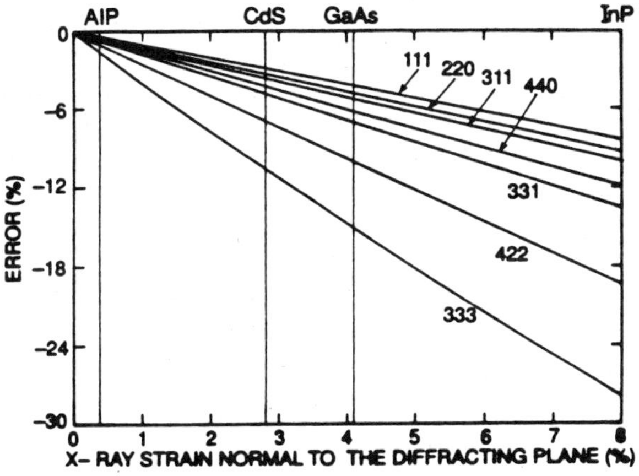


Fig. 1. The error in the x-ray strain for stress-free crystals introduced by the small strain approximation, Eq. 2 using a Si reference crystal and different diffracting planes with $FeK\alpha_1$.

The x-ray strains must be converted to the strains of elasticity theory in order to employ Hooke's law for determination of the stresses. The elasticity normal strains are given by

$$\epsilon_{elasticity} \equiv (d_f - d_{fo})/d_{fo} = (\epsilon_{x\text{-ray}} + 1)d_r/d_{fo} - 1 \quad (4)$$

where $d_{fo}$ is the film d spacing when it is stress-free.

Direct measurement of the angle $\Delta\theta$, between the maximas of the reference peak and the film peak on a rocking curve may differ from $\Delta\theta_B$ due to a) an asymmetric reference peak, as discussed above, b) a dynamical interaction of the x-ray waves when the film and reference crystals are in contact (coupled), as discussed below, c) tilt between the film and reference diffracting planes, and d) peak overlap. The overlap effect is removed when the diffracted intensity from the reference crystal is subtracted from the rocking curve.

Crystals have a multiplicity of planes suitable for x-ray diffraction so normal strains in different directions in a film may be found. The strain normal to the film surface is particularly easy to determine when the diffracting planes are parallel to that surface (symmetric diffraction). This is one principal strain when the axes of principal stress and strain coincide.

Additional normal strains can then be measured using planes inclined to the surface (asymmetric diffraction), from which the other two principal strains are found. Unequal strains normal and parallel to the film surface cause tilt of the asymmetric Bragg planes in the film relative to those in the reference crystal (which is a function of the angle between these film planes and the film surface), so that the film diffraction peaks are shifted by this tilt in addition to the shift due to the d spacing change [5]. Additional tilt between the diffracting planes of the film and a substrate may result in the growth process, so the total tilt should be measured when asymmetric diffraction is used as discussed below.

Asymmetric Rocking Curves from Uncoupled Crystals

The x-ray geometry for a non-skew asymmetric diffraction in the diffraction plane (the plane containing the incident and diffracted beams) is shown in Fig. 2. (In a skew diffraction the normal to the diffracting plane does not lie in the diffraction plane.) When there is tilt between the film and reference crystals, this tilt may be found by obtaining two rocking curves, one as shown in Fig. 2 and one with the diffraction vectors reversed. This reverses the sign of the angular shift of the film diffraction peak due to the tilt. The tilt angle is then one-half of the difference in peak shifts, $\Delta\theta_1 - \Delta\theta_2$, in the two rocking curves, while the strain is related to one-half of the sum of the two peak shifts. The x-ray strain normal to the diffracting plane is then

$$\epsilon_{x\text{-ray}} = \sin\theta_{Br}/\sin\{\theta_{Br} + \Delta\theta_i \pm \xi(\phi_r)\} - 1 \quad (i = 1,2) \qquad (5)$$

where $\xi(\phi_r)$ = the tilt angle and the minus sign is used when the diffraction vectors are as shown in Fig. 2 (incident x-rays at angle $\theta - \phi_r$ from the sample surface), and the plus sign is used for $\theta + \phi_r$ incidence.

When the x-ray strains are small Eq. 5 reduces to

$$\epsilon_{x\text{-ray}} \cong -\{\Delta\theta_i \pm \xi(\phi_r)\}/\tan\theta_{Br} \qquad (6)$$

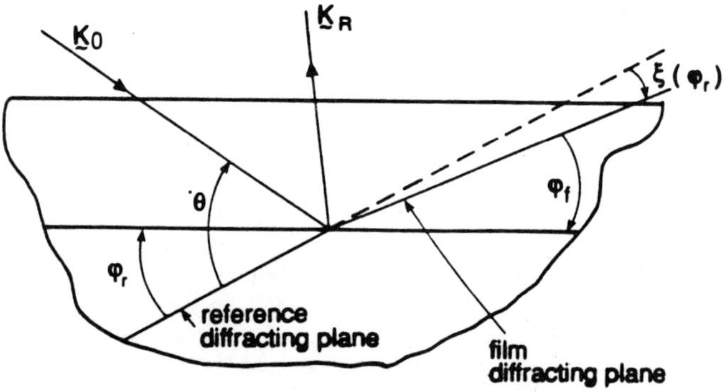


Fig. 2. X-ray geometry for an asymmetric, non-skew reflection with a tilt, $\xi(\phi_r)$ between a film and reference crystal. The incident beam is $K_0$, the diffracted beam is $K_R$, and the angles between the diffracting planes in the film and reference crystal and the film surface are $\phi_f$ and $\phi_r$ respectively.

Procedure for Determination of Principal Strains

The principal strain directions in a strained epitaxial film are dictated by the crystallography of the film and substrate. We treat here the case where both crystals are orthotropic in the surface planes or have perpendicular planes of mirror symmetry which are also perpendicular to the film surface, such as {111}, {100}, and {110} planes of cubic crystals. The surface plane and the mirror planes, or any other two orthogonal planes perpendicular to the surface in the orthotropic case, are then planes of principal strain. Rocking curves are obtained from two different diffracting planes whose normals lie in a principal plane. Normal strains, perpendicular to the two diffraction planes are determined from these rocking curves together with the tilt between film and substrate (using Eqs. 5 or 6). The principal x-ray strains in the plane of the measured normal strains are then given by

$$\epsilon_{perpendicular} = (\epsilon_{\psi_2}/\sin^2\psi_2 - \epsilon_{\psi_1}/\sin^2\psi_1)/(\cot^2\psi_2 - \cot^2\psi_1) \tag{7}$$

and

$$\epsilon_{parallel} = (\epsilon_{\psi_2}/\cos^2\psi_2 - \epsilon_{\psi_1}/\cos^2\psi_1)/(\tan^2\psi_2 - \tan^2\psi_1) \ . \tag{8}$$

where $\epsilon_{\psi_1}$ and $\epsilon_{\psi_2}$ are the x-ray strains normal to the two diffracting planes and perpendicular and parallel refers to the film surface. To maximize the strain sensitivity, one of the diffracting planes should have the minimum possible $\psi$ and the other the maximum. When the film surface is a diffracting plane, $\psi_1 = 0$, and

$$\epsilon_{perpendicular} = \epsilon_{\psi_1} \tag{9}$$

and

$$\epsilon_{parallel} = (\epsilon_{\psi_2} - \epsilon_{\psi_1}\cos^2\psi_2)/\sin^2\psi_2 \tag{10}$$

Repetition of these steps using rocking curves from planes whose normals lie in the third principal plane provides a consistency check on $\epsilon_{perpendicular}$ and gives the third principal strain. For epitaxial films grown on surfaces of lower symmetry, for example surfaces tilted with respect to {100}, {111}, or {110} cubic crystal faces, additional diffracting planes must be used in order to determine the principal strains and their directions.

Examples

Strain measurement using rocking curves of epitaxial InGaAsP layers of differing alloy composition on InP substrates have proven useful in finding the growth conditions which avoid premature failure of the laser due to high strain in the active alloy layer. The rocking curve of Fig. 3a shows the substrate peak and the peak from a single alloy layer of the composition which gives the desired frequency in a laser structure. However, the strain in a laser with a single layer of this alloy causes failure after a few lasing periods. The rocking curve in Fig. 3b is from a five layer laser structure with three layers of the same alloy as that of the single layer in Fig. 3a. The layers below the active layer have been decoupled from the substrate by interface dislocations to reduce layer strains in the five layer structure by over a factor of two and premature failure in this structure is avoided.

The strain tensor in a heteroepitaxial film of (001) Si on $(1\bar{1}02)$ sapphire was found using symmetric and asymmetric rocking curves from the Si film and from a piece of Si wafer as a reference, fixed to the surface of the film (with a viscous grease) [6]. In this case, the x-ray strain is also the elasticity strain. Fig. 4 shows the superimposed symmetric rocking curves taken with the diffraction vectors reversed from which the perpendicular strain and one component of the tilt between film and reference was obtained.

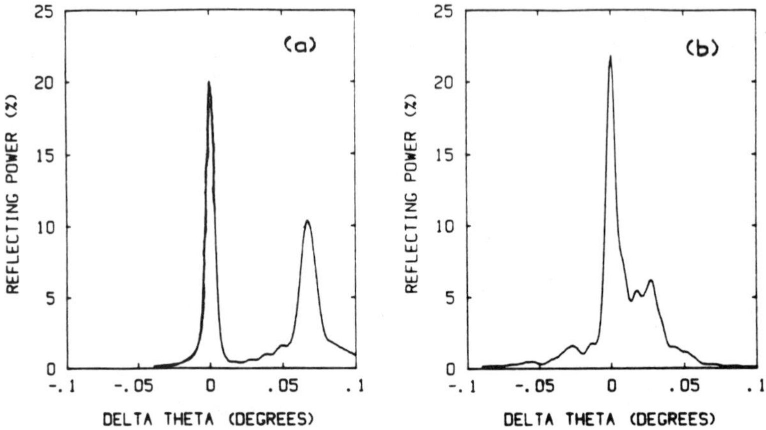


Fig. 3. Symmetric rocking curves of the 004 reflection from a) a single layer of InGaAsP on InP with the alloy content adjusted to that required for the desired wavelength in a laser structure, and from b) a five layer laser structure with the alloy content in three of the layers the same as that in a). Significant relaxation of strain in the active layer of the laser structure by interface dislocations is indicated by the positions of the peaks in a) and b), FeK$\alpha_1$ radiation.

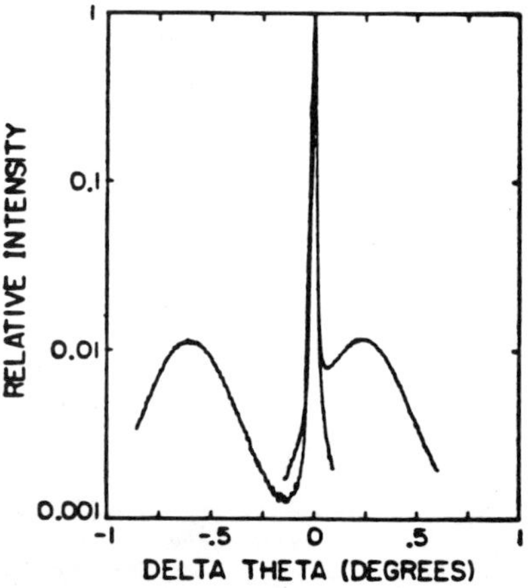


Fig. 4. Superimposed symmetric 004 rocking curves from a (001) Si film grown on (1$\bar{1}$02) sapphire using a stress-free Si reference crystal attached to the surface of the film. The diffraction vectors are reversed in the two topographs (sample was rotated 180 degrees about the normal to the film). The reference crystal peak is at $\Delta\theta$ = 0, the difference in peak shifts is equal to twice the tilt angle between film and reference, and the sum of the two peak shifts is equal to $2\Delta\theta_B$, FeK$\alpha_1$ radiation.

Coupled Crystals

The dynamical theory of x-ray diffraction must be used to properly take into account the boundary condition between coupled crystal layers (layers in intimate contact) [7]. It has been shown that the peak positions in this case are not equal to those for uncoupled layers [8]. While the difference is small when the layer thickness is sufficiently large, it is not negligible for thin layers with small strains. The ratio of $\Delta\theta_{peaks}$ to $\Delta\theta_B$ is shown in Fig. 5 for $Al_{0.25}Ga_{0.75}As$ films of thickness from 200 to 1000 nm on GaAs calculated using the dynamical theory and $FeK\alpha_1$. The peak separation in the calculated curve becomes significantly smaller than $\Delta\theta_B$ as the layer thickness decreases. This example together with more extensive calculations for different epitaxial systems indicate that the ratio approaches unity with increasing film thickness and with increasing strain for a layer of given thickness [9]. Therefore, simulation of the rocking curve using the dynamical theory should be employed to find the x-ray strain in thin layers with small strains.

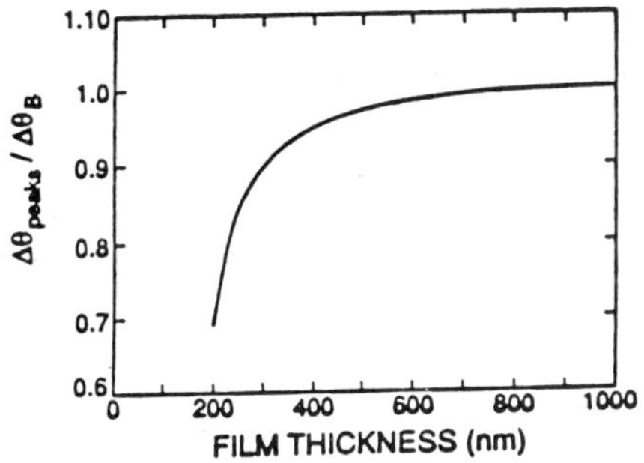


Fig. 5. The ratio of $\Delta\theta_{peaks}$, calculated by the dynamical theory using $FeK\alpha_1$ and 004 diffraction, to $\Delta\theta_B$ as a function of the thickness of coherent $Al_{0.25}Ga_{0.75}As$ layers 200 to 1000 nm thick on (001) GaAs.

MULTILAYER STRUCTURES

The stresses in multilayer structures may be investigated by matching their rocking curves to those obtained by computer simulation. Examples in epitaxial systems and in ion implanted crystals are given in [10], where the structures were modeled by uniform layers of different thickness, x-ray structure factor, and strain (using kinematic or dynamical x-ray calculations). While a layer model which gives a calculated rocking curve that matches well with the experimental curve is not a unique solution, we generally know enough about the actual structure to model it accurately.

A recent extension of the rocking curve technique has been reported which gives the promise of extending it to the study of very thin layers, such as the active layer in quantum laser structures. When a thin crystal epilayer is confined between two multilayer structures which have a mirror symmetry about the confined layer, a phase shift in the x-rays diffracting from the confined layer occurs. A small kink on the flank of the principal peak from the confining layers then may be observed (11,12). The kink position and amplitude depends upon the product of the strain in the confined layer and its thickness. Computer simulation of coherent model systems with large lattice mismatch (such as $In_{0.37}Ga_{0.63}As$ with 100 nm $Al_{0.4}Ga_{0.6}As$ confining layers on GaAs, which have perpendicular x-ray strains of 3.9% and 0.08% respectively) show a kink when the confined layer is thicker than about one nanometer. The active layer thickness in a quantum well laser affects the threshold voltage and band gap and it can be determined by x-ray analysis in highly perfect structures.

## AVERAGE STRESS MEASUREMENT

### Curvature Measurements and Film Stresses, Films on a Crystalline Substrate

Stresses in a film on a crystalline substrate cause bending of the substrate in the plane of the film, and the bending may be related to the average stresses in the plane of the film and the elastic properties of the sample. With a translation stage added to the rocking curve sample holder, the shift in the substrate Bragg peak may be measured at different lateral positions on a film grown on a crystalline substrate. The average principal stresses in polycrystalline or non-crystalline films can then be found from the peak shifts. When the film is less than about one x-ray absorption length thick, the measurements may be made using either the film side or the back side of the substrate.

In order to obtain good angular sensitivity for the peak shifts, the sample stage should be capable of translations of several mm with rotation less than 1E-4 degrees. We have found a sample stage supported by four thin parallel flexures, when translated by a motor mike pushing on the elastic center of the flexure system, meets this requirement.

The radius of curvature of the substrate crystal in the diffraction plane is given by

$$R = \Delta s/\Delta\theta \qquad (11)$$

where $\Delta s$ = separation of the positions on the sample which gave substrate diffraction peaks separated by $\Delta\theta$, and
$\Delta\theta$ = angular shift of the substrate diffraction peak.

A radius of curvature of 1 km gives a $\Delta\theta$ of 1E-4 degrees with a $\Delta s$ of about 1.7 mm.

For a sample free of tractions, the radius of curvature is related to the total force in the film and the elastic properties of the sample. When the film is thin compared to the substrate thickness, the elastic properties of the substrate, its thickness, and the bending moment in the substrate determine its principal radii of curvature. We will consider the case for elastically orthotropic or orthorhombic substrate plates here, specifically cubic crystals with {001}, {111}, or {011} surfaces, and give an example where (001) Si is used.

The principal radii of curvature are then [13]

$$R_1 = (h^3/12S'_{11})/(M_1 - M_2S'_{12}/S'_{11}) , \qquad (12)$$

and

$$R_2 = (h^3/12S'_{22})/(M_2 - M_1S'_{12}/S'_{22}) \qquad (13)$$

where $h$ = the sample thickness ($\simeq$ substrate thickness),
$S'_{11}$, $S'_{22}$, $S'_{12}$ = the substrate elastic constants referred to the orthorhombic symmetry directions, and
$M_1$, $M_2$ = bending moments producing $R_1$ and $R_2$.

Each bending moment is directly related to the corresponding force per unit length in the film, and the average stresses in the film are found when the film thickness is known, giving

$$\sigma_1 = 2M_1/ht \text{ and } \sigma_2 = 2M_2/ht \text{ for } h >> t. \qquad (14)$$

where $t$ = film thickness.

The stresses, in terms of the measured radii are then

$$\sigma_1 = (h^2/6t)(S'_{22}/R_1 + S'_{12}/R_2)/(S'_{11}S'_{22} - S'^2_{12}), \qquad (15)$$

and

$$\sigma_2 = (h^2/6t)(S'_{11}/R_2 + S'_{12}/R_1)/(S'_{11}S'_{22} - S'^2_{12}). \qquad (16)$$

## Stresses in Films on (100), (111), and (110) Substrates

The stresses given above are for plates in pure bending, and it is assumed that the plates are strain-free with no films on them and have no temperature gradients in them. Plates with [100] normals have {100} principal planes, so that

$$S'_{11} = S'_{22} = S_{11} = (c_{11} + c_{12})/[(c_{11} - c_{12})(c_{11} + 2c_{12})], \text{ and} \quad (17)$$

$$S'_{12} = S_{12} = - c_{12}/[(c_{11} - c_{12})(c_{11} + 2c_{12})]. \quad (18)$$

where $c_{11}$, $c_{22}$, $c_{12}$ = elastic modulli referred to the cubic directions.

Plates with (111) normals which are free of shear stress on their surfaces are orthotropic in bending, and

$$S'_{11} = S'_{22} = [c_{11}/\{(c_{11} - c_{12})(c_{11} + 2c_{12})\} - 1/2c_{44}]/2, \text{ and} \quad (19)$$

$$S'_{12} = [(c_{11} - 4c_{12})/\{(c_{11} - c_{12})(c_{11} + 2c_{12})\} - 1/2c_{44}]/6. \quad (20)$$

A plate with (110) surfaces in pure bending has principal curvatures in [001] and [$\bar{1}$10] directions, which are the directions of $\sigma_1$ and $\sigma_2$ respectively. The elastic coefficients in this case $S'_{11}$ and $S'_{12}$ are given by Eqs. 17 and 18, and

$$S'_{22} = c_{11}/[(c_{11} - c_{12})(c_{11} + 2c_{12})] + 1/4c_{44}. \quad (21)$$

The substrate 400 diffraction peak from a traction-free (100) Si crystal, measured at two points with $\Delta s$ = 4 mm is shown in Fig. 6a. The $\Delta\theta$ determined by fitting the measured points with a Gaussian curve is less than the probable error in the fitting of 1E-4 degrees. The same result is obtained from both faces of the crystal, indicating that the translation stage rotates less than 1E-4 degrees with 4 mm of translation. Alternating layers of Ti and Ni were evaporated on 0.2 mm thick (100) substrates, with a total metal thickness of about 40 nm. The substrate peaks measured on one of these samples, at two points 4 mm apart on the film side, is shown in Fig. 6b. The probable error in the peak shift determined by Gaussian fitting is less than 1E-4 degrees.

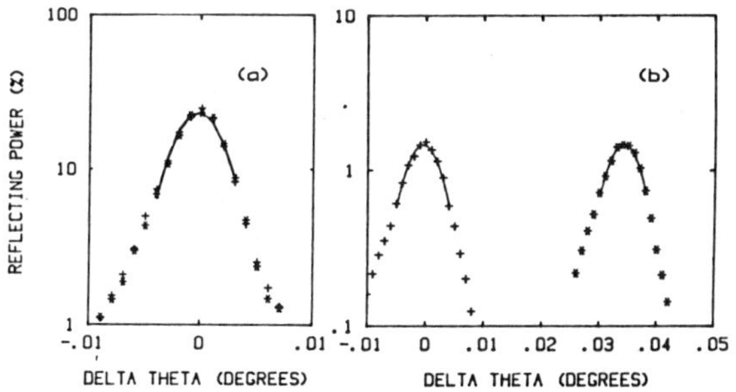


Fig. 6. The 400 FeK$\alpha_1$ diffraction peaks from regions 4 mm apart on (100) Si for a) a stress free crystal, and b) on a crystal with alternating evaporated layers of Ti and Ni. Different plotting symbols are used for the different regions, and Gaussian curves fit to the data from each region are shown as solid lines.

Samples prepared with varying thicknesses of the Ti and Ni layers, of 8 to 12 periods each, were analysed for film stresses after some room temperature ageing. The results are shown in Fig. 7, and indicate an interesting effect of the average concentration in the multilayer [14].

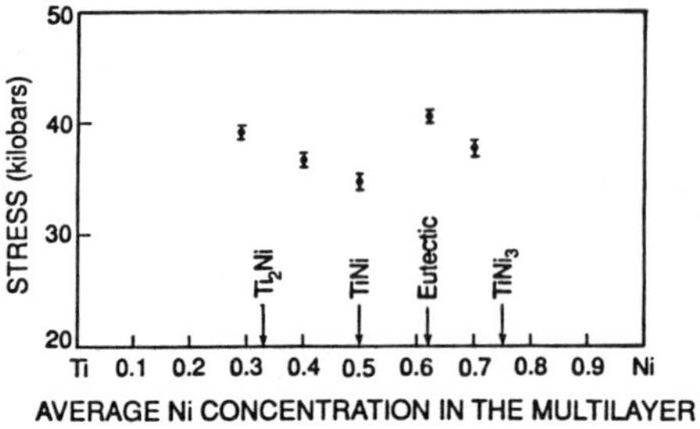


Fig. 7. Stresses determined by substrate curvature measurements in samples with alternating layers of Ti and Ni on (100) Si as a function of the average concentration in the multilayer. The ratio of the total thickness of the layers to the stbstrate thickness was about 1/7000.

ACKNOWLEDGEMENTS

The authors wish to thank Prof. D. S. Wood for checking our calculations of the elastic constants. This work was sponsored by the Semiconductor Research Corporation and the National Science Foundation, Materials Research Groups, grant no. 8811795.

REFERENCES

1. V. Bonse and I. Hartmann, Z. Kristallogr. **156**, 256 (1981).

2. J. W. M. DuMond, Phys. Rev. **52**, 812 (1973).

3. V. S. Speriosu, J. Appl. Phys. **52**, 6094 (1981).

4. W. H. Zachariasen, Theory of X-ray Diffraction in Crystals, (Wiley, New York, 1945).

5. V. S. Speriosu and T. Vreeland, Jr., J. Appl. Phys. **56**, 1591 (1984)

6. Thad Vreeland, Jr., J. Mater. Res. **1**, 712 (1986).

7. C. R. Wie, T. Tombrello, and T. Vreeland, Jr., J. Appl. Phys. **59**, 3742 (1986).

8. P. F. Fewster and C. J. Curling, J. Appl. Phys. **62**, 4154 (1987). An incorrect boundary condition was used in the calculations which introduced a small error in their results. The correct boundary condition is given in [7].

9. C. R. Wie, Y. Choi, J. F. Chen, T. Vreeland, Jr. and C.-J. Tsai, (unpublished) to be presented at the Spring MRS meeting, April 1989.

10. T. Vreeland, Jr. and B. M. Paine, J. Vac. Sci. Technol., **A4**, 3151 (1986).

11. X. Chu and B. K. Tanner, Appl. Phys. Lett., **49**, 1773 (1986).

12. C. R. Wie, to appear in J. Appl. Phys.

13. R. F. S. Hearmon, Applied Anisotropic Elasticity, (Oxford University Press, 1961).

14. Unpublished results, T. Vreeland, Jr., R. B. Schwarz, and K. H. Samwer.

# MECHANICAL PROPERTIES OF NANOMETRE VOLUMES OF MATERIAL: USE OF THE ELASTIC RESPONSE OF SMALL AREA INDENTATIONS

J.B.PETHICA* AND W.C.OLIVER**
*Department of Metallurgy and Science of Materials, Parks Road, Oxford OX1 3PH, England
**Oak Ridge National Laboratory, P.O.Box M, Oak Ridge, TN 37832

## ABSTRACT

A new, differential method for determining the stiffness of a sub-micron indentation contact area is presented. This allows measurement of elastic modulus as well as plastic hardness, continuously during a single indentation, and without the need for discrete unloading cycles. Some of the new experiments that become possible with this technique, especially at the nanometre scale, are described. We show quantitatively that electropolished tungsten reproducibly exhibits the ideal theoretical lattice strength at small indentation loads.

## INTRODUCTION

In recent years there has been a considerable increase in interest in the mechanical properties of small volumes of materials. This is motivated partly by interest in inherently small structures such as thin films, interfaces and composites, and partly by newly available methods of directly measuring local mechanical properties. Prominent amongst these is sub-micron indentation. By sensing the depth of indentation as a function of the applied load, it has been shown that both plastic hardness and elastic modulus of volumes a few hundred nm across can be obtained, [1,2,3,4] There is no need to go through the difficult process of imaging the nanometre scale indentations. In this paper we present a significant further development in the scale at which these tests can be made, and a means of further simplifying the calculation of hardness. We show that at the smallest scale, inherent atomic lattice properties can be observed.

Two problems have to date made it difficult to use depth sensing tests at extremely low loads and depths. The first is the risk of thermal drift in the apparatus, which being unknown, gives uncertainty in the plastic depth. The second is the accuracy with which the shape of the indenter is known, which in turn limits the knowledge of the true contact area. The key to reducing these problems lies in determining the stiffness (or spring constant) of the contact between indenter and surface. For an indenter load P and displacement x of indenter normal to the surface, the stiffness $S = \partial P/\partial x$. The utility of S comes from the fact that for a very wide range of stress distributions under the indenter, S is given by [5,6]

$$S = A E^* \sqrt{\text{Contact area}} \qquad (1)$$

$$\text{where } \frac{1}{E^*} = \frac{1 - \nu_1^2}{E_1} + \frac{1 - \nu_2^2}{E_2}$$

E being the Young's modulus and ν the Poisson's ratio of the indenter and surface (suffices 1 & 2 respectively). The constant A varies at most between $2/(\sqrt{\pi})$ for a flat punch or a spherical (Hertzian) contact, and $(\sqrt{\pi})/2$ for some intermediate stress fields. It has been noted elsewhere [1,3] that S is often constant for a significant part of the unloading process; that is, the indenter plus its 'cap' of plastically deformed material resembles a blunt punch, and the contact area does not change. Therefore, by extrapolating the line of constant S to zero load, the elastic and plastic components of the indentation depth can be separated out . This is the basis of measurement of elastic modulus [3]. Unfortunately, particularly for non-metals, S is often not constant for any significant part of the unloading curve, [4] making the determination of S from these curves subject to uncertainty additional to that due to thermal drift.

We show here that S is directly measurable by simply applying a small oscillatory modulation $\partial P$ to the overall load, and sensing the resulting oscillations $\partial x$ in the displacement. This brings two advantages. Firstly, a true, instantaneous value of S is available during the entire indentation procedure of loading and unloading. Therefore the value of elastic modulus or contact area can be continuously calculated for all penetration depths, without the need for discrete load-unload sequences. Secondly, as the AC frequency used is typically 60-70Hz, the value of S is virtually immune to the influence of thermal drift, since a cycle takes only some 15msec. This is particularly useful at very small depths.

## EXPERIMENTAL

The nanometre indentation system used has been described elsewhere [1,2]. The addition of AC modulation to the load was achieved by adding a small AC current source to the load generating coil. The AC signal was separated from the DC load by a 2.2μF blocking capacitor, which with the resistive components present produced an 80msec time constant for low frequency roll-off. The AC displacement signal was measured with a Brookdeal two channel phase sensitive detector (or 'lock-in'), giving both amplitude and phase angle of the displacement signal. The displacement sensing capacitive bridge system used introduced a 4msec time constant high frequency roll-off into the signal. Together with the machine damping characteristics discussed below, these time constants limited the useful modulation frequency range to between 10 and 150 Hz.

To calculate the stiffness of the actual contact zone, the influence of the indentation system itself upon the dynamic response has to be determined. The relevant components are the mass m of the indenter, the spring constant $K_0$ of the indenter support/guide springs, the stiffness of the overall frame of the machine $K_f$, and the damping constant c due to the air in the gaps of the capacitor plate displacement sensing system. These combine with the contact stiffness S, as shown schematically in Figure 1 to produce the overall response. Many other possible contributions were considered, but after much model fitting to the data, it was found that only the above terms had any significant influence.

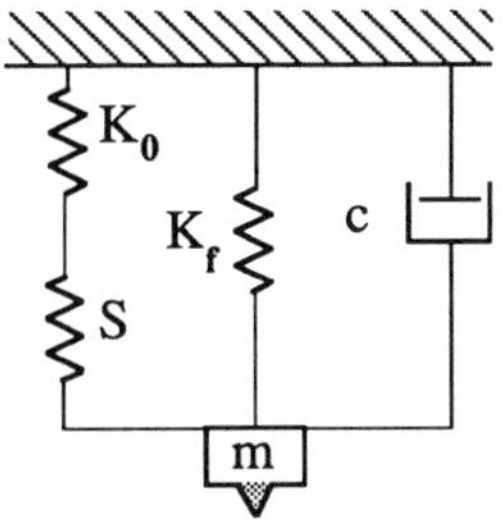


Figure 1. Components of the dynamic model of the indentation system

The system of Fig. 1 is simple to analyse; if the driving force is $F = F_0 \exp(i\omega t)$ and the displacement response is $a = a_0 \exp(i\omega t + \phi)$ then

$$\frac{F_0}{a_0} = \sqrt{\omega^2 c^2 + (K - m\omega^2)^2}$$

and the phase angle $\phi$ between drive and response is given by

$$\tan\phi = \frac{\omega c}{K - m\omega^2}$$

where K, the combined spring constant is given by

$$\frac{1}{K} = \frac{1}{K_f} + \frac{1}{K_0 + S}$$

S may be determined from either of the first two equations, since m, $\omega$, c (below) $K_f$ and $K_0$ are known, and $a_0/F_0$ or $\phi$ are available from the lock-in amplifier.

The damping constant c varies with plate position, being as expected, a minimum when the moving plate is half way between the outer fixed plates. With m, $K_f$ and $K_0$ known from static or DC measurements (4.7g, 6 x $10^5$ N/m & 45 N/m respectively), c was evaluated for various values of ω by simply measuring the response with the indenter well clear of any test surface. Some results are shown in Figure 2.

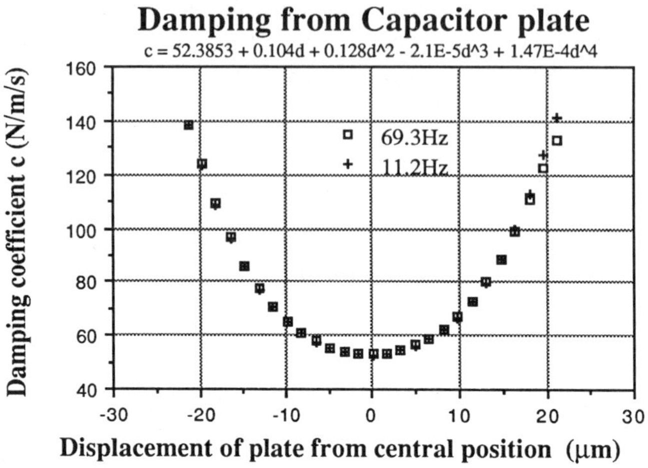


Figure 2

The uniformity of values, independent of the driving frequency, gives confidence in the dynamic model used. In practice, most indents are carried out within a few microns of the plate centre position, where c is approximately constant at ~52 N/m/s, but even outside that region a simple polynomial fit formula gives the appropriate value of c.

The stiffness S can be determined either from phase angle or from amplitude of response. Inspection of the formulae shows that $\phi$ will be most sensitive to small values of S and amplitude $a_0$ will be best for larger values of S. We find remarkably good agreement between these two independent methods for mid range values of S. To check that the measured S is physically reasonable, a special cantilever beam specimen was prepared.
Figure 3 shows the results of measuring K (the combination of spring constants) for various positions along the length of the cantilever. A sufficiently large DC load was used that the contact stiffness S was larger than any of the other spring constants. The cantilever beam spring constant thus acts in place of S in the analysis above. The expected cubic power dependence of stiffness on length of beam is clearly seen on the right hand side. As the indent position approaches the beam root, the beam stiffness begins to exceed the apparatus stiffness $K_f$, and the measured value of K levels out as seen. The figure contains stiffness values determined from the DC unload, and from the amplitude and phase based AC methods. As may be seen, the phase method becomes innaccurate for large K, and the amplitude innaccurate for small K. Both, however, in their appropriate ranges give excellent agreement with the DC value of cantilever spring constant.

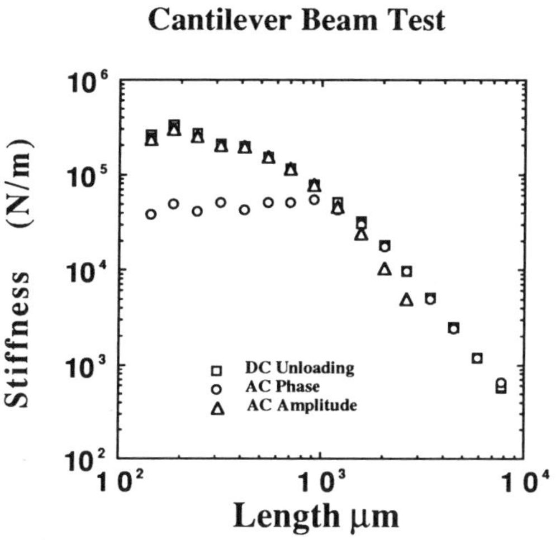


Figure 3.
Variation of total stiffness K with position of test point along cantilever beam

RESULTS

Having established the neccessary techniques, the purpose of this paper is to show some of the phenomena that start to occur as the indent size becomes very small. Typical behaviour of S during indentation, as determined by the AC method, is given elsewhere [6]. The drift independent nature of the AC measurement becomes particularly important in verifying the DC data when the load displacement curve becomes nearly or perfectly reversible. In [6] we reported the observation of purely elastic indentation in sapphire. Here we show such behaviour in a non-brittle material: tungsten. Figure 4 is the load-displacement curve for a fairly large indentation in electropolished annealed tungsten.

This shows the typical plastic loading behaviour, and two cycles of unloading followed by reloading. Hardness and modulus can be readily determined if needed. Note that the unloading cycle is not at first fully reversible. Some inelastic processes are still taking place, particularly on the first unload. The AC stiffness data shows this very clearly. In Figure 5, for a lower load test, S is plotted as a function of time since the beginning of the indent, so the four loading and unloading cycles are seen as a sequence of rises and falls of S, following the rising and falling of load. The maximum and minimum values of S (and hence contact area) continue to rise with each cycle. Note also that S is not quite independent of load even at the top of the loading cycle.

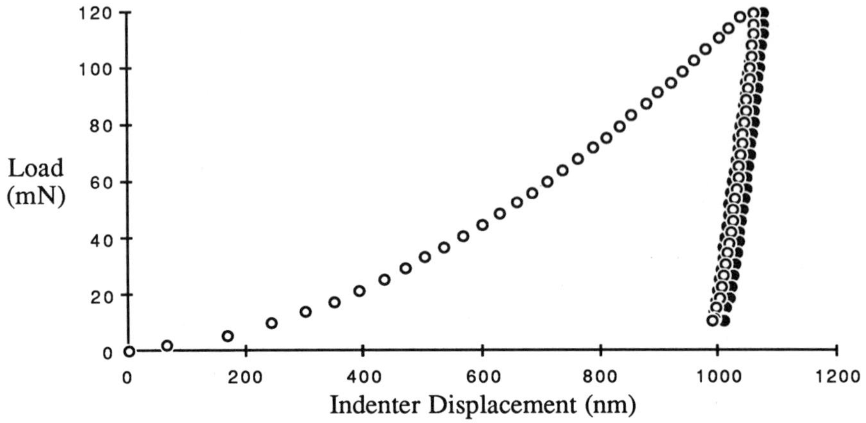


Figure 4. Large load indentation test on electropolished tungsten. The reload-unload cycle after the first is shown with a solid symbol.

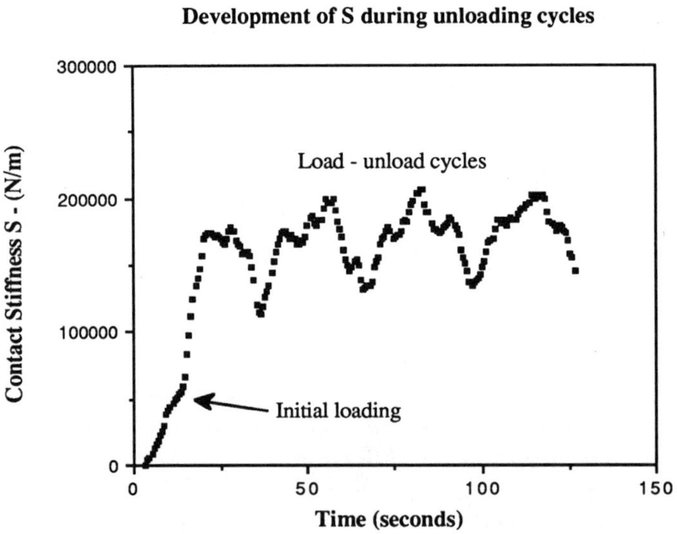


Figure 5. The evolution of contact stiffness S with time, during a test having four load-unload cycles including the original loading. The surface is electropolished tungsten.

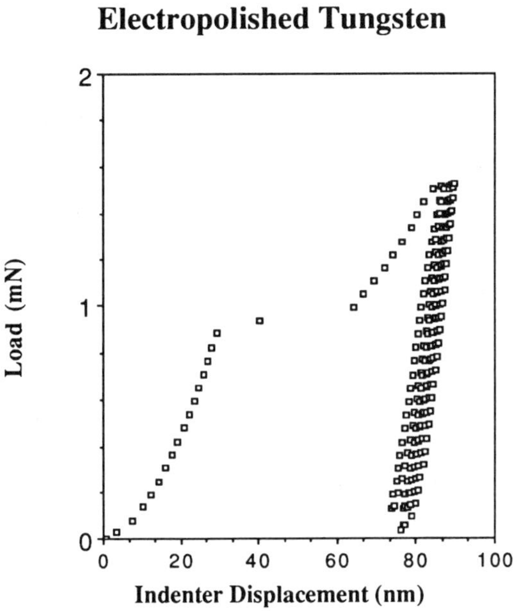


Figure 6. Low load behaviour on electropolished Tungsten, showing sudden "pop-in"

Now if we repeat this measurement at a very much lower load (Figure 6) a marked feature becomes apparent. At a load around 1.1mN the indenter displacement suddenly jumps some 30nm into the surface. The AC cycle data of Figure 5 is actually from the same indent as Figure 6, and the sharp increase in S at the 'pop-in' is clearly seen in the initial loading. Subsequent to this jump-in, the load-displacement behaviour is similar to a normal plastic contact. To understand what is occurring, tests were made with the indenter maximum load smaller still (Figure 7) such that the 'pop-in' load is not reached. As can be seen, the loading and unloading curves are then perfectly reversible. Even over the four cycles of loading and unloading shown on Figure 7 no plastic deformation occurs — the contact is perfectly elastic. The AC measurement confirmed the perfect reversibility, and hence that the data of Figure 7 is drift free.

The purely elastic contact followed by abrupt and extensive plastic flow strongly suggests a critical stress for dislocation nucleation followed by rapid multiplication, that is, a marked yield drop. Consider now the stress at which this occurs. A very important feature of these results is that the load P at which 'pop-in' occurs, is always (we have performed numerous measurements) within ~10% of 1mN. This contrasts with earlier reports of a yield drop in indentation, where a range of loads, up to some maximum value were reported [6,12]. Since the behaviour is purely elastic we may evaluate the maximum shear stress present in the contact using standard Hertzian analysis. Although an effective tip radius could be calculated from the data of Figure 6, there is no need to do so. It is only neccessary to recall that onset of plastic flow in a Hertzian contact occurs at a contact pressure of 1.1Y, where Y is the uniaxial

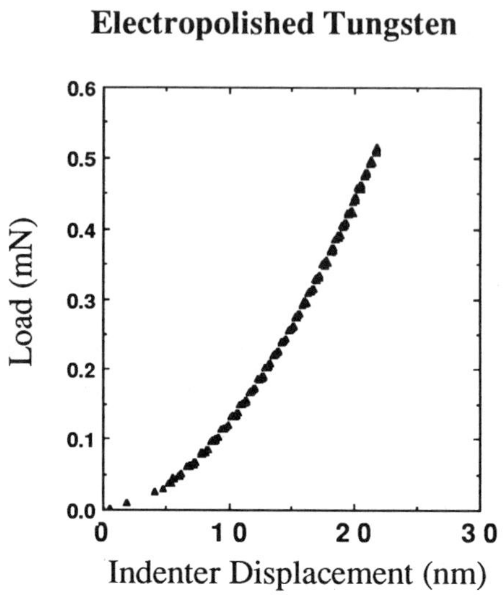


Figure 7. Indentation at a load below the critical value for indenter 'pop-in'. Note the perfect reversibility even over the four cycles of load-unload plotted.

yield stress [7]. Now the contact stiffness S just before 'pop-in' is given by equation (1) above, and so the contact pressure $\sigma_c$ is given by

$$\sigma_c = \frac{P}{\pi S^2 / 4E^{*2}} = \frac{4 P E^{*2}}{\pi S^2} \qquad (2)$$

Using this formula, the value of S at 'pop-in' from the data of Figure 6 (or Figure 5), and using average values of elastic constants for tungsten and diamond (the indenter) we find a maximum contact pressure of $3.15 \times 10^{10}$ N/m$^2$. This means that the onset of yield is occurring at a yield strain (Y/E) for tungsten of 8% — the sudden nucleation process is occurring at or close to the theoretical lattice limit. Therefore in these tests we are probing the ultimate defect-free strength of the lattice.

To further confirm this picture, exactly the same series of tests was run on mechanically polished tungsten, where we expect numerous pre-existing defects. At large loads the curves looked fairly similar (requiring somewhat higher loads for the same depth) to the electropolished surfaces. However, at smaller loads on the mechanically polished, there was no evidence of a sudden 'pop-in' event, and at all times the data showed clearly plastic, irreversible contacts. At the lowest loads used, a slight uneveness of loading curve was seen, pehaps indicating discrete flow onset processes (Figure 8). However, as can be seen from the magnitude of loads in Figure 8, these occurred at considerably lower stresses than present prior to 'pop-in' for the electropolished surfaces. There was no evidence of a single high stress nucleation event,

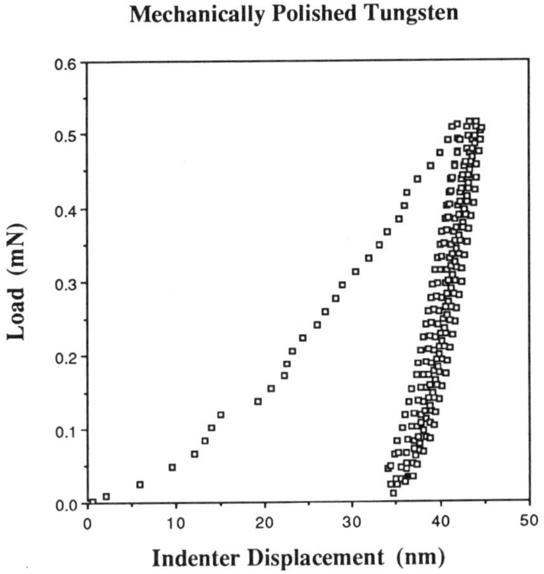


Figure 8.
Low load test on mechanically polished surface.

consistent with the idea that unlike the electropolished surface the mechanically polished surface contains numerous pre-existing dislocation generating sites.

We have earlier reported such theoretical lattice behaviour in sapphire [6] where the defect free nature of contact prior to 'pop-in' was also confirmed by TEM [8]. A similar effect, resembling the breakage of a layer of ice, has also been observed from hard oxide layers such as $Al_2O_3$ on Al [6]. However, the values of stress at 'pop-in' on these ceramics is rather random, and only occasionally reaches the theoretical limits. This is almost certainly due to flaws, as the subsequent inelastic behaviour seems to involve fracture processes [8]. In contrast, the results presented here show a uniform and repeatable attainment of the same theoretical strength in the over thirty different tests that were made. We conclude that given a defect-free surface, a sufficiently sharp indenter used in the manner described here will always enable observation of the lattice theoretical strength.

## DISCUSSION AND CONCLUSIONS

In calculating stresses we have used the contact stiffness S directly in equation (2). Hitherto in sub-micron indentation testing, it has always been neccessary to know the tip shape to allow conversion of the plastic penetration depth into contact area.[2] Even this use of tip shape represented an advance over the need to image every individual indent. However, evaluation of the shape of the indenter is a time-consuming task, requiring TEM imaging of a series of indentations. In addition, with so much effort invested in characterising a given diamond indenter, any damage to the tip due to accident or (say) mishap with a diamond-like substrate is a serious matter. Therefore there is a great deal to recommend the use of S for calculating the overall contact stress where possible; the value of $PE^{*2}/S^2$ (proportional to contact pressure) is continuously available during a test if $E^*$ is known. If $E^*$ itself is unknown or a variable, then the tip shape is still needed. However, we note that, using a test material of well defined elastic modulus, the tip shape may itself be calculated from the indentation data, as plastic depth and contact area ($S^2$) are continuously available. This will provide a very rapid method of characterising tip shape, and hence mitigate many of the risks and difficulties described above. These matters will be reported on elsewhere.

In view of the great usefulness of S it is worth considering its limitations. Firstly, it is difficult to measure large values of S, because the rigidity of the indenter system $K_f$ effectively 'short circuits' the contact stiffness (see Figure 1). Even with improved construction it is unlikely that values of S in excess of $10^8$ N/m could be easily measured. This restricts the indentation contact areas for high modulus materials to the hundreds of square microns range. In present practice, a few square microns is the maximum for good accuracy. A second limitation is that equation (1), the basis of the use of S, only applies for a single asperity contact. Therefore for blunter indenters there is a risk of error, particularly when the test surface is rough. As a result of these two limitations we arrive at the somewhat unexpected conclusion that more information is obtainable (via S) from sub-micron indentation than from the more traditional (i.e. large) scale of hardness testing. This argues for the use of sub-micron indentation testing in a wider range of problems than those restricted purely by dimensional size of the test objects.

Finally we consider processes occurring at the smallest scales of indentation. In principle the sensitivity in S is such that contacts of single atomic dimensions should be measurable. Indeed by going out of contact, and into the Van der Waals attraction range of indenter surface separation, we expect the S sensing method described here to give a useful version of the Atomic Force Microscope.[9] For actual contact, it is an open question whether single atom contacts are mechanically stable, even in principle.[10] This is an interesting question, as there is considerable interest in the electrical properties of extremely small volumes of material. In any case, the results we present here show that at the smallest scales ot testing now accessible, simple macroscopic strength concepts may no longer apply. It has been known for a long time that sufficiently defect free whiskers can show ideal strengths, [11] and it is also known [12] that some thin films on surfaces may allow the bulk below to exhibit near ideal strength. Our results show ideal strength in a quantified, direct manner. A question might now be asked. The surfaces of the tip and tungsten are unlikely to be both perfectly atomically smooth; why then do atomic steps, which when compressed produce locally high stresses, not immediately generate dislocations?

The primary answer must lie in the extremely localised nature of the stress at a surface step, and also in the other possible ways besides bulk defect nucleation in which that stress can be relieved. The high degree of localisation (within an atom or two depth [13]) and the non-uniformity of stress across the surface will make nucleation of an extended defect such as a prismatic loop very unlikely. In any case, for very small loops the nucleation stress will approach the theoretical limit.[12] Even if a series of local or point defects were possible, it is much more likely that a surface (lateral) rearrangement will occur, particularly for facets, or two or more atom steps. Activation energies such as for surface self diffusion are generally much less than bulk nucleation energies. This correlates with the fact that monolayer coverages of some adsorbates at surfaces can influence observed strengths. Clearly, an understanding of mechanical processes at or near to the atomic scale is required.

The methods and results presented here suggest that it is now possible to make direct investigations of mechanical response at the atomic scale, and hence to obtain a new understanding of such central problems as adhesion and crack tip propagation.

References


1. J.B.Pethica, R.Hutchings and W.C.Oliver, *Phil. Mag.*, **A48**, 593, (1983)

2. J.B.Pethica in *Ion Beam Implantation into Metals,* edited by V.Ashworth, W.Grant and R.Procter, (Pergamon Press, Oxford,1982) pp.147-156.

3. M.F.Doerner and W.D.Nix, *J. Mater. Res.,* **1**(4), 601, (1986)

4. S.I. Bulychev, V.P.Alekhin and M.Kh.Shorshorov, *Zavodskaya Laboratoriya,* **9**, 1137, (1975); J.L.Loubet, J.M.Georges and G.Meille, in *Microindentation Techniques in Materials Science and Engineering,* edited by P.J.Blau and B.R.Lawn, (ASTM, Philadelphia, 1986) pp.72-89.

6. J.B.Pethica and W.C.Oliver, *Physica Scripta,* **T19**, 61, (1987)

5. K.Kendall and D.Tabor, *Proc. Roy. Soc. (London)* , **A323**, 321 (1971)

7. D.Tabor, *The Hardness of Metals,* (Clarendon Press, Oxford, 1951)

8. T.F.Page, W.C.Oliver and C.J.McHargue, to appear in *J. Mater. Sci.,* (1989)

9. G.Binnig, C.F.Quate and Ch.Gerber, *Phys. Rev. Lett.,* **56**, 930, (1986)

10. J.B.Pethica and A.P.Sutton, *Jnl. Vac. Sci. & Technol.,* **A6**, 2490, (1988)

11. C.Herring and J.K.Galt, *Phys. Rev.,* **85**, 1060, (1952)

12. N.Gane, *Proc. Roy. Soc. (London)* , **A317**, 367, (1970)

13. J.B.Pethica, *Phys. Rev. Lett.,* **57**, 3235, (1987)

# Ion Channeling Studies of CdTe Films on GaAs

Barry Wilkens, Bellcore, Red Bank, NJ
Ahmet Erbil, Georgia Institute of Technology, Atlanta, Georgia

**Abstract**

Thin films of [111] oriented CdTe have been MOCVD grown onto [111] GaAs substrates. When thicknesses exceed 1000Å the epitaxy is quite good (backscattering minimum yield of approximately 15%) in spite of a 14% lattice mismatch. A narrowing of the Cd angular scan suggests a displacement of some of the Cd atoms in the lattice. A model based on a Te vacancy is presented to describe the data.

## Experiment:

High quality thin films of II-VI compound semi-conductors (e.g. CdTe) can be epitaxially grown on Group III-V materials. such as GaAs, through various UHV techniques. In the present experiment [111] CdTe was MOCVD grown onto [111] GaAs substrates. Ion channeling with MeV He ions was used to study the effects of the large (14%) lattice mismatch between the two materials, on the crystalinity of the film and its interface. Fig. 1 shown the minimum backscatter yield vs film thickness for the series of films measured. It can be seen

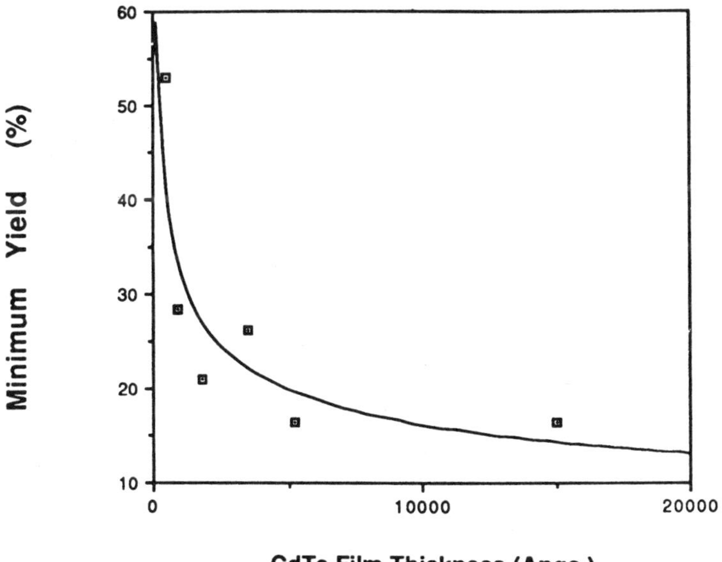


Figure 1) Minimum yield vs film thickness for [111] alignment.

Mat. Res. Soc. Symp. Proc. Vol. 130. ©1989 Materials Research Society

from this figure that after approximately 1000 Å the minimum yield reaches a value under 20%. In the [100] CdTe/ [100] GaAs system it has been shown that the mismatch is taken up at the interface by edge dislocations with a periodicity corresponding to the 14% lattice mismatch.[1] This is most likely what is occuring in the present system, however confirmation awaits TEM results. Angular scans of the films about the [111] axis (fig. 2) show a narrowing in the

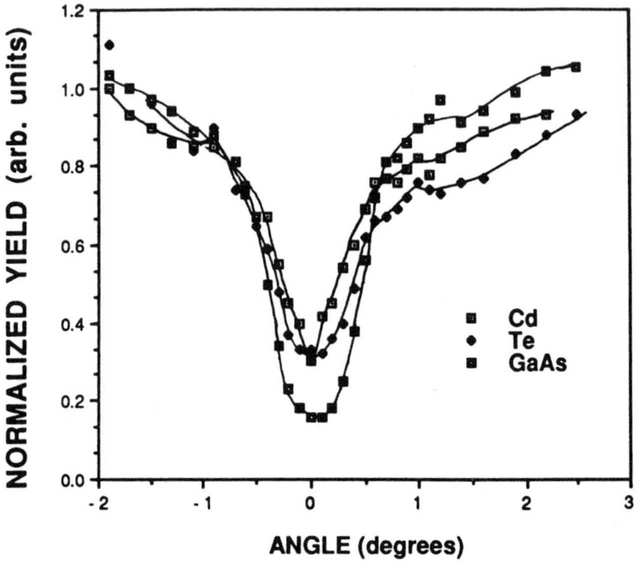


**Figure 2)** Normalized yield vs angle about [111] axis for Cd, Te and GaAs surface backscattered He ions.

Cd angular minimum yield with respect to the Te and GaAs substrate, indicating a small displacement of some of the Cd atoms from their lattice sites. Channeling along the off-normal [211] axis reveals, along with poorer overall channeling, an offset (.15$^{o}$ + .02$^{o}$) of the Cd minimum yield (fig. 3). The Te minimum yield remains aligned with that of the GaAs substrate indicating an effect in the Cd sublattice itself.

**Discussion:**

A model based on Te vacancies gives a geometric description for the angular scan observations. A Te vacancy would allow its four tetragonally bonded Cd atoms to relax or flatten out to a trigonal configuration with their three Te nearest neighbors. This results in Cd atoms moving off lattice sites and protruding into channels by as much as .3Å depending on the "flatness" of the trigonal bond. The lattice displacements are symmetrical about the [111] axis resulting only in a narrowing of the Cd scan as shown in fig. 2.. However, the displacements are

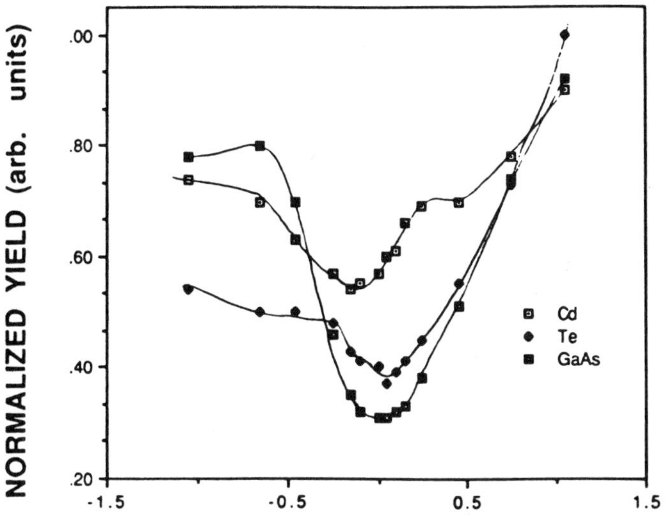


**Fig. 3)** Normalized yield vs angle about [211] axis for Cd, Te and GaAs surface backscattered He ions

assymmetric about the [211] channel with 3 atoms projecting into the [211] channel from one side. On closer inspection of the Zn-blende structure it can be seen that the polarity or stacking order of the Cd and Te atoms can be determined by the direction of the Cd angular minimum offset. With the minimum shifted towards the [111] direction (fig. 3), it can be shown that the Te layer was deposited first onto the GaAs substrate. The Cd offset then is not actually a shift in alignment axis but is merely an assymmetry in the angular yield due to the three Cd atoms (per Te vacancy) protruding into the [211] channel.

**Conclusions:**

We have shown that [111] CdTe/[111] GaAs can be grown with relatively good epitaxy (minimum yield of less than 20%). An offset in the Cd angular scan about the [211] axis was explained by invoking a model based on Te vacancy.

**Acknowledgements:**

The authors wish to acknowledge helpful discussions with H. Farrell, C. Palmstrom, and N. Stoffel.

1. F. Ponce et.al Surf. Sci. 168, 564 (1986)

# RESIDUAL STRESS ANALYSIS OF Al ALLOY THIN FILMS BY X-RAY DIFFRACTION AS A FUNCTION OF FILM THICKNESS

CARLA J. SHUTE*, J.B. COHEN* and D. A. Jeannotte**
*Northwestern University, Department of Materials Science and Engineering, Evanston, IL 60208.
**IBM, IBM-U.S. Technology Products, Hopewell Junction, NY 12533.

## ABSTRACT

Residual stress has been measured as a function of layer thickness in thin films of an Al alloy on oxidized Si by the x-ray "d" versus $\sin^2\psi$ technique. Samples with and without a passivation layer were examined. The results show an increase in residual stress with decreasing film thickness for the passivated samples and indicates that the interface between the metal film and $SiO_2$ may be a region of high stress.

## INTRODUCTION

As the integrated circuit industry pushes for smaller and smaller metal interconnects, the mechanical properties of the thin films used for these connections becomes more and more important to the lifetime of VLSI devices. However, the difficulty of examining these materials increases as their size decreases. One of the major concerns for these materials is the residual stress level present in these films as a result of the fabrication process. These stresses arise due to the thermal expansion mismatch between substrate and film, growth mismatch and lattice mismatch [1,2]. Residual stresses in thin films are typically measured by cantilever beam or curvature techniques [1]. To determine the stress with these techniques, a homogeneous, isotropic, biaxial state of stress must be assumed. These techniques also require special specimen configurations and are destructive in nature. X-ray diffraction techniques, on the other hand, have been developed and in use for many years and offer a non-destructive means of measuring residual stresses in these materials.

## THEORY

The polycrystalline x-ray technique for stress analysis involves measuring the d-spacing of a single reflection for several orientations of a sample to determine the strain along different directions of the sample. Referring to Fig.1, the strain is determined along a direction perpendicular to the diffracting planes from the following Eqn.,

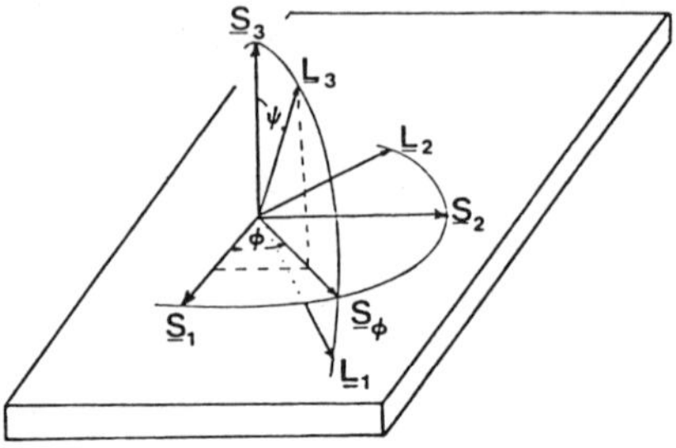


Fig. 1. Coordinate systems and geometry used for polycrystalline x-ray residual stress determination.

$$(\epsilon'_{33})_{\phi\psi} = \frac{d_{\phi\psi} - d_o}{d_o} \tag{1}$$

where $d_o$ is the unstressed lattice parameter and $(\epsilon'_{33})_{\phi\psi}$ is the strain along the $L_3$ direction. Six independent strain measurements must be performed in order to completely determine the symmetric strain tensor. The strains determined for the L or measurement (laboratory) coordinate system can then be transformed to strains in the sample coordinate system, S, and converted to stresses through the use of Hooke's law. The following "stress" equation results for an isotropic material [3]:

$$\begin{aligned}\frac{d_{\phi\psi}-d_o}{d_o} &= \frac{1+\nu}{E}(\sigma_{11}\cos^2\phi + \sigma_{12}\sin2\phi + \sigma_{22}\sin^2\phi - \sigma_{33})\sin^2\psi \\ &+ \frac{1+\nu}{E}\sigma_{33} - \frac{\nu}{E}(\sigma_{11} + \sigma_{22} + \sigma_{33}) \\ &+ \frac{1+\nu}{E}(\sigma_{13}\cos\phi + \sigma_{23}\sin\phi)\sin2\psi.\end{aligned} \tag{2}$$

For the case of a biaxial stress state, this equation reduces to,

$$\begin{aligned}\frac{d_\psi-d_o}{d_o} &= \frac{1+\nu}{E}\sigma_\phi\sin^2\psi - \frac{\nu}{E}(\sigma_{11} + \sigma_{22}) \\ \sigma_\phi &= \sigma_{11}\cos^2\phi + \sigma_{12}\sin2\phi + \sigma_{22}\sin^2\phi.\end{aligned} \tag{3}$$

For a homogeneous stress distribution, $\sigma_\phi$ is constant for all $\psi$ tilts and can be determined from the slope of the "d" versus $\sin^2\psi$ plot. Data at three different $\phi$ tilts is typically taken for the triaxial stress determination [4].

EXPERIMENTAL

The samples for this study consisted of 0.5, 1 and 2 $\mu$ thick films of an Al-Cu alloy deposited on oxidized 100 Si wafers. Samples with and without a 1$\mu$ thick $SiO_2$ passivation layer were examined. The stress analysis was performed on a 12 kW Rigaku rotating anode with Cr target. A graphite diffracted beam monocromator was used to improve the signal to noise ratio. A state of biaxial stress was assumed for these films due to their finite thickness and the "d" versus $\sin^2\psi$ technique was employed.

The films for this study were strongly textured with the {111} planes nearly parallel to the surface of the sample. The 222 peak used for the stress analysis was therefore very weak at intermediate $\psi$ tilts, i.e. between 0 and 65 degrees. To examine the behavior of the "d" versus $\sin^2\psi$ plot, tilts of 0, 2, 18.43, 26.57, 36.27, 45.0, 56.79, 71.57, and 77.0 degrees were used for the 1$\mu$ sample. The more intense peaks at $\psi$ values of 0,2,56.79,67.21 and 71.57 degrees were used for the remaining samples. All data was corrected for Lorentz-polarization and absorption before fitting peak locations. The applied absorption correction was derived for each type of sample configuration to account for any peak shift due to absorption from the thin film layers.

DISCUSSION

Three types of behavior are predicted for "d" versus $\sin^2\psi$. Linear behavior has been described above. If the shear terms, $\sigma_{13}$ and $\sigma_{23}$ are not zero, then $\psi$-splitting will occur for positive and negative $\psi$ tilts due to the $\sin 2\psi$ term in Eqn. 2. The présence of $\sigma_{33}$ causes curvature in the $\sin^2\psi$ plot. This curvature is due to the fact that $\sigma_{33}$ can only exist as a gradient in the surface layers. Since the penetration depth of the x-rays is a function of $\psi$, different averages of this gradient in stress will be sampled at the various $\psi$ tilts. An illustration of these is shown in Figs. 2a,b. Fig. 2c illustrates oscillatory behavior and is not predicted by Eqn. 2. This type of behavior is due to an inhomogeneous stress distribution within the irradiated volume used for the stress analysis. This is caused by an inhomogeneous distribution of plastic deformation or by elastic incompatibility between the grains of the sample [5,6,7].

A plot of "d" versus $\sin^2\psi$ for the 1$\mu$ as deposited film without a passivation layer is shown in Fig. 3. The error bars show the statistical

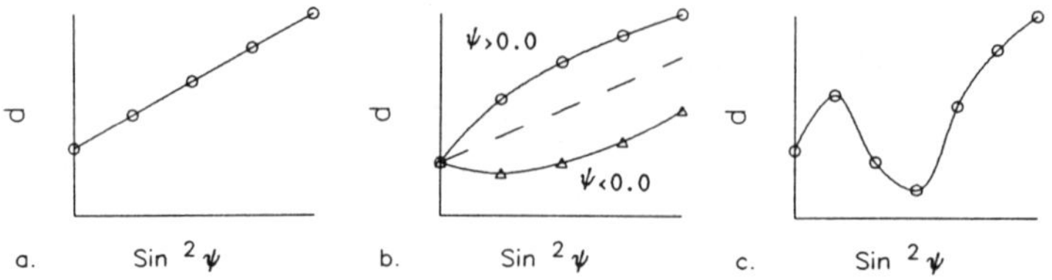


Fig. 2. Types of "d" versus $\sin^2\psi$ plots. Regular behavior which is predicted by Eqn. 2 is shown in a and b. Oscillatory behavior due to inhomogeneous stress distribution is shown in c.

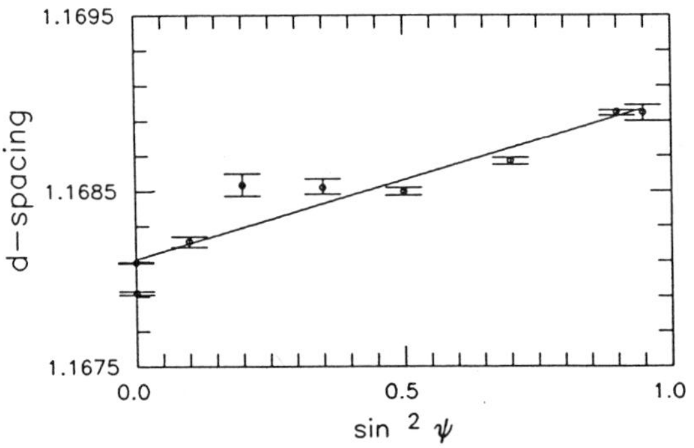


Fig. 3. D-spacing versus $\sin^2\psi$ for unpassivated, $1\mu$ thick Al alloy film.

counting error associated with determining the peak location and subsequent d-spacing of the diffracting planes. As is common with textured materials, this plot shows a slight oscillatory behavior. Noyan and Nguyen [8,9] have recently examined the errors associated with applying conventional stress methods to oscillatory data for textured Cu (311). They conclude that use of the "d" versus $\sin^2\psi$ technique results in a stress value which is representative of the average stress present in the sample.

The results of the stress analyses are plotted in Fig. 4. The instrumental error estimated for a horizontal beam divergence of $1^\circ$ and also including sample displacement error is less than 5 MPa for all samples. The statistical error is 1-2 MPa for all measurements. For a detailed description of these errors and their estimation see references [3,10].

The residual stress for the unpassivated samples ranges from approximately 50-80 MPa and does not vary with film thickness. The passivated samples show a significant variation with thickness however. The

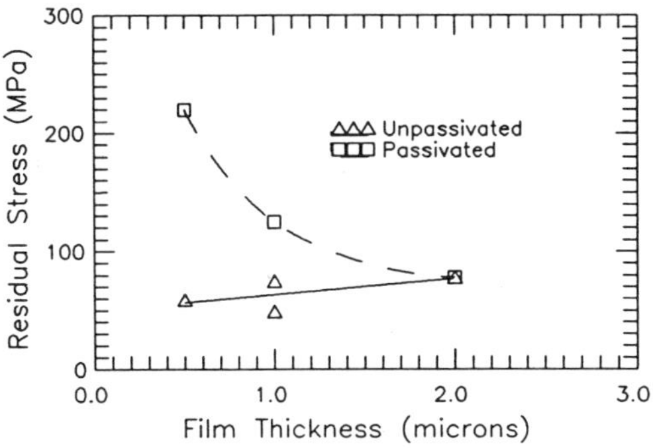


Fig. 4. Residual Stress versus film thickness for passivated and unpassivated samples.

stress value obtained for the 0.5, 1.0, and 2.0$\mu$ passivated samples is 220, 125, and 78 MPa respectively. The x-rays in all cases penetrate through the entire film thickness and the stress values are averages over the entire thickness of the films. These results therefore indicate that there may be an interface region which experiences increased constraint from the sputtered and thermal $SiO_2$ layers. There may also be some recrystallization occurring in the thicker film, resulting in a lower stress value. Examination of the diffraction peaks from these samples reveals that the peaks from the 2$\mu$ passivated sample are sharper than the others and may indicate a more ordered, less plastically deformed material. Line broadening studies to examine this behavior are planned.

As discussed above, the presence of a gradient in $\sigma_{33}$ should cause curvature in the "d" versus $\sin^2\psi$ plot. For these measurements however, the entire gradient was sampled at all $\psi$ tilts and the presence of a gradient could not be verified. Stress analyses using glancing angle geometry to determine the stress at different depths within the film will be performed to examine this.

CONCLUSION

The use of x-ray diffraction to determine stresses in thin films has been demonstrated. This technique eliminates the need for special sample configurations and is non-destructive. The assumptions about the stress state present in the sample can also be verified from the data for the stress analysis. This is not true for cantilever beam or curvature techniques.

The results of this study indicate that the metal film-$SiO_2$ interface may be a region of high stress and may be increasingly more important to the behavior of these materials as the film thickness decreases.

## ACKNOWLEDGEMENTS

This work has been supported by IBM.

## REFERENCES

1. K. Kinosita, Thin Solid Films 12, 17 (1972).
2. A. K. Sinha and T. T. Sheng, Thin Solid Films 48, 117 (1978).
3. I. C. Noyan and J. B. Cohen, Residual Stress: Measurement by Diffraction and Interpretation, (Springer-Verlag New York Inc., 1987), p. 121.
4. R. A. Winholtz and J. B. Cohen, Aust. J. Phys. 41, 189 (1988).
5. H. Dolle, J. Appl. Cryst. 12, 489 (1979).
6. R. Marion, Ph.D. Thesis, Northwestern University, Evanston IL (1972).
7. I. C. Noyan, Ph.D. Thesis, Northwestern University, Evanston IL (1984).
8. I. C. Noyan and L. T. Nguyen in Advances in X-ray Analysis, edited by C. S. Barrett, J. V. Gilfrich, R. Jenkins, J. C. Russ, J. W. Richardson, Jr. and P. K. Predecki (Plenum Publishing Corporation 31, 1988) pp. 191-204.
9. I. C. Noyan and L. T. Nguyen, to be published in Volume 32 of Advances in X-ray Analysis.
10. M. R. James, Ph.D. Thesis, Northwestern University, Evanston IL (1977).

# THERMAL STRAIN MEASUREMENTS IN EPITAXIAL $CoSi_2$/Si BY DOUBLE CRYSTAL X-RAY DIFFRACTION

GANG BAI*, MARC-A. NICOLET*, THAD VREELAND JR.*, Q. YE**, Y. C. KAO**, K. L. WANG**
*California Institute of Technology, Pasadena, CA 91125
**UCLA, Los Angeles, CA 90024

## ABSTRACT

The perpendicular x-ray strain of epitaxial $CoSi_2$ films grown on Si(111) substrates at $\sim 600^oC$ by MBE was measured at various temperatures. Within experimental error margins, the strain decreases linearly with rising temperature at a rate of $(1.3\pm0.1)\times10^{-5}/^oC$ from room temperature up to $600^oC$. Over that temperature range and the duration of a complete measurement ($\sim .5h$ to $\sim 2h$), these strain values remain reversible. At $593^oC$, the x-ray strain is $-0.85\%$, which is about the strain that a stress-free $CoSi_2$ film on Si(111) would have at that temperature. This results show that the stress in the epitaxial $CoSi_2$ film is fully relaxed at the growth temperature. Strains below the growth temperature are induced in the film by the difference in the linear coefficient of thermal expansion of $CoSi_2$ and Si. They were calculated by assuming that the density of misfit dislocations formed at the growth temperature remains constant. The slope of the strain-temperature dependence obtained that way agrees with the measured slope if the unknown Possion ratio of $CoSi_2$ is assumed to be $\nu_{CoSi_2} = 0.35$. A film stress of $\sim 0.8$ GPa at room temperature was calculated using the above value for the Possion ratio, 130 GPa for the Young modulus, and the measured x-ray strain.

## INTRODUCTION

The successful growth of an epitaxial $CoSi_2$ film on a Si(111) substrate holds promise for the fabrication of novel devices such as tunneling structures and metal/semiconductor superlattices. Since the initial experimentation on epitaxial growth of $CoSi_2$ on Si[1], progress in understanding and controlling the growth process and the imperfections has been achieved. Many important structural and electrical properties of the $CoSi_2$ films have been characterized[2–3].

The atomic coordination of the epitaxial $CoSi_2$ film is rotated $180^o$ about the [111] axis with respect to the Si(111) substrate (type-B)[4]. The lattice mismatch, $f$, between the $CoSi_2$ and Si at room temperature is $-1.23\%$. The critical thickness, $t_{cr}$, for pseudomorphic growth of $CoSi_2$ on Si(111) was reported to be $\sim 30\AA$[5]. The growth of $CoSi_2$ on porous Si(111) substrates was explored with the aim of increasing the critical thickness of the film[6]. Double crystal diffractometry shows that epitaxial $CoSi_2$ films grown on Si substrates, with surfaces offset from {111}, are misoriented, i.e., there is a tilt angle between the [111] directions of the $CoSi_2$ film and the Si substrate[7]. Strain measurements on thick films (i.e., the film thickness, $t_f > t_{cr}$) by double crystal diffractometry at room temperature show that the perpendicular x-ray strain $\epsilon^{\perp}$ of epitaxial $CoSi_2$ films has a typical value of about $-1.65\%$ over the range of film thickness $t_f$ from 100Å to 2500Å[7,8] (Perpendicular x-ray strain is defined as $\epsilon^{\perp} = (d_f - d_s)/d_s$, where $d_f$ and $d_s$ are the interplanar spacings of (111) planes in the film and the substrate). This result differs from the value of $\epsilon^{\perp} = -1.23\%$ of a stress-free film at room temperature, which indicates that thick $CoSi_2$ films on Si(111) are under tensile stress at room temperature.

The lattice mismatch between bulk $CoSi_2$ and bulk Si at the growth temperature ($\sim$ 600°C) is estimated to be about $-0.85\%$ from the lattice mismatch at room temperature, $f = -1.23\%$, and thermal expansion coefficients of the two materials, $\alpha_{Si} = 2.7 \times 10^{-6}/^{o}C$ and $\alpha_{CoSi_2} = 9.4 \times 10^{-6}/^{o}C$[9]. Thermal stress can be induced in the $CoSi_2$ film when the sample is cooled down from the growth temperature to room temperature if there are no interface dislocations introduced to release the stress during the cooling.

In this paper, we present the results of perpendicular x-ray strain measurements of thick $CoSi_2$ films on Si at temperatures between room temperature and 600°C by double crystal diffractometry. This gives some insight on the stress state at the growth temperature, the plastic flow below the growth temperature, and the residual stress at room temperature of epitaxial $CoSi_2$ films grown on Si(111) substrates by MBE.

## EXPERIMENTAL METHOD AND RESULTS

Single strained layers of $CoSi_2$ were grown epitaxially on Si(111) substrates by MBE. The surfaces of all substrates were offset from the (111) planes of the Si towards the $[1\bar{1}0]$ direction, with offset angles, $\phi_s$, from 0° to 16°. The growth conditions have been described elsewhere[10]. The $CoSi_2$ was formed at about (600±50)°C. The film thicknesses of the samples used in this study are about 1000Å, which is more than 30 times thicker than the critical thickness of an epitaxial $CoSi_2$ film on Si(111) ($t_{cr} \sim 30$Å). Rutherford Backscattering Spectroscopy and channeling with a 2 MeV $^4He^+$ beam were used to characterize the stochiometry and crystallinity of the epitaxial $CoSi_2$ films. The films have the stochiometry of the $CoSi_2$ phase (Co:Si$\approx 1:2$ within the 10% experimental error) and excellent crystallinity ($\chi_{min} \approx 3\%$).

X-ray rocking curves were taken with Fe $K_{\alpha_1}$ radiation diffracted from the (111) planes of a high quality Si(111) crystal. The sample was mounted on a heating stage in air. The sample temperature T was controlled between room temperature and 600°C. The symmetrical (111) diffraction was recorded and the perpendicular x-ray strain $\epsilon^{\perp}$ (strain perpendicular to (111)) was extracted.

The x-ray rocking curves of sample K310 measured at room temperature and 593°C are shown in Fig. 1. The perpendicular x-ray strain measured at 593°C is $\epsilon^{\perp} = (-0.85 \pm 0.01)\%$. This is the value of $\epsilon^{\perp}$ for a stress-free $CoSi_2$ film at that temperature. The conclusion therefore is that at the growth temperature, the film stress is fully relaxed. The perpendicular elastic strain at room temperature, $e^{\perp} = \epsilon^{\perp} - f$, calculated from the measured perpendicular x-ray strain $\epsilon^{\perp} = (-1.66 \pm 0.01)\%$ and the lattice mismatch $f = -1.23\%$ at room temperature is $-0.43\%$. This strain is induced by the different thermal contraction of the film and the substrate, and the epitaxial constraint.

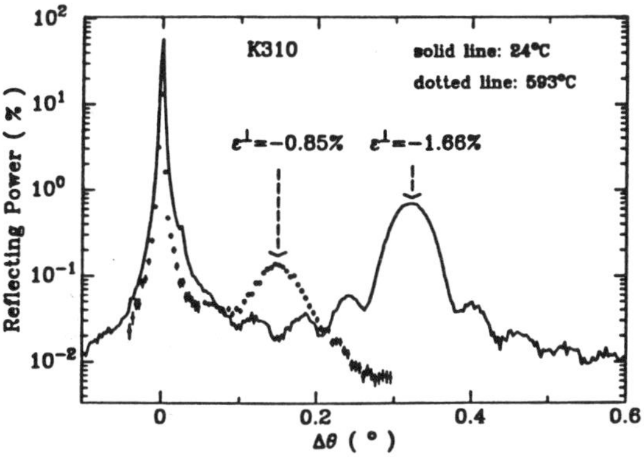


Fig. 1 *Fe* $K_{\alpha 1}$ x-ray rocking curves from the symmetrical (111) diffraction planes. The solid curve is for the measurement taken at room temperature in air. The dotted line is for the measurement taken at 593°C.

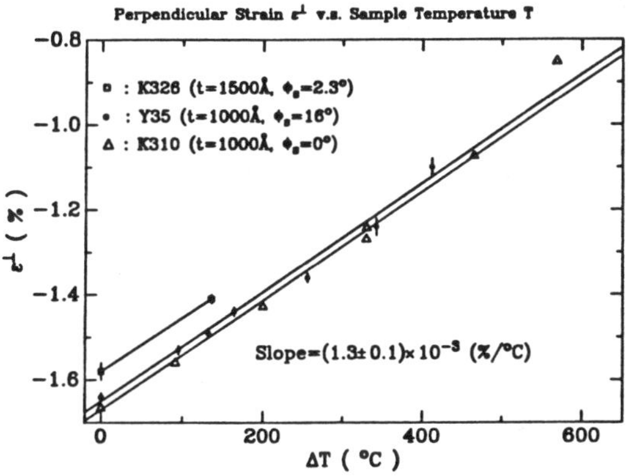


Fig. 2 Dependence of the perpendicular x-ray strain *versus* the temperature at which measurement was performed. Three sets of data are for three different samples. The slopes are same for all samples.

The results of the perpendicular x-ray strain measurements at elevated temperatures are shown in Fig. 2. There are three sets of experimental data corresponding to three different samples which have slightly different perpendicular x-ray strains at room temperature. The variations of the strain values measured at room temperature are probably due to the variations of the growth temperature. The slopes of the three curves are the same, $(1.3 \pm 0.1) \times 10^{-5}/^{o}C$, which is larger than the difference of the bulk thermal expansion coefficients between $CoSi_2$ and Si at room temperature, $(\alpha_{CoSi_2} - \alpha_{Si}) = 6.7 \times 10^{-6}/^{o}C$. This fact can be explained by assuming that the lateral change of the $CoSi_2$ lattice is constrained to be same as that of the Si substrate. The perpendicular change of the $CoSi_2$ lattice is therefore larger than that of a stress free $CoSi_2$ film due to the Possion effect. If one assumes that the misfit dislocations are locked in below the growth temperature, meaning that their numbers and positions are fixed after the growth of the film, the slope of the perpendicular x-ray strain $\epsilon^{\perp}$ versus temperature T is then given by

$$\frac{\delta\epsilon^{\perp}}{\delta T} = (\frac{1+\nu_{CoSi_2}}{1-\nu_{CoSi_2}}) \cdot (\alpha_{CoSi_2} - \alpha_{Si}),$$

where $\nu_{CoSi_2}$ is the Possion ratio of $CoSi_2$. We were unable to find its value in the literature. But if we assume that $\nu_{CoSi_2} = 0.35$ and use the value of ($\alpha_{CoSi_2}$-$\alpha_{Si}$) at room temperature, we obtain a slope of $1.4 \times 10^{-5}/^{\circ}C$, which is within 10% of the measured value ( see Fig. 2). This agreement indicates that the misfit dislocation density remains unchanged in the temperature range and durations of this experiment. A direct proof of this assertion is in fact contained in the additional observation that thermal cycling of a sample does not create a hysterisis in $\epsilon^{\perp}(T)$ of Fig. 2, even for temperature up to 600°C for 2-4 hours.

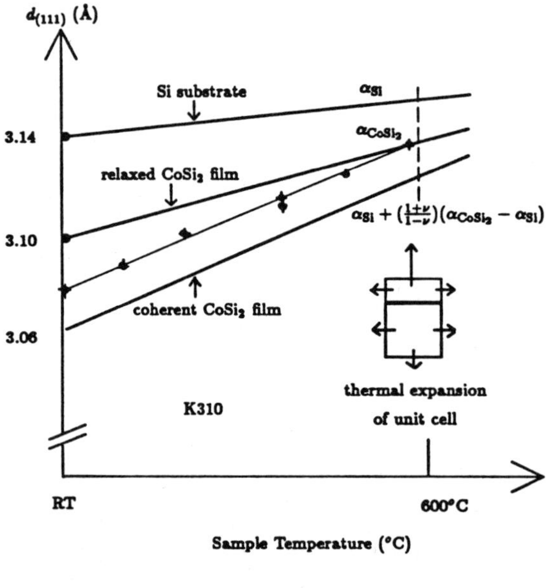


Fig. 3 Schmatic plot of the lattice constants of $CoSi_2$ and Si *versus* temperature. The experimental curve meets the curve for the bulk thermal expansion of $CoSi_2$ at the growth temperature, indicating that the $CoSi_2$ film is relaxed on Si(111) at the growth temperature. The slope of the experimental curve is parallel to the curve for the coherent $CoSi_2$ film on Si, indicating no plastic flow below the growth temperature.

## DISCUSSION

In Fig. 3, the results of the strain measurements are translated into a picture that shows the lattice constants as a function of temperature. It is consistent within the experimental errors and the uncertanties in the knowledge of some parameters (such as the Possion ratio and the bulk thermal expansion coefficients of $CoSi_2$ from room temperature to 600°C) to conclude that our thick epitaxial $CoSi_2$ films ($t_f > t_{cr}$) grown on

Si(111) substrates are free of stress at the growth temperature. At that temperature, the lattice mismatch of $\sim -0.85\%$ is accomodated by misfit dislocations. When the samples are cooled to room temperature, these misfit dislocations are essentially immobile and no additional misfit dislocations are generated. As a result, thermal strains are induced in the $CoSi_2$ films by the differential thermal contraction of the film and the substrate. The $CoSi_2$ films are under tension below the growth temperature. The thermal stresses in the $CoSi_2$ films at room temperature can be calculated from the measured perpendicular strain $\epsilon^{\perp}$ at room temperature,

$$\sigma \approx -\frac{E_{CoSi_2}}{2\nu_{CoSi_2}} \cdot (\epsilon^{\perp} - f),$$

where $E_{CoSi_2}$ is the Young modulus of $CoSi_2$. For example, using the measured x-ray strain of the sample K310 at room temperature, $\epsilon^{\perp} = -1.66\%$, and assuming the elastic constants of $CoSi_2$, $E_{CoSi_2} = 130$ GPa[11] and $\nu_{CoSi_2} = 0.35$, we obtain the tensile stress at room temperature in the $CoSi_2$ film, $\sigma = 0.8$ GPa. We can generalize this result to calculate the thermal stress in the $CoSi_2$ film at any temperature $T$ below the growth temperature $T_g$,

$$\sigma(T) = -\frac{E_{CoSi_2}}{(1 - \nu_{CoSi_2})} \cdot (\alpha_{CoSi_2} - \alpha_{Si}) \cdot (T - T_g).$$

From the experimental results and the previous discussion, we know that the above formula for the thermal stress is valid for temperature from 600°C to as low as room temperature, and for tensile stress as large as 0.8 GPa. This stress is not large enough for misfit dislocation generation or multiplication.

## CONCLUSION

In summary, we have measured the perpendicular x-ray strain in $CoSi_2$ films *in situ* from room temperature to 600°C. The x-ray strain value of $-0.85\%$ measured at 600°C suggests that the thick ( $t_f > t_{cr}$ ) epitaxial $CoSi_2$ films are relaxed on Si(111) substrates at the growth temperature. The slope of the perpendicular strain versus temperature below the growth temperature is larger than the difference of bulk thermal expansion coefficient between $CoSi_2$ and Si. This result is consistent with the model that there is no plastic flow below the growth temperature. Therefore, we conclude that the strain in the $CoSi_2$ film measured at temperatures below the growth temperature is induced by differential thermal contraction of $CoSi_2$ and Si under an epitaxial constraint. The perpendicular x-ray strain of about $-1.65\%$ measured at room temerature for thick samples ( $t_f > t_{cr}$ ) with thickness from 100Å to 2500Å can be predicted from the elastic constants of $CoSi_2$, the bulk thermal expansion coefficients of $CoSi_2$ and Si, and the growth temperature.

## ACKNOWLEDGEMENTS

This work is sponsored by the Semiconductor Research Corporation under contract #87-SJ-100.

## REFERENCES


[1] R.Tung, J.Poate, J.Bean, J.Gibson, and D.Jacobson, Thin Solid Film, 93, 77(1982).

[2] Mat. Res. Soc. Proc., Vol.67, Heteroepitaxy of Silicon I, edited by J.C.C. Fan and J.M.Poate (1986).

[3] Mat. Res. Soc. Proc., Vol.91, Heteroepitaxy of Silicon II, edited by J.C.C. Fan, J.M. Phillips, and B.Y. Tsaur (1987).

[4] R.T.Tung, J.M.Gibson, and J.M.Poate, Phys. Rev. Lett., 50, 429(1983).

[5] J.L.Batstone, J.M.Phillips, and J.M.Gibson, Mat. Res. Soc. Proc., Vol.91, 445(1987).

[6] Y.C.Kao, D.Jamieson, G.Bai, C.W.Nieh, T.L.Lin, B.J.Wu, H.Y.Chen, and K.L.Wang, Mat. Res. Soc. Proc., Vol.91, 445(1987).

[7] G.Bai, D.N.Jamieson, T.Vreeland Jr., and M-A.Nicolet, Mat. Res. Soc. Proc., Vol.102, 259(1988).

[8] D.N.Jamieson, G.Bai, Y.C.Kao, C.W.Nieh, M-A.Nicolet, and K.L.Wang, Mat. Res. Soc. Proc., Vol.91, 479(1987).

[9] M-A.Nicolet, S.S.Lau, VLSI Electronics, Vol.6, 330(1983).

[10] Y.C.Kao, K.L.Wang, E de Fresart, R.Hull, G.Bai, D.N.Jamieson, and M-A. Nicolet, J. Vac. Sci. Technol., B6, 745(1987).

[11] A.H.van Ommen, C.W.T.Bulle-Lieuwma, and C.Langereis, J. Appl. Phys., 64, 2706(1988).

# PRINCIPLES AND APPLICATIONS OF WAFER CURVATURE TECHNIQUES FOR STRESS MEASUREMENTS IN THIN FILMS

PAUL A. FLINN
Intel Corporation, SC9-45, 2250 Mission College Blvd, Santa Clara, CA. 95124 and
Dept. of Materials Science and Engineering, Stanford University, Stanford, CA 94305.

## ABSTRACT

Measurement of the curvature induced in a wafer (or other flat plate) by the stress in a thin film has long been used as a convenient and accurate technique for the determination of the stress. Numerous improvements over the years have led to instruments that provide simple and rapid measurements of stress as a function of the time and temperature for any desired thermal history. A computer controlled instrument using laser scanning will be briefly described and its capabilities and limitations discussed.

Applications of the technique to a variety of thin film materials will be discussed. In addition to the effects of differences in thermal expansion, stresses associated with various deposition techniques, gain or loss of material, phase transformations and flow will be considered. In aluminum based systems, themal expansion, plastic flow and phase transformation play major roles. Refractory metals show, in addition, large stresses associated with the deposition process. In inorganic dielectric systems thermal expansion effects are usually relatively small; deposition effects and the gain or loss of material are the dominant effects. Silica based glasses formed by chemical vapor deposition, for example, show large stress changes due to gain or loss of water, and plasma deposited silicon nitride films show large effects associated with hydrogen. Overall, determination of the stress as a function of time and temperature is a valuable part of the evaluation of a thin film material for use in a VLSI device.

## INTRODUCTION

Stress and strain in thin films have been a matter of practical importance for many years, long before the era of microelectronics, in areas such as electrodeposition[1] and optical coatings[2]. Many experimental techniques were developed. Some are based on direct measurement of force and displacement, esssentially minaturization of traditional mechanical testing techniques[3]. Some are indirect, such as the well known X-ray "residual stress" measurement which is actually an elastic strain mesurement from which stress is inferred[4, 5]. Perhaps the most widely used class of techniques is based on measuring the curvature of a substrate produced by the stress in the film. The curvature may be measured in a variety of ways that fall into two categories: displacement measurements and direct curvature measurements. Displacement may be measured by a mechanical or elctromechanical gage, by microscopic observation, or by optical interference (fringe counting). Curvature may be measured by optical means, as will be discussed in detail later, or, if a single crystal substrate is used, by an X-ray diffraction technique[6]. Once the radius of curvature is known, the stress, $\sigma_f$, in the film may easily be calculated[1, 7]:

$$\sigma_f=[E_s t_s^2]/[6(1-\nu)t_f R] \tag{1}$$

where $E_s$ is Young's modulus for the substrate, $\nu$ is Poisson's ratio for the substrate, $t_s$ is the substrate thickness, $t_f$ is the film thickness, and R is the radius of curvature.

For various orientations of single crystal substrates, the appropriate values of $E/(1-\nu)$ are given by Brantley[8]. Note that only the properties of the substrate and the thickness of the film need be known; the elastic properties and texture of the film are not needed to determine the stress.

Since the early work by Sinha[9], interest in stress in the thin films used in the microelectronics industry has been steadily growing. All of the traditional classes of materials: metals, semiconductors, inorganic insulators and polymers are used in VLSI fabrication. For a variety of reasons, to be discussed later, these films are, quite commonly, in states of high mechanical stress. Not surprisingly, these stresses can result in failure; either immediate, as in the case of complete fracture of a conductor, or, even more

Mat. Res. Soc. Symp. Proc. Vol. 130. ©1989 Materials Research Society

serious, a delayed failure, as in the case of cracked passivation leading ultimately to failure by corrosion. Equipment for measurement of stress in the films used for VLSI fabrication should satisfy several requirements. In addition to the obvious needs for adequate sensitivity and reproducibility, the measurements should be made with the films on their normal substrate, since deposition conditions and substrate-film interactions are an important part of the problem. Since the film stress generally depends in detail on the thermal history, capability of measurement as a function of temperature and time is needed. The wide variety of materials and process conditions of interest results in such a large number of measurements that computer controlled, unattended operation is essential. The wide range of stresses and film thicknesses encountered necessitates an instrument with wide dynamic range; that is, the ability to measure relatively large as well as quite small curvatures.

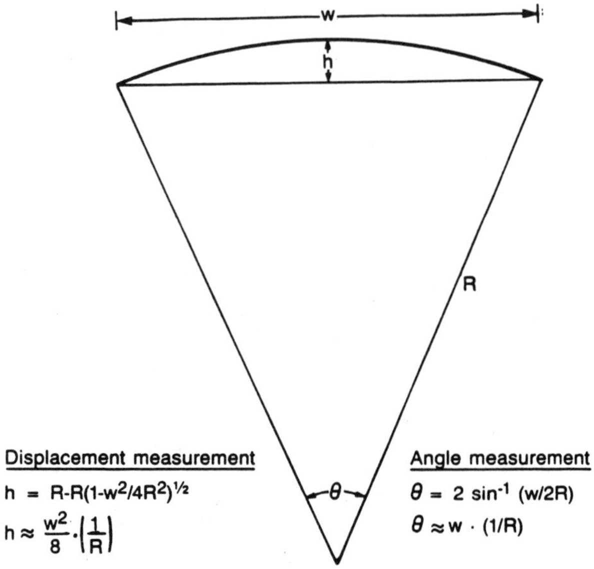


**Figure 1:** Comparison of Displacement and Angle Measurements. For R=100 m and w = 100 mm, h = 12.5 µm and Θ = $10^{-3}$ radians. For a precision of 1%, h must be measured to 125 nm, or Θ to $10^{-5}$ radians.

## BASIC DESIGN CONSIDERATIONS

The necessary sensitivity can be obtained by either displacement or curvature measurement; the relations are illustrated in Figure 1. It is desireable to avoid physical contact with the sample, particularly for work at elevated temperatures; consequently, an optical or X-ray technique is preferable to an electromechanical one. For reasonably large wafers the required precision in displacement measurement can be obtained by an interferometric technique[10], but the apparatus is complex and not easily adapted to fully automatic operation. X-ray measurement of the curvature has the advantage that it is a direct measurement of the bending due to the film independent of the surface profile of the unstressed substrate. Unfortunately, the X-ray method is rather slow, requires expensive equipment, and does not provide as high precision as can be obtained by optical means.

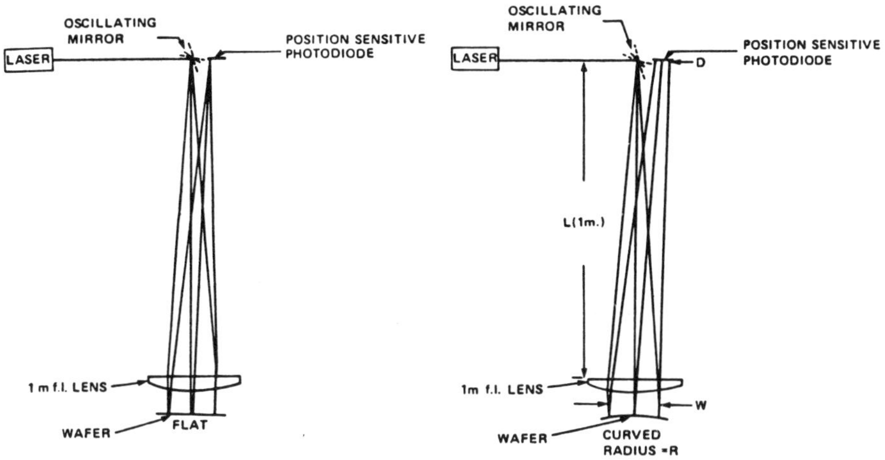


**Figure 2:** a. Reflection from flat surface. b. Reflection from curved surface. For a uniformly curved wafer of radius R, the total displacement D is given by D = 2LW/R.

The technique of laser scanning introduced by Sinha, and used, with various modifications, by others. provides a relatively inexpensive, rapid, easily automated technique. By suitable design, the necessary sensitivity and dynamic range can be obtained[11]. In contrast to the earlier techniques that depend on mechanical translation of the sample[9, 12] or of both laser and detector[13], this design uses an oscillating mirror and lens combination[14] to obtain the necessary scan. The basic principle is shown in Figure 2. The mirror is located at the focus of the lens; as the mirror rotates the various light rays pass through the lens and emerge parallel to each other. If the light is reflected from a flat surface (Figure 2a), the reflected beams reenter the lens parallel to each other and all reach the same point in the focal plane. If the reflecting surface is curved (Figure 2b) the rays reenter the lens at different angles and reach the focal plane at different points. The displacement of the spot in the focal plane is proportional to the slope of the reflecting surface. A position sensitive photodiode can measure the position of the reflected spot with a precision of about 5μm. With the 1 meter path length, the angular sensitivity of measurement is about 2.5 μrad. Greater sensitivity could be obtained by using a longer optical path, but the present equipment has been adequate so far. The bending of a wafer under the force of gravity is easily observed, and for typical films a precision of stress measurement of a few MPa is normally obtained. A more detailed description of the apparatus is given elswhere[11].

## LIMITATIONS

This equipment has a significant inherent limitation; it is a differential technique. The initially present curvature of even high quality silicon wafers is of the same order as the effects under study; consequently, it is necessary to measure the wafer without the film (preferably before deposition) and carry out a point by point subtraction of the reference measurement from the measurement with the film to obtain the effect of the film. Since wafers are not usually curved uniformly, accurate repositioning of the wafer for the before and after measurements is needed. The stress in individual films of a multilayer structure can, in principle, also be determined by a differential method; that is, measurements with and without a particular film. The problem is that one often loses precision by taking a small difference between large numbers. In particular, it is usually difficult to determine the stress of a metal film in the presence of a high stress dielectric film, since the much lower stress in the metal film produces a relatively small change in the already large curvature due to the dielectric film.

In addition, there are several practical limitation to the equipment in its present form. The precision can be limited by temperature measurement problems: the precision of stress measurement, a few MPa, corresponds to the effect of a quite small temperature change (about 2 degrees for an aluminum film on silicon). The present hot stage is only usable to about 600 C, and the sample is in air. Operation at higher temperatures and in a controlled atmosphere should be possible, but at the cost of considerably increased complexity.

## STRESS IN METAL FILMS

In addition to a convenient means of measurement, some degree of understanding of the stress in thin films is needed; otherwise, we would be confronted with a totally unmanageable mass of data. Although metal films exhibit the most complex behavior, they are perhaps the best understood. Depending on the deposition conditions, there may or may not be a large initial stress present in the film as deposited. One critical factor is the temperature of deposition relative to the melting point of the metal; this determines the amount of surface diffusion that can occur as the metal is deposited. If a relatively low melting point metal, such as aluminum, is deposited at room temperature or above, enough atomic motion is possible so that the resulting structure is relatively stress free. In the case of a high melting point metal, such as tungsten, the low atomic mobility at ordinary temperatures results in highly nonequilibrium structures, and an initial state of very high stress. Sputtered tungsten is normally in a highly compressive state[15], while CVD tungsten silicide is highly tensile[16]. If the metal is deposited on a crystalline substrate, an epitaxial relation between the film and the substrate may occur, and the lattice mismatch can result in an initial intrinsic strain, even when the atomic mobility is adequate. After the deposition, during the subsequent thermal history, phase transformations, the effects of differences in thermal expansion, and plastic flow all play a role in determining the state of stress in the film[17].

Conventional aluminum based metallizations deposited on non-crystalline substrates provide a relatively simple case amenable to detailed analysis. If the metal is deposited at a moderate temperature (above about 125 C), the microstructure is relatively stable, and the effect af any thermal history can be accounted for on the basis of thermal expansion and plastic flow in the aluminum governed by thermally activated glide[18]. On the basis of this model, at any given temperature, the stress required for plastic flow can be calculated as[11]:

$$\sigma = \tau \cdot (1 - T/T_F) \tag{2}$$

where $\tau$ is the limiting stress at absolute zero where no thermal activation is present and $T_F$ is the (hypothetical) temperature at which the flow stress would go to zero. $T_F$ is weakly (logarithmically) dependent on the strain rate.

As shown in Figure 3, this equation defines a region in the stress-temperature plane whithin which the metal will exhibit elastic behavior, and the stress change with temperature results simply from the difference in thermal expansion between aluminum and silicon, a value of about -2.3 MPa/K. (The exact position of the boundary depends on the strain rate, but the effect is small. The lines D-E and D'-E' show the effect of an order of magnitude change in strain rate.) Only when the sample is heated or cooled so as to approach a boundary line does substantial plastic flow occur. This is illustrated by the thermal history shown in Figure 4. At the start of this thermal cycle, the sample was at room temperature with a stress of about 200 MPa. (Point A). On heating it followed the elastic curve until it approached the lower limiting line, then followed it approximately up to Point B, about 300 C and -100 MPa. On cooling, there is a short elastic region, then the upper limiting curve is followed to Point C, about 120 C and 200 MPa. The sample was then heated and we see again elastic behavior with heating to Point D. On cooling, there is now a substantial elastic region until the upper limiting curve is approached. Cooling continues approximately along the limiting curve to room temperature (Point E).

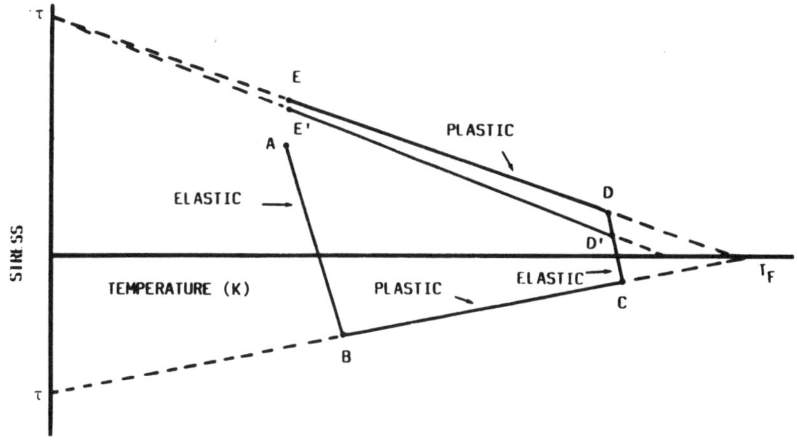


**Figure 3:** Idealized behavior of stress as a function of temperature as given by equation 2.

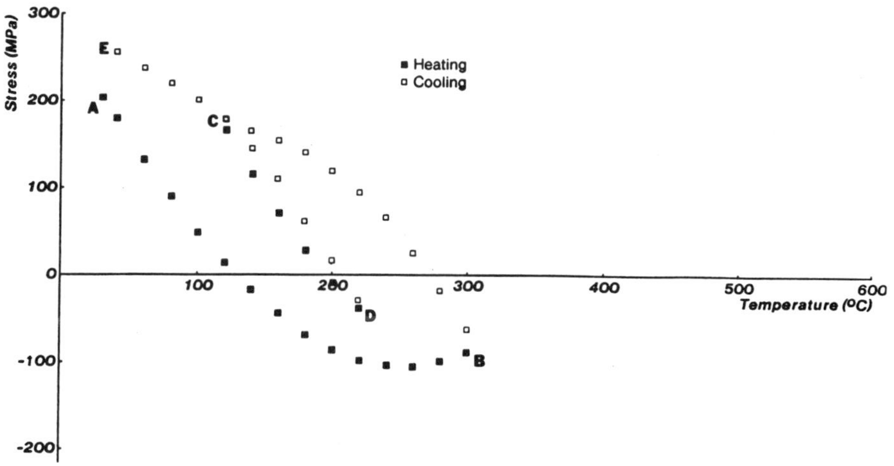


**Figure 4:** Stress as a function of temperature for an Al-1%Si film during thermal cycling

The model can easily be extended to account for the behavior as the limiting curve is approached[11], and a quantitative fit to experimental data can be obtained, as shown in Figures 5 and 6.

As may be seen, the characteristic material parameters are quite sensitive to deposition conditions. Deposition at low temperatures results in relatively low values of $T_F$ and high values of $\tau$, so that the limiting lines are steeply sloped. Higher deposition temperatures lower $\tau$ and raise $T_F$, so that the limiting lines are nearly horizontal. The effect may be at least partly due to grain size effects, since material deposited at elevated temperatures, but in a poor atmosphere, shows behavior similar to that obtained with low temperature deposition. The same model also predicts the relaxation behavior at constant temperature:

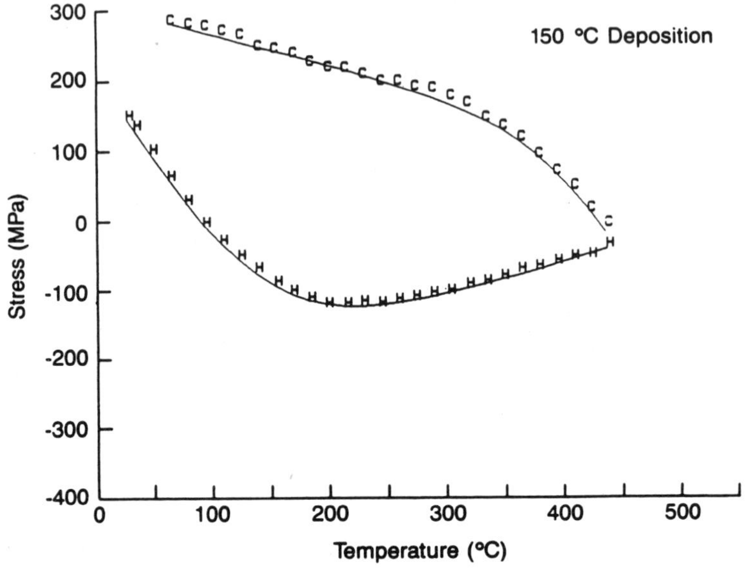


**Figure 5:** Stress vs. temperature for an Al-Si film deposited at 150 C. On heating $\tau$ = 450 MPa, $T_F$ = 800 K; On cooling $\tau$ = 435 MPa, $T_F$ = 1000 K

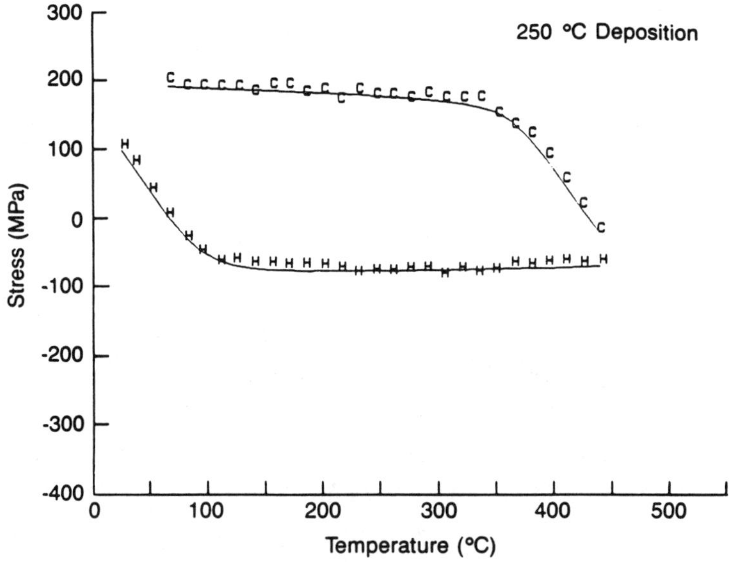


**Figure 6:** Stress vs. temperature for an Al-Si film deposited at 250 C. On heating $\tau$ = 110 MPa, $T_F$ = 1780 K; On cooling $\tau$ = 210 MPa, $T_F$ = 3600 K

$$\sigma = B\{\ln(B/R) - \ln[t + (B/R)\cdot\exp(-\sigma_0/B)]\} \quad (3)$$

where:

$$B = kT\tau/\Delta F$$
$$R = [E/(1-\nu)]\cdot\varepsilon_0\cdot\exp(-\Delta F/kT) \quad (4)$$

and $\sigma_0$ is the initial stress.

This is in excellent agreement with experimental data, as shown in Figure 7.

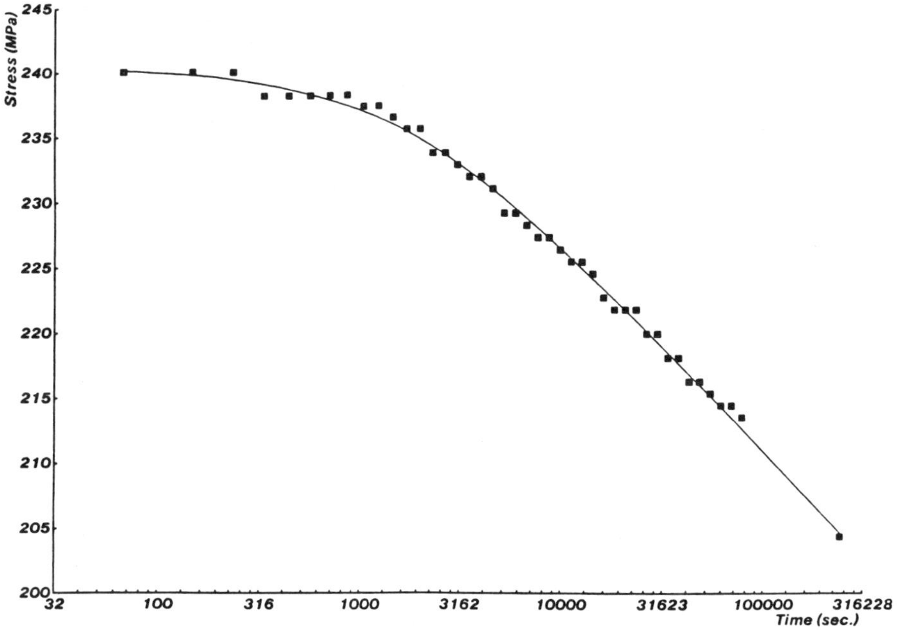


**Figure 7:** Stress relaxation in an Al-1%Si alloy at room temperature.
Solid line given by equation 3

## STRESS IN DIELECTRIC FILMS

Dielectric films formed at sufficiently high temperatures; e. g., thermal oxide, normally have little initial stress, and the relatively small stress observed at room temperature is simply due to the thermal expansion mismatch. The films are quite stable; no hysteresis is observed on thermal cycling, and no time dependent changes are observed at ordinary temperatures. The situation is quite different and much more complicated for the films used in the upper layers of the structure. These films are necessarily deposited at relatively low temperatures, so that highly nonequilibrium structures result. Quite large initial stresses are commonly present. These may be compressive or tensile, depending on the details of deposition, as well as the material composition.

One widely used class of dielectrics is based on the chemical vapor depostion of silicon dioxide, often containing some phosphorous. These glasses are usually under compressive stress as deposited and become more compressive if exposed to a moist atmosphere at room temperature. Typical behavior on heating to a moderate temperature and holding is shown in Figure 8. The effect of thermal expansion is quite small; the main effect is the loss of water. Water is a byproduct of the deposition reaction and some is trapped during deposition. More may be absorbed at room temperature. On heating, the water is driven off resulting in a stress change from compressive to highly tensile. Below about 400 C the basic glass structure is stable and the process is reversible. After dehydration, if the sample is exposed to a moist atmosphere, the stress will again become compressive. Stress changes with time can be accounted for quantitatively by a simple diffusion model[12]. The effect of heating a glass of this type to a somewhat higher temperature is shown in Figure 9. After the increase in tensile stress due to loss of water, the stress begins to decrease. This is the process of "densification", the elimination of the microporosity that accomodated the water. After this treatment, the glass will no longer absorb a significant amount of water. It is important to note that during cooling after densification the stress vs. temperature curve is straight with a slope that con be accounted for by thermal expansion effects. This shows that the densification is different from ordinary glass flow, which occurs at higher temperatures. If the stress change were due to flow it would occur during cooling as well as heating.

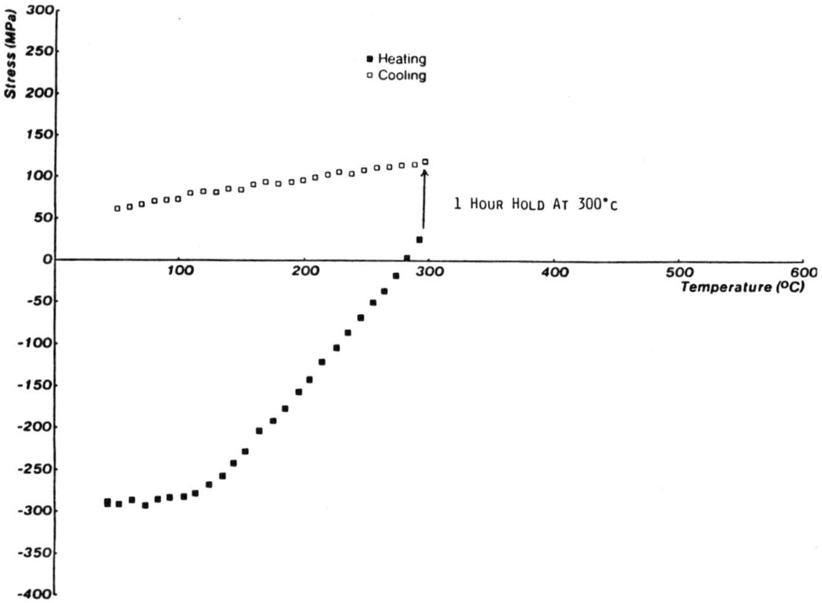


**Figure 8:** Stress change in a phosphosilicate glass due to loss of water on heating.

Another important class of dielectrics is based on silicon nitride or oxynitride. Plasma enhancement of the chemical vapor deposition is necessary to obtain the reaction at sufficiently low temperatures. The resultant films typically have quite high initial stresses, usually compressive, but they may be made tensile by choosing appropriate deposition conditions. These films are stable at room temperature, and have no tendency to absorb moisture, but they do exhibit large, irreversible stress changes on heating above about 400 C, as shown in Figure 10. This is clearly not a stress relaxation effect, since the stress does not simply decrease to zero, but changes sign. The change of stress with time at constant temperature, as may be seen in Figure 11, is well fit by the appropriate solution to the diffusion equation, essentially the same equation as used to describe the behavior of water in silicate based glass[12]:

$$\sigma=\sigma_0-\Delta\sigma\{1-(8/\pi^2)\sum_{n=1}^{\infty}\frac{1}{(2n-1)^2}\exp(-[(2n-1)/2L]^2Dt)\} \quad (5)$$

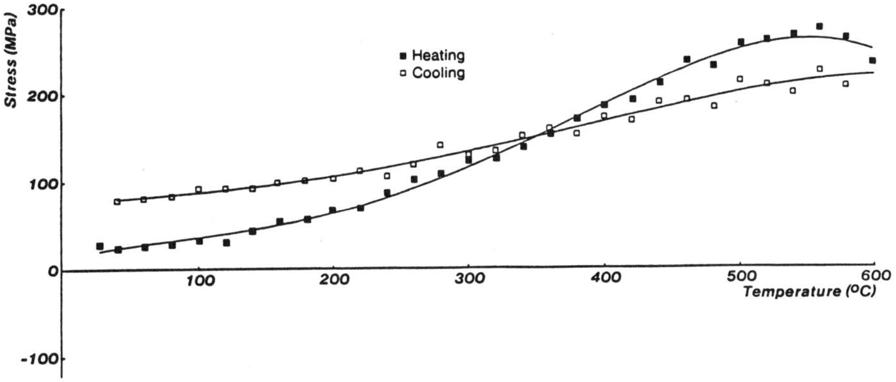


**Figure 9:** Stress changes due to loss of water and densification in BPSG.

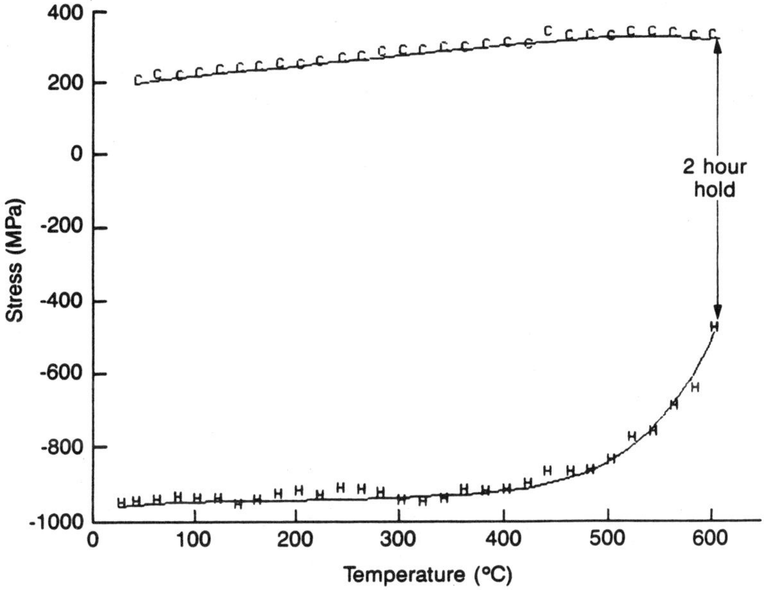


**Figure 10:** Stress change in a PECVD silicon nitride on heating.

The stress change can thus be accounted for by the diffusive evolution of hydrogen trapped in the glass during the plasma deposition. This hydrogen is present in a different form than the bonded hydrogen seen by IR spectroscopy. A substantially higher temperature is required to break these bonds and release this hydrogen[19].

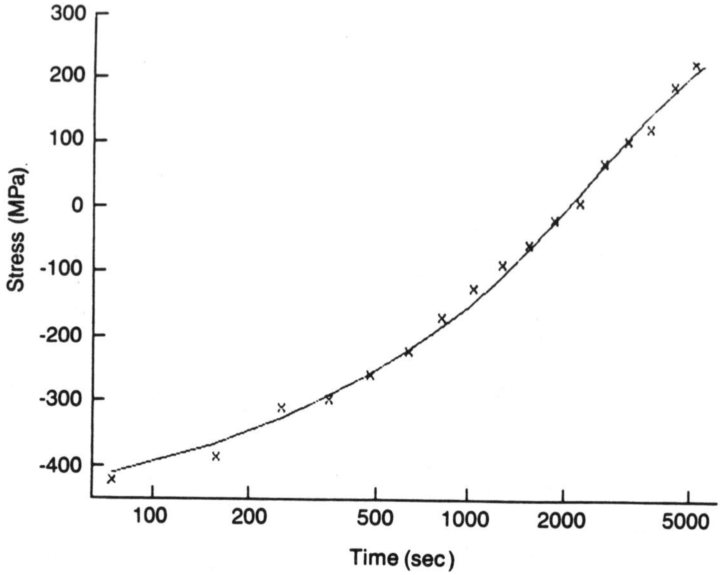


**Figure 11:** Stress change at constant temperature in the PECVD silicon nitride. The solid curve is given by equation 5

## SUMMARY

The stress in various films of importance in the manufacture of VLSI devices can easily be determined as a function of time and temperature by optical measurement of wafer curvature. Differences in thermal expansion between film and substrate play a role in producing film stress, but initial stress developed during deposition, and effects due to structural changes in the film, and gain or loss of material often are much larger than the thermal expansion effects. In general, the stress state of a given film cannot be understood without taking into account its entire history, beginning with a detail knowledge of deposition conditions.

## ACKNOWLEDGEMENTS

Many people at both Stanford and Intel contributed to this work. In particular I wish to thank Professor W. D. Nix, Professor D. Barnett, D. S. Gardner, J. Turlo and F. von Preissig of Stanford, and C. Chiang, J. Wei and Z. Fong of Intel.

## References


1. N. N. Davidenkov, "Measurement of Residual Stresses in Electrolytic Deposits", *Soviet Physics - Solid State*, Vol. 2, 1961, pp. 2595-2598.

2. H. K. Pulker, "Mechanical Properties of Optical Films", *Thin Solid Films*, Vol. 89, 1982, pp. 191-204.

3. M. F. Doerner and W. D. Nix, "A Method for Interpreting the Data from Depth-sensing Indentation Instruments", *Journal of Materials Research*, Vol. 1, 1986, pp. 601-609.

4. M. R. James and J. B. Cohen, "The Measurement of Residual Stresses by X-Ray Diffraction Techniques", in *Treatise on Materials Science and Technology, Vol 19A*, H. Herman, ed., Academic Press, New York, 1980, pp. 1-62.

5. A. Segmueller and M. Murakami, "X-Ray Diffraction Analysis of Strains and Stresses in Thin Films", in *Treatise on Materials Science and Engineering, Vol. 27*, H. Herman, ed., Academic Press, New York, 1988, pp. 143-200.

6. A. Segmueller, J. Angilelo and S. J. LaPlaca, "Automatic X-ray Diffraction Measurement of the Lattice Curvature of Substrate Wafers for the Determination of Linear Strain Patterns", *Journal of Applied Physics*, Vol. 51, 1980, pp. 6224-6230.

7. R. W. Hoffman, "The Mechanical Properties of Thin Condensed Films", in *Physics of Thin Films*, G. Hass and R. E. Thun, ed., Academic Press, New York, Vol. 3, 1966, pp. 211-273.

8. U. A. Brantley, "Calculated Elastic Constants for Stress Problems Associated with Semiconductor Devices", *Journal of Applied Physics*, Vol. 44, 1973, pp. 534-535.

9. A. K. Sinha, H. J. Levenstein and T. E. Smith, "Thermal Stresses and Cracking Resistance of Dielectric Films on Si Substrates", *Journal of Applied Physics*, Vol. 49, 1978, pp. 2423-2426.

10. A. G. van Nie, "A Method for the Determination of the Stress in, and Young's Modulus of Silicon Nitride Passivation Layers", *Solid State Technology*, Vol. 23, 1980, pp. 81-84.

11. P. A. Flinn, D. S. Gardner and W. D. Nix, "Measurement and Interpretation of Stress in Aluminum-Based Metallization as a Function of Thermal History", *IEEE Transactions on Electron Devices*, Vol. ED-34, 1987, pp. 689-699.

12. E. J. McInerney and P. A. Flinn, "Diffusivity of Moisture in Thin Films", *Proceedings of the 20th Annual International Reliability Symposium*, IEEE, 1982, pp. 264-267.

13. J. T. Pan and I. Blech, "In situ Stress Measurement of Refractory Metal Silicides during Sintering", *Journal of Applied Physics*, Vol. 55, 1984, pp. 2874-2880.

14. J. T. Luxon and D. E. Parker, *Low Power Laser Applications*, Prentice-Hall, Englewood Cliffs, N.J., 1985, ch. 9.

15. M. F. Doerner, D. S. Gardner, and W. D. Nix, "Platic Properties of Thin Films on Substrates as Measured by Sub-Micron Indentation Hardness and Substrate Curvature Techniques", *Journal of Materials Research*, Vol. 1, 1986, pp. 845-851.

16. Y. Shioya, T. Itoh, S. Inoue and M. Maeda, "Analysis of Stress in Chemical Vapor Deposition Tungsten Silicide Film", *Journal of Applied Physics*, Vol. 58, 1985, pp. 4194-4199.

17. D. S. Gardner and P. A. Flinn, "Mechanical Stress as a Function of Temperature in Thin Aluminum Films and its Alloys", *Proceedings of the Materials Research Society - Symposium C*, Materials Research Society, 1988.

18. M. F. Ashby and H. J. Frost, "The Kinetics of Inelastic Deformation above 0 K.", in *Constitutive Equations in Plasticity*, A. S. Argon, ed., MIT Press, Cambridge, MA, 1975, pp. 117-147.

19. R. C. Budhani, R. F. Bunshah and P. A. Flinn, "Kinetics of Structural Relaxation and Hydrogen Evolution from Plasma Deposited Silicon Nitride", *Applied Physics Letters*, Vol. 52, 1988, pp. 284-286.

# CHARACTERIZATION OF SURFACE MECHANICAL PROPERTIES AND RESIDUAL STRESSES IN ION IMPLANTED NICKEL

D.J. MORRISON,* J.W. JONES,* G.S. WAS,** A. MASHAYEKHI,** AND D.W. HOFFMAN***
* Department of Materials Science and Engineering, The University of Michigan, Ann Arbor, MI 48109
** Department of Nuclear Engineering, The University of Michigan, Ann Arbor, MI 48109
*** Research Staff, Ford Motor Company, Dearborn, MI 48124

## ABSTRACT

In this study, we have measured changes in hardness and residual stress state in polycrystalline nickel samples which have undergone a variety of ion beam surface treatments including self-implantation, amorphization by ion beam mixing Ni-Al multilayers, and the formation of a two-phase microstructure by elevated temperature aluminum implantation. Hardness-depth profiles of the the near-surface regions were determined using an ultra-low load microindentation system, and residual stress states induced by the surface modifications were analyzed using optical interferometry. These property changes were correlated with the accumulation of surface fatigue damage.

## INTRODUCTION

Ion implantation techniques are frequently used to produce microstructures that result in surface properties that are significantly different from the substrate. Nastasi et al.[1] found increases in the surface hardness of nickel specimens implanted with carbon; and Oliver, McHargue, and Zinkle [2] reported surface hardening in self-implanted copper. In addition to producing changes in surface hardness, ion implantation can induce residual stresses in the surface region [3]. Previous research [4] has shown that the formation of persistent slip band (PSB) features at the surface was significantly inhibited by using ion beam processes to microalloy the surface of nickel specimens with aluminum. In this study, the effects of several Ni-Al system ion beam modifications on near-surface hardness and residual stress were analyzed and compared to the ability of the ion beam modified region to suppress surface fatigue damage in polycrystalline nickel. Previous studies on the effects of thin surface layers on fatigue damage accumulation primarily addressed elastic modulus [5] and stacking fault energy [6] effects.

## EXPERIMENTAL PROCEDURES

Experiments were conducted on three types of nickel specimens. Specimens used for the ultra-low load microindentation tests were cut from a Ni-270 (99.98% Ni) rod. These specimens were annealed in a flowing argon-3% hydrogen atmosphere at 500°C for three hours resulting in a grain size (average linear intercept) of 24 μm. After annealing, the disks were electropolished in a 20% perchloric acid and ethanol solution at -50°C. Cantilever beam fatigue specimens (design similar to ASTM D-671) were fabricated from Ni-200 (99.0% Ni) sheet. These specimens were annealed in the argon-hydrogen atmosphere at 800°C for four hours which produced a grain size of 70 μm. Following heat treating, the fatigue specimens were electropolished in the same manner as previously discussed. Residual stress specimens were made from 0.125 mm thick nickel (99.9% Ni) foil. Disks 19 mm in diameter were punched from the foil, and both sides were mechanically ground on 600 grit SiC paper followed by final polishing on a Syntron vibratory polisher using 0.3 μm alumina as an abrasive. Polishing resulted in a final thickness of about 0.10 mm.

Ion beam modifications listed in Table I were applied to the nickel hardness specimens and cantilever beam fatigue specimens. In addition, the 350 keV modifications were applied to the residual stress samples. All specimens were irradiated in either a 400 kV Varian ion implanter (350 keV modifications) or 1.7 MV Ionex tandem ion accelerator (3 MeV modifications) at pressures less than $5 \times 10^{-7}$ torr. Ion beam mixed multilayer specimens were produced by alternately electron beam evaporating Ni and Al layers in which the thicknesses were adjusted to produce an overall composition of Ni-75 at% Al in the 100 nm thick multilayer region. Evaporations were performed at a base pressure better than $2x10^{-7}$ torr and with the nickel substrate at room temperature. Eleven layers were applied (6 Ni and 5 Al) with Ni as the first and last layer to reduce subsequent oxidation. Ion beam mixing of the multilayers and self-implantation were carried out with a beam of either 3 MeV $Ni^{++}$ or 350 keV $Ni^{+}$ to a fluence of $1x10^{16}$ ions/cm$^2$ at 220°C. A two-phase $\gamma + \gamma'$ near-surface microstructure was produced by implantation of 350 keV $Al^{+}$ to a fluence of $5x10^{17}$ ions/cm$^2$ at 500°C. Beam heating was used to maintain specimen temperature within ± 5°C of the desired temperature.

TABLE I. Ion Beam Surface Modifications of Nickel Specimens

| Modification Type | Ion Irradiation | Fluence (i/cm$^2$) | Irrad. Temp. °C | Surface Layer Thickness | Normalized Hardness (1) | Residual Stress (GPa)(2) | Slip Line Density (mm$^{-1}$)(3) |
|---|---|---|---|---|---|---|---|
| Self-Implantation | 350 keV $Ni^{+}$ | $1x10^{16}$ | 220(4) | 100 nm | 1.9 | -0.4 | 34.8 |
| Self-Implantation | 3 MeV $Ni^{++}$ | $1x10^{16}$ | 220 | 100 nm | 2.3 | note (5) | 30.7 |
| Mixed Ni-75%Al Multilayers | 350 keV $Ni^{+}$ | $1x10^{16}$ | 220(4) | 100 nm | 2.2 | +0.3 | 8.0 |
| Mixed Ni-75%Al Multilayers | 3 MeV $Ni^{++}$ | $1x10^{16}$ | 220 | 100 nm | 2.8 | note (5) | 5.9 |
| Elevated Temp. Al Implanted | 350 keV $Al^{+}$ | $5x10^{17}$ | 500 | 400 nm | 1.9 | -1.3 | 2.2 |

(1) Hardness values at 50 nm normalized to unmodified specimen values at the same depth.
(2) Negative values represent compressive residual stresses and positive values represent tensile stresses.
(3) Slip line length per unit area after $10^6$ cycles. Value for unmodified Ni was 37.1 mm$^{-1}$.
(4) Residual stress specimens irradiated at 24°C
(5) Residual stress measurements not obtained on this specimen.

Hardness-depth profiles of the near-surface regions were obtained using a Nanoindenter mechanical properties microprobe described in references [7] and [8]. For each type of surface condition, at least nine indentations were made normal to the ion beam modified surface. Results of the nine indentations were averaged to produce hardness-depth profiles.

Surface residual stresses were determined by the classical method of measuring the resultant warping or curvature of the substrates [9]. The shape of the nickel foil substrate was determined before and after treatment with reference to an optical flat by observation in monochromatic light. The resulting pattern of Newton's rings was photographed and the fringe spacing measured in orthogonal directions using a computer driven mechanical scanner. Foil curvatures were determined by least squares fitting of parabolas to the recorded fringe spacings. The resulting quadratic coefficients, $K_x$ and $K_y$, are linearly related to the average stress by the formula:

$$\text{Stress} = \frac{E h^2}{3(1-\nu)t} (\Delta K_x + \Delta K_y)$$

where E is Young's modulus of the nickel substrate, $\nu$ is Poisson's ratio, h is the foil thickness, t (<<h) is the thickness of the stressed layer at the treated surface, and the $\Delta$Ks indicate the changes in the quadratic coefficients induced by the treatment. This analysis assumes that the substrate only deforms elastically. If this condition is not met, the equation above overestimates the magnitude of the stress.

The effects of the ion beam surface modifications on fatigue damage accumulation were determined by performing fully reversed cantilever beam fatigue testing on unmodified and ion beam modified nickel specimens. Maximum bending deflection was adjusted to produce a surface total strain amplitude of $\pm 1.2x10^{-3}$. All fatigue testing was performed at room temperature. Periodically during each fatigue test, acetate replicas were extracted from the surface to provide a record of fatigue damage. The replicas were evaporation shadowed with Au-40% Pd at a 30° angle and analyzed using optical microscopy.

## RESULTS

The results of the Nanoindenter studies indicated that all the ion beam surface modifications significantly increased the hardness of the near-surface region. Figure 1 shows normalized hardness-depth profiles for the modifications carried out at 350 keV. Normalized hardness is defined as the hardness of the ion beam modified specimen at a specified depth divided by the hardness of an unmodified specimen at the same depth. Self-implantation and Al implantation produced almost identical hardness profiles. The ion beam mixed multilayer specimen exhibited significantly higher hardness at depths less than about 80 nm, but followed the hardness profiles of the other modifications at depths greater than 100 nm. Significantly higher hardnesses were observed on specimens implanted or mixed with 3 MeV ions. Figure 2(a) compares hardness-depth profiles for the specimens self-implanted at 350 keV and 3 MeV, and Figure 2(b) shows the effect of mixing the multilayers with 350 keV and 3 MeV ions. Hardness values at a depth of 50 nm for the different modifications are also tabulated in Table I.

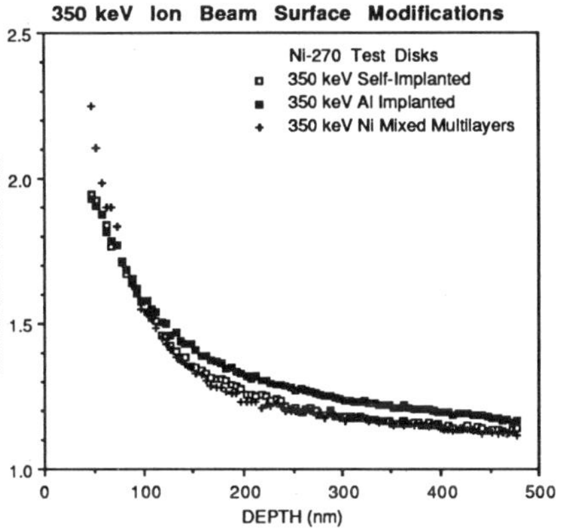


FIGURE 1: (Left) Hardness-depth profiles of 350 keV self-implanted specimen, 350 keV aluminum implanted specimen, and multilayer specimen mixed with 350 keV $Ni^{+}$. Hardness is normalized to the hardness of an unmodified specimen at the same depth. See Table I for irradiation parameters.

FIGURE 2: (Below) Hardness-depth profiles of (a) self-implanted and (b) Ni mixed multilayer specimens that were irradiated with 3 MeV and 350 keV Ni ions to a fluence of $1x10^{16}$ ions/cm$^2$.

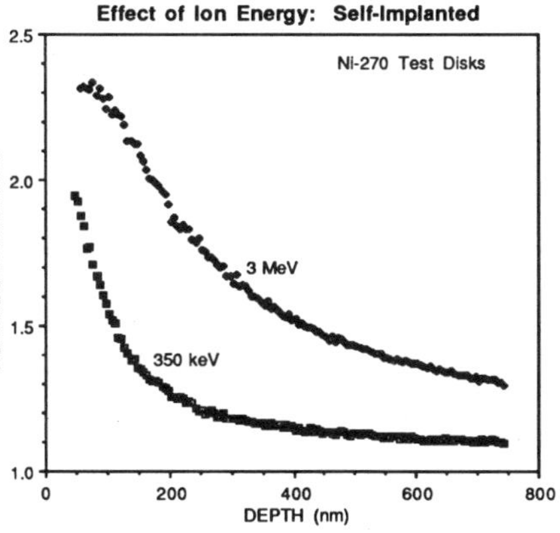


2(a)

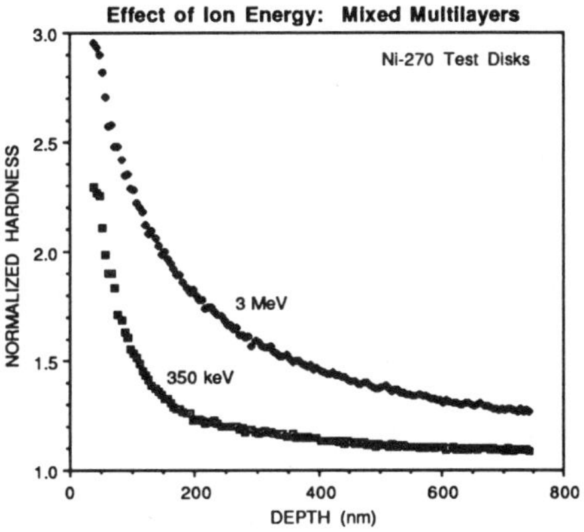


2(b)

Results of the residual stress measurements are presented in Table I. Self-implantation produced a small compressive residual stress while implantation with Al generated larger compressive residual stresses which extended to a depth of 400 nm due to enhanced diffusion of Al in Ni. The unexpectedly large compressive residual stress measured in the aluminum implanted specimen was probably due to plastic deformation of the substrate during irradiation. Residual tensile stresses were observed in the multilayered specimen after the evaporation process. Subsequent ion beam mixing reduced the magnitude of the tensile stress slightly.

The evolution of surface PSB features during fatigue testing is summarized in Table I where the slip line density (line length per unit area) after $10^6$ fatigue cycles is given. Aluminum implantation almost completely eliminated surface PSB features while the ion beam mixed multilayer modification reduced the frequency of PSB surface features by a factor of six. Self-implantation had almost no effect on the suppression of surface PSB features. Increasing ion energy to 3 MeV did not significantly alter the effect of the modification on PSB suppression compared to the same type of modification carried out at 350 keV. Optical micrographs depicting typical surface fatigue damage in an unmodified specimen after $1.5 \times 10^6$ cycles and a specimen implanted with 350 keV Al at 500°C after $1.8 \times 10^6$ cycles are shown in Figure 3.

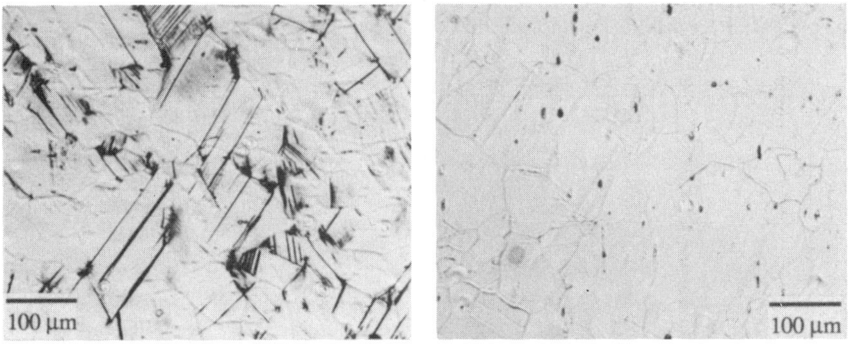


3(a) 3(b)

FIGURE 3. Optical micrographs of flexure fatigue specimens showing surface fatigue damage in (a) unmodified nickel specimen after $1.5 \times 10^6$ cycles and (b) 350 keV $Al^+$ implanted specimen after $1.8 \times 10^6$ cycles.

## DISCUSSION

The near-surface hardness increase produced in the self-implanted specimen indicated that irradiation-induced lattice damage played an important role in surface strengthening. Previous research [10] indicated that nickel self-implantation produced point defects such as vacancies and interstitials which coalesced to form complex dislocation loop structures on {111} planes. These implantation-induced dislocation structures would be expected to inhibit the movement of mobile dislocations, thus providing a strengthening effect. TEM analysis indicated that a two phase $\gamma + \gamma'$ structure was produced by the elevated temperature implantation of Al similar to results reported by Ahmed and Potter [11]. As a result, strengthening would also be provided by precipitation hardening mechanisms in the Al implanted specimen. TEM evaluation of the mixed multilayer specimens indicated that ion beam mixing amorphized the Ni-75 at% Al multilayer region. In this case a hardness increase would be produced as a result of the amorphous structure as well as solid solution strengthening.

Self-implantation of the nickel specimen resulted in a small compressive residual stress due to the defect structure produced during irradiation. The magnitude of the compressive residual stress was greater in the aluminum implanted specimen due to the additional effect of the larger size of the aluminum atom. Tensile residual stresses observed in the mixed multilayer specimen resulted from large tensile residual stresses that are commonly found in evaporated layers [9]. Mixing the evaporated layers with $Ni^{+}$ reduced the magnitude of the tensile stress slightly due to the effect of irradiation induced defects.

The results of the fatigue experiments indicated that high surface hardness was the most important factor in suppressing the the evolution of surface fatigue damage. The self-implanted specimen exhibited high surface hardness after implantation but resulted in practically no suppression of surface damage because of significant softening of the hard near-surface region during cyclic deformation in a manner similar to cyclic softening commonly observed in cold-worked FCC metals [12]. Cyclic softening of the near-surface region in the self-implanted specimen allowed easy penetration of PSBs to the surface. This phenomenon is shown in the hardness-depth profiles in Figure 4(a). A comparison of the profiles acquired before and after fatigue loading clearly shows softening at depths less than 100 nm in the self-implanted specimen. Solid solution and precipitate strengthening mechanisms found in the mixed multilayer and aluminum implanted specimens remained effective during cyclic deformation and provided a barrier to the emergence of subsurface PSB structures. No softening was observed in the near surface region of the mixed multilayer specimen as shown in Figure 4(b). Therefore, the resistance of the microstructure to cyclic softening is an important factor in determining if a surface modification will enhance resistance to fatigue damage. Additionally, if the surface modification increases the yield strain (yield strength/modulus), then the surface may experience only elastic deformation even though the bulk deforms plastically. In this case, no cyclic softening of the near-surface region would occur. Both profiles in Figure 4 indicated cyclic hardening of the specimens at depths greater than 100 nm.

Residual stress appears to play a secondary role in inhibiting the formation of surface PSB structures. The mixed multilayer modification effectively inhibited PSB penetration to the surface in spite of the tensile residual stresses that developed in the near-surface region. While not significantly affecting the evolution of surface slip band features, the tensile residual stresses appeared to accelerate the initiation of a critical fatigue crack. The elevated temperature aluminum implantation induced a compressive surface residual stress and increased fatigue life 25% while the mixed multilayer modification induced a tensile surface residual stress and decreased fatigue life by 30%.

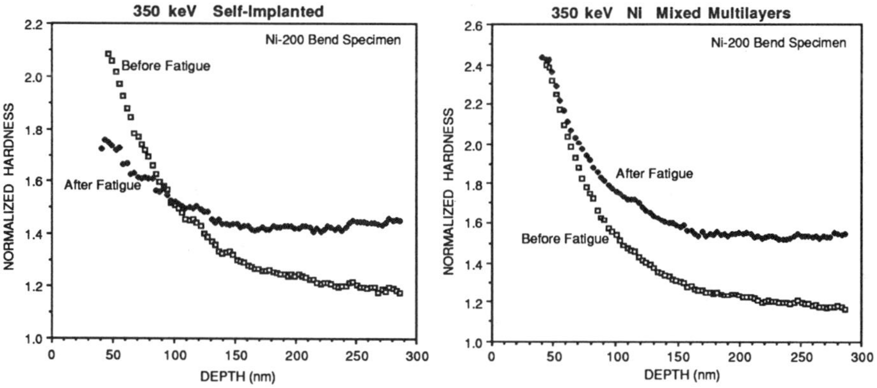


4(a) 4(b)

FIGURE 4. Normalized hardness-depth profiles of the (a) 350 keV self-implanted and (b) 350 keV Ni mixed multilayer flexure specimens showing the effects of cyclic deformation on hardness. Cyclic softening is evident in the near-surface region of the self-implanted specimen. Both specimens cyclically hardened at depths greater than 100 nm.

## CONCLUSIONS

The results of this study indicate that the near-surface regions of nickel substrates can be strengthened by self-implantation, aluminum implantation, and ion beam mixing evaporated Ni-Al multilayers. Even though self-implantation increased near-surface hardness, significant softening of this region occurred during cyclic deformation which resulted in very little suppression of surface slip features. Aluminum implantation at 500°C nearly eliminated the penetration of subsurface slip bands to the surface while the mixed multilayer modification decreased surface slip line density by a factor of six. Compressive residual stresses were produced by elevated temperature aluminum implantation and contributed to a delay in the initiation of fatigue cracks. Tensile residual stresses produced as a result of the evaporated Ni-Al multilayers accelerated crack initiation in the ion beam mixed multilayer specimen.

## ACKNOWLEDGMENTS

The authors acknowledge the financial support of the National Science Foundation under grant DMR 8603174. The ultra-low load indentation studies were performed in the High Temperature Materials Laboratory at Oak Ridge National Laboratory with the assistance of Dr. John Cathcart which is gratefully acknowledged. We also acknowledge the assistance of Klenton Willis in reducing the hardness data.

## REFERENCES

1. M. Nastasi, J. K. Hirvonen, T. R. Jervis, G. M. Pharr and W. C. Oliver, Journal of Materials Research 3, 226 (1988).

2. W. C. Oliver, C. J. McHargue and S. J. Zinkle, Thin Solid Films 153, 185 (1987).

3. E. P. EerNisse and S. T. Picraux, Journal of Applied Physics 48, 9 (1977).

4. D. S. Grummon, J. W. Jones and G. S. Was, Metallurgical Transactions A 19A, 2775 (1988).

5. I. G. Greenfield and A. Purohit, Materials Science and Engineering 46, (1980).

6. E. Y. Chen and E. A. Starke, Materials Science and Engineering 24, 209 (1976).

7. J. B. Pethica, R. Hutchings and W. C. Oliver, Philosophical Magazine A 48, 593 (1983).

8. W. C. Oliver, R. Hutchings and J. B. Pethica, in Microindentation Techniques in Materials Science, ASTM STP 889, edited by P.J. Blau and B.R. Lawn (ASTM, Phila., PA, 1986) p. 90.

9. J. D. Finegan and R. W. Hoffman, in Transactions of the 8th National Symposium on Vacuum Technology, (Pergamon, New York, 1961) p. 935.

10. B. O. Hall and D. I. Potter, in Effects of Radiation on Structural Materials. ASTM STP 683, edited by J. A. Sprague and D. Kramer (ASTM, 1979) p. 32.

11. M. Ahmed and D. I. Potter, Acta Metallurgica 33, 2221 (1985).

12. C. Laird, in Treatise on Materials Science and Technology. Plastic Deformation of Materials. Volume 6 , edited by R. J. Arsenault (Academic Press, San Francisco, 1975) p. 101.

# MEASUREMENT OF THE SOUND VELOCITY IN AlAs BY PICOSECOND ULTRASONICS

D. A. Young,* H. T. Grahn,* H. J. Maris,* J. Tauc,* J. M. Hong,** and T. P. Smith, III**
*Brown University, Providence, RI 02912
**IBM Thomas J. Watson Research Center, Yorktown Heights, NY 10598

## ABSTRACT

Picosecond ultrasonics have been used to measure the sound velocity and index of refraction in a thin film of AlAs. The AlAs layer was grown by molecular-beam epitaxy on a buffer layer of GaAs which was deposited on a semi-insulating GaAs wafer. The AlAs film was capped by a thin layer of GaAs, and a very thin film of InSb. To generate a sound wave effectively a picosecond light pulse was absorbed in the InSb film. As the sound wave propagates through the microstructure it changes the optical properties in the various films, and this produces a change in the optical reflectivity, which is measured. From the round trip time of the acoustic wave we have determined the sound velocity in the AlAs film.

## EXPERIMENTAL

We have measured the sound velocity and index of refraction of AlAs using a novel technique based on picosecond optics.[1] This method is ideal for ultrasonics in nanometer-scale films, and epitaxial films in particular. The samples measured in this work were grown using molecular-beam epitaxy. The substrate was semi-insulating GaAs aligned to within 0.5° of the <100> axis. To ensure high-quality AlAs epitaxy, a 600 nm GaAs buffer layer was grown prior to the growth of the 800 nm AlAs film. A 22 nm GaAs cap layer was grown over the AlAs to smooth the surface and reduce oxidation.

Picosecond optical ultrasonics depends on the interaction of light with an absorbing material. Optical energy is deposited in an absorbing material and this generates a stress which in turn launches a strain pulse into the structure. This strain pulse is then tracked through the sample optically. In our experiments the energy gap of the AlAs (2.2 eV) is larger than the photon energy (2.0 eV) and the absorption is negligible. In addition, the absorption length of the GaAs is too long (about 230 nm) to yield significant stress in the cap layer. For these reasons a very thin strongly absorbing layer (transducer) of InSb was sputtered onto the sample surface. We can safely assume that the stress generated by the light is only in the transducer.

The strain pulse is tracked through the structure by measuring the changes it produces in the index of refraction and absorption coefficient of the layers and the attendant changes in the optical reflectivity of the sample. These changes are measured in a pump-probe technique where the time delay between a pump pulse (used to create the strain pulse) and a probe pulse (used to measure the optical reflectivity) is varied.

## RESULTS AND DISCUSSION

Results for room temperature measurements of the AlAs film described above are shown in Fig. 1. There are a number of oscillatory components to the spectra, and we have performed computer simulations of the optical response to understand them more fully. The calculations are similar to those used to study amorphous superlattices.[2] To calculate the change in the optical reflectivity we used established values for the density and sound velocities of InSb and GaAs, the density of AlAs, and assumed values for the strain dependence of the optical constants of these materials. Fig 2 shows the computed reflectivity if we assume that only InSb has a photoacoustic response (change in n and $\alpha$ due to strain). The structure at $\tau_1$ is produced when a strain pulse is reflected from the AlAs-GaAs interface and returns to the InSb transducer. Thus

$$\tau_1 = 2(d_1/v_1 + d_2/v_2), \quad (1)$$

where $d_1$, $v_1$, $d_2$, and $v_2$ are the thicknesses and sound velocities in the AlAs and GaAs capping layer.

The damped oscillations which begin at $\tau_2$ occur because the strain pulse has had sufficient time to travel through the AlAs and enter the GaAs buffer layer. Therefore $\tau_2 = \tau_1/2$. The oscillations between $\tau_2$ and $\tau_1$ are damped due to optical absorption in the GaAs. As the strain pulse retreats from the sample surface the light beam is more strongly attenuated and the changes in the reflectivity become smaller. The period of these damped oscillations is

$$\tau = \lambda/2vn\cos\theta \quad (2)$$

where $\lambda$ is the vacuum wavelength of the light pulse (602nm and 582nm), v the sound velocity, n the refractive index, and $\theta$ the angle between the lines of propagation of the

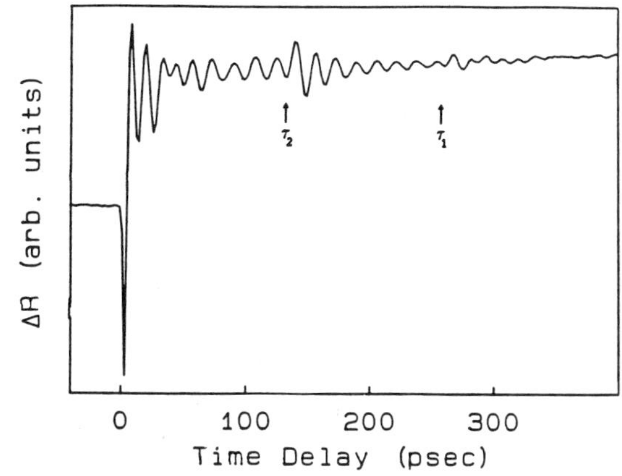


Figure 1. Photoinduced reflectivty changes in an AlAs/GaAs heterostructure. $\tau_1$ is the time at which the strain pulse reflected from the GaAs substrate reaches the transducer, and $\tau_2$ is the time at which the strain pulse first enters the GaAs substrate.

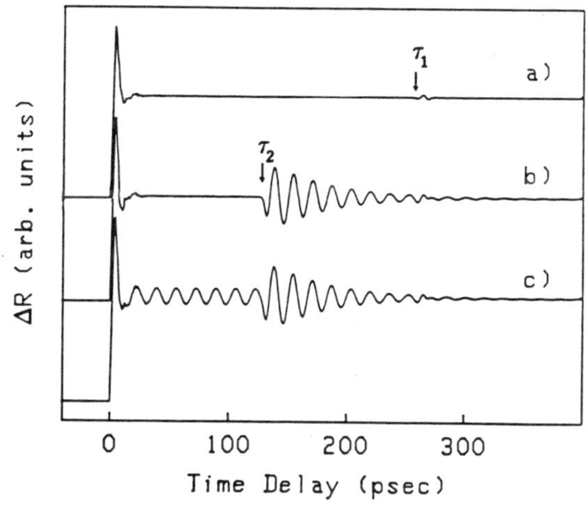


Figure 2. Computer simulations of the changes in the optical reflectivity. Curve (a) includes photoacoustic response of only the InSb transducer, curve (b) includes the GaAs and InSb, and curve (c) includes the photoacoustic response from all three materials.

light and strain pulses. The calculated period (16ps) agrees well with the experimental values.

The oscillations prior to $\tau_2$ are produced by the photoacoustic response of the AlAs layer. Theoretically, they are undamped because there is no optical absorption in the AlAs and their period is given by Eqn. (2). Coupled with the foregoing discussion this allows determination of both the sound velocity and the index of refraction of the AlAs layer. Using Eqn. (1) we found the sound velocity of AlAs to be $(6.44 \pm 0.32) \times 10^5$cm sec$^{-1}$. This is about 10% larger than indirect estimates in the literature.[3] Using Eqn. (2) we found the index of refraction at to be $2.78 \pm 0.20$and$2.83 \pm 0.21$ for $\lambda$ = 602nm and 582nm respectively. These values are slightly smaller than those previously measured in bulk AlAs.[4] This may be due to the small amount of strain between the AlAs and GaAs but this seems unlikely.

SUMMARY

We have discussed a novel technique for picosecond ultrasonic spectroscopy of thin-film materials. These measurements can be used to determine the sound velocity and index of refraction of the component materials. Because of its very high temporal resolution this technique can be used for nanometer-scale films and may be useful in studying the many new epitaxial systems currently being investigated.

ACKNOWLEDGEMENTS

The authors would like to thank T. R. Kirst for his technical assistance. The work at Brown University was supported in part by Department of Energy grant DE-FG02-86ER45267. The work at IBM Research was sponsored in part by the Army Research Office.

## References

[1] C. Thomsen, H. T. Grahn, H. J. Maris, and J. Tauc, Phys. Rev. B **34**, 4129 (1986).

[2] H. T. Grahn, H. J. Maris, J. Tauc, and B. Abeles, Phys. Rev. B **38**, 6066 (1988).

[3] S. Adachi, J. Appl. Phys. **58**, R1 (1985).

[4] R. E. Fern and A. Onton, J. Appl. Phys. **42**, 3499 (1971).

# SILICON NITRIDE FIBERS USING MICRO FABRICATION METHODS

J. KOSKINEN AND H. H. JOHNSON
Department of Materials Science and Engineering, Cornell University, Ithaca, NY 14853.

## ABSTRACT

Amorphous LPCVD and PECVD $Si_3N_4$ films have been patterned into fibers of various shapes using integrated circuit methods. A tensile test apparatus was developed which allows a simultaneous testing of up to 20 fibers. The average tensile strength of LPCVD $Si_3N_4$ fibers with a 30 µm gauge section was 7.0 ± 0.4 GPa. The corresponding value for PECVD $Si_3N_4$ was 2.4 ± 0.2 GPa. The following properties of the materials were measured: atomic composition (RBS, FRES), density, refractive index, internal stress, elastic modulus and micro hardness. A Weibull analysis of the tensile strength values was performed.

## INTRODUCTION

Research on high strength ceramic fibers is essential to the development of high performance composite materials. Fibers of refractory materials are difficult to manufacture to a specific geometric shape.

This work presents methods to manufacture fibers from deposited $Si_3N_4$ films using integrated circuit methods.[1] The films were deposited using low pressure chemical vapour deposition (LPCVD) and plasma enhanced chemical vapour deposition (PECVD). The tensile strength of the fibers was measured using a specifically designed multi-fiber measuring device. This method allows a unique evaluation of the mechanical properties of the film without interference from the substrate.

## EXPERIMENTAL

The $Si_3N_4$ films were deposited on two kinds of samples; pure Si (100) wafers and Si wafers with a 1 µm thick film of thermal oxide in LPCVD and in PECVD a 1.2 µm thick film of polyimide ( Probimide® 287 baked at 250 °C). The deposition parameters are in Table I.

The fiber forms were patterned on the films using an optical projection aligner and reactive ion etching ($CHF_3$). Self supporting fibers were then obtained by dissolving the intermediate film. Hydro fluoric acid buffered with ammonium nitride (6:1) was used to etch the thermal oxide and dichloromethane to dissolve the polyimide layer. The etch rate of the oxide was 90 nm/min as compared to the ≤ 1 nm/min etch rate of LPCVD $Si_3N_4$. The photoresist layer survived the 70 min etch protecting the top side of the fiber. Finally the photoresist was removed by reactive ion etching ($O_2$).

Rutherford back scattering (RBS) and forward recoil spectroscopy (FRES) were used to determine the atomic compositions of the films.

Films deposited on pure Si wafers were used to determine the following properties of the materials: density, refractive index, residual stress and micro hardness. The wafers were weighed and the curvature was measured before and after the deposition. Since the LPCVD $Si_3N_4$ coating was grown on both sides of the

Mat. Res. Soc. Symp. Proc. Vol. 130. 

Table I

Deposition parameters of $Si_3N_4$ films

| | LPCVD | PECVD |
|---|---|---|
| Source gas | $SiH_2Cl_2/NH_3$ | $SiH_4/NH_3$ |
| Flow rate [sccm] | 30/90 | 20/60 |
| Temperature [C°] | 800 | 280 |
| Pressure [mtorr] | 230 | 525 |
| Deposition rate [nm/min] | 4.3 | 30 |
| Substrate | Si; Si/$SiO_2$(thermal) | Si; Si/Polyimide |
| Film thickness [nm] | 410 | 400 |
| Reactor type | Hot wall | Radial flow |
| Plasma | | 30 kHz; 0.12 W/cm$^2$ |

wafer the nitride film was removed on one side by reactive ion etching to obtain the "initial" weight and curvature. The refractive index and the thickness of the films were measured using a Rudoph 43603 Research Ellipsometer (HeNe laser with 632.8 nm wave length). Knowing the thickness of the films the residual stress and the density of the films could be calculated from the curvature[2] and weight data. Hardness was measured using a continuous indentation test[3] where the load and the depth of penetration are constantly monitored as the Berkovitch diamond indenter is drawn in and out of the the substrate at a constant speed of 18 nm/s.

The maximum film thickness was limited by the high internal stress which was tensile in LPCVD and compressive in PECVD $Si_3N_4$. Due to this fact a specific form of the fiber in LPCVD $Si_3N_4$ was developed allowing partial relief of the stress without breaking the fiber as the oxide layer was dissolved, Fig.1. The following factors were taken into account when designing the fiber: the fiber has a narrow gauge section for tensile strength testing; long grip section to prevent misalignment in the tensile test; wider segments of the fiber have holes to permit HF acid to attack the oxide layer simultaneously in many places to minimize the etch time; the stress in the end parts of the fiber has to be relieved before the neck part of the gauge section is disconnected from the substrate; the fibers must be connected to the substrate so that the the whole array of the fibers may later be lifted off and attached to the tensile testing equipment. After etching contraction of the fibers was observed indicating that the stresses are relieved. The transversal area of the fiber was measured by SEM.

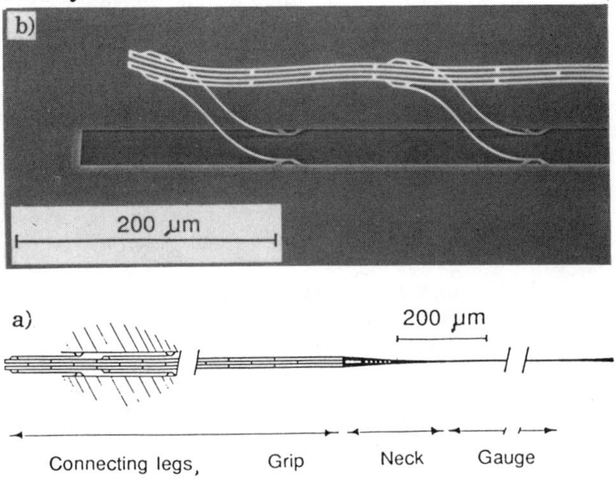


FIG. 1. a) The fiber used in the tensile tests with different parts indicated; b) SEM picture of the fiber end after the oxide etch. The tensile stress is relieved in the fiber which remains attached to the substrate.

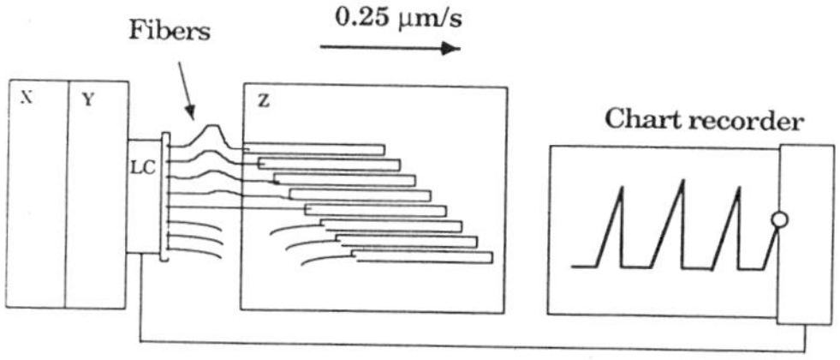


FIG. 2. Schematic view of the tensile test apparatus. X, Y and Z are translation stages and LC is a load cell.

The tensile test apparatus is schematically shown in Fig.2. One end of the fiber was glued to a load cell and the other to a holder mounted on a motor driven translation stage. The fiber holder is a set of steel blades that could be slid relative to each other allowing a sequence of tensile tests in one run.

The pull tests were performed to measure the elastic modulus and tensile strength. The voltage induced by the load cells was recorded as a function of time on a chart recorder. The elastic modulus for the material was obtained by measuring the speed of translation stage ( 0.25 μm/s) and the spring constant of the load cell. The gauge section of the fibers was 30 μm long and 3.9 μm wide. The transversal area was $1.36 \pm 0.07 \cdot 10^{-12} m^2$ in LPCVD and $1.6 \pm 0.1 \cdot 10^{-12} m^2$ in PECVD fibers. In addition 300 μm and 3000 μm gauge sections were used to study the strength values as a function of gauge length in LPCVD $Si_3N_4$ fibers.

## RESULTS

The values of atomic composition, density, refractive index and internal film stress measured for LPCVD and PECVD $Si_3N_4$ films are presented in Table II. The RBS measurements indicated a stoichiometric $Si_3N_4$ composition for LPCVD films but some excess of N in PECVD films. Transmission electron diffraction indicated that all films were amorphous.

Table II

Measured properties of the $Si_3N_4$ films

| | LPCVD | PECVD |
|---|---|---|
| TEM | Amorph. | Amorph. |
| Composition: | | |
| RBS: $Si_1N_x$ | $1.33 \pm 0.03$ | $1.48 \pm 0.04$ |
| FRES: H cont. [at.%] | $5 \pm 1$ | $23 \pm 4$ |
| Density [g/cm$^3$] | $3.1 \pm 0.1$ | $2.67 \pm 0.05$ |
| Refractive index | $1.96 \pm 0.02$ | $1.88 \pm 0.02$ |
| Internal stress [GPa] | $1.2 \pm 0.2$ Tensile | $1.1 \pm 0.1$ Compr. |
| Hardness [GPa] (5g) | $23 \pm 4$ | $23 \pm 4$ |
| Elastic modulus [$10^{11}$ Pa] | $3.0 \pm 0.2$ | $2.4 \pm 0.3$ |
| Avg.tens. strength [GPa]: | | |
| gauge: 30 μm | $7.0 \pm 0.4$ | $2.4 \pm 0.2$ |
| " 300 μm | $6.5 \pm 0.4$ | |
| " 3000 μm | $5.7 \pm 0.3$ | |
| Weibull parameters: | | |
| $S_{0a}$ [GPa $\mu m^{-1/m_{0a}}$] | 8.3 | 3.5 |
| $S_{0b}$ [GPa $\mu m^{-1/m_{0b}}$] | 12.9 | - |
| $m_{0a}$ | 27.9 | 11.6 |
| $m_{0b}$ | 10.5 | - |

Despite the similar silicon and nitrogen composition in LPCVD and PECVD films the materials turn out to have quite different properties. The PECVD $Si_3N_4$ has over four times as much hydrogen, is less dense and has a lower elastic modulus. Both films have very high internal stress which in the LPCVD film is tensile and PECVD compressive. Due to the stress relief of the PECVD $Si_3N_4$ on polyimide crumpling of the film was observed with a local curvature of ≥ 100 μm.

The micro hardness values as a function of the plastic indentation depth are shown in Fig. 3. The location of the interfaces and the hardness of Si are indicated in the same plot.[4]

The tensile strength cumulative Weibull distribution function of the LPCVD $Si_3N_4$ fibers with gauge lengths of 30, 300 and 3000 μm is shown in Fig. 4. In the Weibull estimation the probability of a fiber having strength less than S is given by[5]

$$P(S) = 1 - \exp(-LN(S)), \tag{1}$$

where L is the length of the fiber and N(S) is the probability of one flaw of strength less than S per unit length. In the two parameter Weibull distribution it is assumed that

$$N(S) = (S/S_o)^m, \tag{2}$$

where $S_o$ is a scaling parameter and m a shape parameter. In the Weibull plot the distribution is represented by a straight line with a slope of m. Since the slope of the measured distribution function seems to increase in shorter gauge lengths a multi risk-model is adopted. If two independent populations of different flaws are present the the distribution can be described as a sum of two separate Weibull populations [6] and a Weibull estimate for N(S) is:

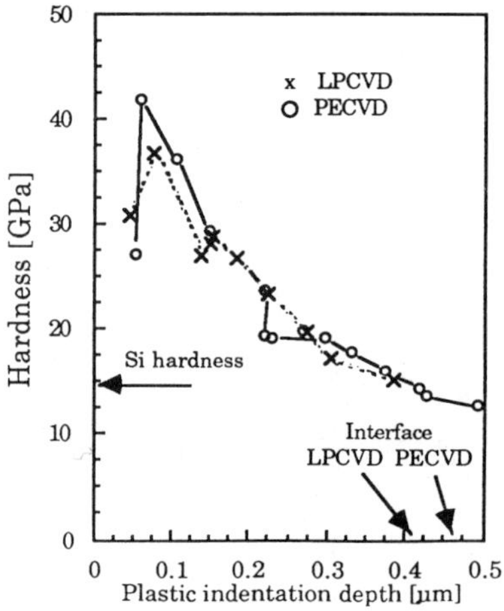


FIG. 3. Hardness as a function of indentation depth as measured with a continuous indentation test. The $Si_3N_4$/ Si interface and the substrate hardness are indicated.

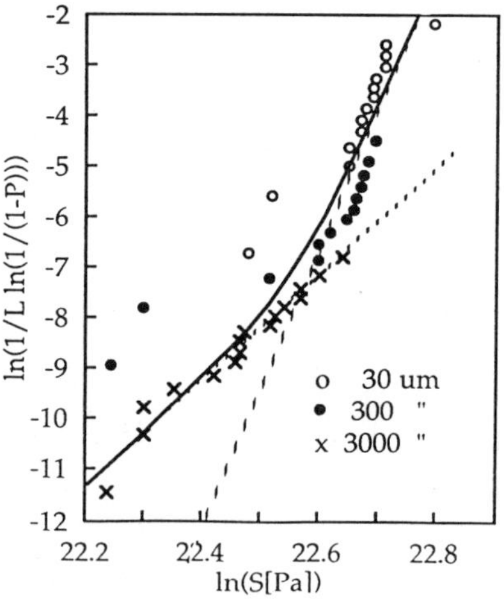


FIG. 4. The tensile strength data of LPCVD $Si_3N_4$ fibers in a Weibull distribution plot. The solid line is a multi-risk estimation.

$$N(S) = (S/S_{oa})^{m_a} + (S/S_{ob})^{m_b}, \qquad (3)$$

where $S_{oa}$ and $S_{ob}$ are scaling parameters and $m_a$ and $m_b$ shape parameters. The Weibull parameters were obtained by least square fitting separately a two parameter distribution to the data points representing a lower and a higher slope. The average tensile strength values and the Weibull parameters are in Table II. The error in the average strength values is mainly due to the uncertainty of the area of the crossection of the fiber.

Most of the LPCVD $Si_3N_4$ fibers shattered when breaking leaving no evidence of the initial breaking point. To show that the fibers did break randomly in the gauge section a set of fibers was tested in water which absorbed the momentum of the breaking fiber and prevented the shattering. Indeed, the fibers broke randomly in the gauge section. The water didn't affect the tensile strength of the fibers. Inspection of the broken fibers showed that the material is brittle and there was no evidence of plastic deformation.

## DISCUSSION

The $Si_3N_4$ films deposited with different methods have quite different properties. The lower density, elastic modulus and higher hydrogen concentration in the PECVD $Si_3N_4$ indicate that the structure of the material is probably characterized by imperfect bonding of Si and N, a lot of Si-H and N-H bonding, and voids trapping unbonded H. The difference in the tensile strength values is even more pronounced. However, the microhardness values are identical within the precision of the measurement for both types of $Si_3N_4$. Therefore it is evident that there is a difference in the internal structure of PECVD $Si_3N_4$ films on pure Si sample and Si/polyimide due to the stress relief and crumpling of the latter film during deposition. When inspected in SEM the fibers have a side wall which has a characteristic roughness. This may be due to the roughness of the photoresist and the variations in the isotropic etch rate of the nitride. The maximum depth of the side wall imperfections was about 70 nm in LPCVD and 150 nm in PECVD $Si_3N_4$. If these imperfections were considered as critical cracks with a length of c the tensile strength can be estimated to be $\sim c^{-1/2}$ using the Griffith brittle fracture equation.[7] Then a relative value of 4.8 GPa is obtained for PECVD $Si_3N_4$ as compared to LPCVD $Si_3N_4$ . This value is clearly greater than the measured value S=2.4 GPa. Thus it is concluded that there is a difference in the tensile strength of the two materials caused by the difference in the internal composition and structure of the materials. On the other hand the strikingly similar hardness values may be addressed to the high but opposite internal stress of the films and possibly the substrate interference.

In the literature there are numerous examples where the two flaw populations observed in the Weibull distribution of ceramic fiber strength can be identified as surface cracks and internal flaws.[8,9] In the case of the present measurement it seems that two separate flaw distributions are superimposed but they remain unidentified.

The tensile strength of the LPCVD $Si_3N_4$ is about half of the value for single crystal whiskers but still stronger than the ceramic fibers produced with other methods.[10,11] Even the PECVD $Si_3N_4$ fibers can be considered strong ceramic fibers. One of the disadvantages of this method to produce fibers is the limited film thickness due to high internal stress. The stress can however be modified by altering the deposition parameters.[12]

The tensile test experiment described has the following advantages in testing mechanical properties of the films: no substrate effects, many fibers can be tested at the same time yielding good statistics and the possibility to perform fatigue tests on the fibers.

## CONCLUSIONS

By introducing integrated circuit methods it is possible: i) to manufacture strong fibers out of a large variety of materials that can be deposited as thin films; ii) to measure the mechanical properties of film without the substrate interference.

Amorphous $Si_3N_4$ fibers were manufactured from LPCVD film with a tensile average strength of 7.0 $\pm$ 0.4 GPa and PECVD $Si_3N_4$ 2.4 $\pm$ 0.2 GPa respectively. The corresponding hydrogen contents were 5 $\pm$ 1 and 23 $\pm$ 4 at. %. Tensile test measurements indicate a difference in the mechanical properties of the nitrides prepared with the two different methods.

The Weibull analysis of the LPCVD $Si_3N_4$ fibers with various gauge lengths indicate a presence of two superimposed populations of flaws causing the failure of the fiber in the tensile test.

## ACKNOWLEDGEMENTS

This research is supported by DARPA and DOE. The National Nanofabrication Facility and the Materials Science Center made critical experimental facilities available. B.Soave provided technical advice and assistance with the depositions. The indentation tests were performed and analyzed by J. Steinwall using the equipment built and developed by the research group of Professor C.Y. Li.

## REFERENCES

1. J.E.Steinwall, H.H.Johnson, Proceedings of Materials Research Society Conference, Reno, Nevada, April 5 - 9, 1988, in print
2. E.W.Hearn, D.J.Werner, D.A.Doney, J. Electrochem. Soc. **133**, 1749 (1986)
3. D.Stone, W.R.LaFontaine, P.Alexopoulos, T.-W. Wu, C.-Y. Li, J. Mater. Res. **3**, 141 (1988)
4. M.F.Doerner, W.D.Nix, J.Mater. Res. **1**, 601 (1986)
5. W.Weibull, J. Appl. Mech. **18**, 293 (1953)
6. R.Olshansky, R.D.Maurer, J. Appl. Phys. **47**, 4497 (1976)
7. eg. R.Moreton, in Strong Fibers, ed. W.Watt and B.V.Perov, (Elsevier, Amsterdam, 1985), p. 464
8. K.K Phani, J. Mater. Sci. **23**, 2424 (1988)
9. K.Goda, H.Fukuaga, J. Mater. Sci. **21**, 4475 (1986)
10. H.M.Schwartz ed., Composite Materials Handbook, (McGraw-Hill, New York, 1984) p. 2.6 - 2.7
11. K.Okamura, M.Sato, Y.Hasegawa, Ceram. Int. **13**, 55 (1987)
12. R.F.Bunshah ed., Deposition Technologies for Thin Films and Coatings, (Noyes Publications, New Jersey, 1982)

# Mechanical Stress as a Function of Temperature in Thin Aluminum Films and Its Alloys

Donald S. Gardner* and Paul A. Flinn**
*C.I.S., Stanford University, Stanford, Calif. 94305
**Intel Corporation and
Dept. of M.S. & E., Stanford University, Stanford, Ca 94305

**Abstract**

Aluminum alloys have virtually replaced aluminum for interconnections in VLSI because of their improved reliability. Mechanical stress is a problem of growing importance in these interconnections. Stress as a function of temperature was measured for thin aluminum films and several aluminum alloys and layered films consisting of silicon, copper, titanium, tungsten, tantalum, vanadium, and $TiSi_2$. Solid-state reactions of the aluminum with the additives and with the ambient during thermal cycling will occur and depending on what compounds have formed and at what temperature, this will determine the morphology and reliability of the metallization. The measurement technique, based on determination of wafer curvature with a laser scanning device, directly measures the total film stress and reflectivity *in situ* during thermal cycling. Changes in stress were detected when film composition and structure varied and were correlated using x-ray diffraction with the formation of aluminides. Other phenomena that contribute to stress changes including elastic behavior, recrystallization, grain growth, plastic behavior, yield strength, and film hardening from precipitates.

## 1 Introduction

Several failure mechanisms associated with mechanical stress are becoming increasingly important in thin films used for interconnections in integrated circuits. Open circuits from metal cracking and voiding, short circuits due to hillocks, and both open and short failures due to electromigration are phenomena related to mechanical stress. Temperature is important because the conductors and dielectrics have relatively large thermal mismatches with silicon resulting in stress. Solid-state reactions and changes in structure normally involving volume changes may also occur during thermal cycling causing variations in stress. More details can be found in our other publications [1,2,3,4,5].

Two types of stresses (or strain) occur in thin films–intrinsic and thermally induced. Intrinsic (or growth) stress is observed in as-deposited films and can be a result of defects or structural mismatch between the film and substrate [6]. Thermally induced stress is introduced by temperature changes after deposition. The origin of these stresses has been reviewed by Campbell [7,8], Kinosita [9], and Hoffman [10].

Studies of stress have helped to achieve a better understanding of hillocks and voids in thin films. Hillock suppression [5,11,12] is one primary advantage of layered films and knowledge of the internal stresses has assisted in explaining the improvements [1,2]. The detection of these changes when film composition and structure change has proven to be very useful because the precise temperature at which precipitate/compound formation, or grain growth occurs can be determined.

## 2 Experimental Techniques

Aluminum was deposited onto (100) oriented 5–20 $\mu\Omega$·cm p-type silicon wafers by magnetron sputtering at room temperature. A thermal oxide 100 nm thick was grown prior to thin-film deposition. This $SiO_2$ layer was grown to prevent epitaxial regrowth of the aluminum on silicon and to prevent dissolution of the silicon substrate in the aluminum. A laser-based optical-lever technique [13,14,4,1] yields the best combination of accuracy, convenience, and speed and was used to measure the wafer curvature *in situ* during thermal cycling. This curvature corresponds to the stress in the total thin film, therefore the film stress is being directly measured. The heating and cooling rates were 5°C/min and the ambient was air. Because the stress-measurement scan requires

Mat. Res. Soc. Symp. Proc. Vol. 130. ©1989 Materials Research Society

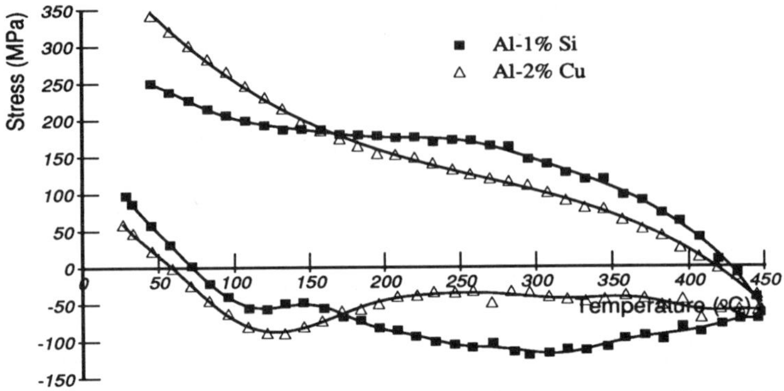


Figure 1. Temperature versus the stress of aluminum and aluminum copper. Here, Al-Cu consists of 2% Cu, and the films were sputter deposited at room temperature.

only a few seconds, heating and cooling can be continuous. The X-ray diffraction measurements were made using a new type of diffractometer called a Generalized Focussing Diffractometer (GFD) designed with increased sensitivity for thin films [15].

## 3 Stress Behavior of Aluminum Alloys

Stress as a function of temperature in aluminum alloys is similar in many ways to stress in pure aluminum films; however, there are several differences. A typical measurement of the stress for pure aluminum and Al-Cu as a function of temperature is shown in Fig. 1 and eight regions of stress were identified, including elastic behavior, recrystallization, plastic behavior, yield strength, film hardening from precipitates and grain growth. An explanation of these different regions of stress have been previously published [1,2,5].

When aluminum alloys are fabricated from elemental targets at room temperature, changes in film morphology can occur and compounds can form after thermal cycling. Figure 2 plots measurements of the stress behavior of Al-Si-Ti films. At approximately 250°C, the change in compressive stress toward zero can be accounted for if a negative atomic volume change occurs when the aluminide phase is formed. The compound $TiAl_3$ has a smaller volume (−5.4%) which will reduce the compressive strain. Additional thermal cycles do not show this change in stress toward zero and, in addition, show a lower compressive yield strength as a result of the different morphology. Further evidence that a variation in stress is the result of a volume change can be seen in Fig. 2 by varying the concentration of titanium in the aluminum film. Larger amounts will increase the size of the stress change because a higher percentage of the film will be forming a compound.

This region or change can be theoretically modeled by calculating the strain from the change in volume by subtracting the volume of each element $V_{element}$ from the volume of the resulting compound $V_{compound}$. This volume strain or dilatation $\Delta$ is hydrostatic and, therefore, mean strain can be determined by dividing by 3. To obtain the resulting stress, Young's modulus and Poisson's ratio are needed for the material to which the strain is applied. Because the precipitates that form are in the matrix of aluminum, they will strain the aluminum matrix. The expression that predicts the change in stress is

$$\Delta\sigma = -\frac{E_{Al}}{3(1-\nu_{Al})}\left(\frac{V_{compound} - \Sigma\, V_{element}}{\Sigma\, V_{element}}\right) \qquad (1)$$

A reduction in volume either reduces compressive or increases tensile stresses. When a layer rather than precipitates form or a compound grows at the surface (such as oxidation), the volume change would strain the forming layer in addition to the aluminum in which event the determination of the modulus should include the materials in the layer.

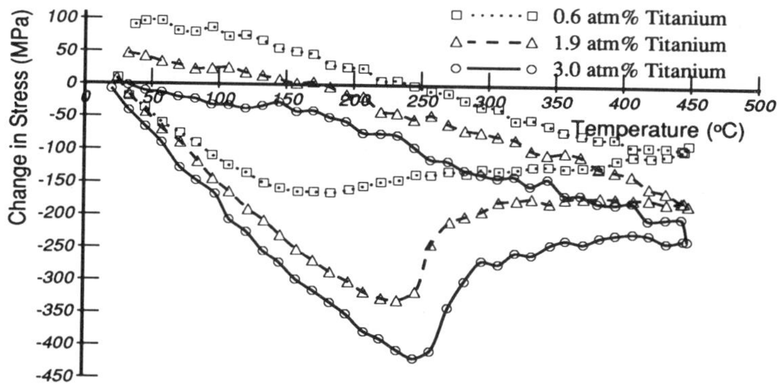


Figure 2. Temperature versus stress of aluminum-silicon-titanium films. These films were fabricated via a multipass deposition, and the deposition conditions were similar.

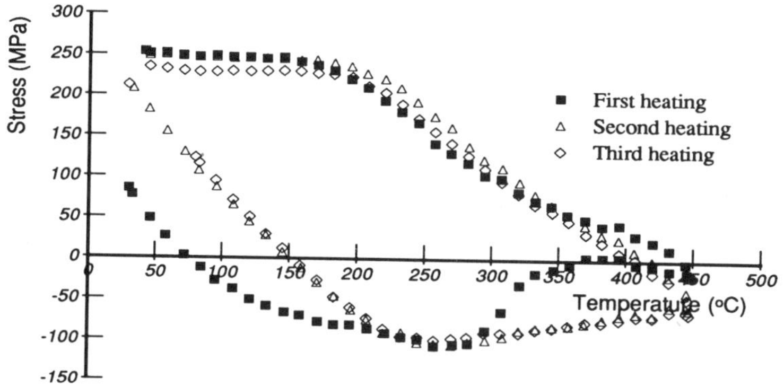


Figure 3. Temperature versus stress in a layered film of aluminum-silicon with titanium.

## 3.1 Thin Films of Aluminum Layered with Titanium

Figure 3 is a plot of temperature versus the total stress in Al-Si layered with titanium. The films were deposited at room temperature and the first thermal cycle exhibited a drop in compressive stress when the aluminide $TiAl_3$ forms. These films consist of three 16 nm layers of titanium alternated with 300 nm of Al-Si which adds to a total of 5.3 atomic percent titanium. This amount should yield a change of 110 MPa [1,2] which is the amount of change in Fig. 3.

During the second and third thermal cycles for the layered films, stress first behaves elastically and then a sharp transition occurs when yield strength is obtained. At this point, plastic deformation occurs. The difference between this curve and that of a pure aluminum film is that the plastic-deformation region is more linear with a slope closer to zero partly because hillocks cannot form.

## 3.2 Thin Films of Aluminum with Vanadium

Films of aluminum with vanadium were also thermally cycled while measuring stress *in situ*. In the stress measurement of Fig. 4, there are two abrupt changes in stress during the first cycle that are not apparent in subsequent cycles [1,2] and these may be the result of the two possible compounds detected at 550°C or an intermediate compound. The first stress change occurs at 370°C and the second at 410°C. Vanadium can form several aluminum-rich compounds and in addition to forming more than one compound during a thermal cycle, it is possible for one compound to transform into other compounds. X-ray diffraction measurements of aluminum layered with vanadium resulted in

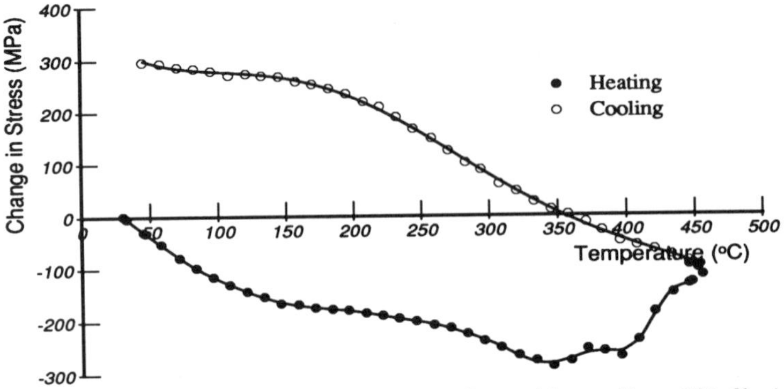


Figure 4. Stress behavior of a layered film of aluminum-silicon with vanadium. This film is 1.04 $\mu$m thick and is composed of two 50 nm layers of vanadium and two 500 nm layers of Al-Si; the Al-Si was the first layer deposited.

the identification of vanadium and aluminum before thermal cycling in Fig. 5a and after thermal cycled to 380°C, $VAl_3$ and possibly $V_7Al_{45}$ was detected in Fig. 5b. After thermal cycling to a higher temperature of 550°C, $VAl_{11}$ and possibly $V_7Al_{45}$ (see Fig. 5c) was detected.

In other work, several phases involving aluminum and vanadium have been detected. The phase $VAl_3$ was reported to form at 450°C when vanadium was chosen as a barrier metal between aluminum and silicon [16,17,18]. Other phases, possibly $VAl_{10}$ or $VAl_{11}$, were also present in some samples [18,19]. A study of the reaction of vanadium with Al-Cu versus Al revealed that $VAl_3$ forms after 450°C 12 hr anneals and that the presence of copper in the aluminum retards the development of $VAl_3$ and can lead to the formation of $VAl_{10}$ [20].

It appears that the less aluminum-rich compound $VAl_3$ first developed during this investigation and then transformed into $VAl_{11}$ and possibly $V_7Al_{45}$. This is being examined using TEM because there are many combinations of molecules that could lead to $VAl_{11}$ or $V_7Al_{45}$ formation from $VAl_3$; for example, one molecule of $VAl_{11}$ could combine with one vanadium atom and 19 aluminum atoms to form two $VAl_{11}$ as

$$VAl_{11} + V + 19Al \longrightarrow 2VAl_{11} \quad (2)$$

which in this case would lead to a slight (−0.2%) reduction in volume.

The temperatures at which the first and second vanadium aluminide compounds form are significant in integrated-circuit manufacturing because a typical $H_2 + N_2$ anneal or low-temperature silicon dioxide deposition takes place over a 350° to 450°C range. If a process is changed such that the final or even intermediate thermal cycle is varied, the resulting morphology and composition of the film may be different.

Another difference observed in Fig. 4 is an increase in the slope of the temperature versus stress curve during heating between 275° and 350°C. This increase, however, does not appear in measurements of Al-Si/V/Al-Si film without vanadium on the top surface [1,3]. The change in slope, therefore, is caused by the formation of a vanadium oxide or nitride which has a larger atomic volume that raises the total compressive stress of the film. In Fig. 6, there is further evidence of the formation of such a compound at the surface in the measurement of reflectivity during thermal cycling. A dramatic drop in reflectivity occurs between 350° and 450°C followed later by a rise. This drop, not observed in the film with a top layer of aluminum, indicates that the oxide or nitride has reached a thickness corresponding to the wavelength of the helium-neon laser within the material.

### 3.3 Thin Films of Aluminum with Tantalum

Stress behavior as a function of temperature was measured in layered films of Al-Si with tantalum. Figure 7 plots the stress in a film composed of two layers each of Al-Si and tantalum. X-ray

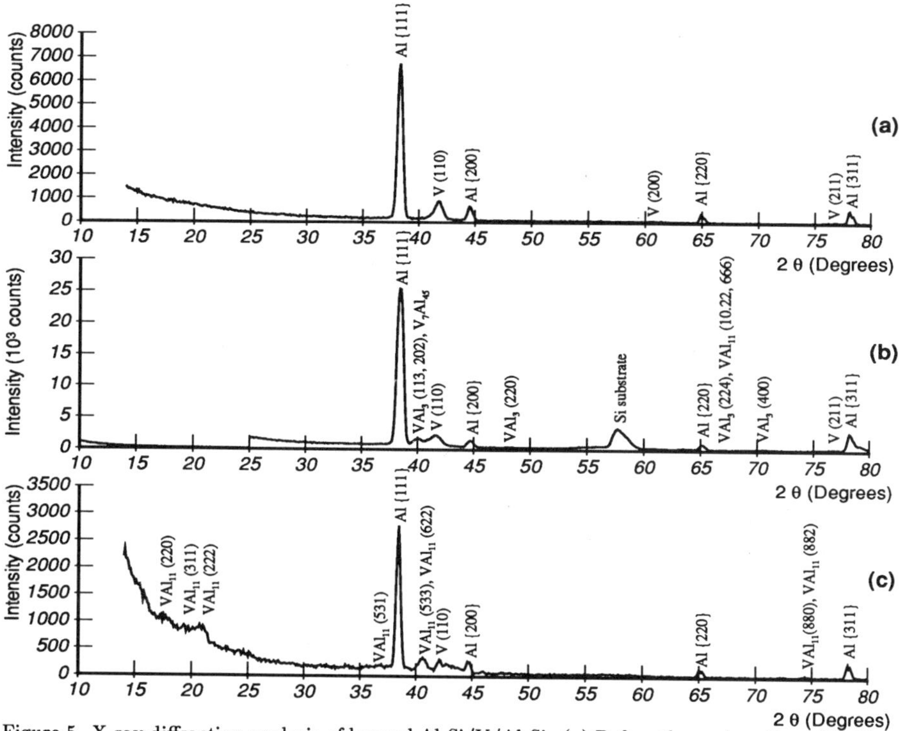


Figure 5. X-ray diffraction analysis of layered Al-Si/V/Al-Si. (a) Before thermal cycling. (b) After thermal cycling to 380°C. (c) After thermal cycling to 550°C.

diffraction measurements of layered films of aluminum with tantalum on $SiO_2$ identified elemental aluminum and the $\beta$ phase of tantalum before thermal cycling [1,3]. Peaks corresponding to $TaAl_3$ and tantalum silicide, probably $Ta_2Si$ or $Ta_5Si_3$ in addition to aluminum and unreacted $\beta$–Ta were identified after thermal cycling to 550°C. All major peaks for $TaAl_3$ was present in the sample. There were very few peaks present for the tantalum rich silicides possibly because of strong texturing. In the stress measurement, a positive change in stress (decrease in compressive stress) observed at approximately 430°C may be the point at which the aluminide or silicide form.

## 3.4 Thin Films of Aluminum with $TiSi_2$

The stress behavior of layered films of Al-Si with $TiSi_2$ was measured and both the first and second thermal cycles of a film that consisted of two layers each of Al-Si and $TiSi_2$ are plotted in Fig. 8. Measurements obtained by x-ray diffraction in Fig. 9a indicated that $TiSi_2$ was amorphous before the first thermal cycle; only aluminum and a small amount of silicon was identified. After thermal cycling twice to 450°C, however, x-ray diffraction identified $TiAl_3$, elemental silicon, and possibly base-center orthorhombic $TiSi_2$ or elemental titanium in Fig. 9b, but not the ternary compound $Ti_7Al_5Si_{12}$.

Crystallization of the amorphous Ti-Si layer would also result in a change in volume and a corresponding change in stress. There were two changes in stress during the initial thermal cycle in Fig. 8, a positive and a negative change. The negative change occurred at the highest temperature in the cycle (450°C), and the sample was kept at this temperature for 30 min. During the second thermal cycle, there was an additional very large negative change at 450°C which is possibly a continuation of any crystallization that may have begun during the first cycle.

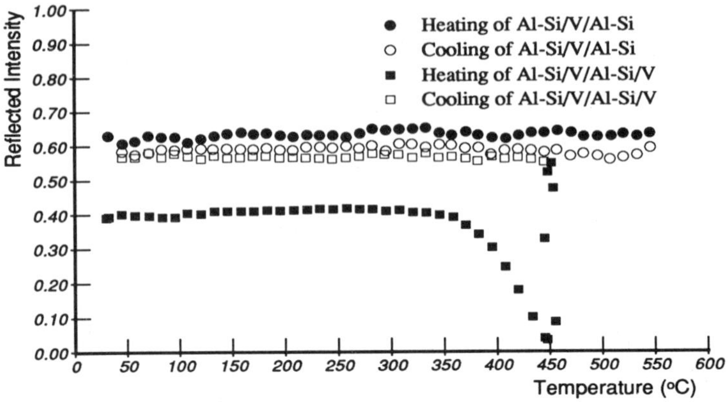


Figure 6. Reflectivity of the layered Al-Si/V/Al-Si/V film and the Al-Si/V/Al-Si single refractory-layer film as a function of temperature.

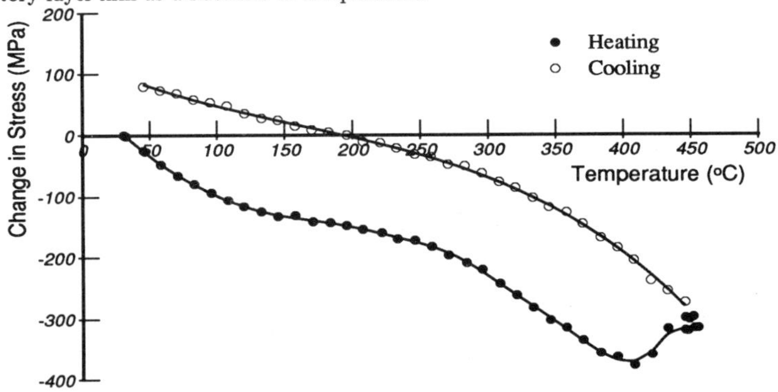


Figure 7. Stress behavior of a layered film of aluminum-silicon with tantalum. This is 1.09 $\mu$m film is composed of two layers of 50 nm tantalum and two layers of 500 nm Al-Si; the Al-Si was the first layer deposited.

Transmission electron microscopy identified several phases in these samples. Precipitates of $TiAl_3$ were found and large silicon nodules were present in the aluminum layers, however, no evidence of any large $TiSi_2$ precipitates were found. This suggests that $TiAl_3$ formed upon first heating between 350° and 400°C and then silicon crystallites and possibly small (less than 20 nm) $TiSi_2$ crystallites formed at 450°C. Such a film would not be desirable for VLSI interconnections since the large silicon nodules create high resistivity paths in the aluminum layers.

## 3.5 Thin Films of Aluminum with Tungsten

The change in stress as a function of temperature in a homogeneous alloy of Al-Si with tungsten is plotted in Fig. 10. Initially, there is an elastic region followed by plastic deformation until the temperature reaches approximately 400°C. At that time, compressive stress is greatly reduced as a result of recrystallization and the formation of an aluminide.

Stress in layered films of Al-Si with tungsten as deposited appears to be high and tensile [1,3]. Most likely, high tensile stresses develop in the tungsten layers of the initially laminar film because thin films of pure tungsten can have very large intrinsic (as deposited) tensile stresses. When aluminum and tungsten interdiffuse, stress may become less tensile because the aluminum would then become part of the layers of tungsten and would plastically deform.

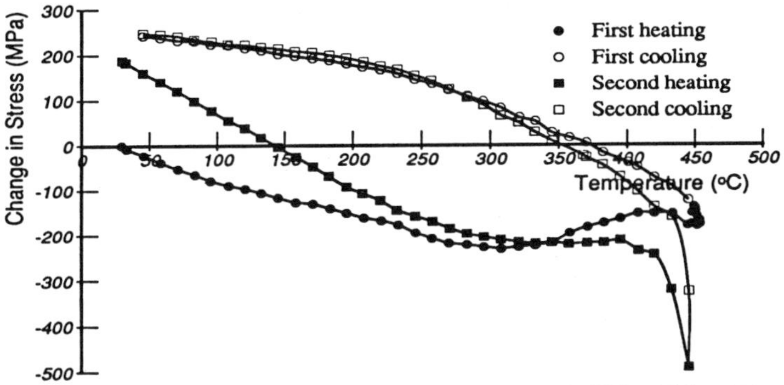


Figure 8. Stress behavior of a layered film of aluminum-silicon with $TiSi_2$. This 1.052 $\mu$m film is composed of two layers of 50 nm $TiSi_2$ and two layers of 500 nm Al-Si; Al-Si was the first layer deposited.

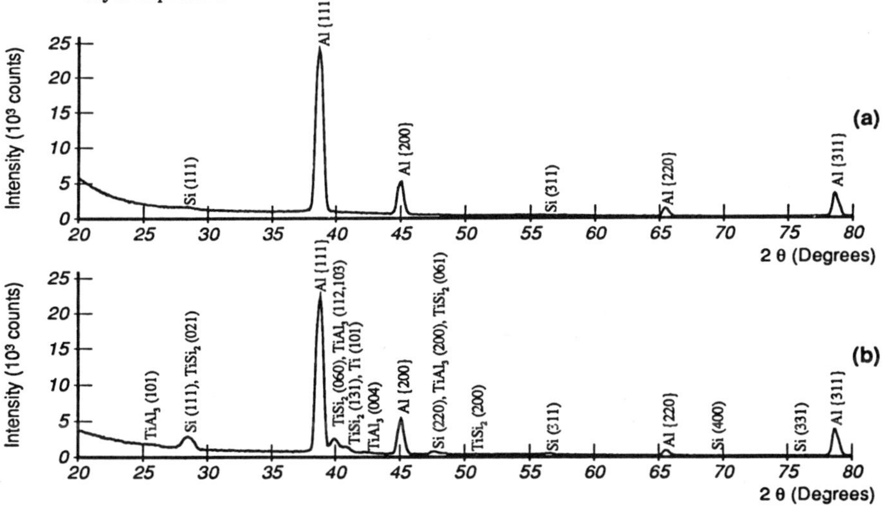


Figure 9. X-ray diffraction analysis of layered Al-Si/$TiSi_2$/Al-Si/$TiSi_2$ using the Seemann-Bohlin mode on the GFD. (a) Before thermal cycling. (b) After thermal cycling twice to 450°C.

## 4 Conclusion

Stress and reflectivity as a function of temperature was measured for aluminum films and several of its alloys consisting of silicon, copper, titanium, tantalum, vanadium, titanium silicide, and tungsten. The temperature at which compounds form was discernible in these measurements. A model based on volume change and its effect on stress explains some of the changes in stress when compounds form. X-ray diffraction measurements identified compounds in all the samples. The effects of oxidation/nitridation of the refractory material at the surface in terms of reflectivity and stress were also examined.

## 5 Acknowledgments

The authors wish to acknowledge Provost J. Meindl, Prof. K. Saraswat, Prof. W. D. Nix, C. Collins, D. Yaney, B. W. Shen and J. Turlo for their valuable discussions and assistance. This work was jointly supported by SRC Contract No. 87-SJ-104 and DARPA Contract No. MDA 903-80-C-0432.

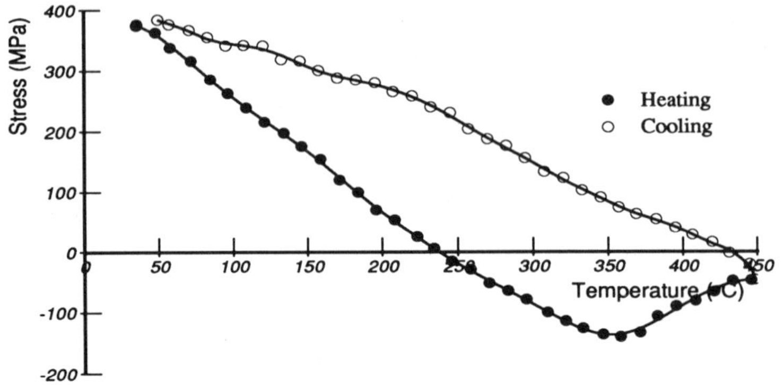


Figure 10. Stress behavior of a homogeneous alloy of Al-Si with 3.6 atm% tungsten.

## References


[1] D. S. Gardner. *Layered and Homogeneous Aluminum Alloys for Interconnection Devices in Integrated Circuits.* Ph.D. dissertation, Stanford University, Stanford, Calif., 1988. Available through U.M.I., 300 N. Zeeb Road, Ann Arbor, Michigan 48106, Order No. 88-15, 003.

[2] D. S. Gardner and P. A. Flinn. *Trans. Electron Devices*, 35(12), pp. 2160–2169, 1988.

[3] D. S. Gardner and P. A. Flinn. *J. Appl. Phys.*, 1989. Submitted for publication.

[4] P. A. Flinn, D. S. Gardner, and W. D. Nix. *IEEE Trans. on Electron Devices*, ED-34(3), pp. 689–699, 1987.

[5] D. S. Gardner, T. L. Michalka, P. A. Flinn, T. W. Barbee, Jr., K. C. Saraswat, and J. D. Meindl. In *1985 Proc. Second Intl. IEEE VLSI Multilevel Interconnection Conf.*, pp. 102–113, 1985.

[6] R. W. Hoffman. In *Physics of Thin Films*, pp. 211–273, 1966.

[7] D. S. Campbell. In *Basic Problems in Thin Film Physics*, pp. 223–230, 1966.

[8] D. S. Campbell. *Handbook of Thin Film Technology*, p. 12.3. Mc-Graw Hill Book Co., New York, 1970.

[9] K. Kinosita. In *Proc. of the Second Colloquium on Thin Films*, Vandenhoeck and Ruprecht, Gottingen, pp. 31–37, 1967.

[10] R. W. Hoffman. *Physics of Nonmetallic Thin Films*, pp. 273–353. Volume 14 of *B*, NATO Advanced Study Institute, Plenum Press, New York, 1976.

[11] D. S. Gardner, T. L. Michalka, K. C. Saraswat, T. W. Barbee, Jr., J. P. McVittie, and J. D. Meindl. *Trans. on Electron Devices*, ED-32(2), 1985.

[12] D. S. Gardner and K. Saraswat. In *Multilayers: Synthesis, Properties, and Nonelectronic Applications*, Materials Res. Soc., 1987.

[13] A. K. Sinha, H. J. Levenstein, and T. E. Smith. *J. Appl. Phys.*, 49, pp. 2423–2426, 1978.

[14] E. J. McInerney and P. A. Flinn. In *20th Annual Proc. Intl. Reliability Physics Symp.*, IEEE, pp. 264–267, 1982.

[15] P. A. Flinn and G. A. Waychunas. *J. Vac. Sci Technol.*, B 6(6), 1988.

[16] K. Nakamura, S. S. Lau, M. Nicolet, and J. W. Mayer. *Appl. Phys. Lett.*, 28(5), pp. 277–280, 1976.

[17] M. Eizenberg, R. D. Thompson, and K. N. Tu. *J. Appl. Phys.*, 53(10), pp. 6891–6897, 1982.

[18] T. G. Finstad, G. Salomonsen, N. Norman, and J. S. Johannessen. *Thin Solid Films*, 114, pp. 271–284, 1984.

[19] A. G. Dirks and J. J. van den Broek. *Scripta Metallurgica*, 21(2), pp. 175–180, 1987.

[20] E. G. Colgan, C. J. Palmstrom, and J. W. Mayer. *J. Appl. Phys.*, 58(5), pp. 1838–1840, 1985.

# MECHANICAL TESTING OF THIN FILMS

C. T. ROSENMAYER*, F. R. BROTZEN*, AND R. J. GALE**
*Rice University, Houston, Texas, 77251
**Texas Instruments Inc., Dallas, Texas, 75265

## ABSTRACT

Metal films used as conductors in integrated circuits must be able to withstand stresses created during their deposition, caused by mismatches in thermal expansion coefficients and by bending which occurs during manufacture and installation. These stresses may be biaxial or uniaxial in nature. This paper discusses testing techniques and gives examples of results of mechanical testing of thin metallic films of aluminum, aluminum-copper(2%), copper, and tungsten in uniaxial and biaxial tension. The film specimens were deposited on and removed from production silicon wafers in order to simulate as much as possible actual device conditions. Basic considerations for the comparison of the two testing methods are also presented.

## INTRODUCTION

It is well known that a substantial fraction of microcircuit failures originates in the thin metallic films that serve as interconnects. The presence of defects brought about by electromigration is a frequent source for these failures. An important additional cause for the deterioration of the films, however, is their inability to withstand the high stresses created partly in the course of the deposition and processing and partly by the large difference in thermal expansion between the metallic films and the non-metallic substrates. As an example, the thermal expansion coefficient of aluminum is about eight times larger than that of silicon. As a result, thermal stresses generated in the metal at usual service temperatures are inordinately high. In view of this, the reliability of VLSI devices depends to a great degree on the mechanical properties of the metallic interconnects, the determination of which presents an interesting and challenging problem.

In the present work, the discussion will be limited to two methods for the determination of stress-strain relations in thin metallic films:

(a) the biaxial bulge-testing technique and

(b) the uniaxial tensile-testing technique.

Examples of film testing by both of these methods and attempts to correlate the results obtained by them will be presented. A background discussion on the techniques and the equipment used in the above methods by earlier investigators will be dispensed with here because it is well covered in the comprehensive reviews of this topic by Hoffman[1,2] and, more recently, by Hardwick [3].

## EXPERIMENTAL PROCEDURES AND EQUIPMENT

### Specimen Preparation

The first step in the preparation of specimens for biaxial and uniaxial testing is the ablation of the films from their substrates. In this study, Al, Al-alloys, Cu and W films of about l µm thickness of the type used in VLSI devices were used. In the case of Al and Al alloys, the films had been sputter deposited on previously oxidized Si-crystal surfaces of {00l} orientation, while in the case of Cu an intermediate layer of Mo of 0.1 μm thickness had been placed between the electro-deposited Cu and the Si substrate. An intermediate layer of phosphosilicate glass (PSG), 1 μm thick, was located between the chemical-vapor deposited W and the Si substrate. Samples were cut from the wafers along the <100> directions using a tungsten-carbide tool . Their dimensions were about 16 mm square for the specimens for the biaxial tests and 10 X 30 mm for those for the uniaxial tests. All wafers were supplied by Texas Instruments Inc. of Dallas, Texas.

The Al and Al-alloy films were ablated by placing the specimens into a teflon reactor at about 100°C, in which they were exposed to a stream of a mixture of fluorine and argon [4]. The Si of the substrate forms gaseous $SiF_4$ which is swept from the system, while the Al remains unaffected. A similar technique is used for the ablation of the Cu films. W films, on the other hand, require a different method for their removal from the Si substrate [5]. Here, liquid HF (48%) is used at room temperature for the removal of the PSG layer and of parts of the substrate. Because of the brittle nature of the W films, the excess acid must be removed with extreme care. W films have the tendency to curl severely during the ablation process, so that a two-hour vacuum-annealing treatment at 800°C of the specimen prior to exposure to HF must be applied. This relieves the stresses in the intermediate PSG layer. Metallographic characterization of specimens was carried out by optical and scanning electron microscopy.

## Biaxial Testing

A bulge test, similar to that developed originally by Beams[6] was used to determine the biaxial stress-strain relations in the films. Details of the testing procedure used in this work were given in an earlier publication [7], but a brief description of the equipment and the procedure is also given here.

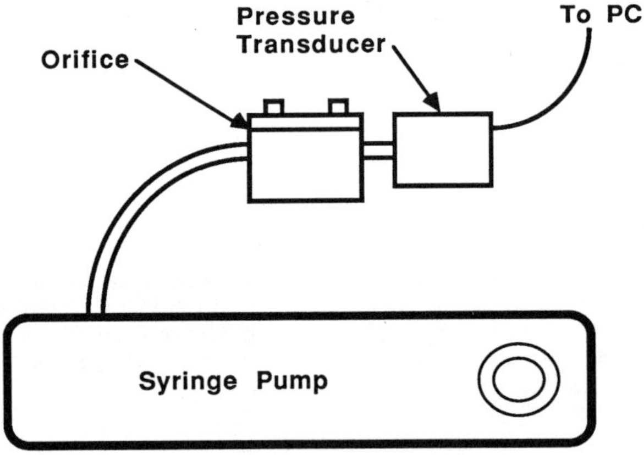


Figure 1. Schematic Diagram of Biaxial Tension Testing Apparatus.

The films were carefully fastened by an O-ring to a specially designed device that consists of a brass block mounted on the top of a syringe pump (Sage Instruments, model 341A). The pump injects glycerol at a precisely known rate into a chamber in the brass block. Care is taken to remove all entrained air from the system. This chamber is connected to a pressure transducer (Honeywell, model 142 PC) and has a sieve-like opening on top. The film lies flat on the top of the sieve and is held down by an O-ring-fitted collar. A schematic diagram of the apparatus may be seen in Figure 1.

The flow rate used was 0.0105 ml min$^{-1}$. Assuming that the bulge of the film formed by the glycerol pressure had spherical shape and that the strain was uniform throughout the film, the meridional strain can be calculated from the displaced volume, while the corresponding stress is derived from the pressure [8,9]. The pressure-volume data were collected by the data-acquisition unit of a microcomputer, then processed and recorded as stress-strain graphs.

Uniaxial Testing

The uniaxial tensile-testing equipment was of the "hard" type, that is, the flow stress is recorded at a constant strain rate. Fig. 2 is a schematic drawing of the equipment. It reveals the constant-displacement drive which is connected to one arm of a balance by a tungsten wire. The drive consists of a synchronous motor which, through a changeable worm-gear assembly, turns a lead screw that generates the linear motion of the connecting wire. In order to obtain uniform deformation, the alignment of the wire must be such as to exert a truly vertical force on the specimen which is suspended from the other arm of the balance. The end of the specimen is linked to a sensitive load cell (Schavitz, model FTA-G-500) with a range of ±500 g. The signal from the load cell is processed by a data-acquisition unit and plotted as a function of the displacement derived from the displacement rate and the time.

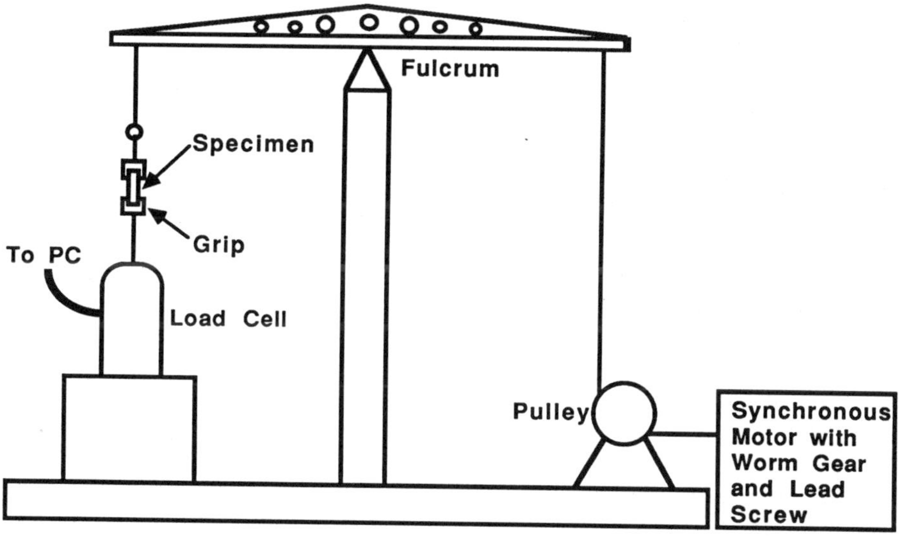


Figure 2. Schematic Diagram of the Tensile-Testing Apparatus

The most exacting feature of the equipment involves the grips.These contain very small ball-and-socket swivels near the end of the specimens, by which the film specimens are allowed to align themselves when the load is applied. To obtain films relatively free of wrinkles and curls, a special process had to be adopted. Small aluminum blocks, 10 X 5 X 1.5 mm are glued with sodium-silicate cement to the two ends of the metal side of the wafers, which measure 10 X 30 mm, Fig. 3. The wafers, consisting of the silicon substrate and the metallic film and resting on the aluminum blocks, are introduced into the fluorine reactor. After the fluorination is completed, the free-standing film, still attached to the two blocks, is glued to the flat portions of the grips with Loctite Super Bonder 495.

The specimens used in the uniaxial tests are rectangular films. However, "dog-bone" specimens, similar to the ones used by Lawley and Schuster [10] and by Springer and Catlett [11], were used in a number of tests. They proved to be less desirable than the rectangular ones which were more easily removed from their substrates.

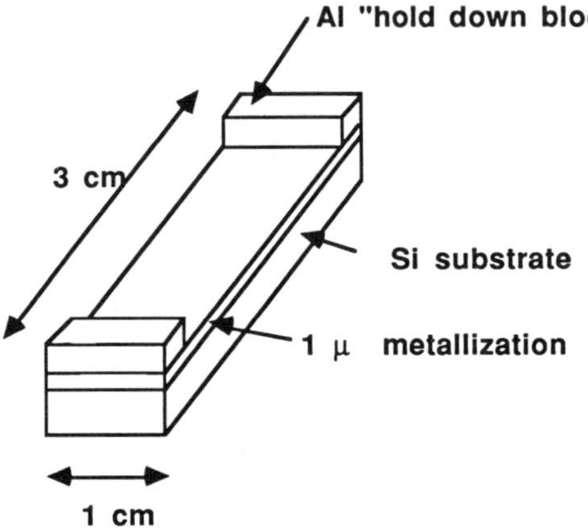


Figure 3. Schematic of tensile test specimens prior to ablation.

BASIC CONCEPTS

In principle, the stress-strain relations of a given material are obtainable from either the uniaxial or the biaxial test. Assumptions, such as the spherical shape of the bulge or the isotropy of the material, that are involved in the determination of these relations are certainly open to question. Yet, by neglecting these imponderables, one can readily obtain a good idea from the biaxial test of Young's modulus, the yield stress and the degree of plastic flow. These material properties can then be translated into uniaxial stress-strain relations. Knowledge of the uniaxial relations is particularly desirable, because they are generally more amenable to calculations of thermal stresses and because integrated circuits are stressed in bending during handling and mounting.

<u>Young's modulus</u>

Consider first the determination of Young's modulus which is predicated on the assumption of elastic isotropy of the films. The slope of the elastic portion of the meridional stress-strain curve in bulge testing is given by[1,9]

$$\frac{\sigma_m}{\varepsilon_m} = \frac{E}{(1 - \nu)}, \qquad (1)$$

where $\sigma_m$ = meridional stress, $\varepsilon_m$ = meridional strain, E = Young's modulus as determined by uniaxial testing, and $\nu$ = Poisson's ratio. $\sigma_m$ is given by

$$\sigma_m = \frac{p\,r}{2\,t}, \tag{2}$$

where p = the fluid pressure, r = the radius of the spherical bulge, and t = the film thickness. $\varepsilon_m$ is determined from the displaced volume and the geometry of the spherical bulge by the relation

$$\varepsilon_m = \frac{\sin^{-1}\left(R/r\right) - \sin^{-1}\left(R/r_0\right)}{\sin^{-1}\left(R/r_0\right)} . \tag{3}$$

Here, $r_0$ = the radius of the initial bulge (p = 0) and R is the radius of the orifice (0.625 mm). The approximation presupposes that $r/2t >> 1$. In view of the small strains involved, the latter inequality is readily satisfied.

The bulge-testing equipment described above does not provide for the direct measurement of the bulge radius. Instead, the initial bulge radius $r_0$ is calculated by assuming that the initial portion of the stress-strain curve has a slope as determined by equation (1). This calculation is performed iteratively by a microcomputer. Once $r_0$ has been determined, increases in r can be easily calculated since the flow rate is precisely known.

The question of the truly spherical shape of the bulge was discussed at length by various investigators [12-15]. While the assumption of sphericity is not generally justifiable, the error introduced in the case of small deformations and, therefore large values of r, is small.

## Yield Stress

The stress at the onset of plastic flow in the bulge test is calculated by using the von Mises distortion-energy criterion. Plastic flow tends to begin at the center of the bulge, where the principal stresses are approximately composed of the two meridional stresses, $\sigma_{m1}$ and $\sigma_{m2}$ and the radial stress $\sigma_r$. Because of symmetry, the two meridional stresses are equal, and $\sigma_r$ is insignificant in comparison to the former two, so that the yield criterion reduces to $\sigma_y = \sigma_m$. Here, $\sigma_y$ is the uniaxial yield stress which, in terms of glycerol pressure, geometry, and stress, is then

$$\sigma_y = \frac{p_y\,r}{2\,t} . \tag{4}$$

The assumptions made here preclude a truly quantitative evaluation of the stress-strain relations in the plastic domain. The total deformation, however, is a semiquantitative indication of the ductility of the film. The question of mechanical stability during plastic flow, so important in the theory of thin shells [16-17], is of little import here in view of the rather low ductility of the films.

## RESULTS AND DISCUSSION

### Uniaxial Tension

Figure 4 demonstrates the raw data of the stress-strain curves obtained for various films by uniaxial testing. The nature of the curves in figure 4 reveals clearly that results obtained by uniaxial testing are not always characteristic. It is often possible, however, to refine the raw results of uniaxial testing through the elimination of the initial portion of hte stress-strain curves. As an example, the uniaxial test results of the Al-Cu(2%) alloy film shown in figure 5 were adjusted by replacing the low-strain part of the curve by a linear plot that starts at the origin and has the maximum slope of the raw curve. This procedure will yield not only values for Young's modulus but will also give an indication of the yield stress. The stress levels are generally higher than those of corresponding bulk materials. This latter observation is in line with earlier investigations[2].

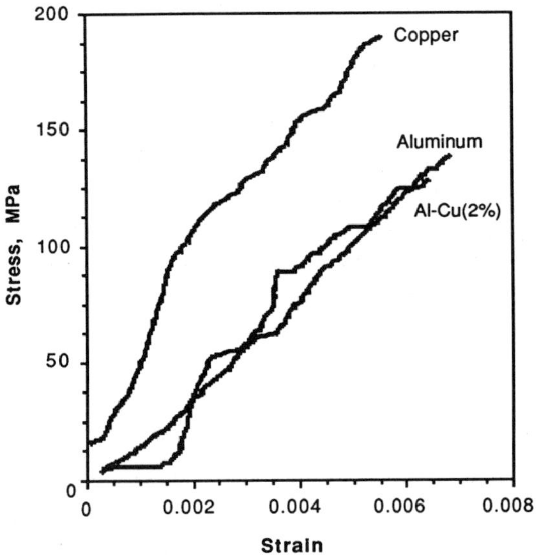


Figure 4. Tensile test results of three thin films. The copper was annealed for two hours at 750° C and cooled at 75° C per hour, the aluminum for two hours at 400° C and furnace cooled, while the Al-Cu(2%) was precipitation hardened for 30 minutes at 250° C and also furnace cooled.

Despite the special procedures for testing described above, problems created by the the very nature of the specimens are often difficult to overcome. Among these are the wrinkling and curling of the specimens as a result of intrinsic and thermal stresses as well as edge defects produced as the specimens are cleaved from the wafer.

Wrinkled specimens tend to generate "S"-shaped stress-strain curves. The initial slope is low as the wrinkles are "ironed" out. Figure 6 is a close-up photograph showing wrinkles in a specimen which is being stressed at 50 MPa. Once the wrinkles are eliminated, the slope represents the elastic response of the specimen. When yielding begins to occur, the slope once again decreases. The yield point is frequently obscured, even when it is normally well defined or is measured by an offset method.

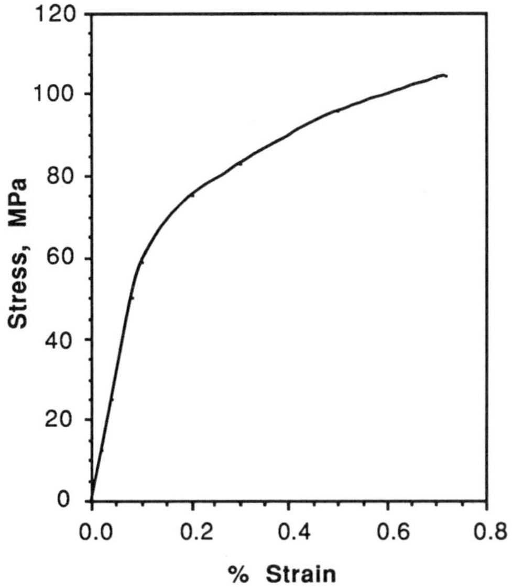


Figure 5. Adjusted stress-strain curve for an Al-Cu(2%) film which was precipitation hardened for 70 hours at 250°C.

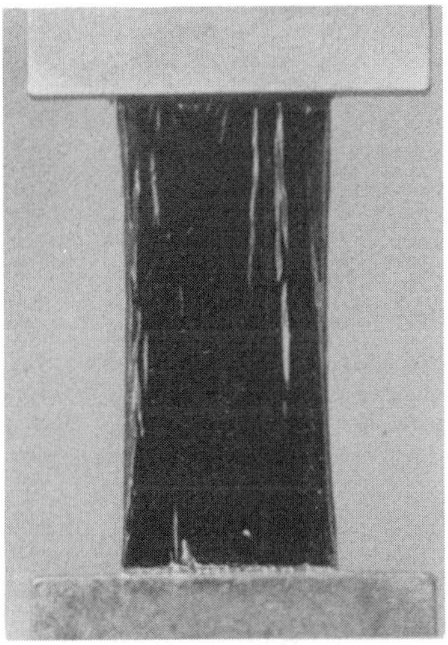

Figure 6. Wrinkles in stressed aluminum specimen

At times, the curling of the specimens is severe enough (in as-received or quickly cooled specimens) to cause the 10 X 30 mm film to roll into a tight scroll 1 mm in diameter or to tear at the point where the specimen is glued to the restraining block. A partially failed specimen may be seen in figure 7. The specimen failed by tearing from left to right. As the tearing proceeded from the left edge, the film began to roll into a tight scroll from the initiation site. Even slightly curled specimens make the task of accurately locating the specimen on the grips difficult, especially when the situation is exacerbated by the presence of wrinkles and by the fragility inherent in the 1 μm thick films. Inaccuracies in mounting prevent the specimen from being evenly loaded. Nonuniform adherence of the film specimens to the grips also results in "S"-shaped stress-strain curves. Initially, only part of the specimen is loaded, which causes a lower than expected slope. The slope eventually increases to becoming equal to the elastic modulus until the yield point is reached.

Figure 7. Failed aluminum specimen which exhibits curling behavior.

In general, the specimen-cleaving process produces microscopically smooth edges. However, If the silicon substrate does not cleave evenly, the metallization may have a jagged edge. These edge defects can be rather sharp, causing considerable stress intensification.

Biaxial Tension

Examples of stress-strain relations obtained by bulge testing are given in figure 8. A listing of the uniaxial Young's moduli , Table I, was obtained by means of (I), using the steepest portion of the slope of the curves in figure 8. The agreement with data from the literature for corresponding bulk materials is satisfactory. The aluminum, aluminum-copper, and copper films all had significant amounts of wrinkles, hence the S-shaped stress-strain curves for these metals. The tungsten films were much smoother, providing curves of a more conventional appearance. The elongation in all cases is rather small, while the stress level is higher than that normally expected from these materials in bulk form.

The bulge-testing methods eliminates the hazards of edge defects, and the thin layer of glycerol on the top of the sieve serves to smooth wrinkles by wetting the bottom of the film and holding it relatively flat. This method tends to be simpler and experimentally much more convenient than uniaxial testing.

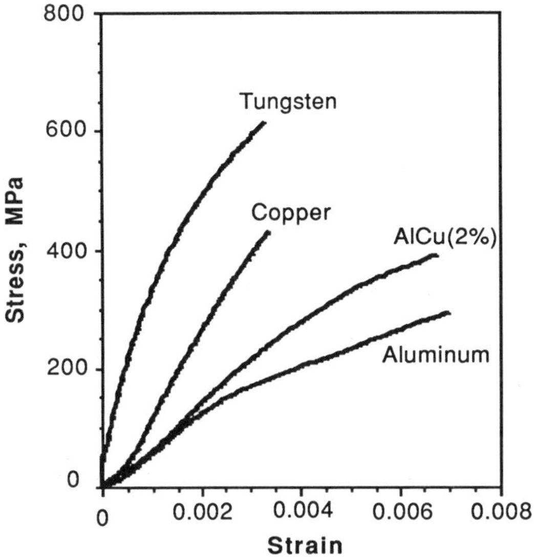


Figure 8. Bulge test results of various metallizations. The tungsten was annealed for 2 hours at 800°C and cooled at 75°C per hour, the copper at 750°C and simularly cooled, the aluminum at 400°C and furnace cooled, and the Al-Cu(2%) was precipitation hardened at 250°C at 30 minutes and also furnace cooled.

Table I
Young's moduli obtained from bulge testing results

| Material | Poisson's Ratio[18] | Young's modulus (Bulge test) GPa | Young's modulus[18] (Bulk) GPa |
|---|---|---|---|
| W | 0.285 | 340 | 392 |
| Cu | 0.352 | 116 | 124 |
| Al | 0.357 | 64 | 70 |

## CONCLUSIONS

Mechanical properties of commercially produced thin-film metallizations can be determined by uniaxial or biaxial testing methods, yielding valuable information on elastic and plastic deformation of these films.

From an experimental point of view, biaxial testing offers numerous advantages over uniaxial testing of films. Problems with reproducibility of results arising in uniaxial testing stem generally from flaws in the specimens, notably wrinkles and a tendency to curl because of internal stresses.

Direct correlation between results obtained by the two techniques is possible but is problematic at times in view of the aforementioned difficulties associated with uniaxial testing.

Elastic moduli derived from bulge-testing results of films of various metallic materials are in good agreement with those of corresponding bulk materials listed in the literature.

The commercially produced films, 1μm in thickness, of Al, Al-Cu(2%), Cu and W proved to have higher strength and lower ductility than the identical materials in bulk.

ACKNOWLEDGMENTS

The authors wish to acknowledge the invaluable assistance by Mr. J. P. Gesenhues in the design and construction of the test equipment.

REFERENCES

1. R. W. Hoffman, in Thin Films, edited by H. G. F. Wilsdorf (Amer. Soc. Met., Metals Park, Ohio, 1964), pp. 99-134.
2. R. W. Hoffman, in Physics of Thin Films, vol. 3, edited by G. Hass and R. E. Thun, (Academic Press, New York, 1966), pp. 211-273.
3. D. A. Hardwick, Thin Solid Films 154, 109-124, (1987).
4. R. B. Badachhape, J. L. Margrave, and F. R. Brotzen, Thin Solid Films 139, L 77-78 (1986).
5. C. T. Rosenmayer, F. R. Brotzen, and D. P. Brady, submitted for publication to Thin Solid Films.
6. J. W. Beams, in Structure and Properties of Thin Films, Edited by C. A. Neugebauer, J. B. Newkirk, and D. A. Vermilyea, (John Wiley and Sons, New York, 1959), pp. 183-193.
7. A. J. Griffin, Jr., F. R. Brotzen, and C. F. Dunn, Thin Solid Films 150, 237-244 (1987).
8. F. R. Brotzen, C. T. Rosenmayer, and R. J. Gale, to appear in Thin Solid Films 166, (1988).
9. S. Jovanovic and C. S. Smith, J. Appl. Phys. 32, 121-122 (1962).
10. A. Lawley and S. Schuster, Rev. Sci. Instr. 33, 1178-1180 (1972).
11. R. W. Springer and D. S. Catlett, Thin Solid Films 54, 197-205 (1978).
12. A. J. Ranta-Eskola, Int. J. of Mech. Sci. 21, 457-465 (1979).
13. T. Tsakalakos, Thin Solid Films 75, 293-305 (1981).
14. R. Papirno, J. Appl. Phys. 32, 1175-1176 (1961).
15. A. Catlin and W. P. Walker. J. Appl. Phys. 31, 2135-2139.
16. G. Gerard and R. Papirno, Trans. ASM 49, 132-148 (1951).
17. W. F. Brown, Jr. and G. Sachs, Trans. ASME 70, 241-251 (1948).
18. G. Simmons and H. Wang, Single Crystals Elastic Constants and Calculated Aggregate Properties: A HANDBOOK, 2nd ed.(MIT Press, Camb ridge, 1971).

# MEASURING THE STRENGTH AND STIFFNESS OF THIN FILM MATERIALS BY MECHANICALLY DEFLECTING CANTILEVER MICROBEAMS

T. P. WEIHS, S. HONG, J. C. BRAVMAN and W. D. NIX

Department of Materials Science and Engineering
Stanford University, Stanford, CA, 94305.

## Abstract

The present authors recently introduced the mechanical deflection of cantilever microbeams as an experimental technique for measuring the strength and stiffness of thin films.[1] The technique utilizes conventional integrated circuit (IC) fabrication to process the samples, a Nanoindenter to deflect the cantilever beams mechanically, and simple elastic beam theory to analyze the data. This paper will review the technique and describe some of its current applications.

## Introduction

When measuring the mechanical properties of thin films, the presence of substrates below the films can affect the measurements[2]. Thus, one would prefer to test the films free of their substrates. However, if the films are approximately 1.0 μm thick and free-standing, they can be damaged if handled directly. To avoid both of these problems, and to offer a measurement of the strength and stiffness of thin films on a local scale, the deflection of cantilever microbeams using a Nanoindenter was introduced recently[1]. This paper will describe the three basic stages of the microbeam technique - the fabrication of samples, the testing of samples, and the analysis of data. In addition, some applications of the technique will be described and several issues regarding the deformation of the beams will be discussed.

## Sample Fabrication

A variety of different materials can be fabricated into cantilever beams for testing. Interconnect metals such as gold, dielectric materials like silicon dioxide, and semiconductors such as silicon have all been tested. For such a diverse group of materials, different IC processing methods are employed.

The first method -- called the "lift-off" technique -- is typically used in the processing of metal beams. It is termed "lift-off" because photoresist is peeled away or "lifted off" the silicon substrate in order to pattern the metal film. The initial step in the technique is the deposition and patterning of photoresist. A positive resist is spun onto a lightly-doped p-type (100) silicon wafer and patterned using typical exposures to UV light and chemicals. The processing leaves photoresist only in sections where the metal film is not desired. In the next step, metal is deposited over the complete top side of the wafer using e-beam evaporation or sputtering techniques. After the deposition, warm acetone dissolves the photoresist that is covered by the metal film and thereby lifts off the unwanted sections of the film. The cantilever beams then appear as thin lines on the silicon. To form the bottom side of the cantilever beams, a heated solution of EDP (66 wt.% ethylendiamine, 23 wt.% $H_2O$, 11 wt.% pyrocatechol) anisotropically etches away the silicon from below the metal lines. The final sample will appear as shown in Figure 1: cantilever beams extending over an open silicon etch pit that is bounded by 4 {111} side walls and a (100) bottom surface.

The second fabrication method, used for fabricating dielectric samples, is more common in IC processing. Silicon dioxide films are grown (wet-thermal oxide) or deposited (low-pressure chemical vapor deposition) on (100) Si wafers. The oxide films are then patterned using standard photolithography processes that include positive photoresist and a wet etchant of buffered hydrofluoric acid. Once the oxide is shaped into thin lines, a heated solution of EDP etches away the Si from beneath the patterned oxide films, thereby forming cantilever beams of $SiO_2$.

The third and final processing technique is used to fabricate silicon beams. It is significantly different than the first two in that we etch the silicon wafer from its back side to form the cantilever beams and not from its front or top side. To begin the processing, silicon nitride is deposited on both sides of the wafer. 3.0 mm squares are then patterned in the nitride on the back side of the wafer using photoresist and dry etching. After the photoresist is removed, liquid KOH anisotropically etches the silicon wafer until the square patterns extend up through most of the wafer. By carefully monitoring the depth of the square etch pit, the etching can be stopped so that only a thin film remains at the top of the wafer. In the final set of steps, these thin films of silicon are patterned into cantilever beams. First, the top of the wafer is patterned with a hardened photoresist. Then, sections of the nitride film are dry etched where silicon is to be removed. Once the silicon is etched to form the beams, the remaining nitride and photoresist layers are removed as well.

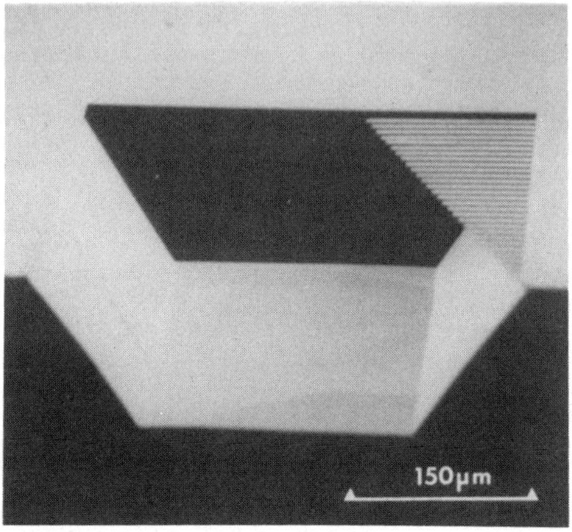


Figure 1: Au Cantilever Microbeams over a Si Etch Pit

## Mechanical Testing

The elastic and plastic properties of the various materials comprising the thin cantilever beams are measured using a Nanoindenter. The instrument mechanically deflects the beams at a constant velocity while continuously recording the applied load and the measured deflection. The rates of loading and unloading are controlled so that the beams bend and recover at a constant velocity between 3 and 6 nm/s. During loading, when a predetermined deflection is reached, the applied force is held constant for 10 s before it is removed. The resolutions of load and displacement measurements are 0.25 μN and 0.3 nm, respectively.

To analyze the load and deflection data recorded for a given test, the physical dimensions of the beam must be known. The width and thickness of a beam are typically measured using scanning electron microscopy and optical microscopy. However, to calculate mechanical properties we do not measure the actual length of the beam. Instead, we measure the distance between the fixed end of the beam and the point of loading which we call the effective length. In Figure 2 the distance between point B and C marks this dimension. For a typical test, it is measured using the X and Y stepper motors on the Nanoindenter which have a 0.5 μm lateral resolution. These motors control the movement of the sample and they prerecord positions A, B, and C. Of course, it would be more accurate to measure the effective length directly by looking for an indentation on the beam at the point of contact. However, the forces applied to the beams are typically too small to form a visible indentation on the top of the beam. As an alternative, we check the position of contact on the beam indirectly by observing the position of the indentation at point A. If the indentation is not at its prerecorded position, then changes in the point of loading on the beam can be inferred.

Figure 2 also shows a schematic of the main electro-mechanical components of the Nanoindenter. A brief description of these components is given here to clarify the test technique and the presentation of data. In simplest terms, the Nanoindenter is a constant-load machine; loads are applied and displacements are measured. We can see from Figure 2 that load is applied to the shaft of the indenter by changing the DC current running through the coil. Changes in the current alter the magnetic field emanating from the coil, and thereby increase or decrease its attraction to the permanent magnet. Note that as the indenter shaft moves up or down, the applied load must offset the resistance of the suspending spring as well as the resistance of the beam being bent. Thus, once the indenter tip contacts a beam and begins to bend it, the applied load is working to deflect two springs in parallel: the suspending spring and the cantilever beam. A final point to notice regarding the electro-mechanical workings of the Nanoindenter is the effect of the relative magnitude of these two springs. If a beam is very compliant because it is thin or long, then its spring constant may be too small, relative to the spring constant of the suspending spring, to be measured accurately.

By continuously recording the load applied to the shaft of the indenter and the resulting deflection of the shaft, the initial contact between the diamond tip and the cantilever beam can be detected. In Figure 3 the total applied load is plotted versus the total tip deflection for two different tests of a silicon beam. In the initial parts of both curves, the slope or spring constants are equivalent, since no beam has yet been contacted. The applied load is deflecting only the suspending spring. At a total deflection of approximately 2.0 μm, the slopes of the two curves change abruptly because the indenter contacts the beams. The plot of data following the contact point then describes the total applied load and the total deflection of the indenter tip. To know the load on the beam, the force carried by the suspending spring must be subtracted, and to know the deflection of the beam, we must subtract the depth to which the diamond tip indents the top of the beam.

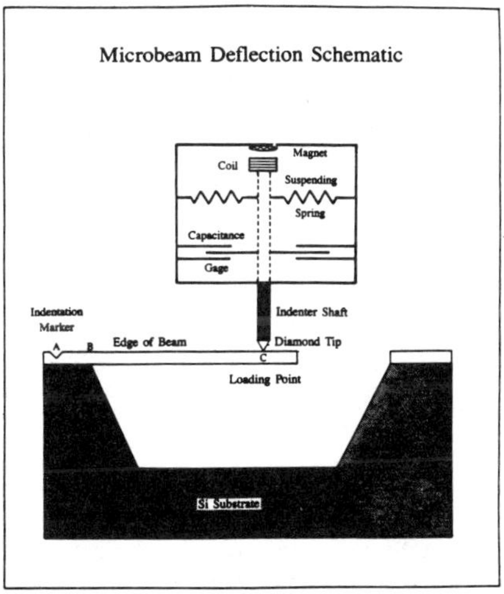


Figure 2: Nanoindenter and Microbeam Schematic

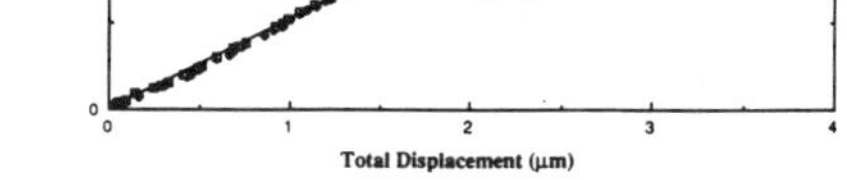


Figure 3: Two Deflections of a Si Beam

## Analysis of Beam Deflections

The total load applied by the instrument deflects both the suspending spring and the beam. To subtract the load on the suspending spring from the total load we simply measure the spring constant and include it in the analysis. To determine the net deflection of the beam, though, more work is required. For the softer, metallic beams, measurable indentations can form on the top of the cantilever beams as the forces are applied. These indentations are often too small to see, but they can be detected given the displacement resolution of the Nanoindenter. For the harder, dielectric materials, the indentations are typically very small and appear to be completely elastic. This is suggested by data presented earlier[1] that shows complete recovery on unloading. However, whether the indentations are small or large, plastic or elastic, their depth should be subtracted from the total deflection data to analyze the deformation of the beams.

To subtract the depth of indentation from the total deflection data, an empirical equation relating the depth of indentation to the applied load is determined for each new beam material. An array of small indentations is made in sections of the material that are supported by the silicon substrate. The indentations vary in depth from 20 nm to 500 nm. To determine the empirical equations, curves are fit to the average values for indentations of similar depths.

Another deflection which must be considered is that of the cantilever beams bending across their width. Since the beams are loaded at a point along their centerline, and since the beams are typically very wide compared to their thicknesses, the centers of the beams will tend to bow downward relative to the edges of the beams. The authors have shown that for an effective length to width ratio of 1.0, the additional deflection of the diamond tip due to curvature across the widths of the beams can be 2.0 % of a beam's net deflection [1]. However, as the length or thickness of a beam increases compared to its width, this deviation decreases significantly. To improve the accuracy of the measurements, this displacement is considered in all analyses.

Once the net deflections and loads for the bending of a beam are known, simple elastic beam theory for plane strain is applied to calculate elastic and plastic properties. Elastic theory says that the applied load, P, should be linearly related to deflection, w, and that the slope of this relationship is determined by the Young's Modulus, E, of the beam material and the geometry of the beam.

$$w = \frac{4Pc^3}{bEt^3}(1 - \nu^2) \qquad (1)$$

In the above equation c is the effective length, b is the width, t is the thickness, and $\nu$ is the Poisson's ratio for the material. The two different deflections of a Si beam which are plotted in Figure 3 support equation 1. The Young's Modulus of the Si can be determined from the slopes of the two curves and their respective geometries. Note that the beam test with the longer effective length has the lower slope, as expected from equation 1. Furthermore, note that the linear nature of the curves in Figure 3 suggest that the two deflections are purely elastic.

In order to measure plastic properties such as yield strength, fracture strength, or strain hardening, beams can be deflected beyond their elastic limit. The onset of yielding or fracture is easily detected through a deviation from linearity in the load-deflection curves. The yield stress, $\sigma_y$, is given by the following equation

$$\sigma_y = \frac{6cP_y}{bt^2} \qquad (2)$$

where $P_y$ marks the load corresponding to the loss of linearity and the onset of plastic deformation. Note that equation 2 assumes the existence of plane stress rather than plane strain. This assumption is made for two reasons. The first is that the plane strain conditions assumed in the elastic analysis for equation 1 are relaxed at the edges of a beam. The second is that because of this relaxation, yielding should start at the edges of the beam where the effective stresses[3] are highest and then move inwards toward the center of the beam.

## Discussion

To date, various metal beams have yielded and shown strain hardening (Figure 4). However, no beam has fractured at its fixed end even though strains in excess of 0.01 have been applied to silicon dioxide and silicon beams (Figure 5). The absence of such fracturing has been attributed to the uniformity of the material. Large planer defects, such as cracks, appear to be absent. The nonlinearities of the curves in Figure 5 do not mark yield or fracture. Instead they are attributed to the indenter tip sliding out along the hard oxide beam as it is bent. Such sliding increases the effective length of the beam and thereby decreases the total spring constant for the test. Consequently, the slopes of the curves decrease. The deformation is thought to be elastic since the unloading curve retraces the loading curve.

When yielding does occur in a cantilever beam, the presence of surface layers such as oxides must be considered. As a cantilever beam is bent beyond its elastic limit, the material at the fixed end of the beam will be the first to deform plastically. The deformation will start at the top and the bottom surfaces of the beam, and at its edges, and then it will spread inwards. If the beam is completely homogeneous, the onset of plastic deformation at the surfaces of the beam will match the strength of the material within the beam. However, if a surface layer such as an oxide is present, then the initial yielding may be delayed and the measured yield strength may not accurately describe the beam material. This concern has been raised as we test metal beams such as Cu and Al which readily form oxide layers when exposed to air at room temperature.

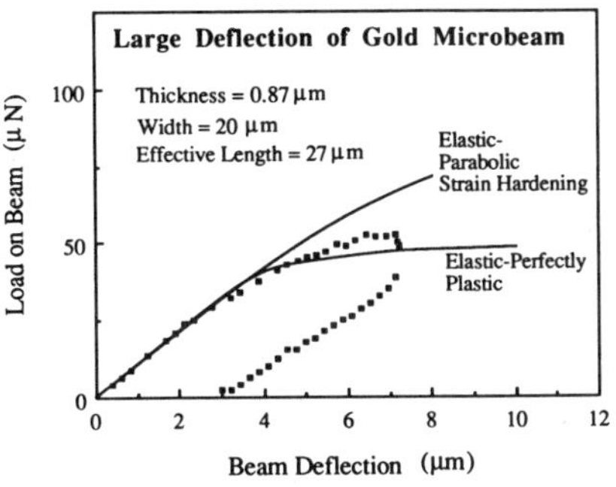


Figure 4: Large Deflection of a Au Beam

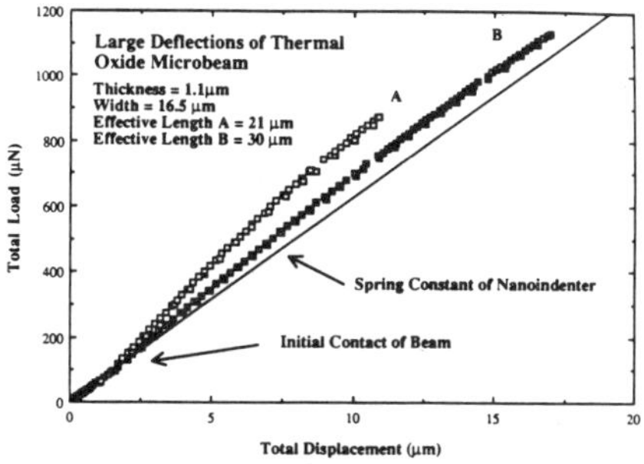


Figure 5: Large Deflection of a $SiO_2$ Beam

To estimate the degree of strain hardening that occurs in a metal beam as it is bent, we compare experimental data with theoretical curves. In Figure 4 the strain hardening of a Au beam is plotted with theoretical data for an elastic-plastic material (i.e. no strain hardening) and a material that strain hardens parabolically [1]. Since the experimental curve for Au lies closest to the curve for the elastic-plastic material, we assume that the thin film does not strain harden significantly during bending. The simple explanation for this is the large ratio of surface area to volume for the Au material. With the large area of free surface, dislocations can leave the beam and avoid strain hardening interactions. For a typical bulk sample that has a lower ratio of surface area to volume, dislocations cannot leave the material as easily. Thus, the dislocations interact more than in the Au beam, and strain hardening is higher.

For comparison of measurement techniques, the hardness and Young's Modulus of the films that form the beams were measured using a Berkovich indenter, the Nanoindenter, and the indentation analysis of Doerner and Nix[2]. Arrays of indentations were made in sections of the films that are supported by the silicon substrates. Properties measured by both the microbeam deflection technique and the nanoindentation technique are presented in Table 1. The values of Young's Modulus measured by the two methods for amorphous and nontextured polycrystalline materials ($SiO_2$ and Au) should be directly comparable. However, in the initial experiments reported here, there was only limited agreement. The moduli measured by indentation for the oxide films are higher than those measured by microbeam deflection and they are higher than literature values reported earlier[1]. Some of this disparity can be attributed to a densification of the oxide films that appears to have occurred upon indentation[1]. In terms of the Au films, preliminary X-ray studies show little texturing in the Au and therefore suggest that the two measurement techniques should yield similar values for Young's Modulus. However, the indentation values are again higher. A possible cause of this difference and the difference in the oxide measurements is the effect of the large pressure under the indenter tip[1,2]. The only strong agreement between the measurements of the two techniques is shown for the plastic properties of the Au films. With an appropriate conversion from yield strength to hardness, $H = 3.2*\sigma_y$, [7], the two different measures of the resistance to plastic deformation compare very favorably.

When single crystal films such as silicon are tested, the microbeam technique measures Young's Modulus for a particular direction in a particular plane, while the indentation technique measures an average Young's Modulus.[4-6] Thus, the two techniques can not be compared directly. But, the Young's Modulus of 163 GPa measured for Si on a (100) plane and in a {110} direction is very similar to the calculated value of 169 GPa[8].

Table 1: Elastic and Plastic Properties Measured for Thin Films

| | Beam Test | | Indentation Test | | |
|---|---|---|---|---|---|
| | E | $\sigma_y$ | E | H | $\nu$ |
| Si | 163 | >3.39 | 188 | 13 | .068(.20)[2] |
| Thermal SiO2 | 64 | >0.57 | 85 | 12 | .16 |
| LTO | 44 | - | 83 | 10 | .16 |
| Au | 57 | 0.34[1] | 74 | 1.0 | .42 |

*All data in GPa
[1]Corrected from initial reported value[1] in which an effective stress was mistakenly listed.
[2]An average value of $\nu$ (0.20) was used for the indentation analysis.

## Conclusion

The deflection of cantilever microbeams has been described as an alternative technique for measuring mechanical properties of thin film material. The various steps involved in sample fabrication, beam deflection, and data analysis have been reviewed and some current applications of the technique have been discussed.

## Acknowledgements

The work presented here was supported by the Defense Advanced Research Projects Agency through the University Research Initiative at University of California at Santa Barbara under the ONR Contract No. N00014-86-k-0753 and the Semiconductor Research Corporation.

## References

1. T.P. Weihs, S. Hong, J.C. Bravman, and W.D. Nix, J. Mater. Res. 5, 931 (1988).
2. M.F. Doerner and W.D. Nix, J. Mater. Res. 1, 601 (1986).
3. G.E. Dieter, *Mechanical Metallurgy*, (McGraw-Hill, San Francisco, 1986), pp. 77-81.
4. L.J. Farthing, T.P. Weihs, D.W. Kisker, J.J. Krajewski, M.F. Tang, and D.A. Stevenson, presented at Fall MRS Meeting, Boston, MA, 1988 (to be published).
5. W.C. Oliver, F.A. List, and R.A. McKee, presented at the Fall MRS Meeting, Boston, MA, 1988 (to be published).
6. J.B. Pethica, presented at the Fall MRS Meeting, Boston, MA, 1988 (to be published).
7. A. Bhattacharya and W.D. Nix, Int. J. Solids Structures, 24, 881 (1988).
8. W.A. Brantley, J. Appl. Phys., 44, 534 (1973).

# THE DETERMINATION OF MECHANICAL PARAMETERS AND RESIDUAL STRESSES FOR THIN FILMS USING MICRO-CANTILEVER BEAMS

S. HONG, T. P. WEIHS, J. C. BRAVMAN AND W. D. NIX
Dept. of Materials Science and Engineering, Stanford University, Stanford, CA. 94305

## ABSTRACT

A method for determining mechanical parameters and residual stresses for thin films is described. Multi-layer cantilever beams (LPCVD $SiN_x$/thermal $SiO_2$) are fabricated utilizing standard IC processing technologies and micromachining of silicon. The elastic response of the beams to imposed deflections is then measured using a Nanoindenter, a sub-micron hardness testing machine. The elastic constants of the nitride films are calculated from the force vs. deflection slope and known elastic constants of the thermal $SiO_2$ and silicon. By measuring the curvature of the multi-layer cantilever beams with a scanning electron microscope after successive etching of the LPCVD nitride films, average and differential stresses in the films were calculated.

## INTRODUCTION

LPCVD (low pressure chemical vapor deposition) $SiN_x$ films are widely used in silicon semiconductor devices as a diffusion barrier and as passivation layers. The structure of thin LPCVD $SiN_x$ over thermally grown $SiO_2$ is particularly important in the fabrication of most MOS devices, since this structure is essential for the LOCOS process. The residual stress in the nitride film is one of the main factors responsible for the generation of point and line defects in the silicon substrate during/after the bird's beak formation. The ratio of the thicknesses in the nitride and oxide layers is critical to the defect formation [1]. To gain a better understanding of the mechanism of defect generation, the stress state and elastic constant of the thin nitride films should be known.

In this paper, a new test method which can determine both the elastic constants and residual stresses of very thin films will be described. The technique is an extension of the single-layer cantilever bending technique [2], and makes use of multi-layer cantilever beam structures. Thin films for which elastic constants are to be determined (LPCVD $SiN_x$ in this paper) are deposited on well characterized films, which can sustain large elastic deformations. Thermal $SiO_2$ ideal for this application [3]. The elastic parameters of the top layer can be evaluated without the yielding of the bottom reference film . Furthermore, the bottom layers can withstand the large stresses that develop in the beams due to significant growth or thermal mismatches between the top and bottom films. This allows the residual stresses to be calculated with a simple elastic model.

After the fabrication of double-layer cantilever beams, the elastic constants of the nitride films are calculated from the force-deflection slope and known elastic constant of the thermal $SiO_2$. The curvature of this double-layer beam provides information on the residual stresses in the films. This test method is useful for measuring films that are either too thin (less than 0.5 μm thick) to test as single layer beams or for films that are easily attacked by the silicon etchant and therefore cannot be fabricated into single-layer beams.

## SAMPLE PREPARATION PROCEDURES

A layer of wet-thermal $SiO_2$ was grown on a lightly doped P-type {100}Si wafer at 1100°C for about 5 hours. The thickness of the $SiO_2$ layer was measured at 1.465 μm using an ellipsometer. Two types of LPCVD nitrides, with different ratios of reactant gas flow rates, were deposited on the $SiO_2$ layer by the chemical reaction of $SiCl_2H_2$ and $NH_3$ at 785°C. The chamber pressure was 360 mtorr. There were no anneals after the depositions. In this paper, the nitride deposited with a gas flow ratio between $SiCl_2H_2$ and $NH_3$ of 1:3, will be refered to as a 'standard nitride'; whereas the nitride deposited with the ratio of 6:1 will be refered to as a 'low stress nitride.' The thicknesses of the standard and low stress nitride were 2140 ± 15 Å and 2870 ± 15 Å, respectively. The thickness measurements were made with both a surface profiler, which directly measures the step height of the nitrides on the oxides, and an ellipsometer, which measures the reference samples of nitrides on bare silicon wafers.

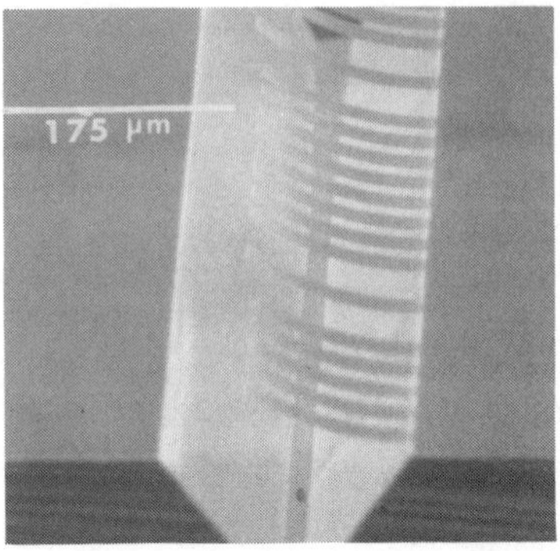


Figure 1: LPCVD $SiN_x$ (2100 Å)/$SiO_2$ (1.465 μm) beams.

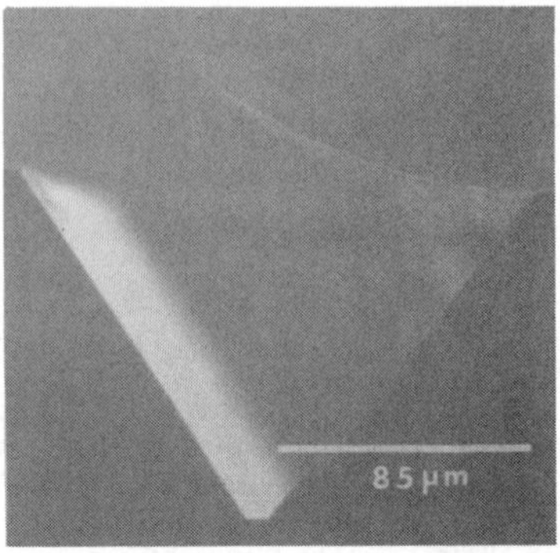


Figure 2: Cross-sectional view of LPCVD $SiN_x$ (2100 Å)/$SiO_2$ (1.465 μm) beams.

The $SiN_x$ layer was patterned by standard photolithography processes and a dry etching process in a gas mixture of $SF_6$ and $CF_3Br$. The underlying $SiO_2$ layer was wet etched in a buffered hydrofluoric acid (BHF) solution. Finally, the exposed silicon was etched in an EDP solution (66 w/o ethylenediamine, 23 w/o DI water, 11 w/o pyrocatechol) at 115°C for 60 minutes, and then rinsed and dried. In this solution, the etch rates of thermal $SiO_2$, standard $SiN_x$, and low stress $SiN_x$ are extremely low (~100 Å/hr, ~35 Å/hr and ~45 Å/hr, respectively) compared to the etch rate of the Si ~60 μm/hr.

For the determination of average residual stress in the nitride, various thicknesses of the nitride films were produced using the dry etching process before the etching processes described above were initiated. The curvatures of the double-layer beams with different nitride thickness were measured in SEM photographs.

Figure 1 and Figure 2 are SEM photographs of multi-layer (standard $SiN_x/SiO_2$) cantilever beams which were fabricated using these procedures. All of these beams curve upwards due to initial compression in $SiO_2$ layer and tension in $SiN_x$ layer. As seen in these figures, the curvatures are smooth and circular in shape.

## DETERMINATION OF ELASTIC CONSTANTS FOR VERY THIN FILMS

### 1. Mechanics for Double-layer Beam Bending

From the assumption that bending strain due to a curvature is continuous throughout the beam thickness, force and moment can be written in a general expressions as follows:

$$F = \int \sigma dZ = \varepsilon_0 \int E dZ + K \int EZ dZ \tag{1}$$

$$M = \int \sigma Z dZ = \varepsilon_0 \int EZ dZ + K \int EZ^2 dZ \tag{2}$$

where $\sigma$ is the normal stress, $\varepsilon_0$ is the axial strain, $Z$ is the distance in a vertical direction from the neutral axis, and $K$ is the curvature generated when a beam is deflected. The position of the neutral axis from the middle of double layer beam, $z_0$ (see Figure 3) can be determined by setting the coupling terms ($\int EZdZ$) to zero. Thus

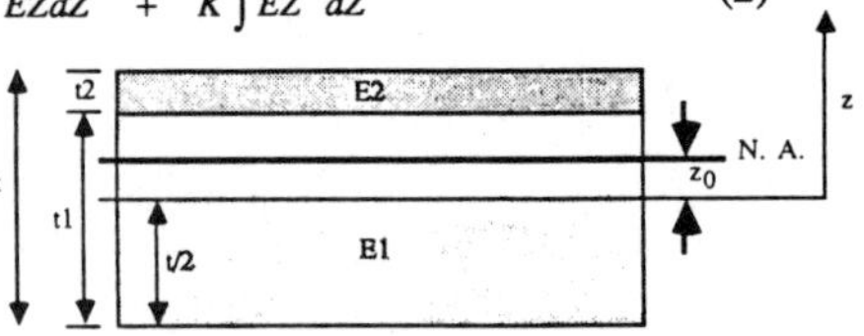


Figure 3: Schematic diagram of double-layer beam cross-section

$$z_0 = \frac{\int_{-t/2}^{t/2} Ez \, dz}{\int_{-t/2}^{t/2} E \, dz} \tag{3}$$

Once the position of the neutral axis is known, the effective stiffness of a double-layer beam per unit width, $(EI)_{effective}$, can be calculated from eq. (2) as follows:

$$(EI)_{effective} = \int_{-t/2}^{t/2} E\left\{z - z_0\right\}^2 dz \tag{4}$$

$$= \int_{-t/2}^{(t_1 - t_2)/2} E_1\left\{z - z_0\right\}^2 dz + \int_{(t_1 - t_2)/2}^{t/2} E_2\left\{z - z_0\right\}^2 dz \tag{5}$$

Equations (3) and (5) suggest that $z_0$ and $(EI)_{effective}$ are functions of the elastic constants and the thicknesses of a multi-layer beam, *e.g.* $E_1, E_2, t_1, t_2$.

## 2. Test Procedures and Data

After the samples were prepared in double-layer cantilever structures, bending tests were performed as described elsewhere[2]. Figure 4 shows data for the bending test of a double-layer cantilever beam which consists of standard $SiN_x$ (2100 Å) and thermal $SiO_2$ (1.465 μm) layers. The width and effective length of the beam were 22.5 μm and 33.1 μm, respectively. The Young's modulus of thermal $SiO_2$, which is the bottom layer in this sample, was measured earlier to be ~ 65 GPa [2]. The slope of the initial straight line in Figure 4 represents the spring constant of the Nanoindenter. When the tip of the indenter hits the beam and starts to deflect it the slope changes abruptly. The slope in the second linear section represents the sum of the spring constants of the machine and the beam. From this slope, it is possible to extract the actual effective stiffness of the beam. This measured value of $(EI)_{effective}$ will be used to determine the elastic constant of the second layer material, which is standard $SiN_x$ in this example.

Since we know all of the physical dimensions of the double-layer beam and the Young's modulus of the bottom layer, the position of the neutral axis can be determined using eq. (3), if any arbitrary number is given for the Young's modulus of the second layer. The calculated position of the neutral axis, in turn, determines the effective stiffness of the beam. Therefore, a set of values for $(EI)_{effective}$ for the beam can be calculated by varying the ratio of the elastic moduli of the first and second layers, $E_2/E_1$. Finally, the Young's modulus of the second layer is determined by comparing the measured stiffness with these calculated values: the correct ratio of $E_2/E_1$ makes the difference between the two stiffnesses vanish. As an example in Figure 5, the difference between the two stiffnesses for the same beam was plotted with the ratio, $E_2/E_1$, varying from 1 to 6. The difference goes to zero when the ratio is ~4.2. Thus, the Young's modulus of the standard

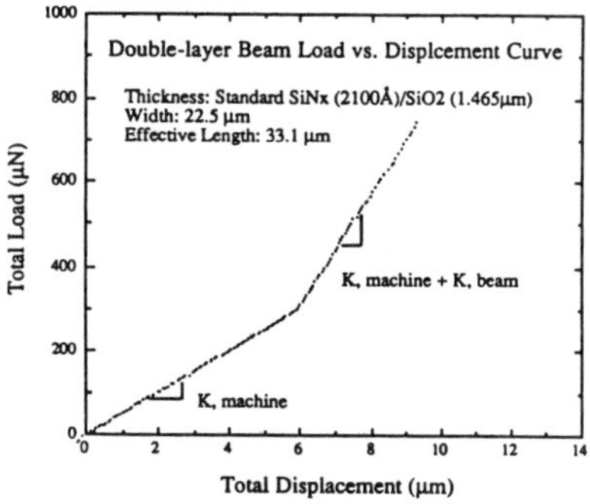


Figure 4: Deflection of a LPCVD $SiN_x/SiO_2$ beam.

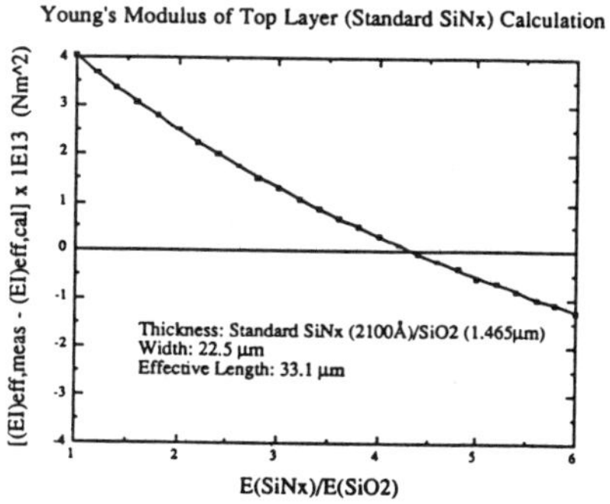


Figure 5: Plot of difference in measured and calculated stiffnesses with $E_2/E_1$.

nitride is ~ 275 GPa, which is ~4.2 times greater than that of thermal $SiO_2$. Following the same test procedures and analysis, the Young's modulus of the low stress LPCVD nitrides was determined to be ~ 235 GPa.

Large deflections of the $SiN_x/SiO_2$ beams, to about 1/4 of their lengths, were also made. The normal stress at the surface section due to this large deflection is approximately 3 GPa (~ 1 % elastic strain). The total normal stress, including the residual stress to be discussed in next section, could be as high as ~ 4 GPa for the standard LPCVD nitride. However, we did not observe any fracture of the nitrides. This suggests that the films, up to the thicknesses we studied, have a very small average flaw size.

## DETERMINATION OF RESIDUAL STRESSES FOR VERY THIN FILMS

### 1. Mechanics for Residual Stress Calculation from A Curvature

In order to calculate the average residual stress in thin films from the curvature of a multi-layer beam, we have adopted the elastic model proposed by P. H. Townsend *et al.* [4]. In their model, the curvature, $K$, and the normal stresses in each film, $\sigma_i$, are equated in terms of physical dimension, $t_i$, material constants, $E_i$, and hypothetical relaxed linear dimensions, $d_i$ . If we modify this model for 2-dimensional cases with the coordinates used in Figure 3, the curvature and the normal stress can be rewritten as follows:

$$K = \frac{\sum_{i=1}^{n} \frac{E_i}{(1-\nu_i)} \gamma_i \frac{t_i}{2} \left[ -\ln d_i + \frac{\sum_{j=1}^{n} \{E_i/(1-\nu_i)\} t_i \ln d_j}{\sum_{j=1}^{n} \{E_i/(1-\nu_i)\} t_i} \right]}{\sum_{i=1}^{n} \frac{E_i t_i}{(1-\nu_i)} \left[ \left\{ \frac{t}{2}\left(\frac{t}{2} + z_0\right) - \frac{t^3}{3} \right\} + \left(\frac{t}{2} - z_0\right)\frac{\gamma_i}{2} - \frac{1}{12}\left(3\gamma_i^2 + t_i^2 + t^2\right) \right]} \tag{6}$$

and

$$\sigma_i = \frac{E_i}{(1-\nu_i)} \left[ -\ln d_i + \frac{\sum_{j=1}^{n} \frac{E_i}{(1-\nu_i)} t_i \ln d_j}{\sum_{j=1}^{n} \frac{E_i}{(1-\nu_i)} t_i} + (z_0 - z)K \right] \tag{7}$$

where t is total thickness, and the coefficients $\gamma_i$ and $\beta_{ij}$ are defined as:

$$\gamma_i = \sum_{j=1}^{n} \beta_{ij} t_j \tag{8}$$

and

$$\beta_{ij} = -1 \text{ (if } i > j\text{)}, \ 0 \text{ (if } i = j\text{)}, \ +1 \text{ (if } i < j\text{)}. \tag{9}$$

Since we know all of the parameters ($E_i$ , $t_i$ ), it is possible to calculate stresses in each film by measuring the ratio of the linear dimensions from curvatures. For the 3-layer case, as in the $SiN_x/SiO_2/Si$ system discussed in next section, it is sufficient to measure two ratios, ($d_2/d_1$) and ($d_3/d_2$), for the stress calculation.

### 2. Calculation Results and Discussion

Using the elastic model discussed above, the residual stresses in LPCVD nitrides, deposited on thermally oxidized silicon substrate, were calculated for various nitride film thicknesses. In order to determine ($d_{SiO2}/d_{Si}$) and the average stress in the $SiO_2$, a laser scanning technique was used. The stress was determined to be ~ -280 MPa. We know that the stress is uniformly distributed throughout the film thickness since single-layer $SiO_2$ beams with various thicknesses show no curvature. For the determination of ($d_{SiNx}/d_{SiO2}$), the curvatures of $SiN_x/SiO_2$ beams were measured from SEM photos. We have used 180.5 GPa for $(E/(1-\nu))_{Si}$ [5], 0.16 [6] and 0.2

for the Poisson's ratios of $SiO_2$ and $SiN_x$, throughout the calculations.

The radii of $SiN_x/SiO_2$ beams are plotted in Figure 6 for various film thicknesses. This plot shows that the radius of the standard $SiN_x$ /$SiO_2$ beam is smaller than the one of low stress $SiN_x/SiO_2$ beam at the same thickness. Therefore, the residual stress in the standard nitride is higher, see Figures 7 and 8. For relatively thick films (>1200 Å), the average stresses in the standard and low stress nitrides are ~1.15 GPa and ~440 MPa, respectively. The difference in residual stress can be explained at least partially by the difference in silicon contents in nitrides. The refractive index, $N_f$, of the standard nitride (2140 Å) was measured at 2.00, which is an accepted value for stoichiometric $Si_3N_4$ [7]. On the other hand, $N_f$ of the low stress nitride (2870 Å) was measured to be ~2.25. Even though this technique does not directly quantify the composition, it indicates that the silicon content in low stress nitride is higher than in stoichiometric films. Many reports in the literrature have shown that the excess silicon relieves tension in nitride films [for example, see 8]. We have also found that the refractive index was not a constant with the film thickness in both nitrides. This suggests that these films are not completely homogenious.

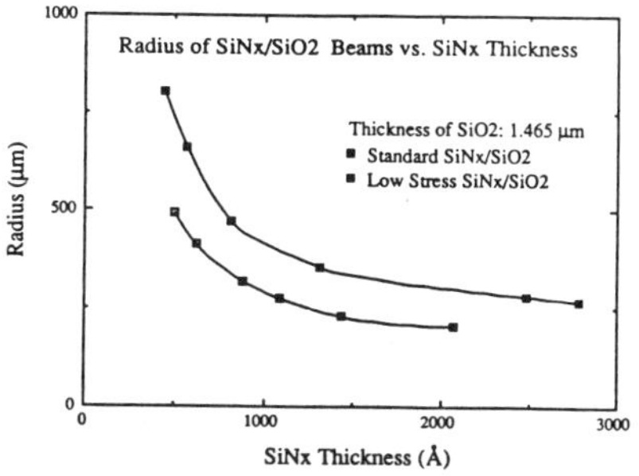


Figure 6: Plot of radius of $SiN_x/SiO_2$ beams for various $SiN_x$ thicknesses.

Figures 7 and 8 show the average stresses, $\sigma_{avg}$, and the differential stresses, $\sigma_d$, for both nitrides at room temperature. The differential stress between two film thicknesses, $T_{n-1}$ and $T_n$, was defined as follow:

$$\sigma_d(T_{n-1}, T_n) = \frac{\sigma_{avg,T_n} T_n - \sigma_{avg,T_{n-1}} T_{n-1}}{T_n - T_{n-1}} \qquad (10)$$

As seen in these figures, the average stresses in both nitrides increase with thickness when the films are very thin, and that they approach constant values (~1.15 GPa and ~440 MPa) after certain thicknesses. The main source of error source in these calculations is the uncertainty in film thickness. We assumed that the nonuniformity in film thickness, after all etching processes, is ± 50Å. This was accounted for by the error bars at each data point. Since the portion of the error range with respect to the total thickness increases as the film gets thinner, the magnitude of the error bar also increases accordingly. However, the general trend can still be clearly seen. The accuracy of this technique largely depends on two main factors: the thickness measurement of top film, and the initial thickness of the bottom layer. If we use a thin bottom layer and accurately measure the top layer thickness, it is possible to detect very small residual stresses.

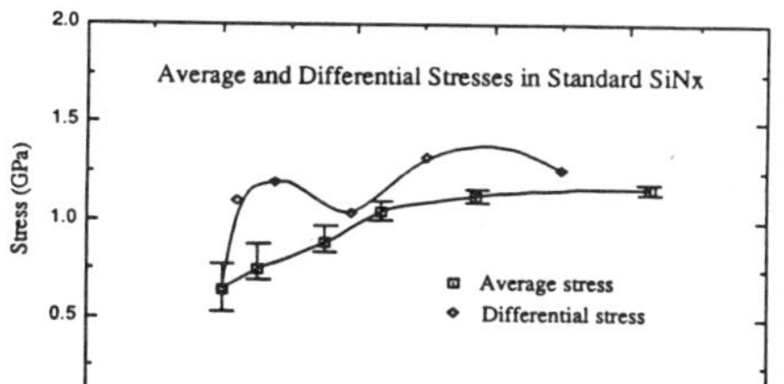


Figure 7: Average and differential stresses in standard nitride film.

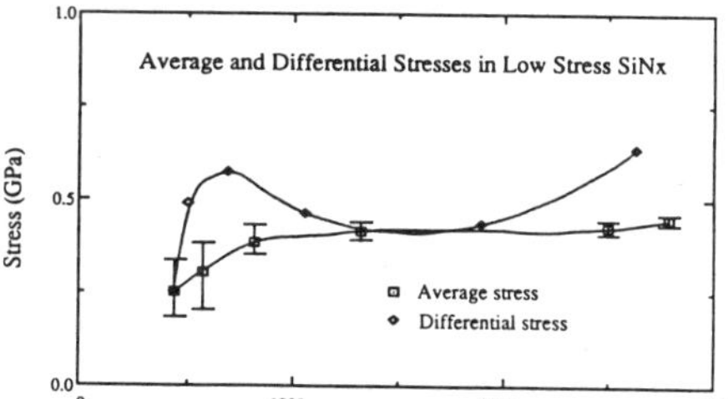


Figure 8: Average and differential stresses in low stress nitride film.

For very thin films, the relatively large change of the average stress indicates the variation of the differential stress. The differential stress increases gradually with a small slope for the regions where the average stress starts to approach a constant value. The trend of the actual stress states was verified by removing the bottom oxide layer with BHF solution, thus making a single-layer nitride beam. Figure 9 shows the cross sectional view of the nitride beams. The curvature of the nitride beams is caused by the non-uniform distribution of the actual stress in the film, and also indicates clearly that, in an average sense, more tension develops in the top portion of the film. However, it is not understood well, at this point, why the variations of the normal stresses are present in these films. One possible explanation is that silicon rich $SiN_x$ grows near the interface, and then the composition composition approaches the proper stoichiometry (a steady value in low stress nitride case) as the film grows thicker.

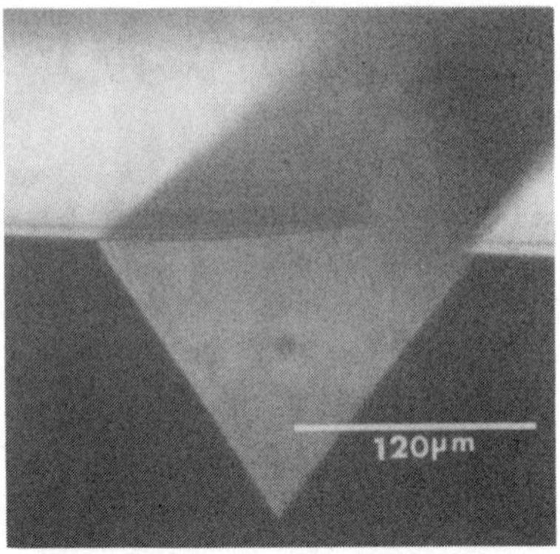

Figure 9: Cross-sectional view of standard $SiN_x$ (2100 Å) beams.

## SUMMARY

A unique technique which can measure both the elastic parameters and residual stresses of very thin films has been developed using multi-layer micro-cantilever beam structures. A film system of LPCVD $SiN_x/SiO_2/Si$ was tested using the technique and showed that:

(1) The Young's moduli of LPCVD standard and low stress $SiN_x$, and PECVD $SiN_x$, were measured at 275 and 235 GPa, respectively.

(2) The average stress at room temperature in both LPCVD standard and low stress $SiN_x$ increases near the interface between $SiN_x/SiO_2$ and approaches constant values, 1.15 GPa and 440 MPa, at larger thicknesses.

## ACKNOWLEDGEMENTS

The authors wish to thank to R. F. W. Pease and D. M. Barnett for fruitful discussions. The financial support of the Center for Integrated Systems at Stanford University, the Semiconductor Research Corporation, and DARPA through the URI at UCSB, is gratefully acknowledged.

## REFERENCES

[1] A. Bohg and A. K. Gained, Appl. Phys. Lett. 33, pp 895-897, Nov., 1978.
[2] T. P. Weihs, S. Hong, J. C. Bravman and W. D. Nix, J. Mater. Res. 3, pp 931-942, Sep/Oct., 1988.
[3] T. P. Weihs, S. Hong, J. C. Bravman and W. D. Nix, These Proceedings.
[4] P. H. Townsend, D. M. Barnett and T. A. Brunner, J. Appl. Phys. 62, pp. 4438-4444, Nov., 1987.
[5] W. A. Brantley, J. Appl. Phys. 44, pp 534-535, Jan., 1973.
[6] CRC Handbook of Chemistry and Physics, edited by R. C. Weast (CRC, Boca Raton, FL, 1985), p.56
[7] K. E. Bean, P. S. Gleim, R. L. Yeakley and W. R. Runyan, J. Electrochem. Soc., 114, pp. 733-737, July, 1967.
[8] P. Pan and W. Berry, J. Electrochem. Soc., 132, pp. 3001-3005, Dec.,1985.

# THE INFLUENCE OF THICKNESS AND WAVELENGTH ON THE MECHANICAL PROPERTIES OF A COMPOSITIONALLY MODULATED CERAMIC THIN FILM

W. C. Oliver, F. A. List and R. A. McKee
Metals and Ceramics Division, Oak Ridge National Laboratory, Oak Ridge, TN 37831

## ABSTRACT

The hardness and modulus of thin (<1 μm) ceramic films have been investigated using the differential mechanical properties microprobe (DMPM). The DMPM consists of a computer controlled ultra-low load indentation system with a special system that allows the elastic stiffness of the contact to be determined continuously during the indentation process. The films were deposited in a molecular beam epitaxy (MBE) system utilizing unique high temperature effusion cells. Two specimens have been studied. The first is a wedge shaped layer of amorphous $Al_2O_3$ on a silicon substrate. The thickness of the layer varies from 0 to 990 nm at the two extreme sides of a 3 in. silicon wafer. The effect of layer thickness on the hardness and modulus as measured with the DMPM has been quantified for this system. The second specimen is made up of 38 repetitions of alternating layers of crystalline $TiO_2$ and amorphous $Al_2O_3$. The wavelength of the composition modulation and the thickness of the film varies uniformly from 0 to 38 nm and from 0 to 1460 nm respectively across the 3 in. diameter of the silicon substrate.

## INTRODUCTION

Multilayer, structured materials research has been cited [1-3] as an area with significant potential for major advances in physics and materials science. Presently active research programs center primarily on metal/metal [4,5], metal/semiconductor[6,7] and semiconductor/semiconductor [8-10] structures with applications in electronic packaging and electronic and optical devices. Unique electrical and optical properties are found in these materials, but, perhaps more importantly, an extremely interesting and unparalleled effect on mechanical properties is also found. A "super modulus" effect has been reported [11] where three-fold increases in the elastic modulus and five-fold increases in the yield strength occur [5,12] and in some cases super-plastic behavior has been observed [13]. Such a combination of strength and ductility properties, if achievable for a ceramic material, would have enormous potential for application. Multilayered, heterogeneous ceramic/ceramic or ceramic/metal composites could evolve as an important new class of materials, but fundamental structure-property research is required for this development. The mechanical properties noted above are biaxial properties. They have been measured using bulge tests. In this work we will be using indentation testing on the DMPM [14] to measure the hardness and effective modulus of the films studied. This type of measurement is more a three dimensional average of the properties because the strain fields produced are roughly spherical.

## EXPERIMENTAL

The films used in this study were grown in a specially designed MBE system. The system utilizes high temperature effusion cells that allow one to generate high quality molecular beams of virtually any element whose melting point is below 2000°C. The system has shutters in front of each cell and one in front of the specimen. The sample can be radiantly heated to 1400°C. The first film studied consisted of amorphous $Al_2O_3$ on a Si (111) substrate. The specimen shutter was manipulated during growth so that the thickness of the film varied monotonically between 0 and 990 nm of the thickness. The growth conditions were as follows: system base pressure $10^{-10}$ Torr, $1 \times 10^{-6}$ oxygen partial pressure, Al flux of 0.02 nm/sec, and a substrate temperature of 400°C. ESCA results indicated $Al_2O_3$ type bonding. In situ RHEED results indicated that the alumina was amorphous. The second film was a layered structure of amorphous $Al_2O_3$ and crystalline $TiO_2$. In this case the specimen and cell shutters were manipulated so that a layered structure was produced. This film varied in total thickness from 0 to 1460 nm and consisted of 38 pairs of layers of amorphous $Al_2O_3$ and crystalline $TiO_2$ of approximately equal thickness. Hence, the wavelength of the compositional oscillation varied from 0 to 38.4 nm. The growth conditions for the amorphous alumina were the same as those for the first specimen. The growth conditions for the $TiO_2$ were the same as those already quoted except that a Ti flux of 0.01 nm/sec was used instead of the aluminum flux. ESCA was used to confirm $TiO_2$ bonding. RHEED results indicated the rutile form of $TiO_2$. TEM results indicated a grain size approximately equivalent to the thickness of the $TiO_2$ layers.

A description of the DMPM, a modified Nanoindenter, is presented in another paper in these proceedings. Indentation experiments were performed on the films at two loading rates - high and low. The high loading rate experiments consisted of loading the indenter such that its velocity was 2 nm/sec to a final load of 120 mN. The low loading rate experiments consisted of loading the indenter such that it's velocity was 0.5 nm/sec to a total depth of 50 nm. As the indentation proceeds, the load on the indenter, the displacement of the indenter, and the elastic stiffness of the contact are measured simultaneously and continuously by the DMPM. An effective hardness and modulus for the material is then calculated using standard techniques [15].

## RESULTS AND DISCUSSION

Figure 1 shows results of the high loading rate experiments for the amorphous alumina specimen. Each curve represents an average of at least three experiments performed at the same thickness. The legend shows the thickness corresponding to each curve. Data below 30 nm indenter displacement should be disregarded due to time constant constraints in the electronics. The shapes of the curves are as expected. At shallow displacements relative to the film thickness, intrinsic film properties are detected; at greater displacements, the substrate properties become effective and the measured values shift from those of the film to those of the substrate. Finally at large displacements the substrate properties dominate the measurement. Clearly, Fig. 1 shows that the hardness of the silicon substrate is approximately 11.5 GPa. The hardness of amorphous alumina is approximately 7.5 GPa.

Figure 1 shows no effects of the film until a film thickness of 93 nm is sampled. To probe thinner films low loading rate experiments were performed for which the DMPM measurements are not affected by electronic time constants for displacements greater than 5 nm. In addition at least nine experiments were performed and averaged at each thickness. The results of such tests on the thin portion of the same amorphous alumina film are shown in Fig. 2. The dip in the data at shallow displacements may be due to using an area function that describes the indenter geometry that is inaccurate at those displacements. This should not change the relative hardness of the film to the substrate determined by the DMPM. One important aspect of Fig. 2 is that the presence of the film is detectable even when it is only 32 nm thick (about 20 atomic layers); however, it is difficult to obtain film properties at such small thickness that are substrate independent. The results of modulus calculations for this film are shown in Fig. 3 and indicate that the modulus of amorphous alumina is approximately 100 GPa. These values of the hardness and modulus determined by the DMPM for amorphous alumina produced by reactive MBE are in agreement with those determined both for amorphous alumina produced by other techniques [16] and by other testing techniques [17].

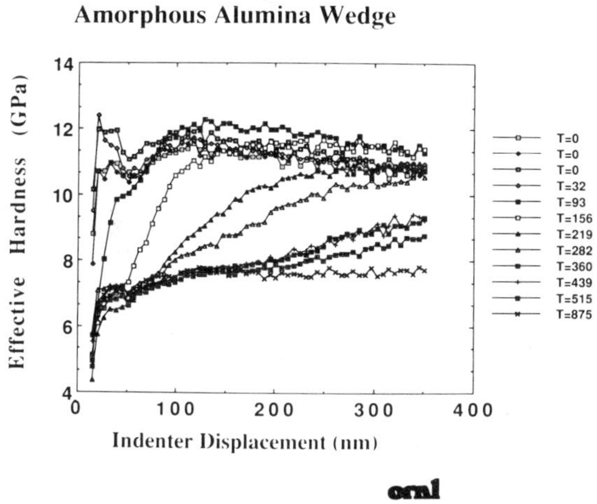


Fig. 1. Differential mechanical properties microprobe (DMPM) results showing hardness of an amorphous alumina film at various film thicknesses (T in nm) as a function of indenter displacement.

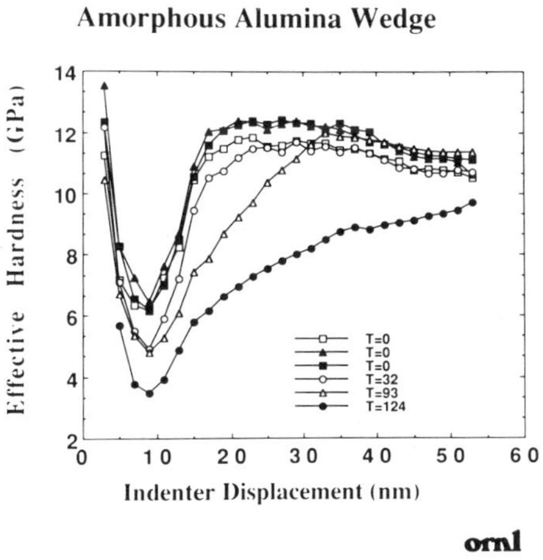


Fig. 2. DMPM results showing hardness of an amorphous alumina film at various film thicknesses (T in nm) as a function of indenter displacement.

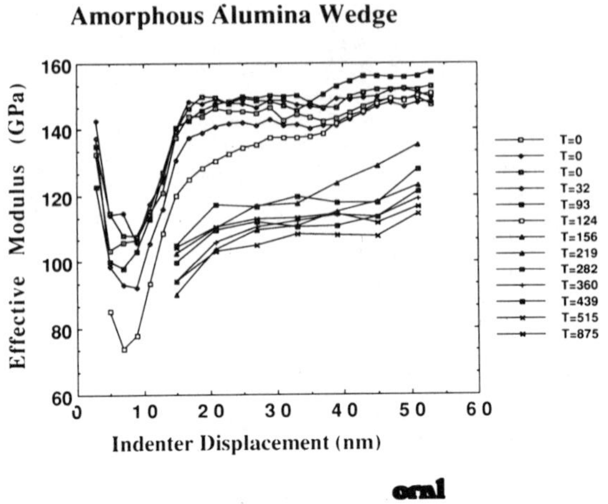


Fig. 3. DMPM results showing effective modulus of an amorphous alumina film at various film thicknesses (T in nm) as a function of indenter displacement.

The final two figures (Figs. 4 and 5) compare the hardness results of the compositionally modulated ceramic (CMC) film to those of the amorphous alumina film. Figure 4, which refers to the low loading rate experiments, shows the hardness curves for both films at several specific depths. The curves for which only a thickness (T) is given are from the amorphous alumina film. The others for which a wavelength (W) is given are from the compositionally modulated film. There appears to be no significant difference in the response of the two films at these indentation depths.

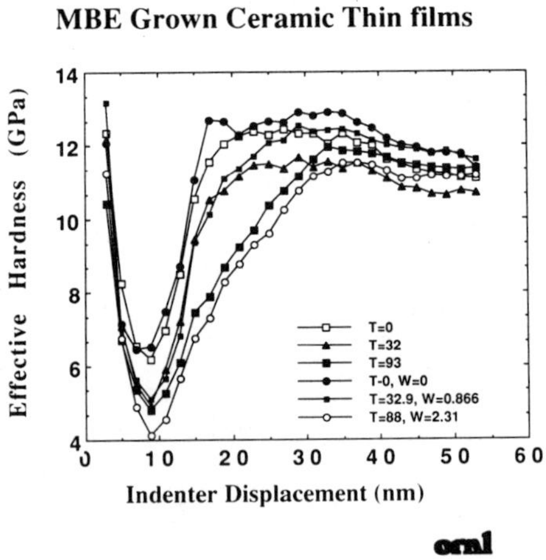


Fig. 4. A comparison of the hardness of a compositionally modulated ceramic (CMC) film and an amorphous alumina film at several film thicknesses (T in nm) and modulation wavelengths (W in nm) as a function of indenter displacement.

Figure 5 shows a comparison of the high loading rate experiments at larger film thicknesses. In this figure the ratio of the hardness of the film to the hardness of the substrate is shown. At T = 360 nm the properties of the films are approximately identical. However, the results for the films of thickness less than 360 nm seem to show a significant difference with the modulated film being harder. Whether this difference simply reflects the different compositions of the two films, or whether it indicates some wavelength or thickness effect on the hardness of the CMC film is not clear from these results. Work is currently in progress to better understand this behavior.

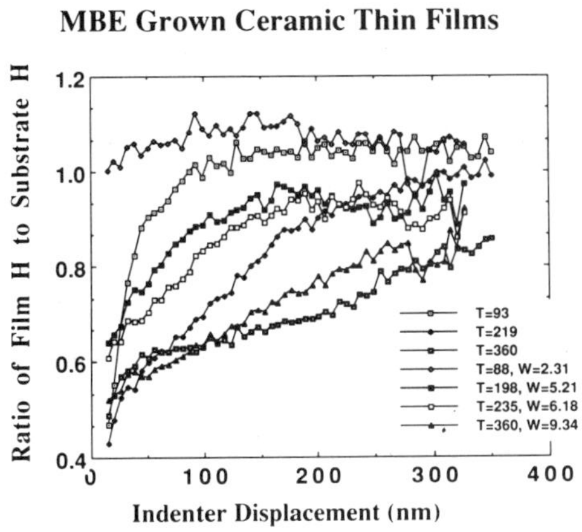


ornl

Fig. 5. A comparison of the relative hardness of a CMC film and an amorphous alumina film at several film thicknesses (T in nm) and wavelengths (W in nm) as a function of indenter displacement.

CONCLUSIONS

1. Amorphous alumina has a hardness of 7 GPa.
2. Amorphous alumina has an effective modulus 30% less than Si. (~100 GPa).
3. These properties are very close to those reported for amorphous alumina films generated by ion beam damaging sapphire.
4. A modulated ceramic film made up of 38 layers of $TiO_2$ and amorphous alumina has properties very similar to those of amorphous alumina.
5. The DMPM can be used to monitor the effective hardness and modulus of a specimen both continuously and simultaneously during an indentation experiment.

ACKNOWLEDGMENTS

Research was sponsored jointly by the U.S. Air Force Office of Scientific Research under interagency agreement DOE No. 1854-A077-A1, USAF No. AFOSR-ISSA-89-0015 and both the Office of Energy Utilization Research, Energy Conversion and Utilization Technologies (ECUT) Program and the Division of Materials Sciences, U.S. Department of Energy, under contract

DE-AC05-84OR21400 with Martin Marietta Energy Systems, Inc. The authors gratefully acknowledge the assistance of Dr. F. J. Walker with the x-ray and RHEED results and Dr. K. Alexander with the TEM results. We appreciate greatly the preparation of the manuscript by Gwen Sims.

## REFERENCES

1. V. Narayanamunti, Phys. Today 24, (October 1984).
2. "Report of Research Briefing Panel on Selected Opportunities in Physics," Hans Frauenfelder and Mildred S. Dresselhaus, Co-chairman, p. 107 in Research Briefings (1984).
3. Phys. Today 21 (January 1985).
4. D. Tench and S. White, Metall. Trans. A 15A, 2039 (1984).
5. D. Baral, J. B. Ketterson, J. E. Hilliard, p. 465 in Proceeding of NATO Advanced Study Institute on Modulated Structure Materials, Elsevier, Oxford, 1984.
6. Sung C. Shin, J. E. Hilliard, and J. B. Ketterson, J. Vac. Sci. Technol. A2, 296 (1984).
7. J.M. Gibson, R. T. Tung, J. M. Phillips, and J. M. Poate, Mater. Res. Soc. Proc., Series 25, edited by G. Baglin, 405 (1984).
8. J. C. Bean, L. C. Feldman, A. T. Fioy, S. Nakahara, I. K. Robinson, J. Vac. Sci. Technol. A2, 436 (1984).
9. V. S. Speriosin, M. A. Nicolet, S. T. Picraux, and R. M Biefeld, Appl. Phys. Lett. 45, 223 (1984).
10. S. S. Pei, N. J. Shah, R. H. Hendel, C. W. Tu, and R. Dingle, in Proceedings of 1984 GaAs IC Conference, to be published.
11. W. M. C. Yang, T. Tsakalakos, and J. E Hillard, J. Appl. Phys. 48, 876 (1977).
12. R. W. Sprugir and D S. Catlett, Thin Solid Films 54, 197 (1978).
13. R. F. Bunshah, R. Nimmagadda, H. J. Doerr, B. A. Mouchan, N. I. Grechanuk, and E. V. Dabizha, Thin Solid Films 72, 261 (1980).
14. J. B. Pethica and W. C. Oliver, in "Thin Films: Stress and Mechanical Properties," Proceedings Symposium C, MRS Meeting Fall 1988.
15. M. F. Doerner, W. D. Nix, J. of Mater. Res. 1(4), 601-09 (1986).
16. W. C. Oliver, C. J. McHargue, S. J. Zinkle, Thin Sol. Films 153, 185-196 (1987).
17. R. W. Hoffman, Mat. Res. Soc. Symp. Proc. Vol. 69, 95 (1986).

# INDENTATION TECHNIQUE TO INVESTIGATE ELASTIC MODULI OF THIN FILMS ON SUBSTRATES

D.S. STONE*, T.W. WU†, P.-S. ALEXOPOULOS†, and W.R. LAFONTAINE‡
*Materials Science and Engineering, Bard Hall, Cornell University, Ithaca, NY 14853; presently at Materials Science and Engineering, University of Wisconsin, 1509 University Avenue, Madison, WI 53706
†IBM Research Division, Almaden Research Center, IBM Magnetic Recording Institute, San Jose, CA 95120-6099
‡Materials Science and Engineering, Bard Hall, Cornell University, Ithaca, NY 14853

## ABSTRACT

Closed-form elasticity solutions are introduced, that predict the average displacement beneath square and triangular, uniformly loaded areas at the surface of a bilayer. The solutions aid in the application of depth-sensing indentation techniques for measuring thin film elastic moduli. The elasticity solutions agree closely with experimental data of Al, Si, 1 μm Al on Si, and 2 μm Cr on Si. The case of poor adhesion between the film and substrate is briefly examined.

## INTRODUCTION

The principles of using depth-sensing indentation tests to measure elastic moduli are discussed by Doerner and Nix [1] and Loubet [2]. According to the former's analysis, the unloading compliance, C, and square root of the projected indent area, S, are related by:

$$C \equiv 1/SE_{eff} = (1-\nu_s^2)/\beta SE_s + (1-\nu_d^2)/\beta SE_d \qquad (1)$$

where $E_s$, $\nu_s$, $E_d$ and $\nu_d$ are the Young's Moduli and Poisson's Ratios of the substrate and diamond indentor, respectively, and $\beta$ is a number near 1. $E_{eff}$ is an effective elastic modulus. The two terms at right represent the specimen and indentor compliance, respectively. Equation 1 excludes the machine compliance, which can be accounted for in the data analysis [1], and it assumes the indentor and substrate are elastically isotropic.

Based on Equation 1 and provided the modulus of the indentor is known, the elastic moduli of bulk specimens can be determined from the slope of C vs. 1/S. However, the compliance measured from a thin film on substrate reflects a combination of both film and substrate properties. This paper reports some elasticity solutions necessary for analysis of unloading data from thin film-substrate systems and compares these solutions with experimental data.

## ELASTICITY MODEL FOR INDENTATION OF BILAYER

The model will be detailed elsewhere [6]. In the model, the average displacement of the surface beneath a region of uniformly distributed pressure is calculated. This approach for determining compliance contrasts with that of King [7], who

Mat. Res. Soc. Symp. Proc. Vol. 130. 

modeled indentation of a bilayer with a rigid punch. Beneath a rigid punch the displacement is uniform, whereas the distribution of pressure is not. While the values of compliance predicted by the different approaches are similar, the present approach is more convenient to use because its results can be presented in closed-form.

Based on Chou and Pagano [8] and the work of Burmister [9], the elastic displacements at the surface of a bilayered half-space can be found in the case where a traction acting normal the surface is of the form:

$$\sigma_z(x,y) = \sigma_o \cos ax \cos by. \tag{2}$$

Any profile of force distributed over a rectangular area at the surface and seperable as $\sigma_z(x,y) = A(x)B(y)$ can be treated as an integral of such terms. The average displacement beneath the indentor can then be determined and the compliance calculated. Through such an analysis, the compliance of a bilayered half-space in response to a Vickers indentor is:

$$C^v_{fs} = \frac{8(1-\nu_f^2)}{\pi^2 E_f S} \int_0^\infty \int_0^{\pi/4} \frac{\sin^2(R\cos\theta)\sin^2(R\sin\theta)}{R^4 \cos^2\theta \sin^2\theta} g(2R\frac{h}{S})\, d\theta\, dR \quad ; \tag{3}$$

$$g(\omega) = \frac{1+4K\omega e^{-2\omega} - KLe^{-4\omega}}{1-(L+K+4K\omega^2)e^{-2\omega}+KLe^{-4\omega}}$$

with

$$K = \frac{1-n}{1+n(3-4\nu_s)}; \qquad L = \frac{(3-4\nu_s)-n(3-4\nu_f)}{(3-4\nu_s)+n}; \qquad n = \frac{G_s}{G_f}.$$

Here, h is the thickness of the surface layer or film. The subscripts s and f refer to the substrate and film, respectively, and $G = E/[2(1+\nu)]$.

For the Berkovitch indentor, the unloading compliance can be calculated by breaking the triangular region into infinitesimal rectangular pieces, distributing the load uniformly over these pieces, and integrating the displacements contributed by each:

$$C^b_{fs} = \frac{12(1-\nu_f^2)}{\pi^2 E_f S} \int_0^\infty \int_0^{\pi} \frac{g(R\frac{h}{S})\, dr\, d\theta}{R^4\cos^2\theta(\cos^2\theta-3\sin^2\theta)} \times$$

$$\left\{\frac{\cos^2\theta+3\sin^2\theta}{\cos^2\theta-3\sin^2\theta} + \cos^2\left(\frac{R\cos\theta}{\sqrt{3}}\right) - \frac{\cos\theta}{\cos\theta+\sqrt{3}\sin\theta}\cos\left[R\left(\frac{\cos\theta}{\sqrt{3}}+\sqrt{3}\sin\theta\right)\right]\right.$$

$$\left. - \frac{\cos\theta}{\cos\theta-\sqrt{3}\sin\theta}\cos\left[R\left(\frac{\cos\theta}{\sqrt{3}}-\sqrt{3}\sin\theta\right)\right]\right\} \tag{4}$$

The integrals in Equations 3 and 4 can be evaluated numerically on a microcomputer. It is generally sufficient to limit the range of R between 0 and 20. In the cases $(E_f,\nu_f) = (E_s,\nu_s)$, h << S, and h >> S, the integral in Eq. 3 may be evaluated directly to obtain an expression like the first term at the extreme right of Eq. 1, but with $\beta = 1.05$.

## COMPARISON OF MODEL WITH THOSE OF OTHER INVESTIGATORS

The values of $\beta$ calculated in this work are compared with those of King [7] and Timoshenko [10] in Table I. The shape of the indentor does not greatly affect $\beta$. Furthermore, the trend of $\beta$ with indentor shape is the same for both traction and displacement boundary conditions. Predictions of $E_{eff}$ vs S/h are shown in Fig. 1. The shapes of the $E_{eff}$ vs S/h curves are relatively independent of indentor geometry.

Table I - Values of $\beta$

| Type of Indent | Displ. B.C.'s | Load B.C.'s | This Work | King[7] | Timoshenko & Goodier[10] |
|---|---|---|---|---|---|
| Circular | X | | ---- | 1.129 | 1.13 |
| Square | X | | ---- | 1.142 | ---- |
| Triangular | X | | ---- | 1.167 | ---- |
| Circular | | X | ---- | ----- | 1.04 |
| Square | | X | 1.05 | ----- | 1.05 |
| Triangular | | X | 1.09* | ----- | ---- |

*Based on numerical integration of Eq. 4.

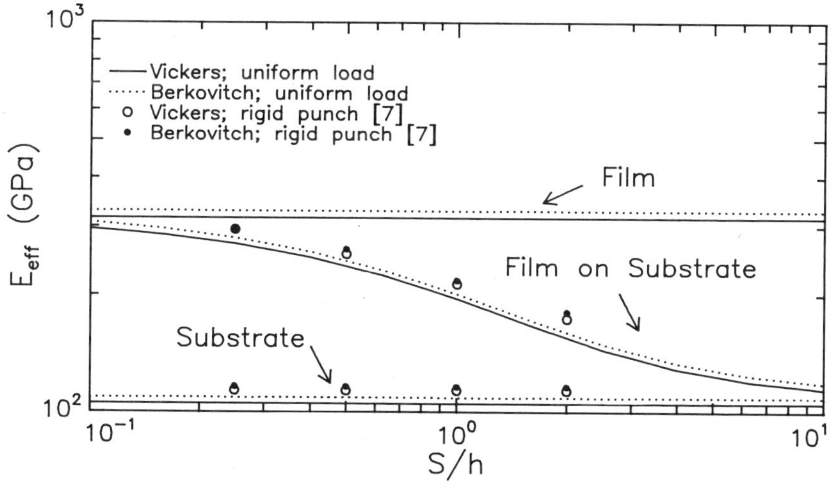


Figure 1. Predictions of $E_{eff}$ vs. S/h. Except where noted, $\nu_s=\nu_f=\nu_d=0.3$, and $E_f:E_s:E_d=4:1:8$ ($E_d=800$ GPa).

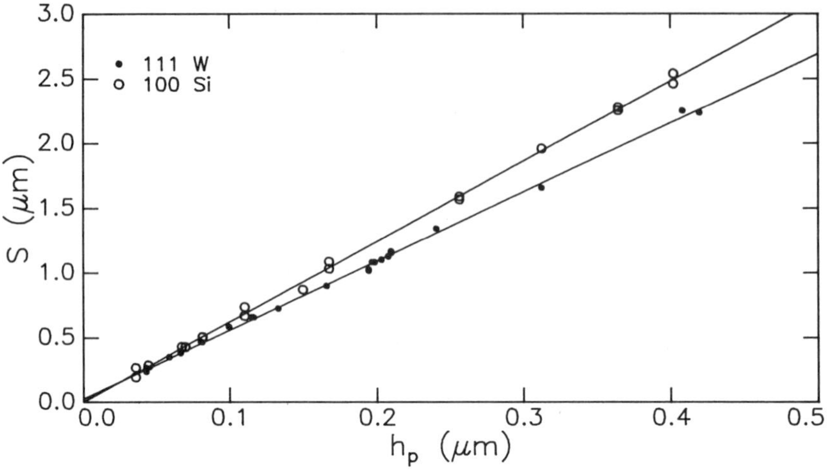


Figure 2. Relationship between the square root of the area (S) and plastic indentation depth ($h_p$) for Si and W.

## COMPARISON BETWEEN THE MODEL AND EXPERIMENTAL DATA

Details of the experimental apparatus are given in references [4] and [5].

In an experiment, C is measured directly from the elastic unloading data [1,2]. Traditionally, the quantity S has either been determined by imaging the indent directly or calculated based on the plastic depth of indentation, $h_p$, and indentor shape [1-3]. This shape is calibrated based on indents of known depth in a reference material [1,3]. We find that the two techniques produce different results. For instance, one of the Berkovitch indentors used at IBM is found to have $dS/dh_p \simeq 6.1$ for Si and $dS/dh_p \simeq 5.3$ for W. See Fig. 2. The direct imaging technique was used to obtain the values of S reported here.

Model predictions are compared in Fig. 3 with experimental data of Al, Si, 1 μm Al on Si, and 2 μm Cr on Si. The values of E and $\nu$ used in the model simulations were based on Reuss average compliances [11]. Note, in Fig. 3, that about 75% of the experimental data fall within 10% of the predicted values. Also there appears to be no systematic deviation in the data to either higher or lower values of $E_{eff}$ than predicted. Such an agreement obviously depend upon how S is measured and E and $\nu$ are chosen for the calculations.

Two sets of data for 1 μm Al on Si are presented in Fig. 3. In one case the film adhered poorly to the substrate: it could be removed by peeling it off the substrate with a piece of adhesive tape. Poor adhesion did not affect the experimental data.

The case of poor adhesion appears at first to be similar to that in which the interface is allowed to slip - no shear tractions can be transmitted across the interface but normal

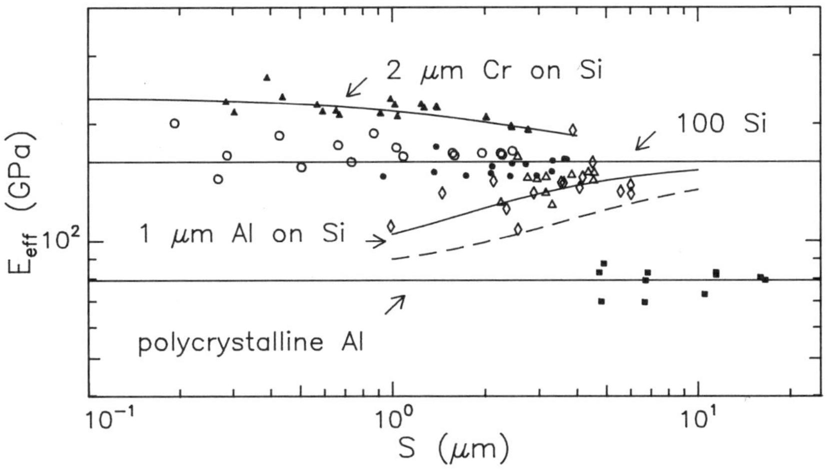


Figure 3. Model predictions compared with experimental data: ▲ - 2μm Cr on 100 Si (IBM); o - 100 Si (IBM); • - 100 Si (Cornell); Δ - 1μm Al on 100 Si (Cornell); ◊ - 1μm Al on Si (poor adhesion; Cornell); ■ - polycrystalline Al (Cornell). Values of E (GPa) and $\nu$ used in the simulations [11]: Cr - 255, 0.186; Si - 159, 0.227; Al - 70.4, 0.347; diamond - 1136, 0.072. All indents were made with Berkovitch indentors, except for polycrystaline Al.

tractions (both tensile and compressive) can be transmitted. In the latter case, the compliance of the film-substrate system can be calculated based on Equations (3) and (4), except with the following function substituted for $g(\omega)$ [12]:

$$g'(\omega) = \frac{F-[2F-1-2\omega]e^{-2\omega} - (1-F)e^{-4\omega}}{F+[(2F-1)2\omega-(1+2\omega^2)]e^{-2\omega}+(1-F)e^{-4\omega}} \; ; \; F = \frac{(1-\nu_s)+n(1-\nu_f)}{2(1-\nu_s)}. \tag{5}$$

The model prediction based on a slipping interface is shown as a dashed line in Fig. 3. The model predicts a lower value of $E_{eff}$ than observed.

## CONCLUSIONS

The elastic unloading of film-substrate systems in response to triangular and square-shaped indentors has been modeled. Model results have been presented in the form of an effective modulus $E_{eff}$. $E_{eff}$ is function of the square root of the projected area, S, divided by the film thickness, h. The model predictions agree well with the experimental data of Al, Si, 1μm Al on Si, and 2 μm Cr on Si. There remain, however, important questions regarding the choices of E, ν, and S. For instance, are the Reuss compliances always the best choices to represent averages of anisotropic elastic behavior? Also, which is better to use, S based on direct imaging or based on indentor shape?

The case of poor film-substrate adhesion has been examined; poor adhesion does not significantly affect the unloading data of 1 μm Al on Si. Whether the unloading data of all film-substrate systems are this insensitive to poor adhesion is presently unknown.

## ACKNOWLEDGEMENTS

The authors wish to acknowledge B. Yost for his help in the experimental work. Also, the encouragement and support of C.-Y. Li have been much appreciated. A portion of this work was supported by the National Science Foundation through the Materials Science Center and the National Nanofabrication Facility (Grant ECS-8619049) at Cornell.

## REFERENCES

1. M. F. Doerner and W. D. Nix, J. Mat. Res. 14 (1986) pp. 601-609.
2. J. L. Loubet, J. M. Georges, and G. F. Meille, in Microindentation Techniques in Materials Science and Engineering, P. J. Blau and B. R. Lawn, Eds., ASTM-STP 889, American Society for Testing and Materials, Philadelphia, PA (1986) pp. 72-89.
3. J. Pethica, R. Hutchings, and W. C. Oliver, Phil. Mag. A 48 (1983) pp. 593-606.
4. T.-W. Wu, C. Hwang, J. Lo, and P. Alexopoulos, in Thin Solid Films, in print.
5. S.-P. Hannula, D. Stone, and C.-Y. Li, Mat. Res. Soc. Symp. Proc. 40 (1985) pp. 217-224.
6. D. Stone, T.-W. Wu, W. R. LaFontaine, and P. Alexopoulos, Unpublished work.
7. R. B. King, Int. J. Solids Structures, 23(12) (1987) pp. 1657-1664.
8. P. C. Chou and N. J. Pagano, Elasticity, Tensor Dyadic, and Engineering Approaches, D. Van Nostrand, New York (1967) pp. 266-282.
9. D. M. Burmister, J. Appl. Phys., 16 (1945) pp. 89-94.
10. S. Timoshenko and J. N. Goodier, Theory of Elasticity, McGraw-Hill, NY (1951) pp. 370-372.
11. J. P. Hirth and J. Lothe, Theory of Disclocations, McGraw-Hill, NY (1968) pp. 761-762.
12. D. M. Burmister, J. Appl. Phys., 16 (1945) pp. 126-302.

# A STUDY OF DEFECTS IN ORDERED TERNARY SEMICONDUCTOR EPILAYERS.

S. McKernan, C. B. Carter, D.P.Bour*†, and J.R. Shealy*
Department of Materials Science and Engineering, Bard Hall, Cornell University, Ithaca, NY 14853
* School of Electrical Engineering, Phillips Hall, Cornell University, Ithaca, NY 14853
† Present address: David Sarnoff Research Center, Princeton, NJ 08543-5300

## ABSTRACT

Ordered $Ga_xIn_{1-x}P$ epilayers grown on GaAs substrates have been studied in the electron microscope. Dark-field images and high-resolution micrographs have directly revealed the presence of an ordering of the Ga and In (111) planes. Several different planar defects have been identified in the ordered structure. There is a very high density of faults parallel to the (001) growth surface. Models for these faults have been constructed in terms of stacking faults and twin boundaries in the ordered structure. Other flat planar defects which occur approximately parallel to (111) and $(11\bar{1})$ planes can also be described in these terms.

## INTRODUCTION

The growth of ternary III-V semiconductors by organo-metallic vapor phase epitaxy (OMVPE) is generally assumed to occur such that the structure is the same as that of the binary semiconductors, but with the two different cation or anion species arrayed randomly on their appropriate sublattice sites. Recently electron microscopy studies [1-9] have shown that ordered states exist for several different ternary compounds, and that the properties of the ordered alloy differ from those of the random one. For example, ordered $Ga_{0.5}In_{0.5}P$ grown by OMVPE is characterized by a bandgap approximately 50meV less than that of the random alloy. The ordered structure of $Ga_{0.5}In_{0.5}P$ is shown schematically in figure 1. The Ga and In forms (111) planes which are Ga rich and In rich alternately.

For the present study $Ga_xIn_{1-x}P$, with x~0.5, was grown on (100) GaAs substrates onto which a 0.5 μm GaAs buffer layer had been deposited. The growth temperature was 625°C and trimethlygallium (TMGa), trimethylindium (TMIn) and phosphine were used as the sources. The "V:III" flux ratio was approximately 400:1, the growth rate was 35nm/min and the layers were grown to a thickness of approximately 0.5 μm. These growth conditions produce lattice matched $Ga_{0.5}In_{0.5}P$ with a bandgap of 1.84 ev, while material grown at 700°C (which did not show this ordering) had a bandgap of 1.89 ev.

Cross-section specimens were prepared for microscopy in the usual manner by cleaving the samples normal to growth surface. Two orthogonal <110> cleaves were epoxied together face-to-face to form the cross section, ensuring that the ordered (111) planes could be observed "end-on". The samples were mechanically polished and finally ion thinned to electron transparency with 3 kV $Ar^+$ ions in a liquid nitrogen cooled stage. Plan-view specimens were also prepared by chemical etching to remove the GaAs substrate and buffer layer, followed by $Ar^+$ ion milling to reduce the thickness of the layer to electron transparency. The electron microscopy was performed in a JEOL 4000EX, the image simulation was made using the TEMPAS program [10] and the image processing carried out using SEMPER [11].

## RESULTS

Selected-area electron-diffraction patterns from the cross-sectional specimens at the $[1\bar{1}0]$ orientation show relatively strong extra reflections occur at $\pm^1/_2(111)$ (A in Figure 2), and weaker ones at $\pm^1/_2(11\bar{1})$ (B in Figure 2). The relative intensity of these two spots in the diffraction pattern varies across the sample so that they may be nearly equal in intensity, or only the stronger $\pm^1/_2(111)$ spots may appear. In cases where the intensities are nearly equal, a clear streak can be seen in the diffraction pattern running approximately parallel to the growth direction and which passes through these extra reflections. Dark-field micrographs from these regions

 

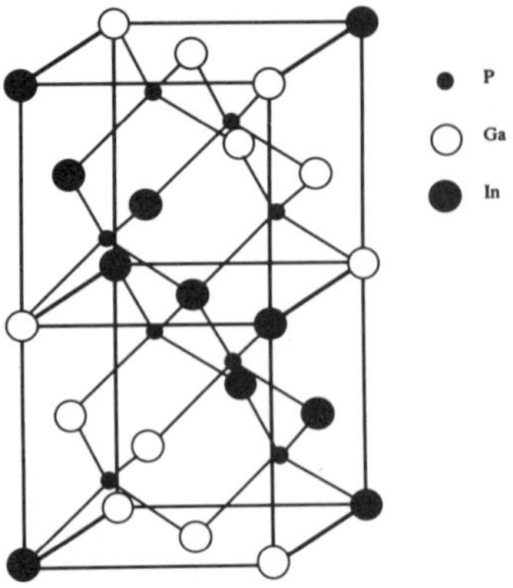


Figure 1 Schematic diagram of the completely ordered unit cell for $Ga_{0.5}In_{0.5}P$

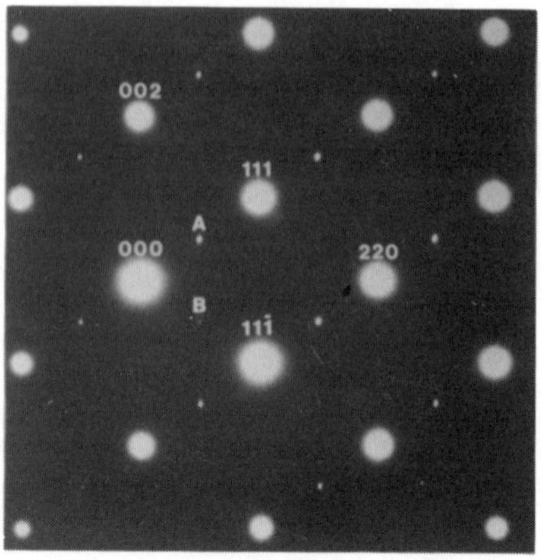


Figure 2 1$\bar{1}$0 electron diffraction pattern for the ordered $Ga_{0.5}In_{0.5}P$

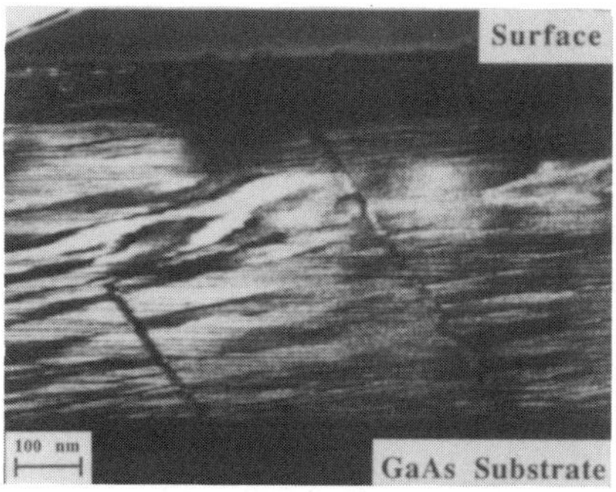


Figure 3 Dark-field micrograph obtained from the ordered reflection at A in figure 2

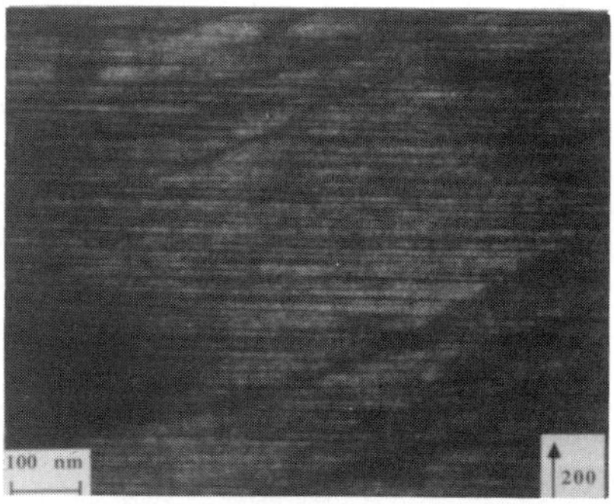


Figure 4 Higher magnification dark-field image of the ordered structure showing the density and sharpness of planar faults

showed domains in which either the (111) planes or the (11$\bar{1}$) planes were ordered and both of these domains contained a high density of defects. Dark-field micrographs from areas of the specimen which showed only the strong ± reflections showed ordering of the (111) planes only, but still a relatively high density of planar defects (see figure 3). These single domain regions were chosen for investigation by high-resolution electron microscopy. Specimens viewed at the orthogonal [110] orientation do not show any extra reflections or streaking.

The major features of the dark-field images are dark, ribbon-like contrast features which are interpreted, from tilting experiments, as inclined planar faults in the ordering of the Ga-rich and In-rich layers. The ribbon-like bands have an apparent width of 100Å wide, but this is a function of the specimen orientation with respect to the electron beam. Two straight parallel-sided features can also be seen, which are closely parallel to {111} planes. They are extremely sharp (less than 50Å wide) when viewed edge-on. These defects occur infrequently and, in this example, are observed to terminate in the epilayer and continue as a ribbon-like defect. This suggests that the defects lying on the (111) planes are closely related to the ribbon-like defects.

When viewed at orientations which are nearly normal to the growth direction, very sharp, dark lines parallel to the layer surface can be seen in the dark field images. At other orientations these lines are less obvious but a general linear texture remains. A higher magnification image of these defects is shown in figure 4, which is a dark-field micrograph obtained using the $^1/_2(111)$ reflection and the electron beam very close to the normal to the growth direction. The faults are seen to be extremely narrow, and closely spaced.

High-resolution images of these specimens obtained with the electron beam oriented parallel to the $[1\bar{1}0]$ direction, show a lattice periodicity which is double that of the (111) "close-packed" planes. In figure 5 the $Ga_xIn_{1-x}P$ epilayer clearly shows this doubled periodicity. The thinnest areas of the specimen do not appear to display this doubling. Image simulations show that the contrast of the ordering is very weak for specimens less than ~200Å thick. For thicker specimens however, the simulations show that the ordering should be easily visible even if the composition of the Ga- or As- rich planes is as low as 60%. Attempts to match the experimental and simulated images have been hampered by a degradation of the lattice images with time. After only a few minutes exposure to the electron beam, the visibility of the ordering in the lattice image is greatly reduced, and the intensity of the superlattice diffraction spots is also diminished. This strongly suggests that the material is becoming disordered during the irradiation by 400 keV electrons.

High-resolution images were obtained, however, from regions of the sample that were heavily faulted. Figure 5 shows such an area, where the doubling occurs in one direction ,then the other alternately. An enlargement of the top right-hand portion of this micrograph is shown in Fig. 6a. The fourier transform, showing the extra reflections in both the [111] and $[11\bar{1}]$ directions, is shown inset. By processing the image so that only the extra, superlattice reflections contribute to the image, the faults in the ordering become slightly more obvious (Fig. 6b). These defects become most apparent when only the reflections from one particular ordering variant are allowed to contribute to the image, as shown in figure 7a and 7b. In this image the contrast has been adjusted so that only the bright peaks in the image are displayed. Here the bright spots in one variant correlate either with the darker regions of the other variant, or with regions where the

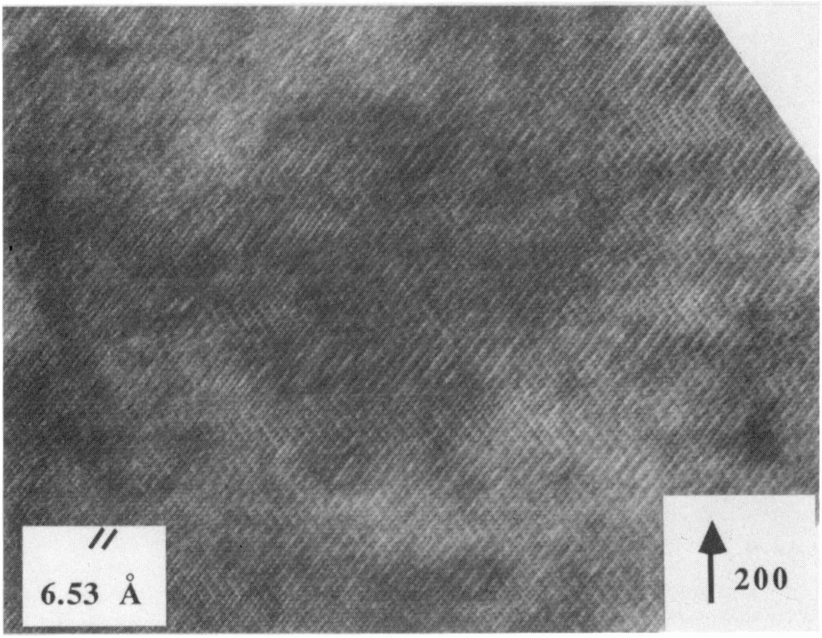


Figure 5 High-resolution lattice image of the ordered phase of $Ga_{0.5}In_{0.5}P$. A doubling of the periodicity of the (111) planes can clearly be seen, throughout the image. The image was recorded from a relatively thick area of the specimen where the doubling appears more pronounced.

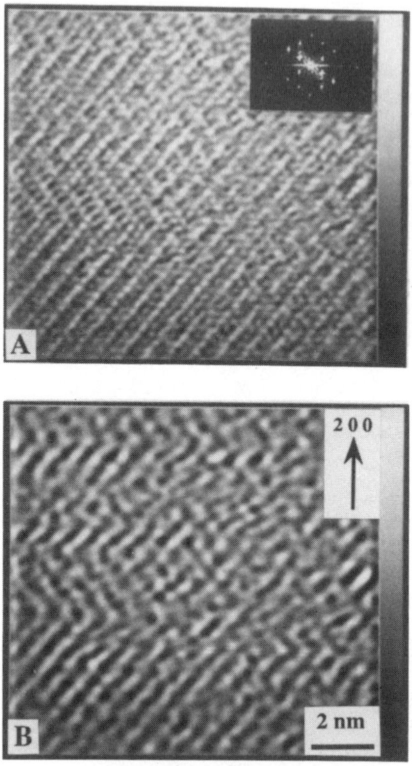


Figure 6 a) enlargement of the top right corner of figure 5. The fourier transform is inset, showing ordering in both of the <111> directions
b) Fourier filtered image of a) with only ordered spots contributing.

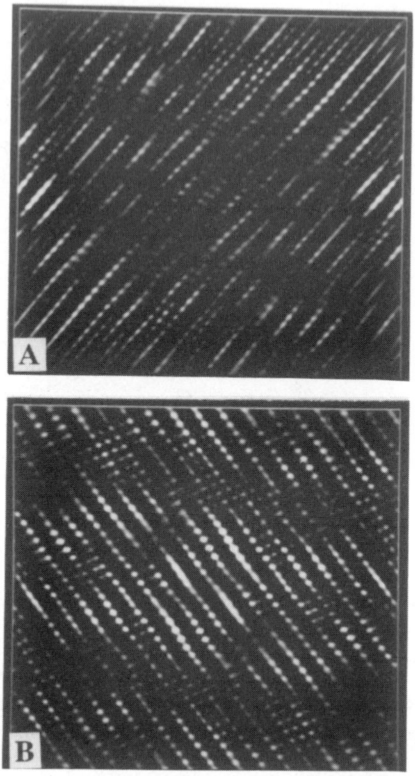


Figure 7 a) Fourier filtered image of 6a, with ordered spots of one variant only contributing to the image
b) Processed image using the spots from the second variant.

periodicity is not doubled in the other image. It can easily be seen, particularly in image 7b, that the lines of bright spots are sometimes offset by a single (111) plane on either side of an area where that variant is weak. It should also be noted that the boundaries of the well-ordered areas are running approximately horizontal, which is the trace of the growth surface. Because of the restrictions applied to the fourier transform in creating these images, the sharpness of these boundaries has not been preserved, and they may actually be less diffuse than they appear.

## DISCUSSION

The (Ga,In)P epilayers are grown on the (001) GaAs surface, which has a 2-fold symmetry; the dangling bonds at the surface lying in [1$\bar{1}$0] and [110] directions on alternate Ga and As layers. It is therefore not surprising that the ordering should occur on only two of the four possible {111} planes. Because of this two-fold symmetry there should be no preference for ordering parallel to either of the favored {111} planes. The existence of domains of ordering parallel to both the (111) and (11$\bar{1}$) planes supports this lack of preference. Although there is no intrinsic reason for one particular variant to be favored during growth, it is possible that

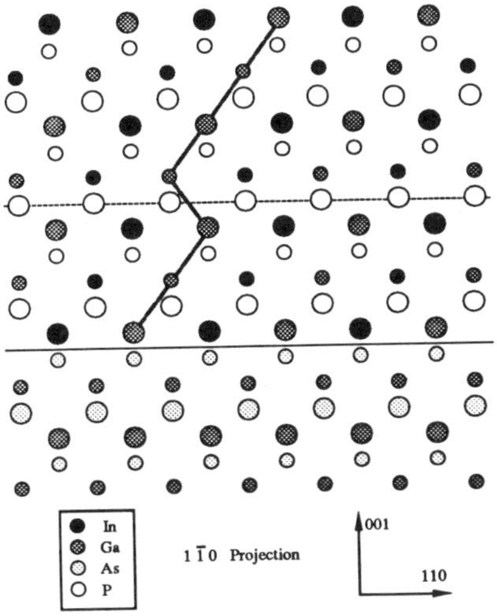


Figure 8 Schematic diagram of a stacking-fault on the growth surface of the ordered epilayer. The step in the ordered plane is illustrated by the shaded lines.

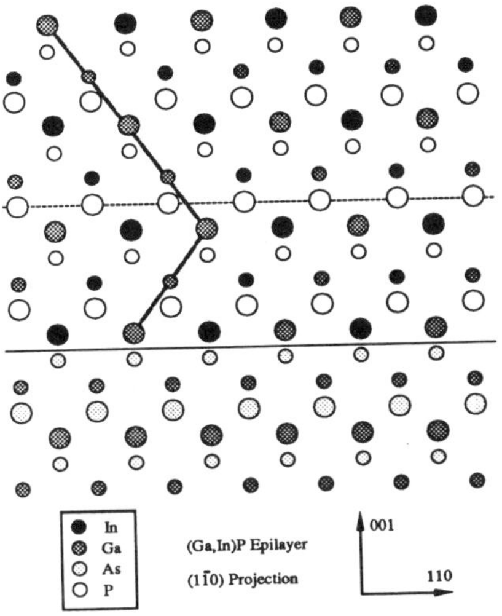


Figure 9 Schematic diagram of a twin boundary parallel to the growth surface of the ordered epilayer. The direction of the ordering in the two twins is illustrated by the shaded line.

appropriate growth conditions or the surface morphologies will favor one variant, either locally within the epilayer, or globally throughout the whole epilayer. As the ordered phase grows, Ga - rich and In - rich rows of atoms are deposited which lie along the [$1\bar{1}0$] direction. There are two equivalent positions for each of these rows, with respect to the previous Ga/In layer, corresponding to ordering of (111) or ($11\bar{1}$) planes. The orientation of the ordering will be determined by the position of the atoms in the Ga/In second layer below the surface. The difference in energy between the two possible sites will therefore be quite small. The difference will be largest when the Ga and In atoms are totally ordered into alternate GaP and InP planes, and will decrease to zero as the enrichment of these planes decreases. The high density of planar defects parallel to the growth surface can then be understood as the statistical result of this small energy difference. This analysis assumes that if the epilayer is already ordered then it will be favorable for the next layer to be ordered also. If, on the other hand, it were energetically favorable for the next layer not to be ordered, then the epilayer would grow with stacking faults in a regular array; producing a [110] ordered state. This type of ordering has already been reported in several ternary semiconductors[6,7]. As the ordered structure has a lower symmetry than the zincblende structure of the substrate, the bond lengths in the otherwise symmetry-related directions need not be preserved. The epilayers would then grow under strain, which would be relieved in a twinned micro-domain structure. This effect, if present, must be extremely small as specimens have been observed in which the entire epilayer thickness (0.5μm) is essentially of a single variant.

Schematic diagrams of a stacking fault and of a twin in the ordered structure are shown in figures 8 and 9. Each of these defects may be made more diffuse by the incorporation of a reduction in the enrichment of ordered (111) planes near the boundary. The interface in figure 8 is a stacking fault with a translation vector which would be a perfect lattice vector in the GaAs structure but is not in the ordered structure. (It may also be viewed as an antiphase domain boundary since the effect is to move Ga atoms to In sites and In atoms to Ga sites.) It is apparent that both of the faults in figures 8 and 9 are intrinsically stoichiometric with equal numbers of Ga and In atoms occurring on either side of the boundary plane. The faults observed on {111} planes can be similarly modelled as stacking faults of twin boundaries. In this case the faults may arise where the nucleii of the ordered phase having different origins or ordering direction coalesce. The defects then propagate through the epilayer as the layer grows. These faults may be either stoichiometric or non-stoichiometric depending on the relative orientation of the fault plane and the ordering direction.

In conclusion, the growth of ordered GaInP by OMVPE produces domains of different ordering orientation, which may be imaged by dark-field and high-resolution electron microscopy. There is considerable faulting in the ordered structure, primarily parallel to the (001) growth surface. The ordered structure is sensitive to electron beam irradiation, so that a determination of the extent of the ordering by high-resolution may be difficult.

## ACKNOWLEDGMENTS

The authors would like to thank Mr R. Coles and Ms M. Fabrizio for technical support. The electron microscopes are part of a central facility provided by the Materials Science Center at Cornell and are supported in part by National Science Foundation. This research has been supported by the Materials Science Center and, in part, by the SRC program for Microscience and Technology at Cornell University. The OMVPE growths were funded by IST under contract number N00014-86-K-0521 and JSEP contract number F49620-84-C-0060

## REFERENCES

1 S.McKernan, C.B.Carter, D. Bour, and J.R.Shealy, Mat. Res. Soc. Symp. Proc., **104**, 637 (1987)
2 S.McKernan, C.B.Carter, D. Bour, and J.R.Shealy, J. Mater. Res. **3**, 406 (1988).
3 S.McKernan, C.B.Carter, D. Bour, and J.R.Shealy, Proc. **46**th annual EMSA meeting, p902 (1988)
4 G.P.Srivastava, J.L.Martins and A.Zunger, Phys. Rev. B **31**, 2561 (1985).
5 J.L.Martins and A.Zunger, J. Mater. Res. **1**, 523 (1986).
6 T.S.Kuan, T.F.Keuch, W.I.Wang and E.L.Wilkie, Phys. Rev. Lett. **54**, 201 (1985).
7 H.R.Jen, M.J.Cherng and G.B.Stringfellow, Appl. Phys. Lett. **48**, 1603 (1986).
8 M.A. Shahid, S. Mahajan, D.E. Laughlin and H.M. Cox, Phys. Rev. Lett. **58**, 2567 (1987)
9 A.Gomyo, T.Suzuki, K.Kobayashi, S.Kawata, I.Hino and T,Yuasa, Appl. Phys. Lett. **50**, 637 (1987).
10 R. Kilaas, Proc. **45**th EMSA meeting, p66 (1987)
11 Synoptics Ltd., 15 The Innovation Centre, Milton Rd., Cambridge, CB4 4BH, UK

# MICRO-INDENTATION AND MICRO-SCRATCH TESTS ON SUB-MICRON CARBON FILMS.

T.W. WU*, R.A. BURN*, M.M. CHEN** AND P.S. ALEXOPOULOS*
*IBM Research Division, Almaden Research Center, San Jose, CA 95120.
**IBM General Product Division, San Jose, CA 95139.

## ABSTRACT

Micro-indentation and micro-scratch techniques were used to characterize the hardness and the adhesion strength of 0.11 $\mu$m thick sputtered carbon films on Silicon substrates. Hardness depth profiles and critical loads were measured using a microindenter under indentation and scratch testing modes, respectively. The carbon film with 6 mtorr argon sputtering pressure shows better practical adhesion (or higher critical load) and slightly higher hardness. The indentation fracture phenomenon observed on the 30 mtorr film is closely related to its poor adhesion. The failure mechanism will also be discussed.

## INTRODUCTION

Adhesion is not just an important problem but also a difficult one. Nowadays, extremely thin films ( < 0.1 $\mu$m) are extensively used both in semiconductor and computer storage technologies; therefore, a device which is capable of measuring this critical property of a sub-micron thick film is highly desirable. The characterization of adhesion strength by scratch testing is well established and commonly used in the hard coatings industry [1-4]. However, because of approximately two orders of magnitude difference in the sample thickness, new specifications are required on the potential testing devices, such as the range and the resolution of the applied load, the sensitivity of the acoustic emission detector, the tip radius and quality of the indenter...etc., and these stringent demands mean conventional type scratch testers are not immediately applicable.

In this study, we have modified the existing microindenter developed at the IBM Almaden Research Center [5] to perform not only the micro-indentation but also the micro-scratch tests. Two sub-micron carbon films with different mechanical properties have been used to demonstrate this added capability. Extensive SEM studies were also carried out to understand the failure mechanisms.

## EXPERIMENTAL PROCEDURE

Two 0.11 $\mu$m thick carbon films were deposited on Si<111> wafers using a DC planar magnetron sputtering system under the same process conditions except for different argon sputtering pressures. The nomenclature, process conditions and some of the sample properties are summarized in Table I. After deposition, the Si wafer was cut into approximately 10 mm square coupons which were mounted on the sample post (2 in Fig.1) for the indentation and scratch tests.

For obtaining the hardness depth profiles, we have performed the continuous load and unload test and utilized the data processing routine described in the previous report [5]. The indentation rate used for the loading and unloading stages were 4 and 15 nm/sec, respectively. The indenter used for hardness measurement was a 3-sided diamond pyramidal or Berkovich indenter [6].

TABLE I: Some Properties of the Sample Films

| Sample ID | Ar-pressure (mtorr) | Residual stress (dyne/cm$^2$) | Mass Density (g/cm$^3$) | Hardness (GPa) | Critical load (g) |
|---|---|---|---|---|---|
| A | 30 | $-2\times10^9$ † | 1.5 ‡ | ~23 | 0.24±0.03 |
| B | 6 | $-6\times10^9$ † | 2.1 ‡ | ~25 | 2.98±0.34 |

† compressive stress, obtained by wafer curvature measurement.
‡ measured by X-ray absorption technique.[11]

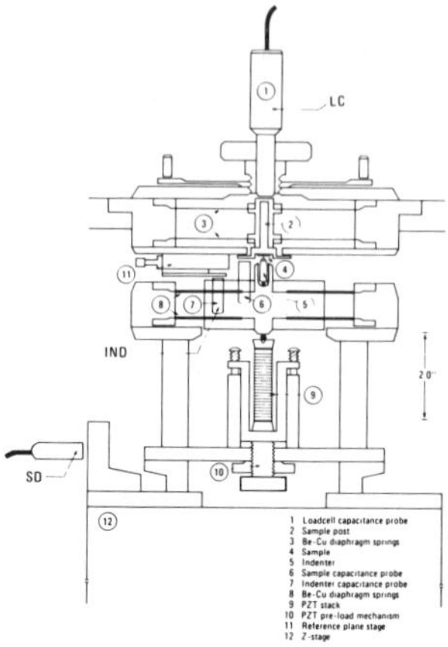


Fig. 1: Schematic diagram of indenter assembly and loadcell assembly. Here IND: indenter probe, LC: loadcell probe and SD: scratch distance probe.

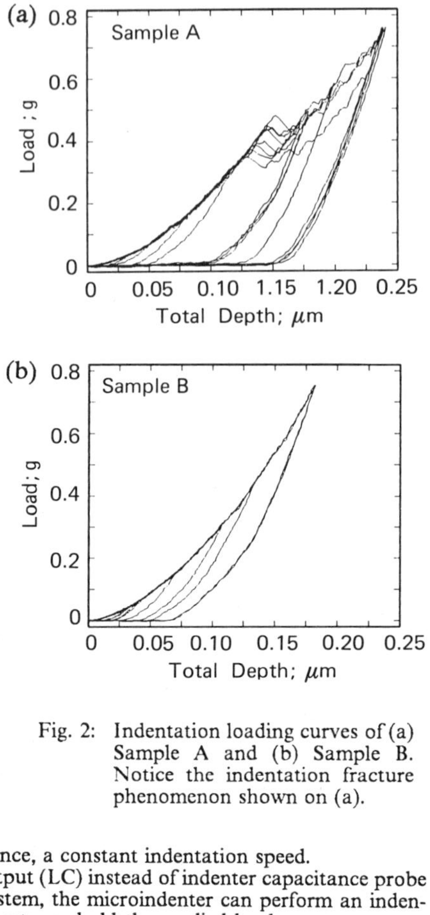


Fig. 2: Indentation loading curves of (a) Sample A and (b) Sample B. Notice the indentation fracture phenomenon shown on (a).

The scratch test can be treated as a combination of two operations, one is the normal indentation process carried out by the indenter assembly, the other is a horizontal motion executed by the precision X-translation stage. Combining the existing capabilities of the microindenter with an additional capacitance probe (SD in Fig. 1) for monitoring the displacement of the X-translation stage, we can perform a scratch test under the following modes:
(1) Indenter-servo mode: We can have any predetermined displacement pattern for the indenter motion during the entire scratch process, for instance, a constant indentation speed.
(2) Loadcell-servo mode: By using the loadcell output (LC) instead of indenter capacitance probe output (IND) as the input signal for the servo system, the microindenter can perform an indentation or a scratch test under a constant loading rate, or hold the applied load constant at any stage. (3) Profilometer mode: By locking in the loadcell output at a small constant load via the servo system, usually it is set at 10 mg, then scanning the indenter across the sample surface by moving the X or Y-translation stage. The recorded locus of the indenter tip is then the surface profile. This is a very useful function for measuring the scratch groove depth.

The current version of the microindenter has a range of applied load from a few mg up to 200 g with a load resolution of 3 mg, and a depth resolution of 1 nm. The scratch speed is from 0.5 $\mu$m/sec up to 50 $\mu$m/sec and the scratch length can be up to several mm.

To perform a scratch test, we first place the indenter about 0.1 $\mu$m away from the sample surface. This step allows us to begin a scratch with a zero applied load. Second, we set the traveling range and speed of the X-translation stage at 120 $\mu$m and 1 $\mu$m/sec, respectively, through the controlling program and then start the motion. Finally, the PZT motor is activated to drive the indenter toward the sample surface at a speed of 15 nm/sec. The outputs from three capacitance probes (i.e. IND, LC and SD in Fig. 1) were recorded during the entire period of the scratch process by a standard data acquisition system. The maximum applied load can be adjusted by selecting the indentation speed, the scratch speed or distance. A conical type diamond indenter with nominal 5 $\mu$m tip radius has been used for the tests.

A JEOL 840 Scanning Electron Microscope was used for examining the fractography of the scratch tracks. Because of the relatively low electrical conductivity of sputtered carbon films, a 10KV accelerating voltage was used.

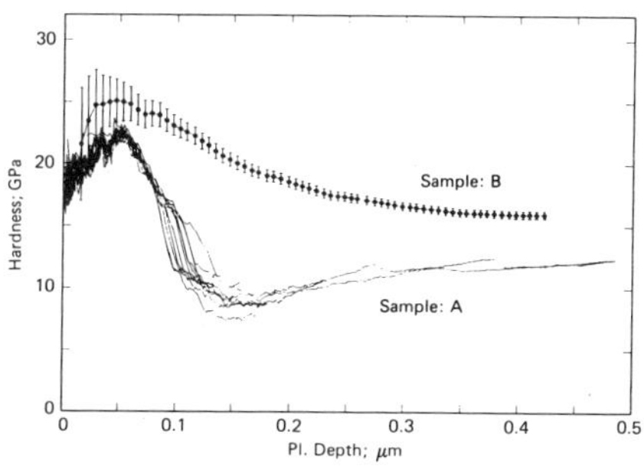


Fig. 3: Hardness depth profiles of Sample A and Sample B.

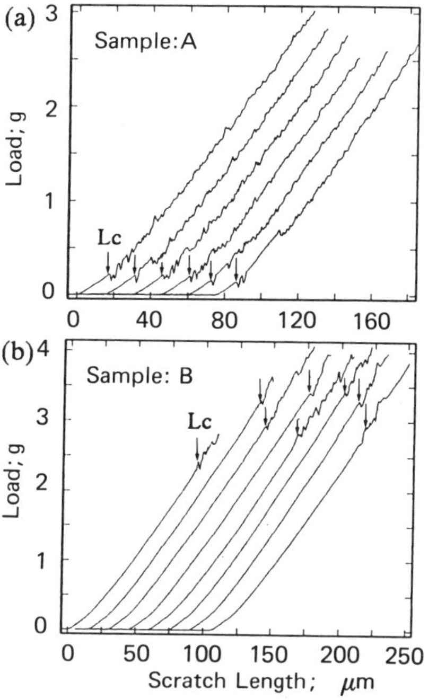


Fig. 4: Scratch loading curves of (a) Sample A, and (b) Sample B. Arrows indicate the positions of the first load drops.

## RESULTS AND DISCUSSION

### Micro-indentation and Micro-scratch Tests

Figs. 2a and 2b show the indentation loading curves of Samples A and B, respectively. Sample A shows failure events during the indenting process as the abrupt drops in load as the applied load reaches roughly 0.36 g. In contrast, Sample B simply shows the normal loading patterns up to 7.5g (only range to 0.8 g shown in Fig. 2b). The hardness depth profiles of Samples A and B are shown in Fig. 3. So as not to obscure the indentation fracture phenomenon which occurs for Sample A, we have omitted plotting the final average hardness values and the standard deviation bars which we have shown for Sample B. The profile of Sample A shown in Fig. 3 is simply a superposition of the individual hardness curves obtained from each indentation. The intrinsic hardness of the carbon coatings can be estimated from the first plateaus in the hardness depth profiles [5], and they are ~23 and ~25 GPa for Samples A and B, respectively. The slightly higher hardness of Sample B may be attributed to its higher mass density as verified by X-ray absorption measurements [11]. As expected, both hardness depth profiles approach the hardness of the Si substrate, which is about 16 GPa, at large penetration depth.

The following information can be acquired after scratch testing. (1) The distribution of the applied load along the scratch length. (2) The distribution of the total and the plastic depth along the scratch length. (3) The critical load, which is the applied load corresponding to the first event of failure during the scratch process, such as film cracking or delamination. The critical load is obtained directly from the loadcell output. (4) The scratch hardness (or scratch resistance); by adopting the concept of the Meyer hardness [7], the scratch hardness can be simply defined as $Hsc= 9.8\times[L\div(0.5\times\pi\times(W/2)^2)]$ [10]. Where, Hsc: the scratch hardness in GPa. L: the applied load in g. and W: the width of the scratch track in μm.

Figs. 4a and 4b are the scratch loading curves of Samples A and B, respectively. The scratch loading curve is the plot of the scratch load vs. the scratch length. For clarity, we have shifted each scratch loading curve 15 μm apart to prevent overlap. Interestingly enough, we have observed that, in both cases, the first abrupt drop on the scratch load occurs very regularly, although for Sample B it has occurred at much higher load than for Sample A. This implies that a characteristic property of each sample has been detected. If we define the critical load, Lc, as the applied load which is associated with the first occurrence of the load drop on a scratch loading curve, then the average critical loads and their standard deviations for Samples A and B can be calculated. The results are also included in Table I. Because of the high accuracy in monitoring the lateral scratch position, for each scratch we can match the applied load to its corresponding scratch pattern along the entire scratch length.

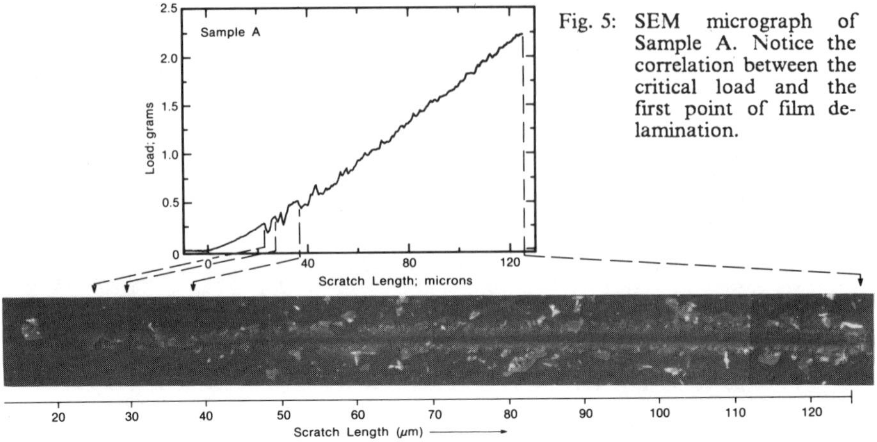


Fig. 5: SEM micrograph of Sample A. Notice the correlation between the critical load and the first point of film delamination.

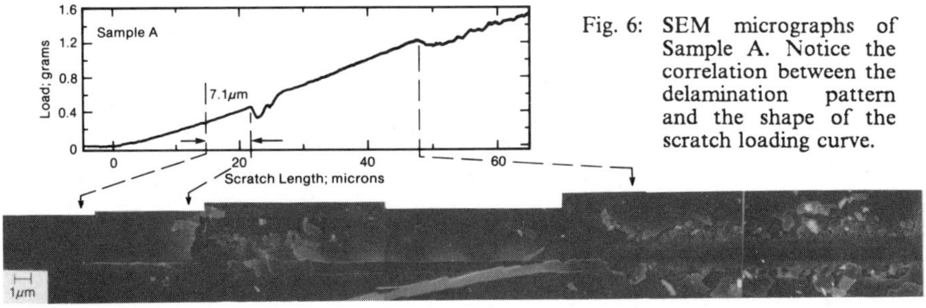


Fig. 6: SEM micrographs of Sample A. Notice the correlation between the delamination pattern and the shape of the scratch loading curve.

## Fractography

Figs. 5 and 6 show the correlations between the scratch loading curves and the corresponding scratch morphology of Sample A. Similarly, Figs. 7 and 8 are for Sample B. The following observations can be made; (1) A smooth scratch loading curve was observed before the applied load exceeded the critical value, Lc, which is in contrast to the zig-zag type curve when it exceeded this value. (2) The initiation of film delamination always occurred at the scratch position where the first drop of the applied load was observed. (3) The zig-zag shape of the scratch loading curve is attributed to consecutive delamination of the carbon films. (4).The size of the load drop is proportional to the scale of the film delamination. (5) On Sample B, a set of uniform tensile type cracks [8] were observed along the scratch tracks before the film delamination initiated (see Fig. 8). This suggests that although the carbon film had been fractured, it was still bonded to the Si substrate until the applied load eventually exceeded Lc. The cracking of the film causes only small changes of the loadcell output and hence induces small kinks or ripple structures on the scratch loading curve.

The explanation for observations (1) and (2) may be argued as follows. Because of the high hardness of the carbon film (ref. Table I) and the low friction between the film and the indenter tip, the indenter slides smoothly on the carbon film up to the load level of Lc, and the failure mechanism up to this stage is simply wearing and thinning down the carbon layer. However, when the applied load reaches Lc, the carbon film starts to delaminate and/or flake out of the track. Failure events such as film cracking and delamination inevitably cause a sudden increase of the local compliance of the sample system, which in turn, results in an abrupt drop of the loadcell output. Items (4) and (5) actually can be thought of in terms of the load drop scaling with the amount of elastic energy released during the corresponding failure event. This also implies that there is a limitation to the sensitivity of our detecting scheme, i.e. direct loadcell output measurement. In other words, if we assume that the released energy can be scaled by the delaminated area, then for small scale delamination, say for which the diameter of the coating flakes is smaller than 0.1 μm, the criterion for Lc determination may not be applicable because the load

Fig. 7: SEM micrograph of Sample B. Notice that the higher critical load corresponds to the later delamination.

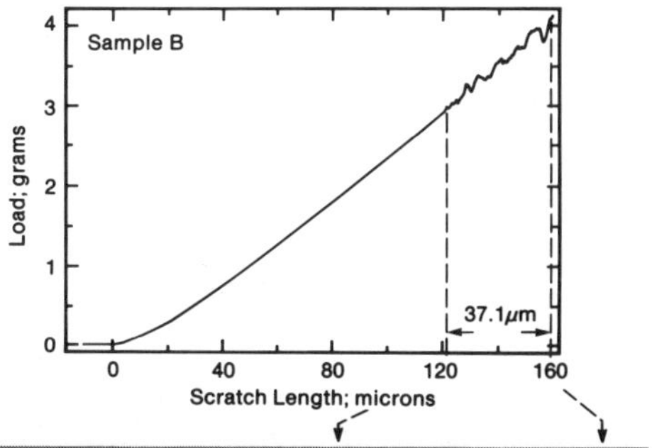


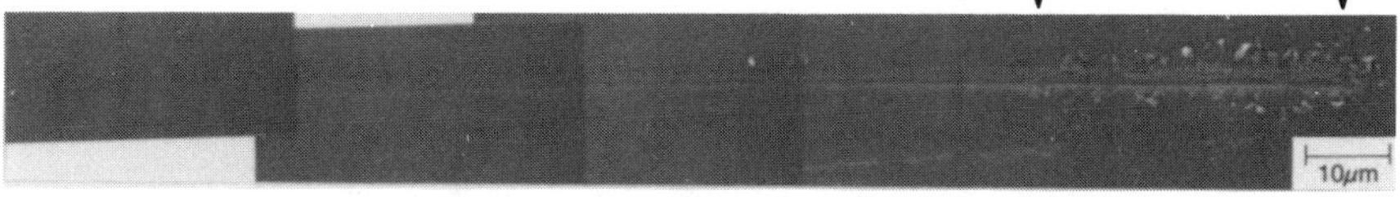


Fig. 8: SEM micrograph of Sample B. Notice that the appearance of the uniform tensile type cracks before film delamination.

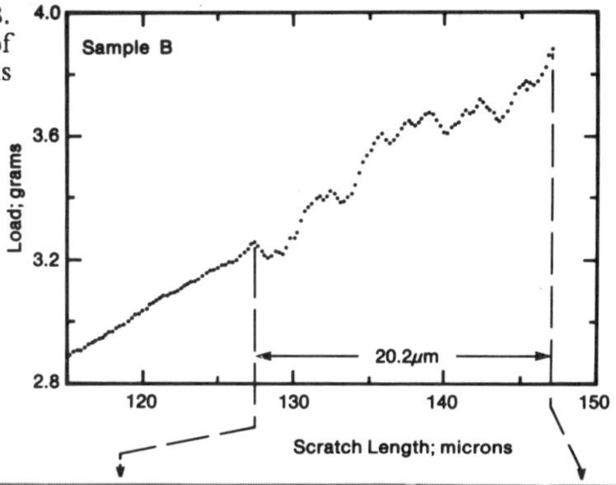


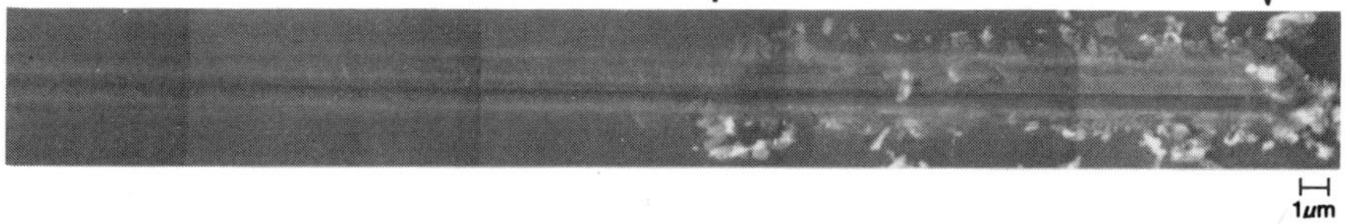


drop becomes so small it may not be discernible from the background noise. A similar problem also exists in devices which use the acoustic emission detection scheme. However, locating the delamination spots by high resolution SEM or increasing the loadcell sensitivity are two possible ways to increase the sensitivity.

Figs. 5 through 8 have clearly demonstrated the one-to-one correspondence between the critical load and the first point of film delamination. Based on this fact, it seems reasonable to conclude that the critical load, Lc, as we defined previously does correlate to the practical adhesion strength [9] of the film/substrate interface. It is worthwhile to emphasize that Lc is a quite complicated function, which includes variables such as the hardness, Young's moduli of coating and substrate, the coating thickness, the friction between the indenter and the coating, indenter tip radius, and even the testing conditions such as the scratch speed and loading rate [3,8]. Therefore care must be exercised when attempting to rank adhesion strength using only the critical loads measurements.

Direct Scratch Depth Measurement

The plastic depth profile along a scratch track can be obtained by the following procedure; (1) Pre-scan: Operating the microindenter in the profilometer mode and then acquiring the original surface profile. (2) Scratch: Carrying out the scratch routine. (3) Post-scan: Setting the microindenter back to the profilometer mode and acquiring the second surface profile along the scratch track created in step (2). (4) Superposition: Overlapping the pre-scan profile with the post-scan one by matching along their common baselines, the separation between two curves is then the plastic depth along the scratch track.

Fig. 9 shows an example of a scratch depth profile for Sample B. On the post-scan profile, the gradual increase of the depth at the beginning confirms that the initial failure mechanism is simply wearing or thinning down the carbon layer. The kinks or ripple structures observed in the range R in Fig. 9 get more and more pronounced along the track and seem to be influenced by the tensile-type cracks. Position S where a sudden increase in depth occurred is exactly correlated to the initiation of the delamination. In addition, the depth measured at position T is about 0.1 μm and it is almost identical to the original film thickness, which also implies the total absence of the carbon film. The evolution of the scratch depth is consistent with the failure mechanisms we have observed by the SEM fractography. Finally, the reverse peak marked P is the pile-up of the materials which has been removed from the track by the indenter, and can be easily seen at the end of the scratch track.

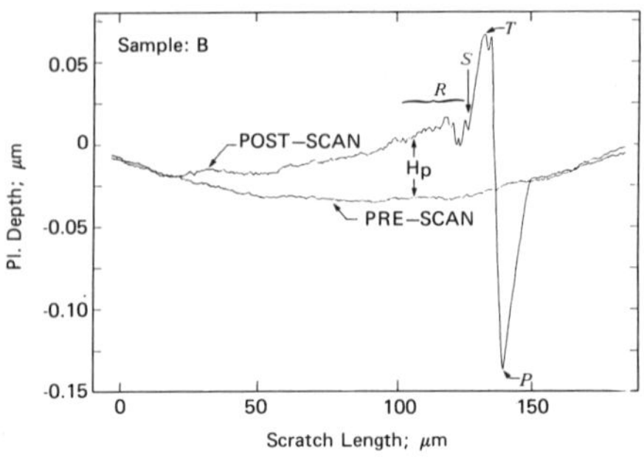


Fig. 9: A typical scratch depth profile of Sample B.

## CONCLUSIONS

After modifying the microindenter, we have demonstrated the capability of performing scratch tests on sub-micron thick coatings under three operating conditions, namely, the indenter-servo, loadcell-servo and profilometer modes. By comparison of the scratch loading curves with the SEM micrographs, we have also shown that the critical load, Lc, defined as the scratch load corresponding to the first load drop, does correlate to the practical adhesion strength of the film/substrate interface.

The Ar sputtering pressure has significant influence on the mechanical properties, including the adhesion, of the sputtered carbon films. The indentation failure and also the early delamination observed on Sample A is attributed to its poor adhesion. On the contrary, Sample B has better adhesion strength so that it can sustain film cracking without debonding. Direct depth measurement can provide the plastic depth profile along the scratch track. This information is very useful in scratch hardness calculation and scratch process modeling.

## REFERENCES

1. M.M. Karnowsky and W.B. Estill, Rev. Sci. Instrum. 35 (10), 1324 (1964).
2. P.A. Steinmann and H.E. Hintermann, J. Vac. Sci. Technol. A3 6 , 2394 (1985).
3. P.A. Steinmann, Y. Tardy and H.E. Hintermann, Thin Solid Film 154 ,333 (1987).
4. A.J. Perry, Thin Solid Films 78 , 77 (1981); 107 , 167 (1983).
5. T.W. Wu, C. Hwang, J. Lo and P.S. Alexopoulos, presented at the 1988 Int. Conf. of Metall. Coatings, San Diego, CA, to be published in Thin Solid Films; IBM Research Report, RJ 6195 (1988).
6. E.S. Berkovich, Fact. Lab., Moscow 16 , 345 (1950); Industr. Diam. Rev. 11 , 129 (1951).
7. D. Tabor, The Hardness of Metals , (Clarendon Press, Oxford, 1951), p. 8.
8. P.J. Burnett and D.S. Rickerby, Thin Solid Films 154 , 403 (1987).
9. K.L. Mittal, in Adhesion Measurement of Thin Films, Thick Films, and Bulk Coatings , edited by K.L. Mittal (STP 640, ASTM, Philadelphia, PA 1978), p. 5.
10. J. Ahn, K.L. Mittal and R.H. MacQueen, ibid. p. 134.
11. G. Lim,M.M. Chen,G. Castillo and R.C.C. Perera, presented at 37th Annual Dever Conf. on Applications of X-ray Analysis, Aug. 3, 1988, to be published in Advanced in X-Ray Analysis.

# HARDNESS AND ELASTIC MODULUS MEASUREMENTS IN CdTe AND ZnTe THIN FILM AND BULK SAMPLES AND ZnTe-CdTe SUPERLATTICES

L. J. Farthing*, T. P. Weihs*, D. W. Kisker**, J. J. Krajewski**, M. F. Tang*, and D. A. Stevenson*
* Department of Materials Science, Stanford University, Stanford, California 94305
** AT&T Bell Laboratories, Homdel, New Jersey 07733

## ABSTRACT

Hardness and modulus values of bulk and epilayer ZnTe and CdTe samples and of ZnTe-CdTe superlattices are reported. Both hardness and Young's modulus values increase with increasing ZnTe content in the ZnCdTe samples. Alloying effects and strains in the superlattice structure are proposed to explain the strengthening.

## INTRODUCTION

Compound semiconductor superlattices are of interest to the materials community because bandgaps can be finely tuned by the proper choice of the semiconductor materials and layer thicknesses. Mechanical properties relate to the changes in electronic and optical properties that may accompany the processing of semiconductor materials into devices, particularly the introduction of dislocations upon thermal processing and the application of films. Furthermore, the mechanical properties are of concern in the integrity of devices during service. In this paper, we report hardness and elastic modulus values for ZnTe and CdTe epilayer and bulk samples and ZnTe-CdTe strained layer superlattices. The ZnTe-CdTe system is of interest because of its promising optical properties [1].

## EXPERIMENTAL

The samples consist of two sets of ZnTe-CdTe strained layer superlattices grown by atmospheric pressure organometallic vapor phase epitaxy (OMVPE) [1], and epilayer and bulk samples of ZnTe and CdTe. The first set of the superlattices consists of 200 cycles of 25 Å ZnTe and 50 Å CdTe with a 25 Å CdTe cap and a 1.0 μm ZnTe buffer layer on a (100) GaAs substrate. The other set of superlattice samples consists of 200 cycles of 25 Å CdTe and 50 Å ZnTe with a 25 Å CdTe cap and a 1.0 μm ZnTe buffer layer on a (100) GaAs substrate. The ZnTe and CdTe have a lattice mismatch of 6%, so these are strained layer superlattices. The (100) ZnTe epilayer and (100) CdTe epilayer are grown on (100) GaAs substrates by OMVPE for thicknesses of 2.0 μm. The bulk samples are a single crystal (100) CdTe sample and a coarse grained ZnTe sample.

X-ray diffraction measurements were made using a Picker diffractometer with a Cu Kα source. Hardnesses were measured by a Vickers hardness tester and a Nanoindenter, which can measure the hardness of very thin films. Important parameters of the Nanoindenter are a displacement resolution of 2 Å and a force resolution of 0.5 mN [2]. The data from the Nanoindenter give hardness values as a function of plastic depth, while Young's modulus values can be calculated from the unloading portion of the load versus depth indentation curves. Most samples were tested for depths between 20 and 1500 nm; however, for the purpose of analysis, only the indentations with depths between 40 and 400 nm were considered.

## RESULTS AND DISCUSSION

Figure 1 shows the hardness of the ZnTe and CdTe epilayer and bulk samples as a function of plastic indentation depth. The hardness values for the ZnTe epilayer and bulk samples are very similar and show the same pattern of decreasing hardness with increasing plastic depth. The large increase in hardness at very small plastic depths is thought to be due to a surface oxide on these samples, which is evidenced by a discontinuity in the indentation curves at small values of plastic depth. Although the hardness values of the CdTe bulk sample are lower than those of the epilayer, the CdTe samples do show a similar pattern in hardness. There is no large increase in hardness at very small depths; indentation curves for the CdTe samples show no evidence of a surface oxide being broken. Both epilayer samples show increases in measured hardness at depths > 400 nm as the much harder GaAs substrates begin to influence the data.

Figure 2 shows the hardness of the CdTe-rich and ZnTe-rich superlattices as a function of depth. As is expected (since ZnTe is harder than CdTe), the unannealed and annealed ZnTe-rich superlattices are harder than the corresponding unannealed and annealed CdTe-rich superlattices, respectively. Both unannealed superlattices are harder than the corresponding annealed superlattices, which is due to layering effects. Strengthening from multilayers arises from the strain in the layers (compressive in the CdTe layers and tensile in the ZnTe layers) and from different moduli in the layers [3]. Since ZnTe is stiffer than CdTe, dislocations in the CdTe layers are repelled by the ZnTe layers, thus requiring larger applied stresses to move dislocations. Both unannealed superlattices show similar patterns in hardness: decreasing hardness with increasing depths < 400 nm. This pattern is thought to arise from annealing that occurs as the superlattices are being grown. Interdiffusion between the ZnTe and CdTe layers occurs to a greater degree in the layers grown first than in the layers grown later. This interdiffusion of the two components decreases the strengthening effects of layers; thus, the hardness values decrease as the indenter reaches the lower layers (i.e., those closer to the buffer layer and substrate).

The annealed superlattices are softer than the corresponding unannealed superlattices, since the layers have interdiffused to form ternary alloys. X-ray diffraction scans of the two annealed samples show no superlattice sattelite peaks, indicating that the superlattice structure has been completely destroyed. There appears to be a difference in the patterns of hardness of the two annealed superlattices, since the hardness of the ZnTe-rich sample is fairly linear with depth while the hardness of the CdTe-rich sample decreases with increasing depth. At closer inspection, however, this apparent difference may be due to a difference in testing: hardness values of the CdTe-rich superlattice range from depths of 18 to 1182 nm, while those of the ZnTe-rich superlattice range from depths of 70 to 1234 nm. Also, the indentation curves at very small depths for the annealed CdTe-rich sample indicate the breaking of a surface oxide. Since we have no data for the annealed ZnTe-rich sample at depths < 70 nm, we do not know if this difference in patterns is real.

Figure 3 shows hardness as a function of ZnTe fraction in the epilayer and bulk ZnTe and CdTe samples, in ZnCdTe samples, and in the ZnTe-CdTe superlattices. Hardness values measured by the Vickers tester and by the Nanoindenter are shown. The lower curve in figure 3 shows Vickers hardness values of the ZnCdTe alloys ranging from 0.5 GPa at 2% ZnTe content to a maximum of 1.0 GPa at 30% ZnTe content. These hardness values show a bowing due to solid solution hardening [4]. This bowing is seen in several ternary semiconductor systems (e.g., HgCdTe [5], GaInSb [4], InAsP [4], etc.). The hardness values of the binary compounds of CdTe and ZnTe are seen at the endpoints of the curves in figure 3. The hardness values of the epilayers of CdTe and ZnTe, 1.1 GPa and 1.5 GPa, compare favorably to values of the bulk CdTe and ZnTe, 0.8 GPa and 1.4 GPa. These values, however, are higher than Vickers hardness values of bulk CdTe and ZnTe, 0.46 GPa and 0.8 GPa. This difference can be explained by the fact that measured hardness typically increases as the applied load decreases [6]. The Vickers measurements were made using a weight of 25 g (245 mN), while the largest load applied by the Nanoindenter at depths of 400 nm was approximately 12 mN. The middle curve in figure 3 shows the hardness of the annealed superlattices. The hardness values of both the CdTe-rich and ZnTe-rich superlattices are 1.7 GPa. If these two values are joined to the hardness values of the binary epilayers, the resulting curve shows a bowing due to solid solution hardening. The upper curve in figure 3 shows the hardness of the unannealed superlattices. The hardness of the CdTe-rich superlattice is 2.2 GPa, while that of the ZnTe-rich superlattice is 2.4 GPa. If these values are joined to the hardness values of the binary epilayers, the resulting curve shows an even more pronounced bowing due to the strengthening by the multilayer structure.

Figure 4 shows Young's modulus values as a function of ZnTe content in the epilayer and bulk ZnTe and CdTe samples and in the superlattices. This figure shows modulus values from the literature [7] and values from the Nanoindenter indentation curves. The lower curve in figure 4 shows calculated modulus values for an isostress model for uniaxial deformation of a (100) multilayer [8]. These moduli for the model are significantly lower than the measured values and the literature aggregate values. This suggests that the modulus extracted from the Nanoindenter data is an aggregate value, not a modulus for the crystallographic orientation of the measured sample. The moduli of the binary compounds are seen at the endpoints of the curves in figure 4. Measured modulus values for the bulk and epilayer CdTe (41.3 GPa and 41.6 GPa, respectively) are very close; however, these values are a little higher than the literature aggregate value (38.1 GPa). Moduli for ZnTe from the literature aggregate value (61.0 GPa) and from the measured bulk (61.3 GPa) and epilayer (61.8 GPa) values are extremely close. The middle curve in figure 4 shows modulus values of the annealed superlattices. The modulus of the CdTe-rich superlattice is 49.8 GPa, while that of the ZnTe-rich superlattice is 52.5 GPa. If these values are joined to the moduli of the binary epilayers, the resulting curve shows a linear increase with increasing ZnTe fraction due to alloying. The upper curve in figure 4 shows moduli of the unannealed superlattices. The modulus of the CdTe-rich superlattice is 51.0 GPa, while that of the ZnTe-rich superlattice is 61.2 GPa. If these values are joined to the moduli of the binary epilayers, the resulting curve shows a slight bowing due to the strain in the multilayers.

## CONCLUSIONS

Hardness and modulus values of bulk and epilayer ZnTe and CdTe samples and of unannealed and annealed ZnTe-CdTe superlattices of two compositions (one third ZnTe content and two thirds ZnTe content) were measured. Increases in hardness in both ZnCdTe samples and annealed superlattices due to solid solution hardening were found. Increases in hardness with increasing ZnTe content in the unannealed superlattices, as compared to the corresponding annealed superlattices, due to the layering effects, were found. Hardness values of bulk ZnTe and CdTe show reasonable agreement with hardness values of epilayer ZnTe and CdTe. A linear increase in modulus values with increasing ZnTe content of annealed superlattices is

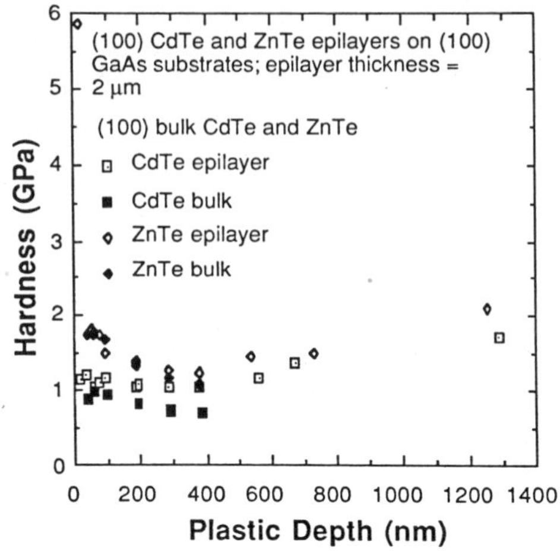


Figure 1. Hardness of epilayer and bulk ZnTe and CdTe as a function of plastic depth

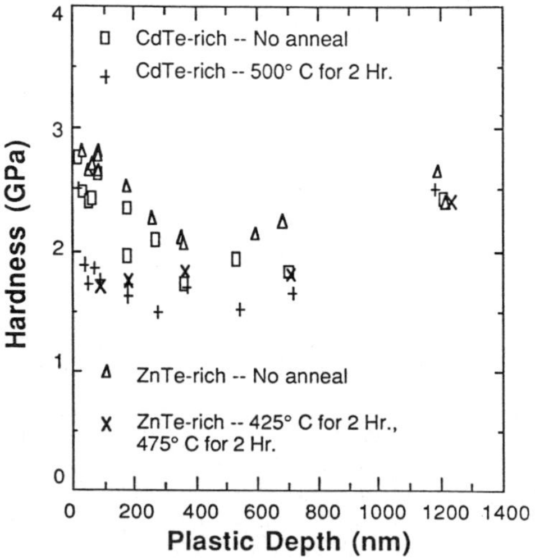


Figure 2. Hardness of ZnTe-CdTe superlattices as a function of plastic depth

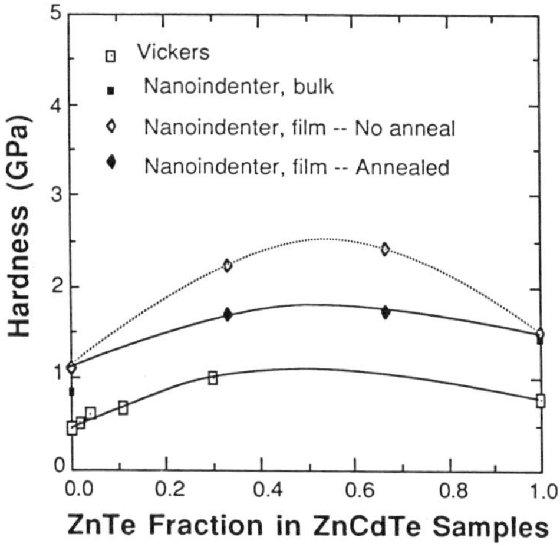


Figure 3. Hardness as a function of ZnTe fraction in epilayer and bulk ZnTe and CdTe and in ZnTe-CdTe superlattices

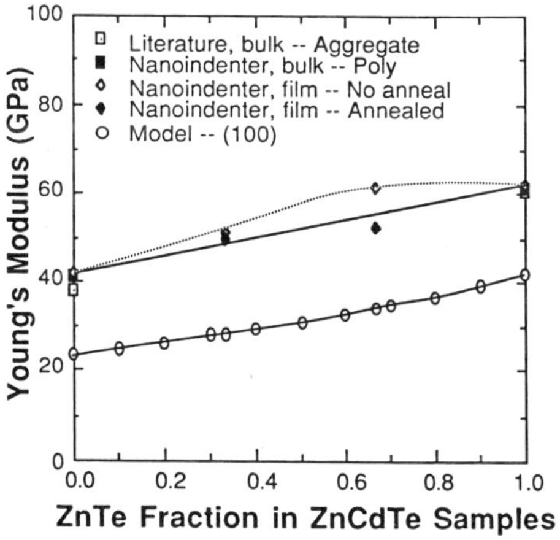


Figure 4. Young's moduli as a function of ZnTe fraction in epilayer and bulk ZnTe and CdTe and in ZnTe-CdTe superlattices

due to alloying and a slight increase (beyond the linear increase) in moduli of unannealed superlattices is due to the superlattice structure. Modulus values of bulk and epilayer samples of ZnTe and CdTe agree with each other and with the literature aggregate moduli for ZnTe and CdTe. The modulus value measured by the Nanoindenter appears to be an aggregate modulus.

## ACKNOWLEDGEMENTS

This work is supported by the Air Force Office of Scientific Research (contract AFOSR 85-0158, Project Task 2306/B1).

## REFERENCES

1. D. W. Kisker, P. H. Fuoss, J. J. Krajewski, R. M. Amirtharaj, S. Nakahara, and J. Menendez, J. Crystal Growth 86, 210 (1988).

2. M. F. Doerner and W. D. Nix, J. Mater. Res. 1, 601, (1986).

3. M. F. Doerner, Ph.D. thesis, Stanford University, 1987.

4. N. A. Goryunova, A. S. Borshchevskii, and P. N. Tretiakov, in Semiconductors and Semimetals, volume 4, edited by R. K. Willardson and A. C. Beer (1968) pp. 3-34.

5. J. G. Fleming, L. J. Farthing, and D. A. Stevenson, J. Crystal Growth 86, 506 (1988).

6. D. Tabor, in Microindentation Techniques in Materials Science and Engineering ASTM STP 889, edited by P. J. Blau and B. R. Lawn (1986) pp. 125-159.

7. G. Simmons and H. Wang, Single Crystal Elastic Constants and Calculated Aggregate Properties: A Handbook, 2nd ed. (The M.I.T. Press, Cambridge, Massachusetts and London, England, 1971).

8. W. D. Nix (private communication).

# GOLD-NICKEL MULTILAYER FILMS: STRUCTURE-PROPERTY CORRELATIONS

S. R. NUTT,* K. A. GREEN,* S. P. BAKER,** W. D. NIX,** and A. JANKOWSKI***
*Brown University, Box D, Providence, RI 02912
**Department of Materials Science and Engineering, Bldg. 550, Stanford University, Stanford, CA 94305
***Lawrence Livermore National Lab, P.O. Box 808, L-350, Livermore, CA 94550

## ABSTRACT

Gold-nickel multilayer films with periods of 1.2 - 4.6 nm were deposited on silicon substrates by magnetron sputtering, and plan and cross-sectional specimens were examined by transmission electron microscopy. The cross-sectional specimens revealed well-defined layering and columnar growth features that extended through the film thickness. Dark striations extending normal to the layers were attributed to diffraction contrast from defect strain fields. Diffraction patterns showed that the films were highly textured and that short-period films had a single fcc structure, while long-period films separated into three fcc structures. High-resolution images of the layer interfaces showed local regions of epitaxy partitioned by regions of disorder.

Indentation tests using a Nanoindenter, a depth-sensing indentation device, were performed to measure the elastic modulus and hardness of the films. No modulus enhancement was detected, and a small variation in hardness was measured.

## INTRODUCTION

Artificially layered bi-metallic thin films can exhibit a remarkable enhancement in elastic modulus known as the "supermodulus effect." The supermodulus effect is the enhancement in elastic modulus (by a factor of two or more) observed in highly textured compositionally modulated bi-metallic films in which the wavelength or bilayer thickness is about 2 nm. This phenomenon was first reported by Hilliard and co-workers in 1977 [1], and has subsequently been confirmed by other groups for several different bi-metal systems [2-3]. A satisfactory explanation of the phenomenon has been elusive, although two theories have been proposed. One theory is based on singularities in the electronic energy caused by interactions between the Fermi surface and the Brillouin zones associated with the composition modulation, and another theory is based on higher order elastic effects arising from the large coherency strains associated with the epitaxial layers [3]. Both of these theoretical approaches have been recently reviewed [3-4].

Many of the systems which exhibit a supermodulus effect are fcc metals in which the lattice parameters of the two components are widely different. The films are generally deposited in vacuum, and the resulting structure is a highly textured strained layer superlattice in which the layers are alternately in tension and compression. The biaxial alternation between compressive and tensile stresses can be described as a strain wave, since the component with the larger lattice parameter is compressed, and the component with the smaller lattice parameter is stretched. However, the details of how the strain is accommodated within the layers is presently unknown.

Our approach to this problem has been to combine detailed microstructural characterization of compositionally modulated Au-Ni films with measurements of the elastic and plastic mechanical response. The objective has been to identify structural features associated with the modulus enhancement using TEM, and to investigate the elastic-plastic properties of the films by light-load indentation measurements. We have selected the gold-nickel system for our study because of the large lattice parameter mismatch (0.4 nm versus 0.35 nm), and because of previous reports of modulus enhancement in Au-Ni films [1].

Mat. Res. Soc. Symp. Proc. Vol. 130. ©1989 Materials Research Society

EXPERIMENTAL

Compositionally modulated gold-nickel films were deposited onto <100> silicon substrates by dc magnetron sputtering. Details of the experimental apparatus and procedures have been reported previously [5]. Briefly, the system includes multiple sources that are individually shielded, and the silicon substrate is mounted on a rotating copper table that is heated during deposition. The total thickness of the film is monitored continuously and a feedback loop to the power supply is used to maintain a constant deposition flux. Layer thickness is controlled by varying the frequency of the rotating table. Using this system, films with periods ranging from .82 to 4.6 nm were fabricated. For some films, a gold buffer layer was deposited on the silicon substrate and annealed at 350C prior to depositing the multilayer. This procedure was used to improve the adhesion of the film and facilitate the preparation of cross-sectional TEM specimens.

Structural characterization of the Au-Ni films was accomplished by TEM. Four films having periods of 1.2, 2.0, 2.8, and 4.6 nm were examined. Both cross-sectional and plan-sectional specimens were prepared using conventional methods. Plan-sectional specimens were back-thinned by dimple grinding and ion milling. Cross-sectional specimens were prepared by methods similar to those described by Bravman and Sinclair [6]. All specimens were ion-milled in a holder cooled by liquid nitrogen to minimize specimen heating. The films were examined using a Philips EM 420T operated at 120 kV and a JEOL 4000-EX operated at 400 kV.

The mechanical properties of Au-Ni films were measured using an indentation device called a Nanoindenter (Microscience, Inc., Norwell, MA). The Nanoindenter is a depth-sensing indentation device with displacement resolution of 0.3 nm and load resolution of 0.25 μN. The instrument is capable of obtaining useful data from indentations as small as 20 nm in total depth. Tests were computer controlled and were conducted by two similar methods. In the first method, the indenter was lowered toward the material very slowly until the surface was detected. The indenter was then pressed into the material at a velocity between 3 and 6 nm/s until the programmed displacement for the indentation was reached. The load was held constant for 10 seconds and then removed at a constant rate. The second method was the same as above except that instead of holding the maximum load for a fixed time, the maximum load was held until the indenter velocity dropped below 0.1 nm/s, thereby minimizing errors that might have been introduced by rate effects in the first test.

RESULTS

Compositionally modulated Au-Ni films with periods of 1.2, 2.0, 2.8, and 4.6 nm were examined in plan- and cross-section using TEM. The plan-view specimens of all four multilayer films showed nodular growth features (Fig. 1). The growth nodules were generally 20-70 nm in diameter, although the nodules were about 40 nm in the film having a 4.6 nm period. Selected area diffraction patterns showed rings typical of structures with small equiaxed grains. There was no evidence of preferred orientation.

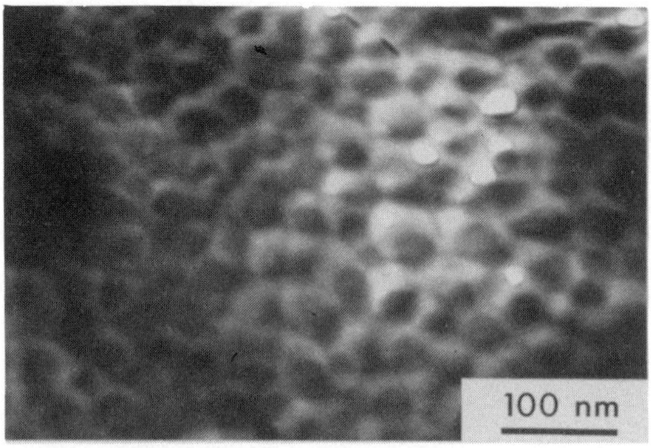


Fig. 1 Plan view TEM image of columnar growth nodules in Au-Ni film with period = 4.6 nm.

The cross-sectional specimens showed discrete alternating layers of gold and nickel that exhibited a convex curvature (Fig. 2). The curvature of the layers was more pronounced in the bottom layers, and the layers became progressively flatter as the film thickness increased. This phenomenon is typical of many vapor-deposited materials and is probably accentuated by the polycrystalline gold buffer layer on which the layers are deposited, which is microscopically rough after annealing. The boundaries between the columnar growth features are apparent as cusps formed by layers inclined in opposite directions. Perhaps the most notable feature in these films is the dark striations extending through the layers in the growth direction. These striations exhibit diffraction contrast and are apparently associated with a local strain field. Unfortunately, this contrast is convolved with structure-factor contrast from the different layers and moire fringes from overlapping growth nodules, thus preventing unambiguous interpretation. It is interesting to note however that there appear to be fewer of the anomalous striations in the film with the longest period (4.6 nm), than in films with shorter periods.

Selected area diffraction patterns from the cross-sectional specimens revealed structural changes in the films as the period was increased from 1.8 to 4.6 nm. Figure 3 shows the diffraction patterns from all four of the films examined. All patterns show strong arcs from the textured film superimposed on a <110> spot pattern from the silicon substrate. (The overall film thickness was only 200 nm, and consequently SAD patterns from the film alone were not possible.) The relative spacings of the different reflections from the Au-Ni film correspond to a *single* fcc-type structure with a lattice parameter of 0.367 nm, approximately midway between the lattice parameters of gold and nickel, which are 0.408 and 0.352, respectively. The lattice parameter of the compositionally modulated structure is slightly closer to that of nickel, the less compliant of the two metals.

The strongest arcs occur just outside the (forbidden) 200 reflection from silicon, and these arcs correspond to {111} planes with an interplanar spacing of 0.212 nm. These planes are oriented normal to the growth direction, resulting in a strong {111} texture. The patterns in Figures 3b, 3c, and 3d all show subsidiary maxima, *i.e.*, less intense arcs inside and outside of the {111} arcs. These reflections, usually termed "satellite" peaks in x-ray diffraction patterns, arise from the periodicity of the layers, and thus are more closely spaced in the longer period structures. The absence of subsidiary maxima in the pattern from the 1.2 nm period structure suggests that this film lacks the strong periodicity found in the other films. In principle, the layers in this film are each 0.6 nm thick, corresponding to slightly less than 3 monolayers of {111} planes. A small degree of atomic-scale roughness and/or intermixing can be expected to have a relatively large effect in such cases, both on the diffraction pattern, and on the mechanical response of the structure. Furthermore, not all regions in images from this film showed discrete layering, implying localized intermixing either during fabrication or during preparation of TEM specimens.

In Figure 3d, the intensity of the subsidiary maxima is asymmetric about the {111} reflection, and there is a set of three strong closely-spaced arcs of equal intensity, two of which are *inside* the {111} arc. Weaker subsidiary maxima are located inside and outside the three strong ones. If one assumes that the three intense arcs are caused by three different interplanar spacings, the values for these three spacings are 0.208, 0.219, and 0.232 nm. When compared with the {111} spacings for *bulk* gold and nickel, which are 0.2355 and 0.20345 nm, respectively, the values measured from the pattern suggest a structural change in the material, in which the individual layers now exhibit lattice spacings that resemble those of the bulk components. If true, this conclusion would indicate a partial loss of coherency that would be manifest in the elastic behavior. Undoubtedly, the interpretation is clouded by the simultaneous presence of the subsidiary maxima from the periodic layers, the Bragg reflections from the {111} planes, and also, in the case of Figure 3d, the forbidden 200 reflection from the silicon substrate. However, there is supporting evidence in a different type of reflection, discussed below.

Most of the strain in these compositionally modulated films must be accommodated *in the plane* of the layers, and it is important to know how atoms are displaced in these directions. One advantage of TEM observations made in cross-section is that this kind of information, *i.e.*, lattice periodicities normal to the growth direction, is accessible in simple selected area diffraction patterns. Looking again at Figure 3, perpendicular to the {111} arcs are the {220} reflections, which represent lattice planes oriented *parallel* to the growth direction. The interpretation of these reflections is facilitated by the absence of superlattice reflections in this region of the pattern. The patterns from films with modulation periods of 1.2 and 2 nm (Figures 3a and 3b) show a single arc corresponding to a 220 spacing of 0.136 nm. (The 220 spacings

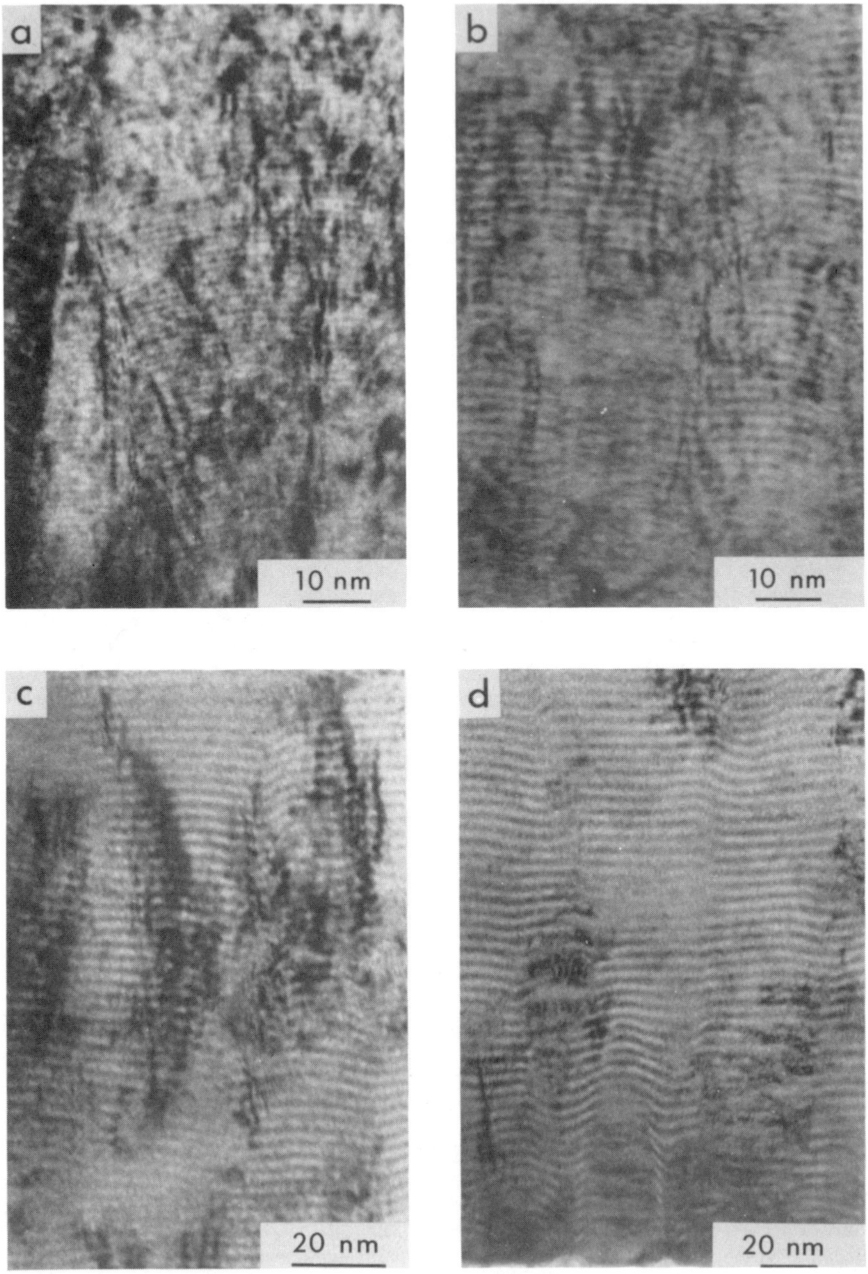


Figure 2. Gold-nickel films viewed in cross-section. Periods are (a) 1.2 nm, (b) 2.0 nm, (c) 2.8 nm, and (d) 4.6 nm. The growth direction is towards the top of the page.

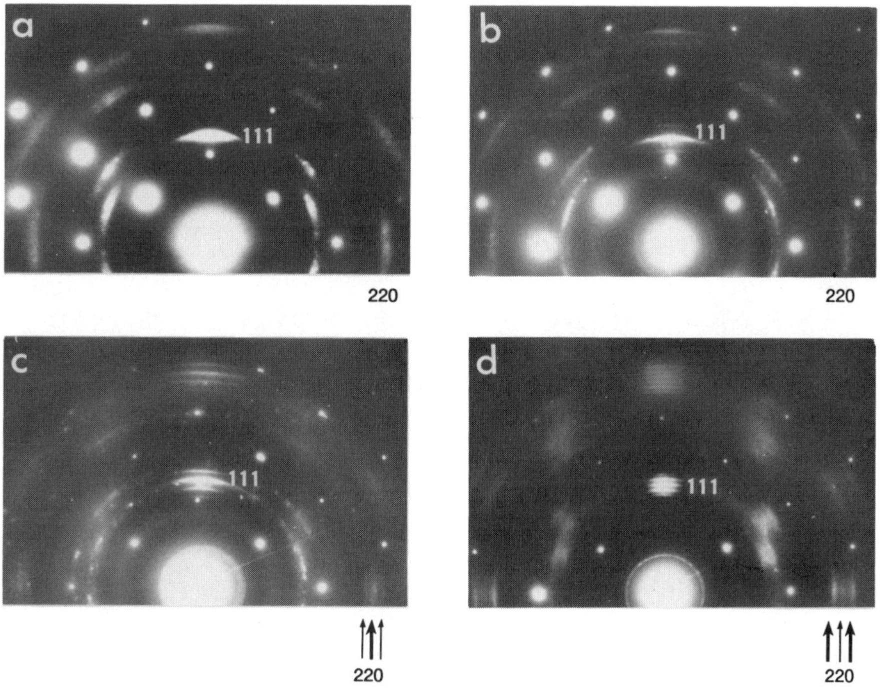


Figure 3. Selected area diffraction patterns from the four Au-Ni films (viewed in cross-section) having periods of (a) 1.2 nm, (b) 2.0 nm, (c) 2.8 nm, and (d) 4.6 nm. The patterns are oriented such that vertical is parallel to the growth direction.

for bulk gold and nickel are 0.1442 and 0.1246 nm respectively, and the average of these values is 0.1344 nm.) In Figure 3c, the 220 reflection is accompanied by two slightly weaker arcs on either side, corresponding to lattice spacings of 0.141 and 0.127 nm. These reflections also appear in Figure 3d, an SAD pattern from the film with a period of 4.6 nm. However, in this pattern the inner and outer arcs are strong and the middle arc is weak. The strong arcs correspond to spacings that are almost identical to the 220 spacings for bulk gold and nickel, implying that the compositionally modulated structure is no longer commensurate and has separated into two (or perhaps three) structures with different lattice parameters. The spatial distribution of these structures or phases within the film is presently unclear, although dark field images with the different 220 reflections might elucidate this issue.

High-resolution images of the films reveal additional structural details. Close examination of images from the Au-Ni films with periods of 1.2 and 2.8 nm (Fig. 4) reveal localized regions that are apparently defect-free. In these small regions, estimated to be no more than 5-10 nm, the lattice fringes are continuous across the interface and parallel to the layers (Fig. 4b). However, the layers are not atomically flat, and there are undoubtedly defects associated with the curvature of the layers. Two interesting features to note are that first, lattice fringes are highly perturbed (and the structure is presumably defective) where the dark striations extend through the layers (Fig. 4b). Secondly, the {111} fringes tend to be continuous in the plane of the layers (Fig. 4a) despite the layer curvature and despite steps in the composition profile (Fig. 4b). Thus, one possible explanation of the dark striations noted in Figure 2 is that they are associated with the layer curvature and are effectively a kind of low-angle tilt boundary. However, these observations are not conclusive, and modulated films with flatter layers are necessary to determine the nature of these defects.

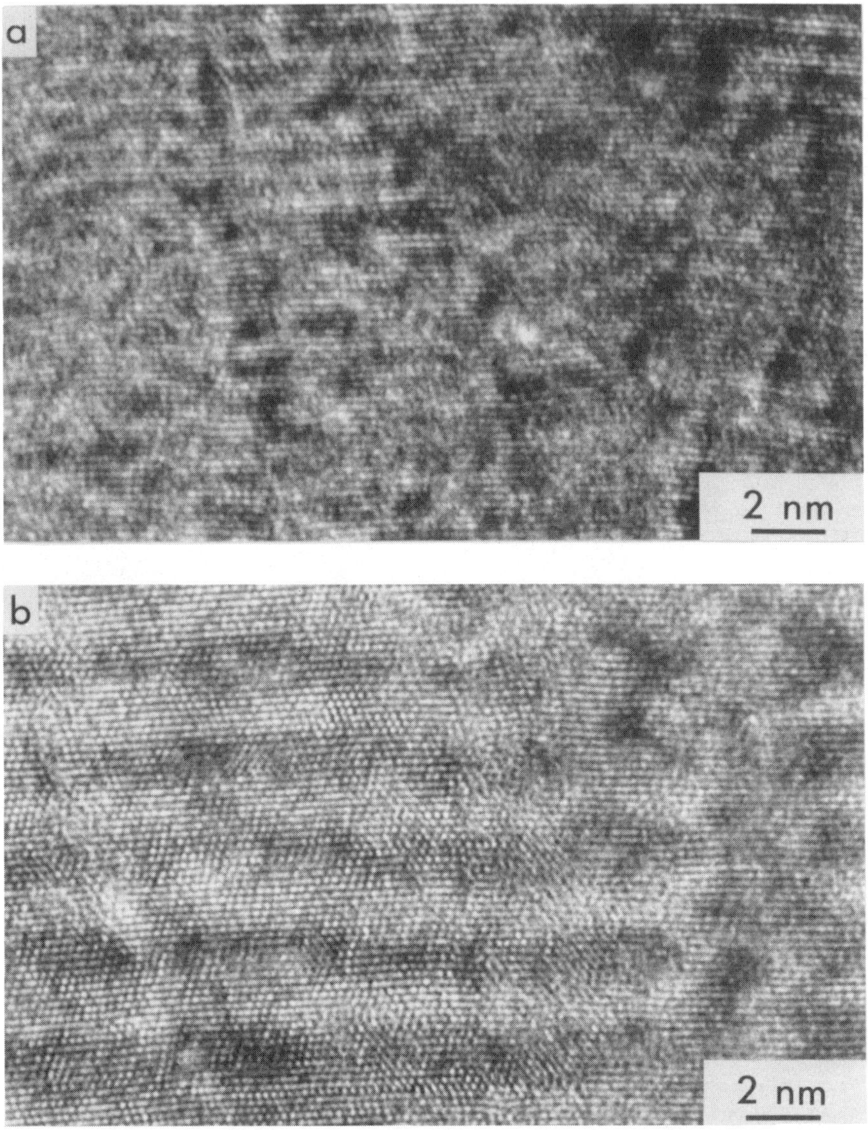


Figure 4. High-resolution images of Au - Ni films with periods of (a) 1.2 nm and (b) 2.8 nm.

Indentation tests using a Nanoindenter revealed *no significant variation of elastic modulus* with modulation period. Measured moduli for all films were near 140 GPa, compared with values of approximately 80 GPa and 140 GPa for bulk gold and bulk nickel, respectively. Although no modulus enhancement was detected, there are compelling reasons to believe that the supermodulus effect should not be observable using our measurement technique. First, the supermodulus effect is an anomalous increase in the *in-plane* stiffness of the film, and indentation measurements sample moduli averaged over *all* directions in the material. Secondly, the indentation process plastically deforms the layers, altering the structure that might be responsible for the elastic effect.

The hardness of the Au-Ni films showed variations with both plastic depth and modulation wavelength. The variations with depth were consistent with trends reported for other multilayer films [7]. The variations of hardness with modulation period showed a local maximum of over 8 GPa for periods of about 2.5 nm, compared with hardness values of about 7 GPa for films with shorter or shorter periods. Similar hardness behavior has been reported in other compositionally modulated films [7].

## CONCLUSIONS

Compositionally modulated Au-Ni films with periods of 1.2 - 4.6 nm were examined by TEM and tested for modulus and hardness. The short-period films exhibited a highly textured fcc structure with a lattice parameter that was midway between gold and nickel. Thus, despite the distinct composition modulation, the lattice spacings in these films were simlilar to those expected for solid solutions in accordance with Vegard's law. As the period was increased above 2 nm, a single strained structure could no longer be sustained, and a phase separation took place in which two additional fcc structures (bulk gold and bulk nickel) appeared. In films with longer periods (4.6 nm), these bulk structures were prevalent, although the strained structure was still present. The distribution of the different structures within the films could not be determined.

No modulus enhancement was detected by indentation measurements, and this effect may only be detectable by measurements which primarily sample the *in-plane* elastic properties of the film. Distinct structural differences were observed in the films, and these may result in substantial variations in elastic properties as a function of composition repeat distance.

## ACKNOWLEDGEMENTS

The authors are grateful for support from the Office of Naval Research, contract N00014-86-K-0125 (SRN), the Department of Energy contract W-7405-Eng-48 (AFJ), and initial funding from Lockheed Palo Alto Research Laboratory. The high-resolution microscopy was performed using the facilities at the Center for High Resolution Electron Microscopy, Center for Solid State Science, Arizona State University, supported by the National Science Foundation (Grant DMR-83-0651) and the State of Arizona.

## REFERENCES

1. W. M. C. Yang, T. Tsakalakos, and J. E. Hilliard, J. Appl. Phys. 48, 876 (1977).
2. T. Tsakalakos and A. F. Jankowski, Ann. Rev. Mater. Sci. 16, 293 (1986).
3. R. C. Cammarata, Scripta Metall. 20, 479 (1986).
4. R. C. Cammarata, Mat. Res. Soc. Symp. Proc. Vol. 103, 315 (1988).
5. A. F. Jankowski, D. M. Makowiecki, M. A. McKernan, S. R. Nutt, and K. A. Green, Mat. Res. Soc. Symp. Proc. Vol. 103, 327 (1988).
6. J. Bravman and R. Sinclair, J. Elec. Micr. Tech. 1, 53 (1984).
7. M. F. Doerner, PhD. Dissertation, Stanford University (1987).

PART II

# Heteroepitaxy

# MECHANICS OF ELASTIC DISLOCATIONS IN STRAINED LAYER STRUCTURES

L. B. FREUND, A. BOWER and J. C. RAMIREZ
Division of Engineering, Brown University, Providence, RI 02912

## ABSTRACT

Application of the elastic continuum theory of dislocations to modeling of phenomena associated with elastic strain relaxation in strained layer epitaxial heterostructures is discussed. The concept of critical thickness for onset of strain relaxation in a strained epitaxial layer is first reviewed, and some extensions to periodic arrays of dislocations and to multiple layers are described. Then, two issues are addressed that arise when the assumptions underlying the critical thickness concept are not met. One issue concerns the nucleation of dislocations at the growth surface of an epitaxial film, particularly the influence of surface irregularities on the activation energy for surface nucleation. A second issue concerns the kinetics of glide of a threading dislocation as it lays down an interface misfit dislocation when the layer thickness exceeds the critical thickness. A generalized driving force for the glide process is defined, and a relationship between this force and the glide speed is proposed.

## INTRODUCTION

It is generally accepted that the elastic continuum theory of dislocations provides a useful framework for the study of strain relaxation in strained layer heterostructures of interest in microelectronics and micro-optics. The situation is somewhat paradoxical, in the sense that the existence of a dislocation derives directly from the discrete nature of a crystalline lattice and that the physical processes of strain relaxation to be modeled are inherently non-elastic. Fortunately, systems of scientific interest and of practical importance have a number of features that permit a convergence of the two views. For one thing, the dislocation densities in the systems of interest are sufficiently low so that the development of models involving isolated dislocations, or interactions of small numbers of dislocations, is justifiable. In addition, the deformation fields of lattice dislocations decay quite slowly with distance from a dislocation line on the scale of lattice dimensions. Finally, the mechanical forces acting on dislocations, and thus controlling their configurations, are representable in terms of global energy variations that are not particularly sensitive to the character of the lattice.

While the case for use of continuum theories of dislocations is quite strong, a major shortcoming is that the details of the structure of the dislocation core are completely overlooked in the continuum approach. The unbounded elastic energy associated with the core of an ideal elastic dislocation is an awkward feature of the continuum models. It is addressed, typically, by assuming the existence of a cutoff radius. Roughly, this radial distance from the dislocation line is determined either as the smallest dimension for which the linear theory of elasticity makes sense or as a radial dimension outside of which the energy of the dislocation based on linear elasticity equals the total energy of the dislocation as calculated on the basis of some detailed discrete model.

In this article, several problems are discussed on the basis of the simplest models of elastic dislocations. Although the materials under consideration are single crystals, the effects of elastic anisotropy are neglected. Furthermore, the differences in mechanical properties between substrates and layers are neglected; common properties are assumed in all cases. While these effects may be quantitatively significant in some cases, they are qualitatively unimportant in most cases.

Fig. 1. Schematic representation of a strained layer on a substrate with a threading dislocation.

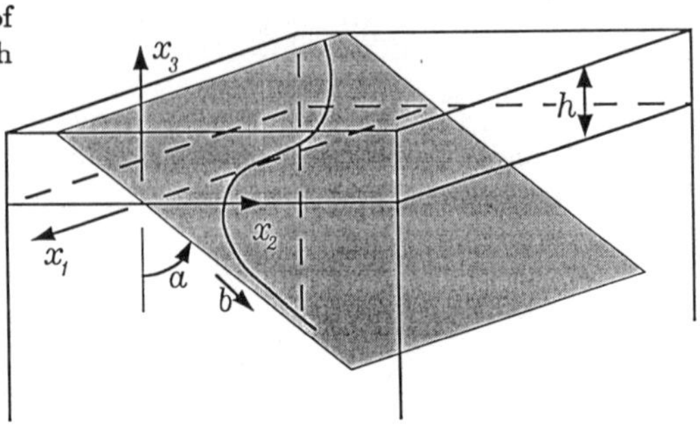


## CRITICAL THICKNESS

Consider a thin, uniform layer that is grown epitaxially on a substrate with a like crystal structure but a different lattice spacing. If the mismatch in lattice spacing of the unstrained crystals is not too great, the layer assumes a uniform strain to eliminate the lattice mismatch. The state of stress in a uniform film is then a spatially uniform membrane tension or compression. Consequently, the stress equilibrium equations are satisfied identically, and there is no traction on the plane interface of the layer and substrate.

If the layer is grown pseudomorphically on the substrate, then any crystalline defects in the substrate are preserved in some way in the layer. The simplest defect, and perhaps the critical defect for purposes of a discussion of critical thickness, is a dislocation whose line penetrates the surface of the substrate and continues through the layer as a threading dislocation; see Fig. 1. If the traction vector on the glide plane due to the strain in the layer has a component in the direction of the dislocation Burgers vector, however, the dislocation will tend to bow out laterally in a direction corresponding to relief of layer stress. When the layer is relatively thin, the bowing is limited because the highly curved configuration of the dislocation line required for glide is energetically unfavorable and the proximity of the free surface would induce an image force to pull the dislocation out of the crystal. As the layer becomes relatively thicker, the curvature constraint becomes less severe and the free surface becomes more distant. Glide within the layer first becomes energetically possible at the critical thickness. Thereafter, as the threading dislocation glides along its slip plane, it leaves behind an expanding segment of interface misfit dislocation. The term *critical thickness* is reserved for the *lowest* layer thickness at which an interface misfit dislocation *may be formed* through glide of a single isolated threading dislocation.

The critical thickness concept is due to Matthews et al[1]. A particularly simple and intuitive way to obtain an estimate is outlined here. The development is based on the fundamental concept of interaction energy of a stressed body containing a dislocation. The interaction energy is defined as the net work done in forming the dislocation in the stressed body. If $\sigma_{ij}^L$ is the equilibrium stress field due to the applied loading and $\sigma_{ij}^D$ is the equilibrium stress field due to the dislocation itself in the absence of any applied loading, the interaction energy is

$$W(S) = \tfrac{1}{2}\int_S b_i n_j \sigma_{ij}^D \, dS + \int_S b_i n_j \sigma_{ij}^L \, dS \qquad (1)$$

where $S$ is the glide area with unit normal $n_i$ and $b_i$ is the Burgers displacement vector of the dislocation. The change in $W$ with any possible variation in $S$ is of particular significance in dislocation mechanics. A change in configuration of a dislocation may occur or not, depending on whether the associated change in the interaction energy is negative or positive, respectively. The key step in applying the concept is to recognize that the glide of the threading dislocation a unit distance down the glide plane, thereby creating a segment of misfit dislocation of unit length, is equivalent to removing a slab of material of unit thickness in the $x_1$–direction from the body far ahead of the threading dislocation, dislocating it under two dimensional conditions to create the misfit segment, and then inserting this slab at some distance behind the threading dislocation[2]. The approach is attractive because it includes the actual mechanism of misfit dislocation formation as a feature, the calculation needed to estimate critical thickness is based on two dimensional fields which can be analyzed exactly within the framework of elastic dislocation theory, and the calculation is independent of the unknown curved shape of the threading dislocation.

For the particular case shown in Fig. 1 of an isotropic layer with extensional strain $\epsilon_o$ (positive or negative) and a Burgers displacement of magnitude $b$ parallel to the $x_2, x_3$–plane, the work required to form a misfit dislocation in a slab of unit thickness at depth $h$ from the free surface on the glide plane inclined at angle $\alpha$ to the $x_3$–axis is[2]

$$w_{disl} = \frac{\mu b^2}{4\pi(1-\nu)} \left\{ \ln\left(\frac{2h}{r_o}\right) - \tfrac{1}{2}\cos 2\alpha - \frac{(1-2\nu)}{4(1-\nu)} \right\} \tag{2}$$

where $\mu$ and $\nu$ are the common shear modulus and Poisson's ratio, and $r_o$ is the dislocation core cutoff radius. This expression includes the work of the tractions on the small cylindrical surface created by artificially cutting out the core region. On the other hand, the work that is recovered from the elastic field of the strained layer in this process is

$$w_{layer} = -\frac{2(1+\nu)}{(1-\nu)} \mu \epsilon_o b h \sin\alpha\,. \tag{3}$$

If the net work per unit distance $w_{disl} + w_{layer}$ is positive then the threading dislocation will regress along the glide plane, whereas if it is negative then progressive glide is spontaneous. Thus, the condition $w_{disl} + w_{layer} = 0$ defines the critical thickness $h_c$. The relationship between $h_c$ and $\epsilon_o$ is

$$\frac{b}{8\pi(1+\nu)h_c \sin\alpha} \left\{ \ln\left(\frac{2h_c}{r_o}\right) - C \right\} = \epsilon_o \tag{4}$$

where $C = \frac{1}{2}\cos 2\alpha + (1-2\nu)/4(1-\nu)$. The result is shown graphically in Fig. 2 with $r_o = b/4$, where the result from Matthews et al[1] is shown for comparison. The problem may be analyzed in precisely the same way if the Burgers vector is not parallel to the $x_2, x_3$ plane. The result (4) is not new, of course, but the approach used to get it offers potential for further development. The critical thickness idea has been verified by observations on many systems, provided that dislocations are indeed present and that conditions for glide prevail. Concerning the assessment of observations on strain relaxation and implications for critical thickness, it has been pointed out by Gourley et al[3] that x-ray diffraction methods for measuring average lattice spacing are not usually sensitive enough to discern the onset of relaxation. Instead, direct observation of dislocations by photoluminescence or electron microscopy appears to be far more effective.

It is important to recall several tacit assumptions in the critical thickness concept. First, the existence of a threading dislocation is presumed; the idea is not applicable in cases in which dislocations must be nucleated before their glide can be considered. In addition, it is assumed that the dislocation is an ideal elastic dislocation; that is, the dislocation moves reversibly through the layer material and it can adjust its configuration instantaneously to minimize energy. On the other hand, the continuation of the dislocation into the substrate must be held back by some intrinsic drag. If the dislocation were not

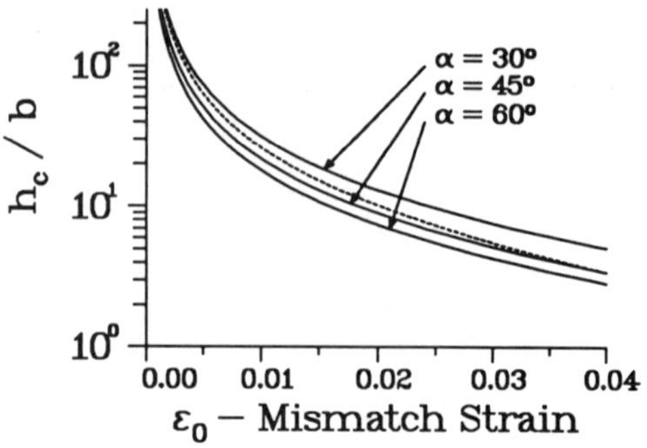


Fig. 2. Critical thickness vs. mismatch strain according to (4) for three values of angle $\alpha$. The dashed curve is the result from Matthews, Mader and Light[1] for $\alpha = 30°$ and the same value of cut-off radius.

restrained, it would simply move along to keep the entire dislocation straight and no interface misfit would be formed.

## RELAXATION DUE TO AN ARRAY OF MISFIT DISLOCATIONS

The matter of glide of threading dislocations in the layer to form interface misfit dislocations can be pursued a bit further in a simple way for distributions of parallel misfit dislocations. The formation of a misfit dislocation relaxes the stress in the strained layer in its vicinity. The formation of another misfit dislocation in the interface on the same glide system is more likely to occur on a parallel glide plane at some distance away where the stress is at a higher level than in the immediate neighborhood of the already formed misfit dislocation. Thus, interface misfit dislocations on a given slip system are expected to be distributed over the interface if threading dislocations or sources are distributed. Furthermore, it is expected that the density of interface misfit dislocations increases, or the spacing between them decreases, as the thickness of the layer increases.

If it is assumed that the spacing of parallel misfit dislocations on the interface is uniform, then the foregoing argument can be modified slightly to obtain a relationship between the spacing of the dislocations and the thickness of the layer beyond the critical thickness. In the present case, there is some nonuniform mechanical field associated with the array of interface dislocations that will determine whether or not a particular threading dislocation will glide. The influence of each misfit dislocation is not included individually but, instead, the effect of this mechanical field on a threading dislocation is represented in an average way. If $b_i$ is the component of the Burgers displacement that lies in the interface and perpendicular to the misfit dislocation line, then the average strain reduction in the layer due to an array of misfit dislocations with spacing $\ell$ is $b_i/\ell$. Here, $b_i = b\sin\alpha$. Therefore, if $h_c$ and $\epsilon_o$ in (4) are replaced by $h\,(\geq h_c)$ and $\epsilon_o - b\sin\alpha/\ell$, respectively, then a relationship between $\ell$ and $h$ is obtained for given $\epsilon_o$ and $b$. This relationship provides an estimate of the misfit dislocation spacing in the interface as a function of thickness beyond the critical thickness, assuming that glide dislocations are available to form the misfit dislocations. The relationship is shown graphically in Fig. 3 in the form of $\ell/b$ versus $h/h_c$ where the influence of initial layer strain $\epsilon_o$ is represented through the parameter $h_c/b$ from Fig. 2. The density of dislocation lines at the interface of an $In_{0.2}Ga_{0.8}As/GaAs$ system was determined recently by photoluminescence microscopy by Gourley et al[3]. The dependence of dislocation density or spacing on thickness beyond critical thickness had the form of the result in Fig. 3 . However, the actual density reported was less by two orders of magnitude than the prediction of the simple model. Data on the spacing of interface misfit dislocations versus layer thickness were also reported by Petruzzello et al[4] for growth of the $ZnSe/GaAs(001)$ system. They found that the onset of relaxation occurred for a thickness of about twice the theoretical critical thickness, but that thereafter the spacing

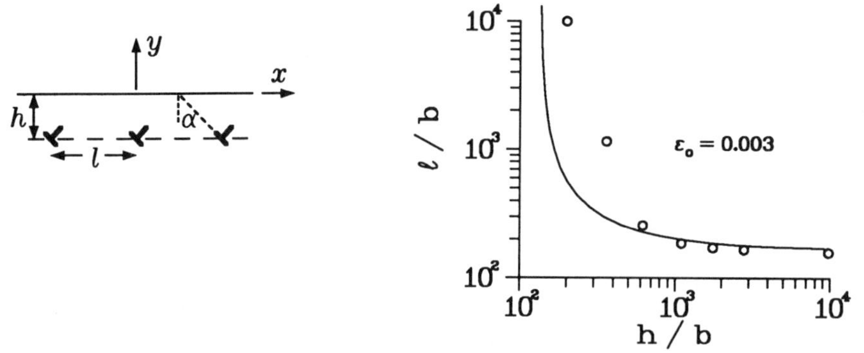


Fig. 3. Dislocation spacing vs. thickness beyond critical thickness for a regular array of evenly spaced dislocations. The data points are for the $ZnSe/GaAs(001)$ system as reported by Petruzzello et al[4].

versus thickness variation agreed quite well with the simple picture outlined here.

The present discussion is based on only the average strain relieved by a regular array of dislocations. It should be noted that the actual nonuniform strain distribution due to a periodic array of dislocations of the same orientation and burgers displacement can be obtained. The details are not included here but the results are summarized in the Appendix.

## MULTIPLE STRAINED LAYERS

The stability of defects in multiple strained layers with parallel interfaces may be pursued in the same way. The details are more complicated because the number of interfaces at which misfit dislocations can form is greater. The matter of critical thickness has been pursued a bit further for the case of an elastic strained layer between two elastic half spaces by actually calculating the equilibrium shape of a dislocation passing from one half space to the other, threading the intervening layer. The approach used was to approximate the generally curved dislocation line by many straight segments so the configuration could be specified by the finite number of vertex positions for the piecewise linear approximation. The equilibrium position was then determined by minimizing the total energy of the set of segments with respect to variations in the vertex positions, accounting for interactions among all segments. A representative equilibrium shape is shown in Fig. 4, where it must be noted that the detailed shape in the unstrained material scales with an assumed stress magnitude representing the intrinsic lattice resistance of the unstrained material to dislocation motion. The critical thickness for a given mismatch strain in the layer was defined in this case as that thickness beyond which a stable equilibrium position could not be found. The results agreed very well with the predictions of the elementary approach outlined above, that is,

$$\frac{b}{4\pi(1+\nu)h_c \sin\alpha}\left\{\ln\left(\frac{h_c}{r_o}\right) - C\right\} = \epsilon_o \tag{5}$$

or with the original estimate of Matthews and Blakeslee[5] for this configuration. For most practical purposes, the critical thickness of an embedded strained layer is about twice the critical thickness for a single layer with a free surface, other parameters being equal.

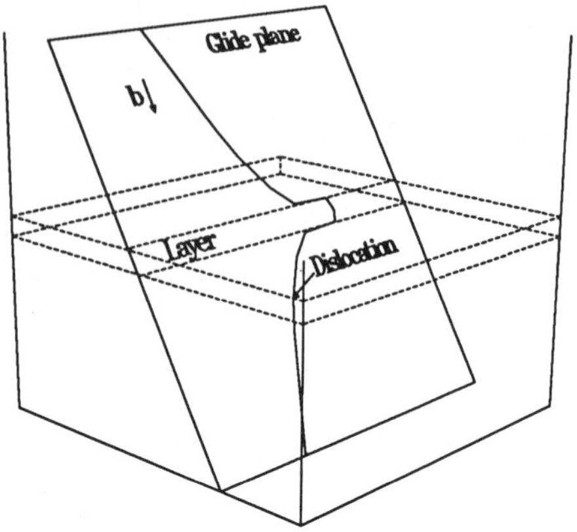


Fig. 4. A computed equilibrium configuration of a dislocation threading a strained layer between two unstrained half spaces.

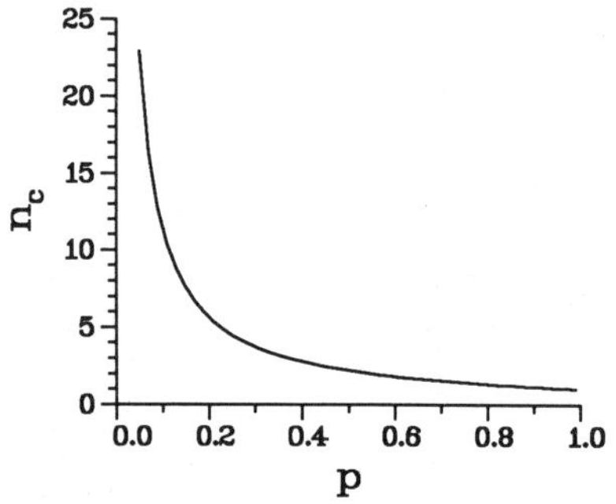


Fig. 5. Number of layers for instability vs. $p$ according to (7).

Critical thickness estimates for a strained layer superlattice, that is, periodic multiple strained layers, bonded to a substrate have been discussed to some extent[6]. As a simple illustration, consider a $BABAB\ldots$ superlattice grown on material $A$ with layers of equal thickness. The first $B$ layer grown onto the $A$ substrate must be grown to a thickness less than its critical thickness $h_{Bc}$. The next $A$ layer grows without mismatch strain, and its presence tends to stabilize the layer of $B$ material. The next layer of $B$ material cannot exceed its critical thickness, and so on. Suppose that there are $n$ layers of material $B$, each with strain $\epsilon_o$ and thickness $h_B$, and that there are $n-1$ strain-free layers of material $A$ each with thickness $h_A$. The direct application of the work argument above leads to the conclusion that $n$ can be increased until its critical value $n_c$ satisfies

$$\frac{b}{8\pi(1+\nu)n_c h_B \sin\alpha}\left\{\ln\left(\frac{2(n_c-1)h_A+2n_c h_B}{r_o}\right)-C\right\}=\epsilon_o \tag{6}$$

whereupon the superlattice layer becomes unstable with respect to the substrate in the sense that interface misfit dislocations can be formed spontaneously. If $h_A = h_B = ph_{Bc}$, then (6) reduces to

$$\ln p(2n_c-1)=(n_c p-1)\left\{\ln\left(\frac{2h_{Bc}}{r_o}\right)-C\right\}. \tag{7}$$

Clearly, $n_c = 1$ is a solution of (7) if $p = 1$. A graph of $n_c$ versus $p$ is shown in Fig. 5. The solution of (7) is essentially $n_c = 1/p$ and is thus virtually independent of the value of $h_{Bc}/b$. Other specific examples can be worked out in much the same way.

## DISLOCATION LOOP NUCLEATION WITH SURFACE IRREGULARITIES

The critical thickness idea is based on the presumption that dislocations are available for glide when mechanical conditions in the strained layer become favorable for glide, but this is not always the case. Consequently, issues of dislocation nucleation and multiplication must be addressed in considering strain relaxation. A number of investigators have reported that the dislocations which ultimately contribute to the process of strain relaxation are formed by means of half-loop nucleation at the growth surface of the strained layer. The energetics of this nucleation process must therefore be examined to understand the conditions necessary for surface nucleation.

The essential features of thermally activated loop nucleation are discussed by Hirth and Lothe[7] and the underlying ideas can be stated in terms of the interaction energy (1). For simplicity, consider first the process of homogeneous nucleation of a circular loop of radius $R$ in a uniform applied stress field. The self-energy of the dislocation loop (the first term in the interaction energy expression) is positive and proportional to the circumference of the circle $2\pi R$, and it is only weakly dependent on the curvature of the dislocation line as $\ln(R/b)$. The work of the applied stress field (the second term in the interaction energy expression) is negative and proportional to the area of the circular loop $\pi R^2$. Thus, $W(R) \sim R\ln(R/b)$ when $R/b \to 1^+$ but $W(R) \sim -R^2$ as $R/b \to \infty$. Consequently, $W(R)$ has an absolute maximum for one intermediate value of $R$, say $R_{eq}$. The condition $W'(R_{eq}) = 0$ is satisfied at this equilibrium size. However, $W''(R_{eq}) < 0$ so that the equilibrium configuration is unstable, that is, the loop tends to grow spontaneously for $R > R_{eq}$ and to contract to vanishingly small size for $R < R_{eq}$. The equilibrium energy indicates the amount of energy that must be supplied from non-mechanical sources to nucleate a dislocation loop. In the terminology of the statistical mechanical theory of thermally activated nucleation processes, this energy is the activation energy for the process.

The activation energy for surface loop nucleation is considered here. This matter has been considered previously[8] by halving the various mechanical quantities appearing in the above picture of loop nucleation in an unbounded body. Hirth and Lothe[7] note that surface nucleation is more favorable than bulk nucleation because image effects at the free surface lower the apparent activation energy, the free energy of surface steps effectively can lower the activation energy, and partial dislocations can be nucleated from a surface in materials with low stacking fault energy.

If the growth surface of a strained epitaxial layer remains flat then it appears that the activation energy is quite high. It has been noted, however, that the growth surface often becomes wavy as the layer thickness increases. This introduces stress concentrations in the valleys of the wavy surface or, in fact, at any abrupt change in orientation of the surface. Because the stress in a valley may be significantly greater than the average stress associated with the nominal mismatch strain, the activation energy for nucleation of a dislocation loop there may be less than for nucleation from a flat surface with the same nominal mismatch strain. An estimate of the magnitude of this effect is given here for the particular case of dislocation nucleation at the root of a cylindrical notch of semi-circular cross section in the surface of the stressed layer.

The configuration considered is shown in Fig. 6, where the glide plane is shaded. The stresses in the material prior to formation of the dislocation have been calculated by means of the elasticity solution by Ling[9] for a semicircular notch in a two dimensional elastic solid subject to remote tension, where the surface of the notch and the otherwise plane surface are traction free. The stress field due to the rectangular $c \times d$ dislocation loop is then added, using the solution given by Comninou and Dundurs[10] for the displacement due to an angular dislocation in a half space. Since the latter solution does not take into

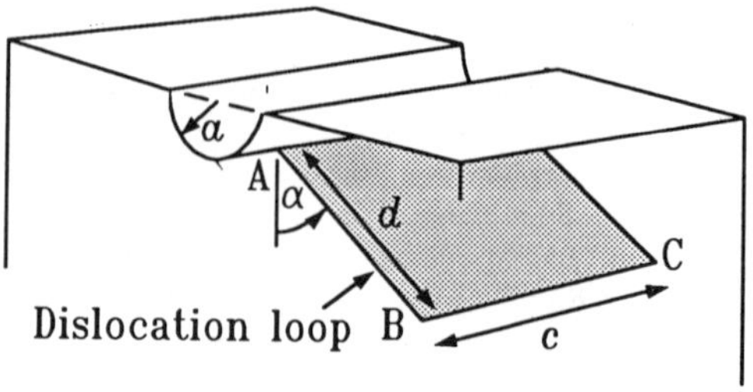


Fig. 6. Schematic representation of the formation of a rectangular dislocation loop (shown shaded) at the root of a surface irregularity.

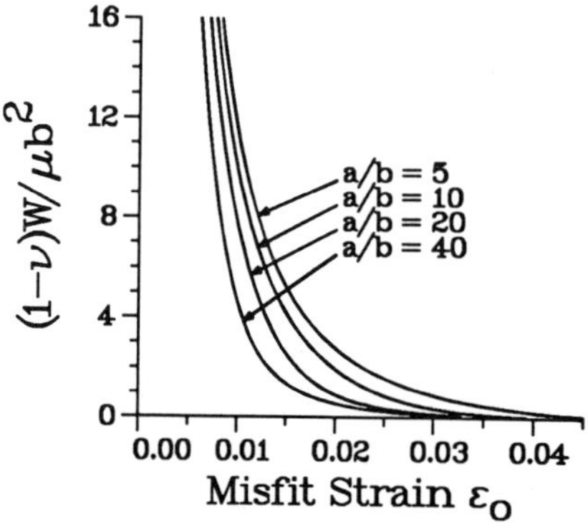


Fig. 7. Activation energy $W$ for surface loop nucleation vs. mismatch strain for four values of the notch radius.

account the effects of the notch, the stress due to the dislocation loop has been estimated by assuming that the free surface is tangential to the notch at its root. Calculations have been carried out for the following values of system parameters: loop aspect ratio $c/d = 2$, glide plane angle $\alpha = 30°$, dislocation core radius $r_o = b/4$, Burgers displacement parallel to the loop edge $AB$, Poisson's ratio $\nu = 0.33$.

The variation of the calculated interaction energy $W(d/b)$ with the size of the loop has the features anticipated in the discussion above. That is, the interaction energy increases with loop size $d/b$ when $d/b$ is small, it reaches a maximum corresponding to an unstable equilibrium configuration some value of $d/b$, and it then decreases with further increase of $d/b$. If the absolute maximum of the interaction energy is identified as the activation energy, then the model shows that the interaction energy depends on mismatch strain $\epsilon_o$ as shown in Fig. 7 for four values of the notch size $a/b$. The presence of the notch has some effect on the activation energy for loop formation. Qualitatively, if the loop can be expanded to its equilibrium size within the region of the stress concentration of the notch, then the presence of the notch has a strong effect on activation energy. However, if the equilibrium size of the loop is large compared to the size of the notch, then the presence of the notch has little effect on the process of loop nucleation.

The amount of energy equal to the activation energy must be supplied from a non-mechanical source for nucleation to be possible. This source is assumed to be the random thermal activity and, on the basis of statistical mechanical arguments, the energy supply is typically assumed to be some multiple of $kT$ where $k$ is Boltzman's constant. Estimates of the multiple factor have been given in the range from 50 to 100 for this process, but such estimates can be viewed as crude at best.

Fig. 8. Observed elastic strain following relaxation during growth vs. initial mismatch strain for $Ge_xSi_{1-x}/Si(001)$ for four values of final film thickness from Bean et al[11].

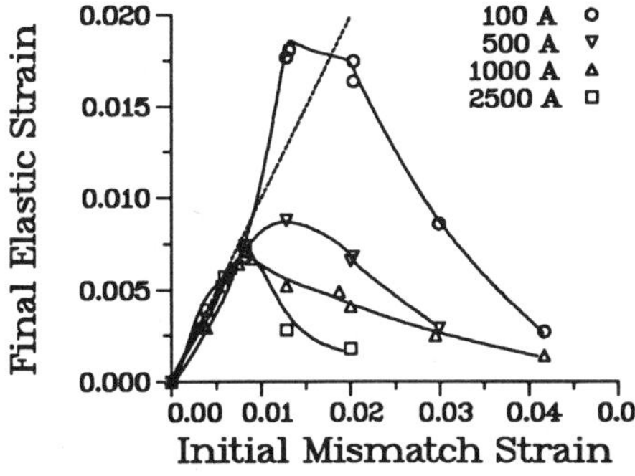


GENERALIZED FORCE FOR GLIDE

Strained layer structures can be fabricated at temperatures for which the dislocation mobility in semiconductor materials is extremely low and strain relaxation is a time dependent process. One of the most complete data sets in which these features are evident is that on the $Ge_xSi_{1-x}/Si(100)$ described by Bean et al[11]. They grew layers of the $GeSi$ alloy on $Si$ under carefully controlled conditions at a temperature of $550\,C$ for values of $x$ in the full range from zero to one, or initial mismatch strain from 0 to 4.2 %. The layers were grown to four different final film thicknesses over the full range of $x$, and the final relaxed elastic strains were measured by means of x-ray diffraction methods. The results are summarized in Fig. 8.

Whereas the final elastic strain for any thickness would be independent of the value of $x$ once interface coherence was lost, as outlined above in the discussion of strain relaxation by arrays of dislocations, the data show a relatively strong dependence. This dependence is apparently due to the absence of numerous threading dislocations initially and to the low mobility of the dislocations that are present. Thus, under realizable growth conditions, the presumption that dislocations can move reversibly through the material to take on minimum energy configurations may not be valid. It then becomes necessary to view dislocation motion as a kinetic process in modeling elastic strain relaxation under conditions that prevail during growth. Direct observations of dislocation motion in $GeSi$ films at various temperatures have been reported by Hull et al[12], and these observations provide a basis for development of kinetic models.

Assume that a strained layer growing on a substrate contains a threading dislocation. When the layer thickness reaches the critical thickness, the dislocation becomes unstable in the sense of elastic equilibrium theory. The threading dislocation then has the potential for laying down an interface misfit dislocation by means of glide. The question posed here is the following: What is the kinetic driving force for the formation of an interface misfit dislocation, and what is the rate of misfit dislocation formation?

First, a definition of the generalized force representing the driving force for motion of the dislocation along its glide plane in the strained layer is developed. The definition follows naturally from a work balance for the process based on the elastic continuum theory of dislocations.

Consider the situation represented schematically in Fig. 9 which is the glide plane of the dislocation when the layer thickness $h$ has exceeded the critical thickness $h_c$. The threading dislocation $AB$ is gliding along its slip plane, leaving behind an expanding segment of interface misfit dislocation $BC$. Suppose that there is an intrinsic resistance to this glide in the crystal lattice. If the threading dislocation moves in a self-similar way along the layer an infinitesimal distance $\delta\lambda$ then the work done on the gliding core region

Fig. 9. Schematic diagram of the glide plane showing a slightly bowed threading dislocation for $h < h_c$ and a gliding dislocation for $h > h_c$ with glide rate $\dot{\lambda}$. The glide segment $AB$ leaves behind the interface misfit dislocation segment $BC$.

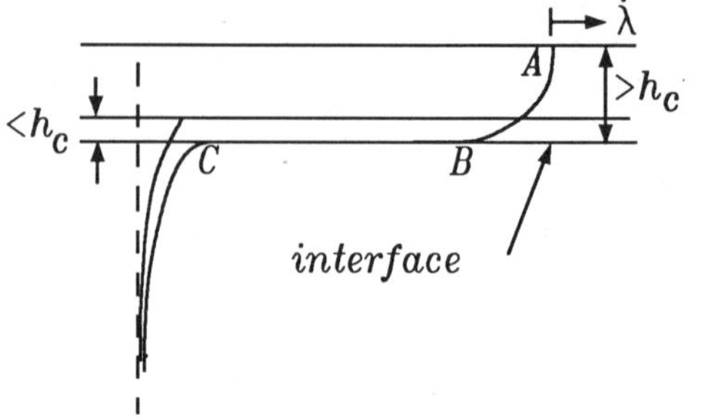


has the form $G\delta\lambda$ where $G$ is the *generalized force* that is work–conjugate to the kinematic parameter $\lambda$. Because $\lambda$ has dimensions of length, $G$ has the physical dimensions of force. An expression for $G$ in terms of system parameters is derived on the basis of the interaction energy argument introduced above.

In the present case, the change in the elastic field corresponding to glide of the threading dislocation a small distance $\delta\lambda$ along the glide plane is equivalent to removing a slab of material of thickness $\delta\lambda$ parallel to the $x_1, x_3$–plane in Fig. 1 far ahead of the gliding part of the dislocation $AB$, introducing a *two dimensional* dislocation into the slab, and then inserting the dislocated slab *behind* the gliding dislocation to extend the interface misfit dislocation $BC$ a length $\delta\lambda$. In carrying out this process of introducing the dislocation in the slab, some mechanical work must be done. The work done to displace one side of the glide plane of the two dimensional dislocation a distance $b$ with respect to the other side is $\delta W_{disl} = w_{disl}\delta\lambda$ where the work per unit distance $w_{disl}$ is defined in (2).

In addition, a traction acts on the glide plane due to mismatch elastic strain in the layer, and this traction does *negative* work $\delta W_{layer} = w_{layer}\delta\lambda$ as the dislocation is formed, where the work per unit distance $w_{layer}$ is defined in (3). The strain $\epsilon_o$ appearing in these expressions is commonly viewed as the initial mismatch strain in the layer. However, it may also be viewed as the average *elastic strain* remaining at some stage of strain relaxation.

The balance

$$\delta W_{layer} + \delta W_{disl} + G\delta\lambda = 0 \tag{8}$$

states that the work recovered from the elastic free energy in the layer must be adequate to form the misfit dislocation *and to sustain glide*. Then $\delta\lambda$ appears as a common factor and (8) reduces to

$$G = \frac{2(1+\nu)}{(1-\nu)}\epsilon_o bh \sin\alpha - \frac{b^2\mu}{4\pi(1-\nu)}\left\{\ln\left(\frac{2h}{r_o}\right) - C\right\}. \tag{9}$$

This expression defines the generalized driving force for glide of the threading dislocations for instantaneous thickness $h$ and elastic strain $\epsilon_o$. It is evident from (8) that the generalized force is related to the interaction energy by $G = -\partial W/\partial\lambda$ for self–similar growth. It is noted that the generalized force derived above on the basis of overall energy variations is similar in form to the excess stress parameter introduced on heuristic grounds by Dodson and Tsao[13] and Tsao and Dodson[14] in discussing the net plastic strain relaxation in strained layer heterostructures.

According to Alexander[15], virtually all observations of dislocation motion in covalent crystals lead to the conclusion that the process of glide is a thermally activated process

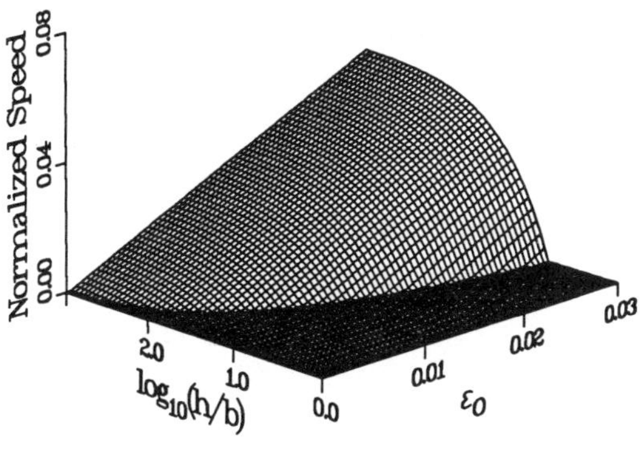


Fig. 10. Surface of normalized speed $(\dot{\lambda}/v_o)\exp(Q/kT)$ over the $h$–$\epsilon_o$ plane according to (11) with $r_o = b/4$ and $\alpha = 30°$.

and that the glide velocity $v$ is given by a rate law of the form

$$v = v_o\,(\tau/\mu)^m\ e^{-Q/kT} \tag{10}$$

where $v_o$ is a material constant, $\tau$ is the resolved shear stress on the glide plane, $m$ is a constant of order unity, $Q$ is an activation energy, $k$ is Boltzman's constant, and $T$ is absolute temperature.

A rate law for glide of a threading dislocation, corresponding to (10), should have the feature that $\dot{\lambda} \sim G^m$ and it should reduce to (10) as $h \to \infty$. A kinetic equation relating $\dot{\lambda}$ and $G$ having these properties is

$$\dot{\lambda} = v_o\left(\frac{G}{\mu bh}\right)^m e^{-Q/kT}\,. \tag{11}$$

The surface of $(\dot{\lambda}/v_o)\exp(Q/kT)$ over the $h$–$\epsilon_o$ plane is shown in Fig. 10 for $m = 1$ to illustrate the thickness dependence of speed. The kink in this surface delineates the region in parameter space in which the dislocation can move according to the kinetic law from the region in which it cannot move. Indeed, the curve in the $h$–$\epsilon_o$ plane defined by this kink is the common relationship between critical thickness and elastic strain.

Alexander[15] has summarized results of measurements on $Si$ and $Ge$, both doped and undoped, over the temperature ranges of 400-900 $C$ and 300-700 $C$, respectively. The activation energy for dislocations on the principal glide plane of undoped $Si$ is slightly stress dependent, with values ranging from about 2.25 $eV$ for a resolved shear stress of 10 $MPa$ to about 2 $eV$ for a resolved shear stress of 500 $MPa$. The values when an electrically active dopant is present are somewhat less. The activation energy results for undoped $Ge$ vary from about 2.2 $eV$ to 1.45 $eV$ as stress varies from 10 to 150 $MPa$.

For the particular case of glide of a 60° dislocation in $Si$ at 710 $C$ and a resolved shear stress of 9.8 $MPa$, Alexander[15] gives $Q = 2.25\,eV$ and $v = 0.37\,\mu m/s$. For $m = 1$, these parameters imply that $v_o = 8.7 \times 10^2 \mu m/s$. The in situ observations reported by Hull et al[12] were made on a layer with $x = 0.3$ so that $\epsilon_o = 0.0126$, at least initially, and a thickness of $h/b \sim 80$. With these values of the parameters involved, $G/\mu bh = 0.015$. For an activation energy of 2 $eV$ (essentially, the lowest value reported for $Si$ at high stress levels) and temperature of 545 $C$, (11) implies a glide speed of $\dot{\lambda} \sim 5.1\,\mu m/s$. The

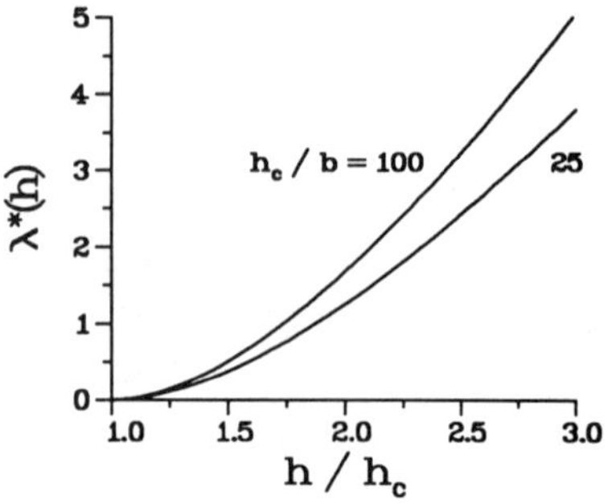


Fig. 11. Normalized glide distance $\lambda^*(h)$ vs. $h/h_c$ according to (12) for two values of critical thickness corresponding to mismatch strain of about 0.4% and 1.3%.

speed observed by Hull et al[12] under these conditions is about $5\,\mu m/s$. The fact that the estimate is in qualitative agreement with the observation provides encouragement for further development of the line of reasoning. The seemingly excellent quantitative agreement should be viewed as fortuitous, however.

The kinetic relationship (11) can be interpreted in other ways. For example, suppose that the process of growth of a strained layer at constant rate and at a fixed temperature is considered. Then $\lambda$ depends on time only through the time dependence of $h$. Thus, $h$ can be viewed as a time-like parameter and $\dot{\lambda} = \dot{h}\, d\lambda/dh$ where $\dot{h}$ is a constant. The relation (11) then becomes a differential equation for glide distance as a function of growth thickness $h$. The initial condition is that $\lambda = 0$ until $h = h_c$, where $h_c$ is the critical thickness defined by $G = 0$ for a given $\epsilon_o$. The differential equation is readily integrated to yield

$$\lambda^*(h) = \left(\frac{h}{h_c} - 1\right) \ln\left(\frac{8h_c}{b}\right) - C\left(\frac{h}{h_c} - 1\right) - \tfrac{1}{2}\ln\left(\frac{h}{h_c}\right)\ln\left(\frac{64h_c h}{b^2}\right) + C\ln\left(\frac{h}{h_c}\right) \quad (12)$$

where $\lambda^*(h)$ is the nondimensional glide distance $4\pi(1-\nu)\lambda(h)\dot{h}/bv_o \exp(-Q/kT)$. Plots of $\lambda^*(h)$ versus $h/h_c$ are shown in Fig. 11 for $h_c/b = 25$ and 100. If the material properties $Q = 2\,eV$ and $v_o = 8.7 \times 10^2\,\mu m/s$ are assumed then the ordinate unit is $\sim 10\,\mu m$ if $T \sim 500\,C$ and is $\sim 10\,mm$ if $T \sim 800\,C$, roughly, if $\dot{h} = 4\,\mu m/s$. Other similar applications of the kinetic relation may also be illustrated.

## CRYSTAL PLASTICITY

The kinetic relation (11) provides the basis for consideration of strain relaxation from the continuum plasticity point of view. Consider a volume of material $V$ with bounding surface $A_o$ and outward unit normal $\overline{n}_i$ subject to a *uniform* surface traction $\overline{\sigma}^*_{ij}\overline{n}_j$. Suppose that the response of the material is linear elastic except for microscopic slip (dislocation glide) producing a displacement discontinuity $b_i$ across well-defined internal planes of area $A_{slip}$ and normal $n_i$; see Fig. 12. Assume that this slip gives rise to a *uniform* macroscopic plastic strain $\overline{\epsilon}^p_{ij}$. The connection between $\overline{\epsilon}^p_{ij}$ and the slip process has been established by means of the principle of virtual work by Rice[16]. If $\sigma^*_{ij}$ is the internal (possibly nonuniform) stress field in equilibrium with the uniform applied stress $\overline{\sigma}^*_{ij}$, then the balance of internal and external virtual work requires that

$$\overline{\sigma}^*_{ij}\overline{\epsilon}^p_{ij} V = \int_{A_{slip}} \sigma^*_{ij} n_j b_i \, dA \quad (13)$$

where the equilibrium and the kinematic fields are independent of each other.

Fig. 12. A macroscopic material volume subject to uniform surface traction $\overline{\sigma}^*_{ij}\overline{n}_j$. The material is elastic except for slip on certain planes which corresponds to a macroscopic plastic strain.

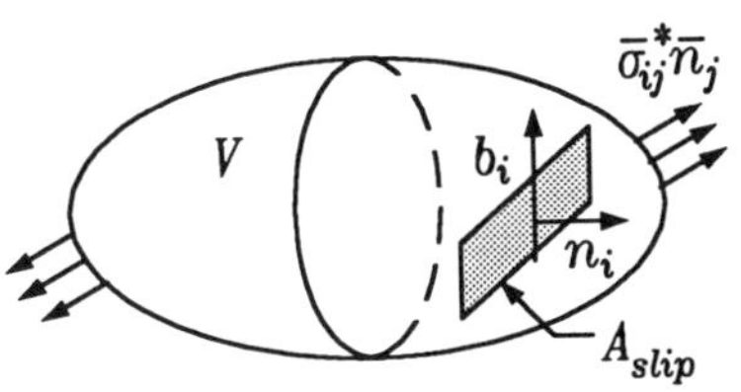


If the material is elastically homogeneous then $\sigma^*_{ij}$ is uniform and equal to $\overline{\sigma}^*_{ij}$, so that

$$\overline{\epsilon}^p_{ij} = \frac{1}{V}\int_{A_{slip}} \tfrac{1}{2}(n_i b_j + n_j b_i)\, dA \tag{14}$$

where the form of (14) is dictated by the fact that the right side of (13) depends only on the symmetric part of $b_i n_j$ due to symmetry of the stress tensor.

For each slip system, $n_i$ and $b_i$ are constant but $A_{slip}$ changes with time. Thus, time differentiation of (14) yields

$$\dot{\overline{\epsilon}}^p_{ij} = \frac{1}{2V}(n_i b_j + b_i n_j)\dot{A}_{slip} \tag{15}$$

where the right side should include one term of the form shown for *each* slip system. For glide of the type shown in Fig. 12, $n_i = (0, \cos\alpha, \sin\alpha)$, $b_i = (0, -\sin\alpha, \cos\alpha)$ and $\dot{A}_{slip} = h\dot{\lambda}/\cos\alpha$. If there are $N$ *mobile* dislocations in the volume $V$ then (15) implies that

$$\dot{\overline{\epsilon}}^p_{22} = NV^{-1}bh\dot{\lambda}\sin\alpha = \rho_m b\dot{\lambda} \tag{16}$$

where $\rho_m$ is proportional to the number of mobile dislocations per unit area of interface. Again, it is noted that each glide system contributes a term of the form appearing in (16). The dependence of $\dot{\lambda}$ on $h$, $T$ and macroscopic plastic strain through $\overline{\epsilon}^e_{22} = \epsilon_o - \overline{\epsilon}^p_{22}$ is given by (11). A corresponding physically based evolution equation for $\rho_m$ in terms of the macroscopic variables is not yet at hand. Some implications of (16) have been discussed by Dodson and Tsao[13].

ACKNOWLEDGMENTS

This work has been supported by an IBM Research Award in Materials Science, by the Office of Naval Research, through contract N00014-87-K-0481 to Brown University, and by the National Science Foundation, through grant DMR87-14665 to Brown University. This support is gratefully acknowledged.

REFERENCES

1. J. W. Matthews, S. Mader and T. B. Light, *J. Appl. Phys.* **41**, 3800 (1970).
2. L. B. Freund, *J. Appl. Mech.* **54**, 553 (1987).
3. P. L. Gourley, I. J. Fritz and L. R. Dawson, *Appl. Phys. Lett.* **52**, 377 (1988).
4. J. Petruzzello, B. L. Greenberg, D. A. Cammack and R. Dalby, *J. Appl. Phys.* **63**, 2299 (1988).
5. J. W. Matthews and A. E. Blakeslee, *J. Crystal Growth* **27**, 118 (1974).

6. R. Hull, J. C. Bean, F. Cerdeira, A. T. Fiory and J. M. Gibson, *Appl. Phys. Lett.* **48**, 56 (1986).
7. J. P. Hirth and J. Lothe *Theory of Dislocations*, (McGraw-Hill, 1968).
8. J. W. Matthews, in *Epitaxial Growth, Part B*, edited by J. W. Matthews (Academic Press, 1975), p. 560.
9. C.-B. Ling, *J. Math. Phys.* **26**, 284 (1947).
10. M. Comninou and J. Dundurs, *J. Elasticity* **5**, 203 (1976).
11. J. C. Bean, L. C. Feldman, A. T. Fiory, S. Nakahara and I. K. Robinson, *J. Vac. Sci. Tech.* **A2**, 936 (1984).
12. R. Hull, J. C. Bean, D. J. Werder and R. E. Leibenguth, *Appl. Phys. Lett.* **52**, 1605 (1988).
13. B. W. Dodson and J. Y. Tsao, *Appl. Phys. Lett.* **51**, 1325 (1987).
14. J. Y. Tsao and B. W. Dodson, *Appl. Phys. Lett.* **53**, 848 (1988).
15. H. Alexander, in *Dislocations in Solids*, edited by F. R. N. Nabarro (Elsevier, New York, 1986), Vol. 7, p. 113.
16. J. R. Rice, *J. Appl. Mech.* **37**, 728 (1970).
17. N. I. Muskhelishvili, *Some Basic Problems in the Mathematical Theory of Elasticity*, translated by J. R. M. Radok (Noordhoff, Groningen, 1953).

## APPENDIX: STRESSES DUE TO A DISLOCATION ARRAY

We consider the stresses associated with a regular array of dislocations, spacing $\ell$, at a depth $h$ below the surface of a half space, as illustrated in Fig. 3. The stresses are most conveniently expressed in terms of complex potentials, following the complex variable formulation of Muskhelishvili[17], so that

$$\sigma_{yy} + \sigma_{xx} = 2\left\{\Phi(z) + \overline{\Phi(z)}\right\}, \quad \sigma_{yy} - \sigma_{xx} - 2i\tau_{xy} = 2\left\{z\overline{\Phi'(z)} + \overline{\Psi(z)}\right\}$$

The stresses at a point $z = x + iy$ due to the dislocation array can be expressed in terms of three potential functions

$$\Phi(z) = \frac{\beta}{\ell}\pi\cot\zeta$$

$$\Psi(z) = \frac{\bar{\beta}}{\ell}\pi\cot\zeta + \frac{\beta}{\ell}\left\{\frac{(z-2ih)}{\ell}\pi^2\mathrm{cosec}^2\pi\zeta - \pi\cot\zeta\right\}$$

$$\Theta(z) = \begin{cases} -\overline{\Phi(\bar{z})} - z\overline{\Phi'(\bar{z})} - \overline{\Psi(\bar{z})}, & \mathrm{Im}(z) < 0; \\ \Phi(z), & \mathrm{Im}(z) > 0. \end{cases}$$

where $\zeta = \pi(z - ih)/\ell$ and $\mathrm{Im}(z)$ denotes the imaginary part of $z$. The complex constant $\beta$ is related to the Burgers vector of the dislocation by $\beta = \mu(b_y - ib_x)/4\pi(1-\nu)$ where $\mu$ is the elastic shear modulus and $\nu$ is Poisson's ratio. The potential function $\Theta(z)$ has been expressed using a half plane continuation, so the stresses due to this potential are given by

$$\sigma_{yy} + \sigma_{xx} = 2\left\{\Theta(z) + \overline{\Theta(z)}\right\}, \quad \sigma_{yy} - i\tau_{xy} = \Theta(z) - \Theta(\bar{z}) + (z - \bar{z})\overline{\Theta'(z)}$$

# Strain relief mechanisms and nature of misfit dislocations in GaAs/Si heterostructures

S.Sharan and J. Narayan
Department of Materials Science and Engineering,
North Carolina State University,
Raleigh, N. C. 27695-7916

J. P. Salerno and J. C. C. Fan
Kopin Corp.
Taunton, Mass 02780

## Abstract

The nucleation and glide of misfit dislocations in GaAs/Si system is investigated using transmission electron microscopy. GaAs epilayers of different thicknesses were examined by electron microscopy (plan and cross-section) and the elastic strain remaining in the film has been related to the average spacing of the misfit dislocations at the interface. A model is developed based on minimum energy considerations to determine the strain-thickness relationship. The theoretical predictions of strain relaxation are compared with experimental observations using high resolution electron microscopy.

## Introduction

Epitaxial growth of GaAs on Si has been the subject of increasing interest since such a system provides the potential for utilizing the benefits of GaAs (high mobility and direct band gap/optical phenomena) in conjunction with Si for advanced integrated device structures. Various methods have been utilized to grow epitaxial GaAs layers on silicon substrates. These include vapor phase epitaxy (VPE) [1] on Ge coated Si substrates as well as direct growth by molecular beam epitaxy (MBE) [2] and organometallic chemical vapor deposition (OMCVD) [3] on Si substrates.

Despite several potential applications for GaAs on Si ranging from GaAs integrated circuits to monolithic integration of GaAs photonic and Si electronic devices, the quality of the epitaxial layer limits their use. The large misfit (4%) between GaAs and Si is responsible for a high density of misfit dislocations at the interface and a large number of emergent dislocations propagating through the epilayer. The misfit problem is further aggravated by the difference in thermal expansion coefficients of the two phases ($\alpha_{GaAs} = 6.8 \times 10^{-6}\ °C^{-1}$ ; $\alpha_{Si} = 2.56 \times 10^{-6}\ °C^{-1}$) which leads to large tensile stresses in the epilayer.

The lattice mismatch between a heteroepitaxial film and its substrate can be accommodated by elastic strain and/or misfit dislocations. Theoretical calculations based on energy balance between the growing coherent film and the film with misfit dislocations indicates that pseudomorphic growth takes place until a critical thickness is reached [4-9]. Misfit dislocations are formed to relax the lattice mismatch upon exceeding the critical thickness. For semiconductor films of diamond cubic and zinc-blende structures, experimental observations show a much smaller relaxation of the elastic strain as a function of epilayer thickness. The epilayer has residual elastic strain even at thicknesses much larger than the equilibrium critical thickness[10-12]. It is realized that dislocation nucleation and glide can be impeded by various processes. There is a large Pierels-Nabarro frictional stress in semiconductors, which can impede the glide of dislocations toward the interface. The good quality of semiconductor substrates and the perfection of growth methods preclude the formation of misfit dislocations from the glide of preexisting threading dislocations. Hence, it is envisaged that the misfit dislocations should nucleate at the free surface and glide to the interface.

Diamond and Zinc blende semiconductors contain perfect dislocations with Burgers vector of the type a/2<110>. Since pure edge (90°) dislocations are sessile, it is envisaged that 90° misfit dislocations are formed by the interaction of 60° glide dislocations. The development of dislocation substructure in GaAs/Si epilayers, as a function of the epilayer thickness can be understood as follows: nucleation of glissile 60° dislocations at the free surface above a critical thickness; glide and interaction of these dislocations on the {111} glide planes to the epilayer-substrate interface; and reaction of two 60° dislocations to form a 90° dislocation at or near the interface (a/2 [101] +a/2 [01$\bar{1}$]⟶a/2 [110]).

It has been shown in the $Ge_{1-x}Si_x$/Si system and boron doped Si (B)/Si systems [13,14] that misfit dislocation loops nucleate at the free surface. These half loops, which are predominantly unfaulted, glide and expand on inclined {111} slip planes toward the epilayer/substrate interface. The 60 ° dislocations reach the interface on each of the four inclined {111} glide planes and form an array of misfit dislocations lying along the [110] and [1$\bar{1}$0].

In this paper the nature of misfit dislocations and their spacing at the interface for the GaAs/Si system are identified. The residual elastic strain in the epilayer for differnt thicknesses is correlated with the dislocation spacing. The nature and number densities of dislocations are determined from high-resolution Transmission Electron Microscopy observations. The strain relaxation in the epilayer as a function of the epilayer thickness is modelled by considering the energy required to nucleate half loops from the free surface. The results of the analysis using this model are compared with the experimental results obtained for the GaAs/Si system.

## Experimental Results

The GaAs epilayers of different thicknesses used for the electron microscopic observations were grown by OMCVD, on tilted silicon substrates. The detailed procedure for growing the GaAs films has been described elsewhere [15]. TEM studies were performed on plan and (110) cross-section samples using a Hitachi H-800 electron microscope for analytical characterization and the JEOL 200-CX for atomic imaging. The JEOL 200-CX has a spherical aberration coefficient, $C_s$ = 1.4mm and a point to point resolution of 2.6 Å. The imaging was done with the electron beam along the [110] axis.

Figure 1 shows a high resolution electron micrograph of the interface between the GaAs/Si for a 1000Å GaAs epilayer. The misfit dislocations at the interface are indicated by the arrows. Majority of the misfit dislocations observed at the interface were of the 60° type (a/2<101> | <110>). The spacing between the dislocations is quite uniform and found to vary from 60Å to 65 Å. There are regions where two 60° dislocations are spaced very closely (Figure 1). These glide dislocations are thought to react to form 90° dislcoations. The only defects observed in the plan view consist of misfit and threading dislocations. There is a distinct change in the misfit dislocation type and spacing upon increasing the thickness of the epilayer. Figure 2 shows a high resolution electron micrograph from a 2000Å thick GaAs epilayer. The misfit dislocations at the interface are also very uniformly spaced but they are predominantly of 90° type. The

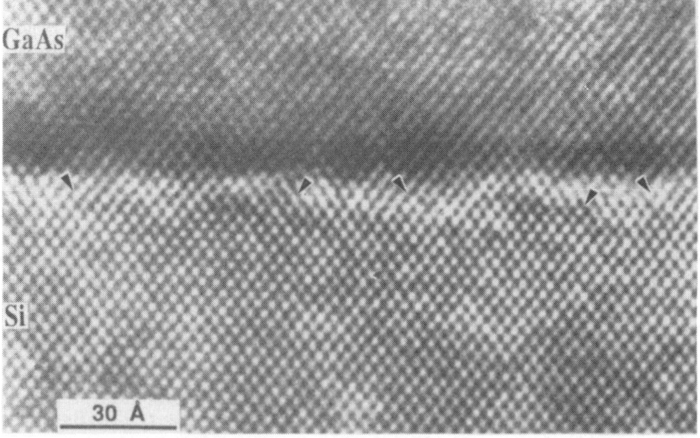


Figure 1. High resolution electron micrograph of the GaAs/Si interface for a 1000 Å thick GaAs epilayer. The dislocations are shown by arrows. All the dislocations are of the mixed (60°) type. The imaging direction is [110].

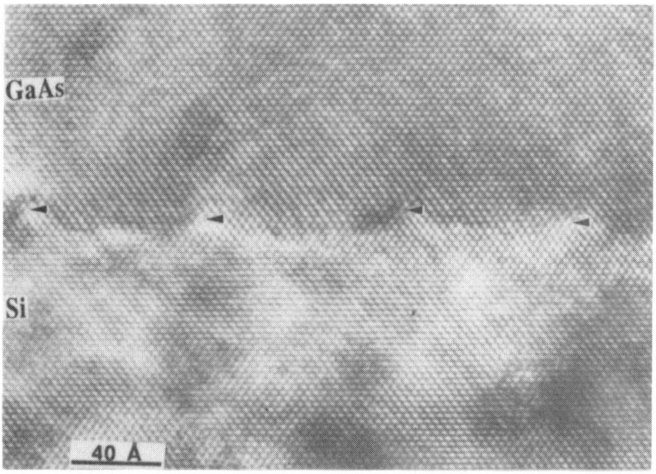


Figure 2. High resolution electron micrograph of the GaAs/Si interface for 2000 Å thick GaAs epilayer. All the dislocations shown are pure edge (90°) type

average spacing of the 90° dislocations is ~ 80 to 85 Å. Further increase in epilayer thickness to 3500 Å and then to 5000 Å, has not been responsible for much change in the dislocation type and spacing.

## Analysis of nucleation of misfit dislocations

In our analysis, we have considered the nucleation and glide of glissile half loops, with the 60° dislocation segment in <110> direction with Burgers vector a/2 <101>. The only nucleation sources considered in the analysis are those at the free surface, which require less activation energy than homogeneous nucleation inside the crystal. The activation energy for the creation of a half loop at the free surface is determined by calculating the energy of a half loop. The total energy of the misfit dislocation configuration consists of self energy of the half loop, work done in expanding the half loop and the interaction energy with the other half loop dislocations in the epilayer. The interaction energy with only the first nearest neighboring half loops is included in the energy of the configuration, because it decreases rapidly with increasing distance of separation.

The total energy of a semicircular glide half loop with radius r, is given by

$$E_{loop} = E_{self} - E_W \qquad (1)$$

where $E_{self}$ is the self energy of the half loop and $E_W$ thework done when the half loop expands in the presence of the resolved shear stress Γ. The self energy is given by [16]

$$E_{self} = \frac{(2 - \upsilon)\mu b^2}{8(1 - \upsilon)} r [\ln(\frac{\alpha r}{b}) - 2] \qquad (2)$$

where μ is the shear modulus, b the Burgers vector of the half loop, υ the Poisson's ratio and α is related to the dislocation core cut-off radius (the distance from the core center below which elastic theory does not apply), $r_0$. The value of α is taken to be 4 for covalent semiconductor structures. The work expended by the shear stress in the film is given by

$$E_W = (\pi r^2/2)\,\Gamma.b \qquad (3)$$

In equation (3), Γ is the resolved shear stress in the glide plane calculated from

$$\Gamma = \sigma \cos\phi \cos\lambda \qquad (4)$$

where $\sigma$ is the stress acting along [110] given by $2\mu \{(1+\upsilon)/(1-\upsilon)\}\varepsilon$, $\phi$ the angle between the slip plane normal and the stress axis and $\lambda$ the angle between the stress axis and the slip direction. The energy of the half loop reaches a maximum for a critical radius $r_c$, and decreases with further increase in the loop radius. The critical radius and the maximum energy for the nucleation of a half loop are obtained by setting $dE_{loop}/dr = 0$.

The interaction of a glide loop with its neighbor of the same radius and separated by a distance d, is given by [17]

$$E_{int} = \frac{\mu b^2}{2(1-\upsilon)} \frac{r_c}{k} [\{(\upsilon - 4)\,\mathbf{K} + \mathbf{E}\}k^2 + 2(3-\upsilon)\,(\mathbf{K}-\mathbf{E})] \qquad (5)$$

where

$$k^2 = \frac{4r^2}{d^2 + 4r^2}$$

and **K** and **E** are complete elliptic integrals of the first and second kind, respectively.
Thus, the total energy associated with the dislocation configuration at the critical radius of the loop; i.e. the activation energy is given by

$$E_{disln} = E_{loop}|_{r=r_c} + E_{int}|_{r=r_c} \qquad (6)$$

For $h \le h_c$ the film is fully coherent and the elastic strain in the film, $\varepsilon$, is given by $\varepsilon = f$, where f is the lattice mismatch and $f = (a_1 - a_2)/a_2$ where $a_1$ is the lattice parameter of the epialyer and $a_2$ that of the substrate. For $h \ge h_c$, the misfit dislocations at the interface relieve some of the lattice mismatch and the coherent energy remaining in the film is given by,

$$E^r_{coh} = \frac{2\mu(1+\upsilon)}{(1-\upsilon)} \varepsilon^2 h \qquad (7)$$

where $\varepsilon$, the elastic strain remaining in the film becomes $\varepsilon = f - \delta$; $\delta$ represents the strain which has been accommodated by the dislocations so that $\delta = b_m/d$; with $b_m$ the component of the Burgers vector of the dislocation that accommodates the lattice mismatch and d the average spacing between the misfit dislocations at the interface. For the case of a 60° mixed dislocation $b_m = b/2$ giving $d = b/[2(f-\varepsilon)]$

Thus, for a film thickness $h > h_c$, the areal energy density of the film is

$$E_{film} = E^A_{loop} + E^A_{int} + E^r_{coh} \qquad (8)$$

where $E^A_{loop}$ and $E^A_{int}$ are the areal energy densities, obtained by averaging the loop and interaction energy terms over an area covering a length $2r_c$ and width $d_{eq}$, and $d_{eq}$ is the spacing between misfit dislocations for full relaxation i.e $d_{eq} = b/2f$. The stable energy configuration of the system for each film thickness ( $h>h_c$) is determined by minimizing the film energy with respect to e, i.e.

$$\partial E_{film} / \partial\varepsilon = 0 \qquad (9)$$

The solution to eqn (9) provides the elastic strain $\varepsilon$ remaining in the film as a function of the film thickness, h.

## Discussion

The existence of a critical thickness for an epitaxial film on a given substrate has been predicted by various authors [4-9] . However, the equilibrium critical thickness models have not been able to account for the observations that epilayers with thicknesses greater then the equilibrium critical thickness can be grown coherent with the substrate. For a thickness of 1000 Å, a residual elastic strain of about 0.006 is present in the film . The dislocation spacing in the 1000 Å sample is larger than the equilibrium spacing, which indicates the presence of residual

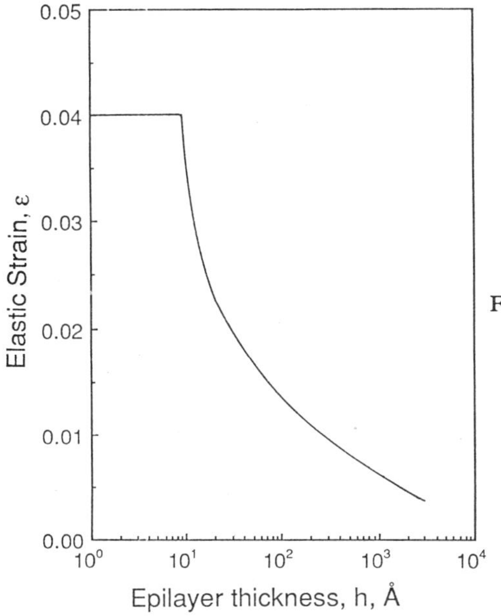


Figure 3. Plot of elastic strain remaining in the film as a function of the epilayer thickness for the GaAs/Si system (f =4 %).

elastic strain. Fig. 3. shows a plot of the elastic strain remaining in the film as a function of the epilayer thickness. In the initial stages of epitaxial growth, when the critical thickness has just been exceeded, the density of dislocations increases rapidly with increasing thickness since a large amount of strain energy is available for nucleating misfit dislocations. This continues until a substantial portion of the lattice mismatch has been relaxed. Most of the strain is relieved when the film reaches a thickness of 2000 Å. Further relaxation of strain beyond a thickness of 2000 Å occurs slowly. Dodson and Tsao [18] have suggested a phenomenological model for strain relaxation to match the experimental data for a $Si_{0.5}Ge_{0.5}$/Si heterostructure. A time dependent solution for the strain relaxation in the form of a single nonlinear ordinary differential equation which includes the glide velocity of the dislocation is suggested. However, for dislocation glide velocities much larger than the growth rates, the dislocations would reach the epilayer/substrate interface almost instantaneously. Thus, the velocity of dislocations should not play an important role in the strain relaxation. In the present analysis, we have used a minimum film energy criterion to obtain the strain relaxation as a function of the epilayer thickness and found a good agreement with experimental results. In our model, it is assumed that only misfit dislocations at the interface relieve the strain due to the lattice mismatch. The relief of the strain due to lattice mismatch by emergent dislocations is expected to be insignificant, since the inplane component of the Burgers vector of the emergent dislocation is very small.

In the 1000Å thick epilayer, the misfit strain has not been fully relieved, since the dislocation spacing is larger than the equilibrium dislocation spacing. For epilayer thicknesses greater than 2000Å , the misfit dislocations are spaced closer than the equilibrium dislocation spacing. It has been shown by X-ray diffraction studies [19] that epitaxial GaAs films less than 2500 Å thick were under a state of compressive stress and the stress state became tensile for thicknesses greater than 2500 Å. This has been explained on the basis of the fact that the larger thermal expansion coefficient of GaAs compared to that of Si, causes the GaAs film to be under a state of biaxial tension, when the heterostructre is cooled from the growth temperature to ambient conditions. The GaAs grows stress free at the growth temperature. Both GaAs and Si possses their bulk lattice parameters at this temperature and the interface is semicoherent. Upon cooling to the ambient, the GaAs epilayer contracts more than the Si substrate and comes under a tensile stress. Hence the stress relief in the GaAs film is a function of the growth temperature. Also, some of the misfit dislocations which are not situated exactly at the interface, do not contribute the entire Burgers vector to relieving the lattice mismatch.

## Conclusions

In the early stages of epitaxial growth, the epilayer is commensurate and the elastic strain in the layer takes the value of the lattice mismatch ($\varepsilon = f$). This regime of growth continues until the equilibrium value of the critical thickness is reached. Upon further increase in the thickness, dislocation sources at the free surface are activated and glissile 60° dislocations glide towards the interface to relieve the elastic strain in the epilayer. However, the strain relaxation continues as a function of the epilayer thickness, and full strain relief occurs only at thicknesses much greater than the equilibrium critical thickness. A model has been developed to calculate the elastic strain remaining in the film as a function of epilayer thickness, for 60° half loop nucleation at the free surface and expansion in the slip system. The most stable film configuration and, thereby, the $\varepsilon$-h relationship, were determined by minimization of the film energy. The theoretical predictions correlated well with experimental observations of GaAs/Si system. The strain in the film has been related to the average misfit dislocation spacing obtained by High resolution TEM for different GaAs epilayer thicknesses.

Part of this work was supported by NSF-ERC at N. C. State University.

## References

1. R. P. Gale, J. C. C. Fan, B-Y. Tsaur, G. W. Turner and F. M. Davis, IEEE Electron Dev. Lett., EDL-2 (7), 169 (1981).
2. B-Y Tsaur and G. M. Metze, Appl. Phys. Lett., 45 (5), 535 (1984).
3. M. Akiyama, Y. Kawarada and K. Kaminishi, Jpn. J. Appl. Phys., 23 (11), L843 (1984).
4. J. Narayan, S. Sharan, A. R. Srivatsa and A. S. Nandedkar, Mat. Sci. and Enggng., B1, 125 (1988).
5. J. W. Matthews in Dislocation in Solids, Ed. F. R. N. Nabarro , North-Holland, Amstersam, (1979).
6. R. People and J. C. Bean, Appl. Phys. Lett., 47, 322, (1985); R. People and J. C. Bean, Appl. Phys. Lett., 49, 229, (1986).
7. M. H. Grabow and G. H. Gilmer, Initial Stages of Epitaxial Growth, Eds. R. Hull, J. M. Gibson and D. A. Smith, MRS Proc., 94, 15, (1987).
8. B. W. Dodson and P. A. Taylor, Appl. Phys. Lett., 49, 642, (1986).
9. A. S. Nandedkar and J. Narayan, to be published
10. J. C. Bean, L. C. Feldman, A. T. Fiory, S. Nakahara and I. K. Robinson, J. Vac. Sci. Technol., A2, 436 (1984).
11. P. M. J. Maree, J. C. Barbour, J. F. van der Veen, K. L. Kavanagh, C. W. T. Bulle-Lieuwma and M. P. A. Viegers, J. Appl. Phys., 62, 4413 (1987).
12. J. Petruzello, B. L. Greenberg, D. A. Cammack and R. Dalby, J. Appl. Phys., 63, 2299 (1988).
13. E. P. Kvam, D. J. Eaglesham, D. M. Maher, C. J. Humphreys, J. C. Bean, G. S. Green and B. K. Tanner, Proc. Mater. Res. Soc., 104, 623 (1987).
14. J. Narayan, B. C. Larson and W. H. Christie, Laser - Solid Interactions and Laser Processing, Eds. S. D. Ferris, H. J. Leamy and J. M. Poate, MRS Boston, p. 440 (1978).
15. J. W. Lee, J. P. Salerno, R. P. Gale and J. C. C. Fan, Proc. Mater. Res. Soc., 91, 33 (1987).
16. J. P. Hirth and J. Lothe, Theory of Dislocations, (1972).
17. E. Kroner, Ergeb. agnew. Math., 5, 162 (1958).
18. B. W. Dodson and J. Y. Tsao, Appl. Phys. Lett., 51, 1325 (1987).
19. T. Yao, Y. Okada. H. Kawanami, S. Matsui, A. Imagawa and K. Ishida, Proc. Mater. Res. Soc., 91, 63 (1987).

# THE STABILITY OF $Si-Si_{1-x}Ge_x$ STRAINED LAYER HETEROSTRUCTURES.

D.C.HOUGHTON*‡, J-M.BARIBEAU*, K.SONG* and D.D.PEROVIC†.
*National Research Council of Canada, Ottawa, Ontario, K1A OR8, CANADA.
†University of Toronto, Toronto, Ontario, Canada, M5S 1A4.
‡British Telecom Research Labs, Martlesham Heath, Ipswich, IP5 7RE, UK.

## ABSTRACT

The structural stability of strained layer superlattices (SLS's) is addressed using an equilibrium model and then compared to the stability of single strained layers. Relaxation mechanisms are described for various superlattice geometries. The application of a critical thickness/strain criterion to define stability limits was found to be very useful in predicting the detailed relaxation process. The competition between relaxation by misfit accommodation at the base of the SLS and at individual strained interfaces is considered for the initial condition of full coherency and after partial relaxation. Experimental data for the Si-Ge strained layer system are presented; as-grown by MBE and after annealing in the temperature range 500°C - 900°C. The extent of relaxation and the detailed dislocation structure within the SLS's were determined by X-ray rocking curve analysis, Nomarski interference microscopy and transmission electron microscopy. The abrupt changes in relaxation behaviour indicate that rigid boundaries between stable and metastable structures do exist, as predicted by the equilibrium models.

## INTRODUCTION

Strained layer epitaxy has recently attracted considerable attention, due to the discovery of novel optoelectronic properties in strained semiconductors and the developments in heteroepitaxy, notably GaAs on Si. Many theoretical suggestions have been put forward [1-6] over the past 25 years to describe the stability of such strained heterostructures. If working devices are to be fabricated from strained material, their structural stability during anneal cycles similar to those experienced during implant activation and contact sintering must be determined. Although experimental data exist for strain relaxation in several material systems, their interpretation is hampered by the poor resolution of many experimental techniques in detecting small strains [7]. The precise point at which plastic relaxation begins is difficult to identify in epilayers grown at low temperatures, since large barriers to nucleation and propagation of misfit dislocations exist [5]. Annealing studies are thus useful since the introduction of thermal energy allows the structure to approach equilibrium.

The upper limit to structural stability is an important parameter since the position of this boundary determines the driving force available for plastic relaxation processes. In this work an expression is derived for the stability of strained multilayers using an energy minimization approach [1,2]. Experimental data are presented which confirm the abrupt transition between stable and metastable structures defined by a critical thickness/strain criterion.

## THE STABILITY OF STRAINED LAYER SUPERLATTICES.

Two possible mechanisms [1] for strain relaxation in SLS's are illustrated in Fig.1. In Fig.1a the SLS adopts the natural lattice parameter of the unsupported SLS. In Fig.1b the layers tend toward their unstrained lattice parameters leaving no residual strain in the multilayer. The residual strains after the mechanism in Fig.1a has reached completion can be readily shown to be a partitioning of the alloy strain energy between the two components A and B. Assuming equal elastic constants for A and B and isotropic behaviour, the in-plane strains in a free standing

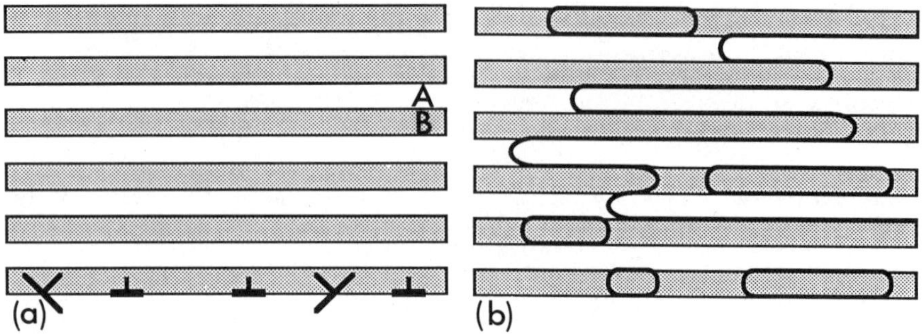


Fig.1 Schematic illustration of two relaxation mechanisms,(a) misfit dislocation array at the base of a multilayer, (b) paired dislocation segments and loops in individual strained layers.

superlattice are, $\varepsilon_B = \varepsilon_f t_A/(t_A+t_B)$ , $\varepsilon_A = - \varepsilon_f t_B/(t_B+t_A)$ (1)

where $\varepsilon_A$, $t_A$ and $\varepsilon_B$, $t_B$ are the strains and thicknesses in the A and B layers respectively. $\varepsilon_f$ is the lattice mismatch between the unconstrained materials which may be conveniently defined as, $\varepsilon_f = (a_B - a_A)/a_A$. The alloy layers are assumed to be initially coherently strained to fit the A substrate. The A layers are initially unstrained. As a superlattice relaxes from its strained condition on thick substrate to its unsupported lattice parameter the new in-plane lattice constant, shared by all the SLS layers is $a_A(1 - \varepsilon_A)$. The misfit strain may be assumed to be accommodated by an array of non-interacting dislocations at the lowest B/A interface. This is the most energetically favourable location for single misfit dislocations since here they reduce strain in the B layers through the entire multilayer. The energy per unit length of an isolated dislocation may be written as [8],

$$E_d = \frac{G\, b^2(1-\nu\cos^2\theta)}{4\pi(1 - \nu)} \left[ \ln(4h/b) \right] , \tag{2}$$

distance from the free surface,b the burgers vector, $\theta$, the angle between the dislocation line and its Burgers vector and where the core parameter has been taken as ~ 4 . A small strain, $\delta$, introduces both misfit dislocations with a spacing $(b\cos\lambda)/\delta$ and partitions the elastic strain between the A and B layers. The energy per unit area, $E_\delta$, of this partially relaxed SLS is,

$$E_\delta = \frac{2\,\delta\, E_d}{b\cos\lambda} + N K \left[ t_A \delta^2 + t_B (\varepsilon_f - \delta)^2 \right] , \tag{3}$$

where $K = 2G(1+\nu)/(1-\nu)$ [9], and $\lambda$ is the angle between the Burgers vector and the direction in the interface normal to the dislocation line. The energy minimum is found at $\delta_{min}$ , where the strain energy just balances the dislocation energy.

$$\delta_{min} = \frac{t_B\, \varepsilon_f - Gb(1-\nu\cos^2\theta)/8\pi NK(1-\nu)\cos\lambda \left[ \ln(4N(t_A+t_B)/b) \right]}{(t_A + t_B)} \tag{4}$$

If the dislocation energy is ignored, or for a freestanding superlattice, $\delta_{min}$ becomes simply the equilibrium in-plane strain in the A layers $\varepsilon_A$ . The onset of relaxation is defined when $\delta_{min} \geq 0$ , which gives an expression for the equilibrium critical thickness of a multilayer of N periods of B, thickness, $Nt_B$, with misfit strain, $\varepsilon_f$, and total thickness $N(t_A+t_B)$. The relation has the familiar form of Matthews and Blakeslee [1] for a single strained layer, modified here for the superlattice case,

$$Nt_B = \frac{b\,(1 - \nu\cos^2\theta)}{8\pi(1+\nu)\varepsilon_f\cos\lambda} \left[ \ln(4N(t_B+t_A)/b) \right] , \qquad (5)$$

The curves in Fig.2 are calculated for various SLS geometries using $\cos\theta = \cos\lambda = 1/2$, appropriate for $60^0$ mixed dislocations, and illustrate the increase in stability of a superlattice with increasing A spacer layer thickness. The stability enhancement is more evident at large strains. At 4% mismatch and $(t_A + t_B)/t_B = 10$ the effective critical thickness for buried strained layers is three times greater than that for an uncapped epilayer. At 1% mismatch this factor is only ~ 2, for the same geometry. The experimental data presented in the next section are for the Si/SiGe system where the SiGe layers are initially strained (B layers) and Si are unstrained (A layers).

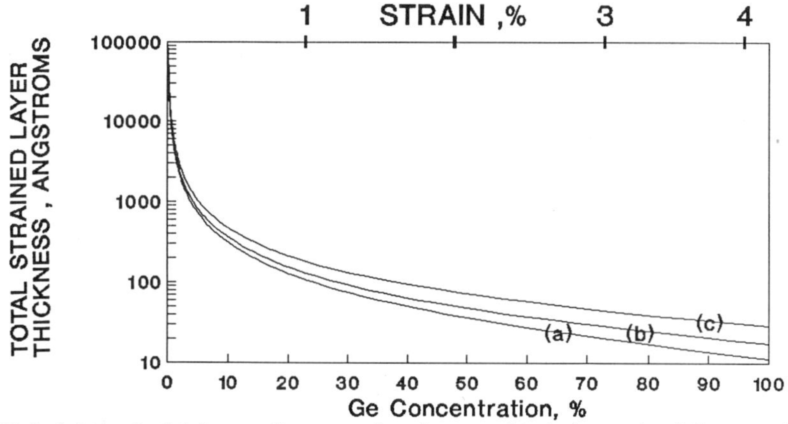


FIG 2.Critical thickness for superlattices or buried strained layers where, $(t_A+t_B)/t_B$ is 1, 2 and 10 for curves (a),(b) and (c) respectively.

Superlattices with layer thicknesses greater than $Nt_B$ in equation (5) are metastable and may relax toward the condition described by equation (1), if kinetic barriers to the nucleation and motion of misfit dislocations can be overcome. Note that the relaxation criterion defined by equation (5) is only weakly dependent on the thickness of the spacer layers. Thus a SLS or a buried strained layer (SiGe capped with Si) is only slightly more stable than an uncapped strained layer of the same thickness.

If the relaxation process depicted in Fig. 1a goes to completion, further relaxation by single dislocations is unlikely since this actually increases the strain energy of the multilayer from its minimum by partitioning more strain into the A layers. A relaxation mechanism which reduces the layer strain to zero is shown schematically in Fig.1b. For this mechanism to operate the individual alloy layers must exceed a different critical thickness calculated at the partially relaxed condition where the strain in the alloy layers is $(\varepsilon_f - \delta)$. This individual layer critical thickness is approximately twice the equilibrium critical thickness for uncapped epilayers, suggested by Matthews and Blakeslee [1], because opposing pairs of dislocations are required to relieve strain in the B layers without simultaneously straining the A layers. The energy per unit area associated with misfit dislocations in this case is $4N\delta E_d/b\cos\lambda$. The critical thickness ($h_s$) for secondary relaxation (Fig.1b) of individual layers can be found by displacing the multilayer from its energy minimum in the partially relaxed state described by equation (4) and following the energy balance as described previously. The individual layer critical thickness may be expressed as,

$$h_s = \frac{b\,(1-\nu\cos^2\theta)}{4\pi(1+\nu)(\varepsilon_f - \delta)\cos\lambda}\left\{\ln\,(4h_s/b)\right\} \qquad (6)$$

The curves in Fig.3 represent the critical thicknesses for individual layer relaxation by 60° glide dislocations and edge dislocations. Also indicated in Fig.3 are the strain and layer thicknesses of SLS 346 from Table I.

TABLE I

| | N | $t_{SiGe}$,nm | $t_{Si}$,nm | x |
|---|---|---|---|---|
| 346 | 40 | 6.2 | 10.3 | 0.19 |
| 349 | 40 | 7.3 | 10.9 | 0.25 |
| 570 | 10 | 4.5 | 33.9 | 0.05 |

In the coherent, as-grown structure the strain in the alloy layers is $\varepsilon_f$, but as relaxation at the base of the SLS proceeds, the alloy strain reduces to $(\varepsilon_f - \delta)$ which tends to $\varepsilon_B$ ($\varepsilon_{SiGe}$) as shown in Fig.3. Superlattices with alloy layers much thicker than their individual critical thickness, $h_s$, and whose spacer layers are thicker than the alloy layers (so that the alloy strain remains ~ $\varepsilon_f$, eq (1)) are more likely to relax internally by the mechanism shown in Fig.1b. The relaxation paths of Fig.1 are, however, competitive; as the alloy strain is reduced by either mechanism less strain energy is available for subsequent relaxation. The driving force for either process will thus be reduced by the progression of the other and may even become zero when the effective critical thickness is not exceeded. This observation and the likelihood that the SLS critical thickness will be exceeded by a larger margin than the individual layer critical thickness (Fig. 3) suggests that for most SLS geometries, misfit accommodation at the lowest strained interface will dominate the early stages of relaxation.

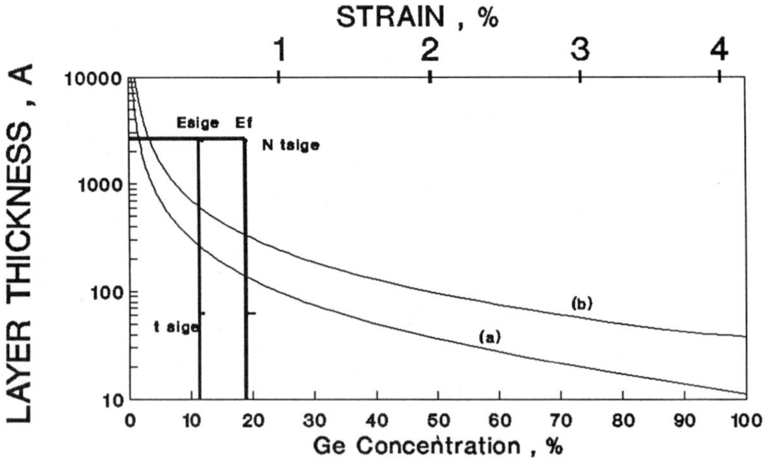


FIG 3. Individual layer critical thicknesses for relaxation by (a) 90° dislocations ,(b) 60° type dislocations.

EXPERIMENTAL DATA

The X-ray rocking curves in Fig.4 are from a 40 period Si-$Si_{1-x}Ge_x$ superlattice (SLS 346) in the as-grown [10] condition and annealed at temperatures in the range (600-900C) for 30 mins. The superlattice peaks are seen to shift toward the substrate peak and broaden progressively with increasing anneal temperature. Inspection of the critical strain/thickness curves in Fig.2 and Fig.3 using the dimensions for SLS 346 from Table I reveals that this SLS is metastable with respect to relaxation at its base

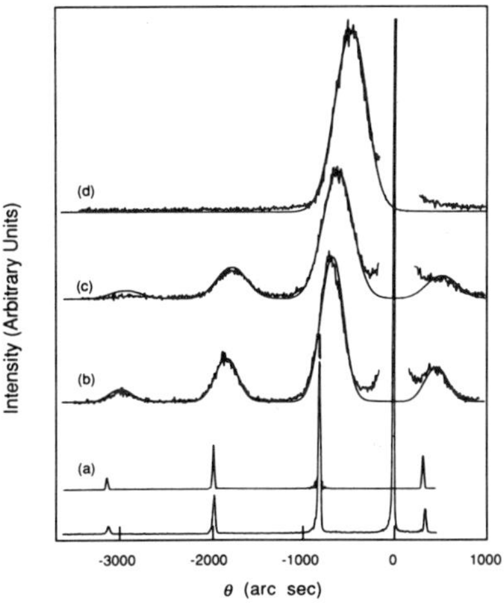


FIG 4.X-ray rocking curves for SLS 346 as-grown (a) and annealed at 600°C(b),800°C(c)and 900°C(d)

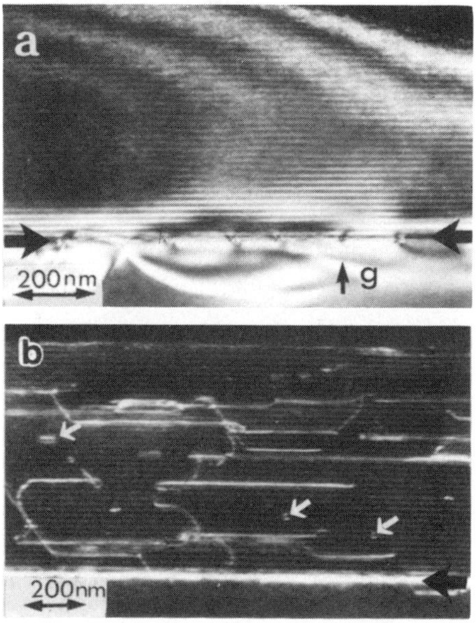


FIG 5.Dark field (g=400) XTEM micrographs of SLS's 346 (a) and 349 (b) annealed at 800C for 30 min.

but stable to further relaxation at interfaces through the SLS. The XTEM micrograph in Fig.5a of annealed SLS 346 confirms these predictions. An array of misfit dislocations, predominantly 60° type, is present at the lowest strained layer/substrate interface (arrowed) but the superlattice itself was found to be virtually dislocation free. SLS 349 in Fig.5b was closely similar and had an identical thermal history to SLS 346 (Fig.5a), with the exception of slightly thicker and higher Ge concentration alloy layers. In this superlattice (SLS 349)each individual alloy layer just exceeded its critical thickness as defined by equation (5). Fig.5b clearly shows that a high density of misfit dislocations and prismatic loops (arrowed) have formed (on (110) planes) during the anneal, throughout the SLS. The abrupt change in relaxation behaviour between these two SLS's is convincing support for the predictions of the equilibrium model described above. Two metastable regimes are defined; the first unstable with respect to relaxation at the base of the SLS (eq 5) and the second where relaxation occurs both at the base and throughout the SLS (eq 6).

The superlattice (SLS 570) in Fig.6 was designed to be stable with respect to both internal (Fig.1b) and lowest interface (Fig.1a) relaxation. On annealing this structure remained coherent with the substrate and misfit dislocation-free, within the resolution of Nomarski and plan view TEM, less than $100/cm^2$. In contrast to the X-ray rocking curves for SLS 346 in Fig.4,

SLS 570 in Fig.6 retained its structural perfection through the annealing cycle at 800C . No evidence for strain relaxation (peak shift) or increase in dislocation density (peak broadening) was observed. The XTEM micrograph in Fig.6a confirms the structural integrity of the annealed SLS, which was indistinguishable from the as-grown structure.

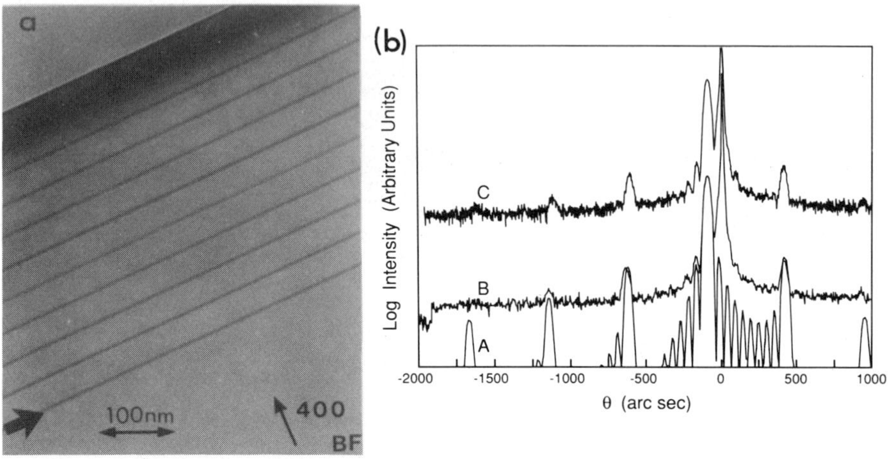


FIG 6.(a) Bright field XTEM micrograph and (b) X-ray rocking curves from SLS 570 kinematical simulation (A), as grown (B), annealed 800°C 30 min (C).

CONCLUSIONS

An equilibrium model has been used to predict the stability limits of strained layer heterostructures. Three stability regimes exist; stable, metastable with respect to misfit dislocations at the base of the SLS and metastable with respect to both relaxation at the base and through the SLS. Experimental data from the Si/Ge system confirm that abrupt changes in relaxation behaviour occur as predicted by the model.

ACKNOWLEDGEMENTS
The authors are grateful to Chris Tuppen and Chris Gibbings at BTRL for useful discussions.

REFERENCES

1.J.W.Matthews and A.E.Blakeslee, J.Cryst Growth 27,118 (1974).
2.J.W.Matthews, J.Vac. Sci. Technol. 12,126(1975).
3.Jan H. van der Merwe and W.A.Jesser, J.Appl. Phys. 63,1509 (1988).
4.R.Hull,J.C.Bean,F.Cerdeira,A.T.Fiory and J.M.Gibson, Appl. Phys.Lett. 48,56 (1986).
5.J.Y.Tsao and B.W.Dodson, Appl. Phys. Lett. 53,848 (1988).
6.R.People and J.C.Bean, Appl. Phys. Lett. 47,322 (1985), erratum ibid 49,229 (1986).
7.I.J.Fritz Appl. Phys. Lett.51,1080(1987)
8.J.P.Hirth and J.Lothe, Theory of Dislocations (Wiley-Interscience,New York,1982) 2nd ed.
9.S.P.Timoshenko and J.N.Goodier,Theory of Elasticity,3rd ed (McGraw-Hill,New York,1970)
10.D.C.Houghton,D.J.Lockwood,M.W.C.Dharma- Wardana,E.W.Fenton,JM.Baribeau and M.W.Denhoff, J.Cryst.Growth 81,434(1987).

# EFFECT OF HIGHLY STRAINED SUPERLATTICES ON THE THREADING DISLOCATION BEHAVIOR IN GaAs GROWN ON SILICON

N. A. El-Masry, J. C.L. Tarn
Materials Science and Engineering Department, North Carolina State University, Raleigh, North Carolina 27695-7907
N.H.Karam* S.M.Bedair
Electrical and Computer Engineering Department, North Carolina State University, Raleigh, N.C.27695-7911
* Currently with Spire Corp. Patriots Park, Bedford MA 01730

## ABSTRACT

GaAs films grown on silicon substrates suffer from a high density of threading dislocations. One technique to reduce these dislocations is the use of strained layered superlattices (SLS) InGaAs-GaAsP as buffer. We found that the effectiveness and interactions of the SLS with the threading dislocations strongly depend on the dislocation type and the strain field of the superlattices. Because of the high dislocation density in GaAs/Si, the SLS acts as a medium for dislocation interactions and annihilations. Highly strained SLS (~ 2%) is required to bend the dislocations and keep them bent at the strained interfaces. We found that the combination between annealing and a highly strained superlattice coupled with selective epitaxy is an effective approach to reduce the threading dislocations in GaAs grown on Si. Transmission electron microscopy is used to study these effects.

## INTRODUCTION

The growth of GaAs on Si substrates could lead to monolithic integration of optoelectronic devices with sophisticated Si electronic circuits and has received increasing attention in recent years. One difficulty that still impedes the progress of achieving device quality GaAs films on Si substrates is the high defect density in the GaAs epilayer. These defects result, in part, from stress generated due to the combination of lattice and thermal expansion coefficient mismatches between GaAs and Si. Considerable efforts have been made to suppress the propagation of defects to the epitaxial films. These approaches include thermal annealing [1], thermal strained layer [2], cyclic annealing [3], rapid thermal annealing [4] and the use of strained-layer superlattices (SLS) [5]. We have reported the effectiveness of InGaAs-GaAsP strained-layer superlattices (SLS) in blocking the threading dislocations in GaAs epilayers [5,6] . These SLS's have the advantage of accommodating a high level of strain (up to 2%) elastically without creating misfit dislocations. Further, we have studied the interactions which take place between the threading dislocations and the strain fields of the InGaAs-GaAsP superlattices [7]. By using these techniques, the dislocation density of GaAs on Si was reduced. However, further reduction in dislocation density is required from the viewpoint of developing the aforementioned defect sensitive device.

In this study, it is demonstrated that the threading dislocation density of GaAs on Si can be greatly reduced by using the combination of SLS and intermittant thermal annealing during or after SLS growth. The advantage of using the combination SLS/intermittant annealing is improving the SLS efficiency significantly, where the energy provided by annealing is used to force the dislocations to glide and interact with the SLS interfaces.

## EXPERIMENT

The GaAs/Si samples used in this study were grown by the molecular beam epitaxy (MBE) growth technique. Thermal annealing was employed under an over pressure of Arsine in the metalorganic chemical vapor deposition (MOCVD) reactor at 830 °C for 50 minutes followed by slow cooling (10°C/minute). The SLS used in this study was grown at 650°C by MOCVD as previously described [7]. Transmission Electron Microscopy (TEM) was used for the structure characterization. Foils for TEM were thinned by mechanical polishing followed by standard ion-thinning techniques. Cross-sectional TEM observations were performed on a Hitachi-800 (200 KV), using the conventional two-beam diffraction technique.

## RESULTS AND DISCUSSION

InGaAs-GaAsP strained-layer superlattices have been used as a buffer layer to reduce the dislocation density in GaAs grown on Si substrates. To demonstrate the effect of strain in InGaAs-GaAsP on the superlattice effeciency in bending threading dislocation three groups of SLS were grown on the same wafer. The composition was varied such that the strain is different in each group. The layer thicknesses were selected to be close to $h_{max}$ for each composition, where $h_{max}$ is the maximum layer thickness that accomodates the misfit strain elastically. Mathews and Blakeslee's Model [8] for $h_{max}$ has been used to calculate the layer thicknesses of the grown SLS. Each group was separated by 800Å GaAs buffer layer. SLS1 with the lowest strain was grown first and the group SLS3 with the highest strain grown last. A micrograph of the above structure is shown in Fig.1. From the figure one can observe that the separation distance between threading dislocations is a critical factor in the SLS efficiency. For high dislocation density the bent distance of a dislocation is longer with a superlattice having higher strain level (dislocation A in the figure).

According to Mathews and Blakeslee's model of mechanical equilibrium [8] the force acting on the threading dislocations from the coherency stresses is the Peach-Koehler Force ($F_{pk}$).

$F_{pk} = C\varepsilon hb$ $\quad\quad$ $\varepsilon$ = misfit strain

h = the strained layer thickness $\quad\quad$ b = strength of the dislocation line

C = constant dependent on the shear modulus and Poisson's ratio in the layers (for isotropic film)

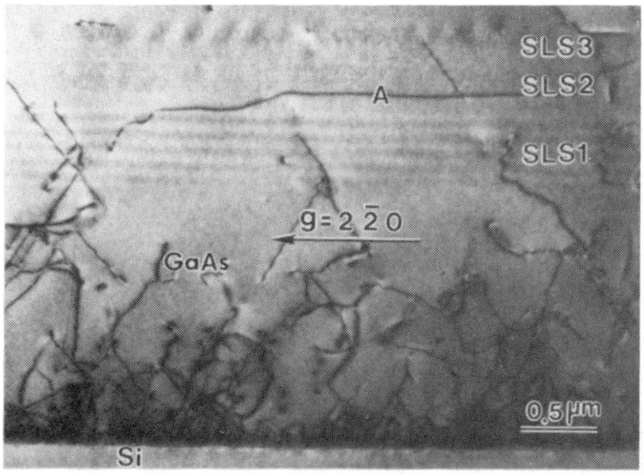


Fig.1. Cross-sectional bright-field TEM micrograph illustrating the bending effect of different strain levels on threading dislocations.

From this relation one can state that as the misfit strain and layer thickness increase, for a given SLS, the gliding force which acts on the the threading dislocation increases. Thus SLS with layer thickness (h) closer to ($h_{max}$) was selected to favor dislocation bending in GaAs/Si. The misfit strain should be kept high as well in order to sustain the maximum gliding force ($F_{pk}$) offered by a given SLS.

Our previous work [7] in utilizing GaAsP-InGaAs SLS to reduce threading dislocations indicated that in areas of low dislocation density almost all threading dislocations are blocked and bent along the SLS interfacial planes. The bent dislocations in the low density regions can propagate along the SLS interface a distance of several microns without distrubance, this occurs in the absence of a strain field generated by a neighboring dislocation. There is a critical separation distance between the threading dislocations below which the local relaxation offered by the dislocations will relax the SLS. The residual strain ($\varepsilon$) will be reduced to $\varepsilon = (f - \delta)$, where $\delta$ is the strain relaxed locally by the dislocation and f is the misfit strain. The critical separation distance can be estimated as:

$S_{crit.} = b\perp / f$, where $b\perp$ is Burger's vector edge component of the threading dislocation in the strained film plane. The calculated values of $S_{crit}$ and f are as follows:

| | | | |
|---|---|---|---|
| f | 0.0075 | 0.0108 | 0.0144 |
| S(A°) | 267 | 185 | 139 |

From the table it becomes obvious that the higher the misfit strain in the SLS the shorter is the distance allowed between the dislocations threading the superlattice, i.e. the higher the dislocation density that the SLS can handle. Thus, the effectiveness of the superlattice in bending dislocations is improved by utilizing higher strain level, thicker layers (closer to $h_{c,max.}$) and large number of periods of the SLS.

Annealing is a thermal process that mobilizes the dislocations and activates their slip mechanism. Combining SLS's with annealing prior and during the SLS growth can activate the dislocation slip systems and force the unbent dislocations to glide and enhance their interactions with the strained interfaces of the SLS. Two types of SLS-annealing combination structures were studied. Structure (A): annealing during the growth for two groups of 5-periods highly strained SLS [$In_{0.2}Ga_{0.8}As$-$GaAs_{0.65}P_{0.35}$ (80 A°/layer)] separated by 800 A° GaAs. The growth was interrupted and annealed in-situ at 830°C for 10 minutes then slowly cooled down to SLS growth temperature 650°C. This was followed by four groups of 5-periods SLS having the same strain level and a 1 μm GaAs cap layer. Structure (B): annealing after the SLS growth, where three groups with the same SLS as in structure (A) were grown with a 1 μm GaAs cap layer followed by post annealing at 830°C for 30 minutes. The TEM micrographs of cross-sectional samples which have been intermittantly annealed during and post the SLS growth are illustrated in Fig. 2 and 3 respectively. It is obvious that the interaction between SLS and threading dislocations is enhanced by intermittant annealing. Also, it is clearly indicated that the threading dislocations are strongly confined within the annealed two series of SLS and that gliding along the SLS interfaces occured. By this intermittant annealing process the grown-in threading dislocation density can be dramatically reduced. This will allow the gradual reduction of the dislocation density and the subsequent SLS to powerfully bend the threading dislocations without the local relaxation of the SLS strain by the high threading dislocation density.

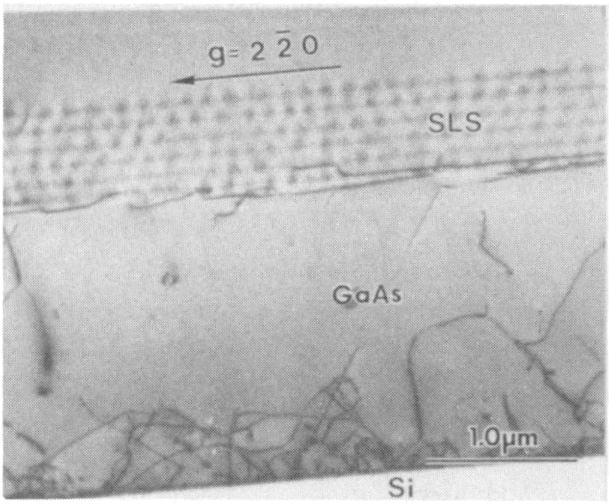


Fig.2. Cross-sectional bright field TEM image of the intermittantly annealed the SLS during growth in dislocations reduction.

Slow and fast cooling after annealing were also employed for the previously mentioned structures after SLS growth. Fig. 3(a) is the cross-sectional TEM image of the specimen which was annealed and slowly cooled after the SLS growth, while Fig. 3(b) shows the effect of fast cooling of the same structure after annealing. It is apparent from the above results that post-annealing of the SLS shows significant defect confinement within the SLS, where the whole SLS stucture during annealing can be considered a dislocation sink. However, considering the higher thermal expansion coefficient mismatch between GaAs and Si, fast cooling after annealing induces a higher thermal strain rate that creates more misfit dislocations within the SLS compared to slow cooling. Therefore, annealing coupled with slow cooling is required to reduce the residual thermal stresses in the grown structures.

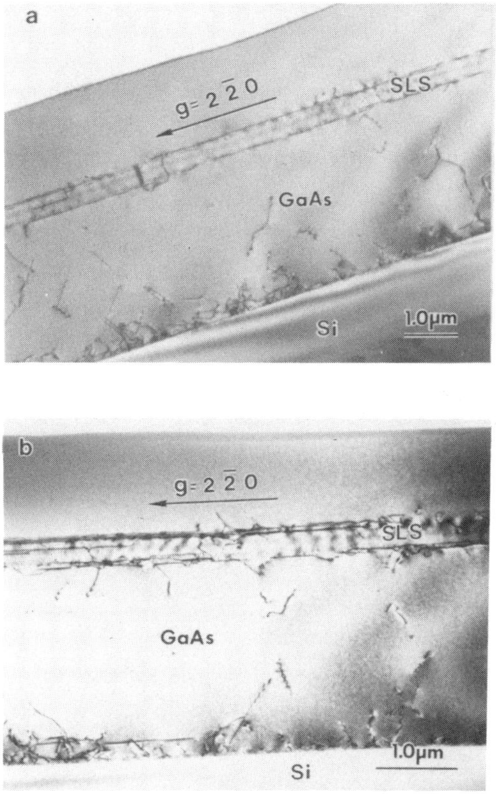


Fig.3. Cross-sectional bright-field TEM images illustrating the effectiveness of the post-annealing after the SLSs growth in dislocations reduction. a) slowly cooled b) fast cooling.

In conclusion, we have found that the effectiveness of SLS in blocking threading dislocations is improved by increasing the strain (ε), the layer thicknesses (close to $h_{max}$) and the number of the strained layers. The SLS coupled with intermittant annealing during or after the SLS growth permits a remarkable reduction of threading dislocation density. The intermittant annealing provides the energy for the interaction between threading dislocations and the strain of SLS, therefore the efficiency of these SLS's can be significantly improved.

It has been shown that the conventional annealing followed by slow cooling is an effective technique to minimize the thermal stresses in GaAs/Si. However, new dislocations can be generated due to thermal stresses if fast cooling is employed after annealing. Further work is in progress to optimize the SLS/annealing parameters in order to achieve device quality GaAs epilayers grown on Si substrates.

REFERENCES

1. C. Choi, N. Otsuka, G. Munns, R. Houdre, H. Morkoç, S.L. Zhang, D. Levi and M.V. Klein, Appl. Phys. Lett. 50, 992(1987).
2. J.W. Lee, J.P. Salerno, R.P. Gale and J.C.C. Fan, Mat. Res. Symp. Proc. 91, 33(1987).
3. H. Okamoto, Y. Watanabe, Y. Kadota and Y. Ohmachi, Japanese J. Appl. Phys. Lett. 50, 992(1987).
4. N. Chand, R. People, F.A. Baiocchi, K.W. Wecht and A.Y. Yao, Appl. Phys. Lett. 49, 815 (1986).
5. N.A. El-Masry, J.C.L. Tarn, T.P. Humphreys, N. Hamaguchi, N.H. Karam and S.M. Bedair, Appl. Phys. Lett. 51, 1608 (1987).
6. N.A. El-masry, J.C.L. Tarn and N.H. Karam, J. Appl. Phys. 64, 3672(1988).
7. J.W.Mathews and A.E.Blackeslee, J. Cryst. Growth 27, 118 (1974).
8. J.W. Mathews, S. Mader and T.B. Light, J. Appl. Phys. 41, 3800 (1970).

# STRAIN RELIEF BY TILTING OF EPITAXIAL GaAs FILMS ON Si

L.J. SCHOWALTER*, E.L. HALL**, N. LEWIS**, AND SHIN HASHIMOTO***
*Physics Dept. and Ctr. Integrated Electronics, Rensselaer Polytechnic Institute, Troy, NY 12181
**GE Research & Development Ctr., Schenectady, NY 12301
***Physics Dept., State University of New York at Albany, Albany, NY 12222

## ABSTRACT

In this paper, a general formula is calculated relating the misalignment of a heteroepitaxial overlayer with respect to the substrate to the defects at the interface. High-resolution TEM studies are presented which show that the density of dislocations along the interface of GaAs grown on vicinal Si(100), with Burgers vector perpendicular to the interface, is consistent with the amount of tilt observed in ion-channeling studies. Arguments are also presented which demonstrate that the tilt observed is an effective form of strain relief but that this mechanism can only operate when the heteroepitaxial growth is nucleated by island growth. One significant conclusion is that the lower bound observed on the threading dislocation density in GaAs on Si is due to this mechanism of strain relief by tilting.

## INTRODUCTION

There is substantial interest in the commercial development of GaAs on Si. A continuing problem, however, is the relatively high number of dislocations (about $10^7$ cm$^{-2}$) which thread through the epitaxial GaAs layer. We argue here that this dislocation density may be an intrinsic property of the strain-relief mechanism during the growth of the GaAs layer (which has a 4% larger lattice spacing than the underlying Si substrate). When a heteroepitaxial layer, such as GaAs on Si, initiates growth by forming islands on the substate, there exists a new mechanism for strain relief of lattice mismatch during growth. By introducing dislocations with an edge-like component of their Burgers vectors perpendicular to the interface, epitaxial islands can change their tilt with respect to the substrate so long as there exist side walls which are free to move. In the case of a vicinal Si(001) substrate in which the (001) axis is deliberately tilted away from the surface normal, the GaAs islands can relieve strain by tilting back toward the surface normal through this mechanism. The amount of tilt which occurs will depend on the aspect ratios of the islands when they coalesce. This implies that there will be some amount of statistical fluctuation in the amount of tilt across the film which will result in the formation of low-angle grain boundaries (LAGB). These LAGB will then act as sources of "threading dislocations." The strain-relief mechanism presented here should be equally effective in relieving strain in any lattice mismatched system in which the epitaxial layer is nucleated by island growth.

In this paper, we first relate the tilt in a coherent epitaxial layer to the defects at the overlayer/substrate interface. In the second part of the paper, a high-resolution transmission electron microscopy (HRTEM) study is presented which relates the observed dislocation densities at the GaAs/Si interface to the observed misalignment of the GaAs with respect to the Si substrate. Finally, we present arguments about the preferential introduction of defects which result in a tilt of the epitaxial film and discuss why this will lead to a lower bound on the threading dislocation density.

## RELATION OF TILT TO DEFECTS AT THE INTERFACE

It has been observed in the growth of a number of lattice-mismatched heteroepitaxial systems on substrates which are misoriented from an exact, low-index, crystallographic direction that the epitaxial layer is slightly misoriented with respect to the substrate (see, for example, Ref. 1-6). If one considers how two crystals with different lattice spacings will line up across an atomic step at their interface, it is readily apparent how one mechanism for tilt can arise just due to geometric considerations. This mechanism is clearly explained in Nagai's work [4] and would require tilt to be observed even if there were no defects at the interface. Additional tilt or even tilt in the opposite direction of that predicted strictly by the lattice mismatch will result when dislocations with an edge component perpendicular to the interface occur. This fact has been qualitatively noted by previous authors [1]. In what follows below, we present a prescription for calculating the tilt based on the number and type of dislocations at the interface.

 

In coherent heteroepitaxy, the misorientation of the epitaxial layer with respect to the substrate can be related to defects along the substrate interface in the following way. Consider the diagram shown in Fig. 1. For simplicity, we will consider the substrate and epitaxial layer to have cubic symmetry and we will also consider the substrate normal to be nominally oriented in the [001] direction. More complicated geometries could be considered by straightforward extension of the procedure described here. Let $\gamma$ denote the misalignment of the surface normal $\hat{n}$ with respect to the [001] direction as shown in Fig. 1. In addition, we consider the substrate to be infinite in extent below the interface plane so that the rigid substrate approximation is valid. Now construct a parallelogram in the epitaxial layer by following along the {100} planes in the horizontal and vertical directions in the substrate. These lines are bent at the substrate/overlayer interface so as to continue to follow the {100} planes of atoms in the overlayer. Of course, near the interface in the overlayer, the {100} planes may be curved due to bond distortions near the interface. However, the range of such "curved" distortions will generally be short and can effectively be ignored by picking the parallelogram far enough away from the interface. Likewise, questions about which {100} plane to follow in the epitaxial layer, when a defect is encountered in the overlayer or at the interface, can be avoided by choosing the parallelogram large enough so that the distinctions become unimportant.

Let $L_z$ and $L'_z$ represent the vertical spacings between the lines drawn in the substrate and epitaxial layer, respectively, as shown in Fig. 1. Let $\theta_1$ be the angle by which these planes are bent as they cross the interface. We define $L_x$, $L'_x$, and $\theta_2$ in a similar way as shown in the figure. Now

$$L'_z = \frac{L_z}{sin\gamma}\frac{sin(\gamma-\theta_1)}{cos(\theta_2-\theta_1)} , \tag{1a}$$

and

$$L'_x = \frac{L_x}{cos\gamma}\frac{cos(\gamma-\theta_2)}{cos(\theta_2-\theta_1)} , \tag{1b}$$

If $a_x$, $a'_x$, $a_z$, and $a'_z$ are allowed to represent the lattice spacings of the horizontal and vertical planes in the substrate and overlayer, then $N_z$ and $N'_z$ can be chosen such that $L_z = N_z a_z$ and $L'_z = N'_z a'_z$. The difference between $N_z$ and $N'_z$ represents the net number of crystal planes which had to be cut out of the overlayer as a result of defects (dislocations) at the substrate/overlayer interface. If we define $B_\perp$ as the total length of horizontal lattice planes which effectively have been cut out of the overlayer due to defects at the interface per unit length of interface, then

$$\frac{Na_z}{sin\gamma}B_\perp = (N_z - N'_z)a'_z \tag{2}$$

since the interface will have length $Na_z(\sin\gamma)^{-1}$. Using this definition, we obtain

$$a'_z = a_z\left[\frac{sin(\gamma-\theta_1)}{sin\gamma \quad cos(\theta_2-\theta_1)} + \frac{B_\perp}{sin\gamma}\right] \tag{3a}$$

Likewise,

$$a'_x = a_x\left[\frac{cos(\gamma-\theta_2)}{cos\gamma \quad cos(\theta_2-\theta_1)} + \frac{B_{//}}{cos\gamma}\right] \tag{3b}$$

where $B_{//}$ is defined in analogy to $B_\perp$. Eqs. (3) can now be used to calculate the necessary $B_{//}$ and $B_\perp$ due to defects along the interface in order to account for the observed misalignments between the epitaxial overlayer and the substrate or, alternatively, the misalignment can be calculated if $B_{//}$ and $B_\perp$ can be measured directly by some technique. Note that our definitions of $B_{//}$ and $B_\perp$ are similar to the way the total Burgers vector per unit length of interface would be defined. However, we do not wish to include the screw component of any dislocation which is encircled by our construction of the parallelogram in our definition since the screw component will not result in any net removal of material in the epitaxial layer. Thus, $B_{//}$ and $B_\perp$ will not be the same as the respective Burgers vectors per unit length of interface if there are screw or mixed dislocations along the interface.

As mentioned earlier, we are considering the simple situation where $a_x = a_z = a$. Let f be the bulk lattice mismatch between the substrate and the overlayer and let $\epsilon'_z$ and $\epsilon'_x$ represent the vertical and horizontal strain in the epitaxial layer. In addition, we will assume that $\theta_1 = \theta_2 = \theta$. This does not have to be the case, even if the epitaxial layer has cubic symmetry, since stress in the epitaxial layer could introduce tetragonal distortions. However, this effect will be second order if $\theta$ is small and would also be negligible if the effective shear modulus for the epitaxial layer is sufficiently large. Finally, we note that $\epsilon'_z$ is proportional to $\epsilon'_x$ when the only stress applied to the thin epitaxial layer is across the interface the substrate. (The strains in the epitaxial layer are assume to be constant everywhere in the approximation of a substrate of infinite extent.) Thus, we will write $\epsilon'_z = -A\epsilon'_x$ where $A$ is related to Poisson's ratio $\sigma$ by

$$A = \frac{2\sigma}{1-\sigma} = \frac{2C_{12}}{C_{11}}, \tag{4}$$

for a (001) film where $C_{11}$ and $C_{12}$ are the elastic constants for a cubic crystal. Now Eqs. (3) simplify to

$$a'_x = (1 + f + A\epsilon'_x)a = \frac{sin(\gamma-\theta)}{sin\gamma}a + \frac{B_\perp}{sin\gamma}a, \tag{5a}$$

and

$$a'_x = (1 + f - \epsilon'_x)a = \frac{cos(\gamma-\theta)}{cos\gamma}a + \frac{B_{//}}{cos\gamma}a. \tag{5b}$$

The above equations can be used to eliminate the strain to find $\theta$ in terms of the substrate misorientation $\gamma$ and the defects along the interface. The resulting expression can be further simplified by considering the case where the tilt $\theta$ is very small (as is generally the case) and keeping only first order terms. In this case, we obtain

$$\theta = [B_\perp cos\gamma - A(f - B_{//})sin\gamma - f sin\gamma](cos^2\gamma - A sin^2\gamma)^{-1} \tag{6}$$

It is interesting to note that Eq. (6) reduces to $\theta = -f(A+1)(cos^2\gamma - Asin^2\gamma)^{-1}$ when there are no defects along the interface. On the other hand, when strain has been eliminated in the epitaxial layer by defects at the interface, Eq. (6) reduces to $\theta cos\gamma = B_\perp - Af(1 - cos\gamma)tan\gamma - ftan\gamma$.

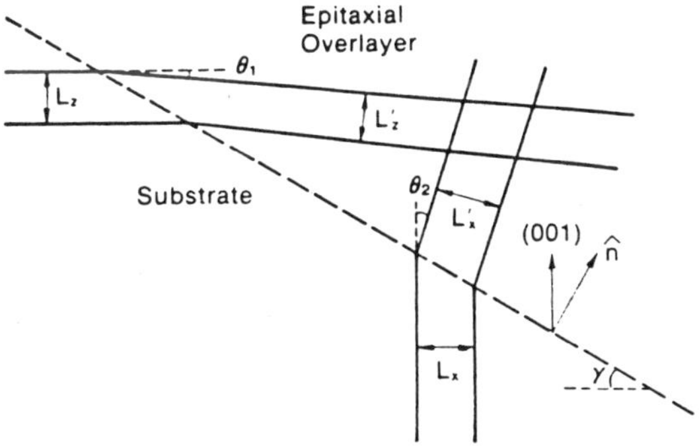


Fig. 1: A schematic diagram of a cross section of an epitaxial layer on a substrate defining $L_z$, $L_{z'}$, $L_x$, $L_{x'}$, $\theta_1$, and $\theta_2$. The interface is shown as a dashed line with a surface normal direction $\hat{n}$.

## ANALYSIS OF DEFECTS AT THE GaAs/Si INTERFACE

We have compared the above results to GaAs grown on Si by use of HRTEM. The layers were grown on vicinal Si(001) substrates by molecular beam epitaxy where the substrate surface normal was tilted 3.7° away from the [110] azimuthal direction. The growth has been described elsewhere [2]. The HRTEM images of the GaAs/Si interface were obtained by looking along the $(\bar{1}10)$ axis perpendicular to the tilt azimuth. A typical image is shown in Fig. 2. In Fig. 3, a blow-up of a particular region of the interface is shown in which two dislocations can be discerned. A Burgers vector analysis shows one of the dislocations to be an edge dislocation with Burgers vector 1/2[110] (this is labeled type-I in conformance with previous literature [7]) while the other dislocation can be identified as a 60° dislocation with a Burgers vector of either $1/2[10\bar{1}]$ or $1/2[01\bar{1}]$ (in this projection it is not possible to distinguish between these two possibilities) which is called [7] a type-II dislocation. We were able to closely examine 171 nm of the interface using images similar to Fig. 2. In these regions, we found a total of 19 dislocations, corresponding to a dislocation density of $1.1 \times 10^6$ cm$^{-1}$. This value corresponds closely to the dislocation density observed with better statistics in lower resolution, bright-field TEM images. (It should be noted that the dislocations are quite regularly spaced as would be expected at this high density.) Of the 19 dislocations, 14 were identified as type-I while the other five were identified as type-II. For all the type-II dislocations identified, the [001] projection of their Burgers vectors were $-(1/2)a$; none were identified with a [001] component aligned in the opposite direction. Each type-I dislocation contributes $(\sqrt{2}/2)a$ to $B_{//}$ and nothing to $B_\perp$ while each type-II dislocation observed contributes $(1/2)a$ to $B_\perp$ and $(\sqrt{2}/4)a$ to $B_{//}$. It should be noted that it is only the edge component of the type-II dislocation which is effective in relieving strain so only this component should be projected onto the interface plane in calculating $B_{//}$ not the total projection of the Burgers vector as some authors have done [1,7].

The observed frequency of the two types of dislocations can be used to calculate $B_{//} = 0.037$ (3.70%) and $B_\perp = 0.79\%$ using $a = 0.543$ nm for the Si substrate. Now $A = 0.91$ for GaAs [8], $f = 3.96\%$, and $\gamma = 3.7°$ for the Si substrates used, so the tilt angle $\theta$ can be calculated from Eq. (6) to be 0.30° while from Eq. (5b), the strain $\epsilon'_x = 0.2\%$ (compressive). We have measured [2] both the tilt and strain by ion channeling experiments with 2 MeV $He^+$ ions. These measurements indicate that tilt is actually 0.2° and that the strain is −0.1% (tensile). Note that the agreement is good given the rather low statistics on dislocation type. Given the measured values of strain and tilt, theoretical values of $B_\perp = 0.6\%$ and $B_{//} = 4.1\%$ are calculated. This would correspond to 3.8 type-II dislocations in 171 nm and 16.3 type-I in the same interval.

## MECHANISM FOR TILT

When the epitaxial layer initiates growth by forming islands on the substrate, there exists a new mechanism for strain relief of lattice mismatch during growth which has not been properly recognized before. An epitaxial layer can change its lattice spacing with respect to the substrate by tilting its crystal axis with respect to the substrate. This effect can be seen with Eq. (5b). The lattice spacing in the epitaxial layer $a'_x$ can be increased with respect to $a_x$ for fixed $B_{//}$ by increasing $\theta$ up to a maximum of $\gamma$. Similarly, $a'_x$ can be made smaller by taking negative $\theta$ (i.e., causing the epitaxial layer's (001) axis to tilt further away from the substrate's surface normal). On the other hand, Eq. (5a) shows that $B_\perp$ must be increased if $\theta$ is increased. We can express the strain in terms of $B_{//}$ and $B_\perp$ where we make the approximation that f and $\epsilon'_x$ are both small;

$$\epsilon'_x = \frac{f - B_\perp sin\gamma - B_{//}cos\gamma}{cos^2\gamma - Asin^2\gamma} \quad . \tag{7}$$

The cost in energy of making a network of dislocations to give some value of $B_\perp$ will be approximately equal to making a network of edge dislocations with the same magnitude of $B_{//}$. Therefore, if the mechanisms for introducing the two kinds of defects were the same, we would expect $B_\perp$ to vanish in thermal equilibrium if the overlayer is a uniform sheet on top of substrate.

When the mechanism of how individual dislocations get introduced is considered, it becomes even less reasonable to consider the introduction of $B_\perp$ in a uniform overlayer. As Fig. 4 shows, an individual edge dislocation with Burgers vector perpendicular to the interface (shown as a missing plane) will result in an expansion of the epitaxial layer on

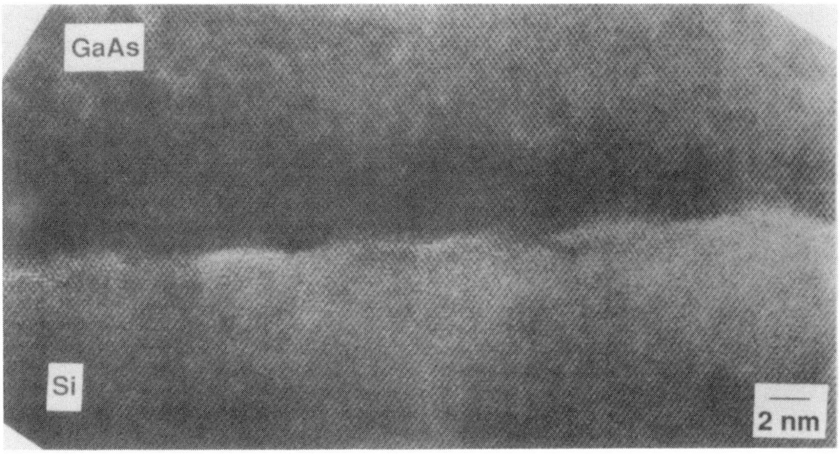


Fig. 2: High-resolution TEM image of a GaAs/Si(001) interface looking down the [1$\bar{1}$0] direction. The interface normal is tilted 3.7° toward the [110] direction (which is directed toward the left in the figure).

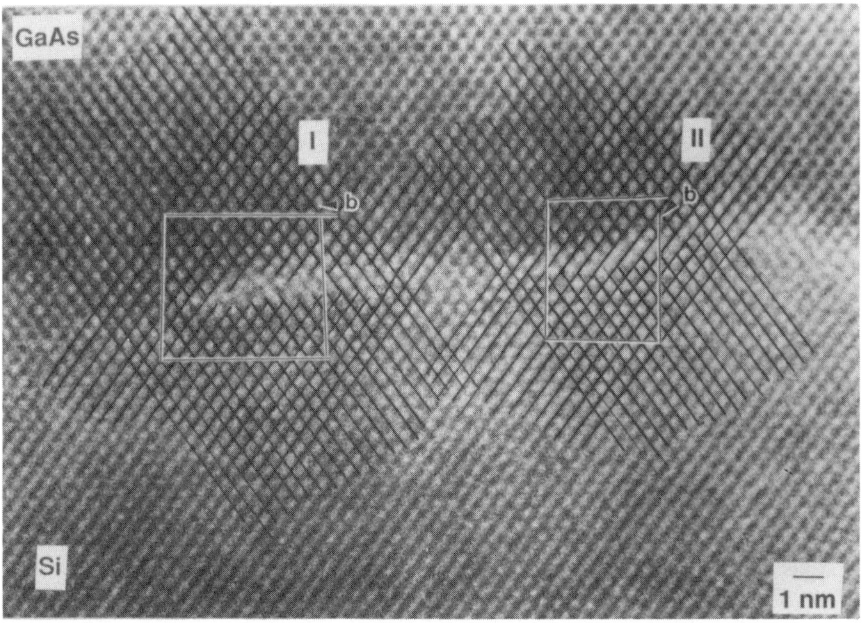


Fig. 3: Blow up of a region of Fig. 2 showing two dislocations. One of these dislocations (type-I) has Burgers vector 1/2[110], which is almost parallel to the interface, and the other (type-II) is a 60° dislocation with a component of its Burgers vector perpendicular to the interface.

one side of the defect but it will result in a corresponding compression of the overlayer on the other side. Thus, it is only when a network of dislocations are introduced, with a net $B_{\perp}$, that the overall strain energy of the epitaxial layer be reduced. This is quite different from the situation for introducing dislocations which contribute to $B_{//}$. In this latter case, each individual dislocation, with an edge component of its Burgers vector, will result in a local reduction of the strain in the epitaxial layer regardless of the distance to the next dislocation.

If the epitaxial layer nucleates as islands, the situation is quite different. Now the side walls of the islands are free surfaces. Any calculation of the strain within the island must then satisfy the boundary condition that can be no stress along these free surfaces. In this situation, introducing dislocations which contribute to $B_{\perp}$ can be energetically attractive since the distance over which the strain field of the dislocation must be integrated [10] is substantially reduced by the side wall of the island. This is qualitatively shown in Fig. 4 where the dotted line represents the side wall of the island. As shown in the figure, the dislocation, as shown, is able to expand the lattice of the overlayer without having to introduce compression further away from the defect because of the side wall.

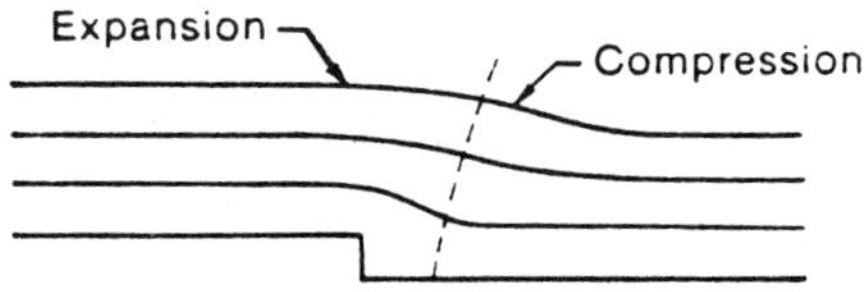


Fig. 4: A schematic diagram showing an edge dislocation with Burgers vector perpendicular to the interface. As explained in the text, the dislocation is ineffective in relieving strain if the epitaxial layer uniform over the entire substrate. However, if the epitaxial layer consists of islands (the edge of the island is shown as a dashed line) which are free to relax, then the dislocation shown effectively relieves strain in the epitaxial overlayer.

## CONCLUSIONS

We have shown that lattice mismatch between a substrate and overlayer will result in a tilt if the substrate is misaligned ($\gamma \neq 0$). This tilt will be modified by the introduction of defects. In particular, the introduction of defects which $B_{\perp}$ will strongly affect the tilt, much more than defects which only contribute to $B_{//}$. It is important to recognize that this mechanism, by which defects which contribute to $B_{\perp}$ are introduced, requires a free surface at the sides of the epitaxial island. There is relatively little advantage (ie. strain relief) to have dislocations with an edge component of their Burgers vector perpendicular to the interface propagate into the epitaxial island a distance along the interface that is greater than the island is thick. This implies that the amount of tilt that the epitaxial layer undergoes will not be uniform across the substrate but rather will depend on statistical fluctuations in the aspect ratios of islands as they coalesce. This variation in misorientation of the GaAs will result in low angle grain boundaries (or sub-boundaries) in the epitaxial layer which would serve as a source for threading dislocations. This result may explain the apparent lower bound on threading dislocations observed in epitaxial GaAs on Si of about $10^7$ $cm^{-2}$ [11] as being an intrinsic property of the large lattice mismatch between the two materials and the fact that GaAs on Si nucleates as islands. One possible solution is to use an appropriate buffer layer on the Si substrate which lattice matches to the GaAs [2].

ACKNOWLEDGEMENTS – This work was supported in part by the New York Science and Technology Foundation under grant no. SSF(87) - 24. We also gratefully acknowledge useful discussions with J. Ayers and Prof. S.K. Ghandhi. The work of S. Divakaruni in identifying defects in HRTEM was also appreciated.

References

1. R.J. Matyi, J.W. Lee, and H.F. Schaake, J. Electronic Materials 17, 87 (1988).
2. L.J. Schowalter, S. Hashimoto, G.A. Smith, W.M. Gibson, N. Lewis, E.L. Hall, and P.W. Sullivan in Epitaxy of Semiconductor Layered Structures, edited by R.T. Tung, L.R. Dawson, and R.L. Gunshor, (Mat. Res. Soc. Proc. Vol. 102, Pittsburgh, PA, 1988).
3. J. W. Lee, J. P. Salerno, R. P. Gale, and J. C. C. Fan in Heteroepitaxy on Silicon II, ed. by J.C.C. Fan, J.M. Phillips, and B.-Y. Tsaur, MRS Proc. Vol. 91, Pittsburgh, PA 1987, p. 33.
4. H. Nagai, J. Appl. Phys. 45, 3789 (1974).
5. J. Kleiman, R.M. Park, and H.A. Mar, J. Appl. Phys. 64, 1201 (1988).
6. S.K. Ghandhi and J.E. Ayers, Appl. Phys. Lett. 53, 1204 (1988).
7. N. Otsuka, C. Choi, Y. Nakamura, S. Nagakura, R. Rischer, C. K. Peng, and H. Morkoc, in Heteroepitaxy on Silicon, ed. by J.C.C. Fan and J.M. Poate (MRS Proc. Vol. 67, Pittsburgh, PA, 1986), p. 85.
8. S. Adachi, J. Appl. Phys. 58, R1 (1985).
9. L.J. Schowalter, to be publ. Heteroepitaxy on Silicon: Fundamentals, Structures, and Devices, ed. by H.K. Choi, (MRS Proc. Vol. 116, Pittsburgh, PA).
10. J.W. Matthews, Materials Science Series: Epitaxial Growth, Part B, ed. J.W. Matthews (Academic Press, New York, 1975) p. 559.
11. J.S. Harris, Jr., S.M. Koch, and S.J. Rosner in Heteroepitaxy on Silicon II, loc. cit., p. 3.

# ONSET OF MISFIT DISLOCATION GENERATION IN AS-GROWN AND ANNEALED $Si_{1-x}Ge_x$/Si FILMS

M.P. Scott, S.S. Laderman, T.I. Kamins, S.J. Rosner, K. Nauka
Hewlett Packard Co., Palo Alto, CA 94303-0867
D.B. Noble, J.L. Hoyt, C.A. King, C.M. Gronet, J.F. Gibbons
Stanford Electronics Labs, Stanford, CA 94305

## ABSTRACT

X-ray topography and transmission electron microscopy were used to quantify misfit-dislocation spacings in as-grown $Si_{1-x}Ge_x$ films formed by Limited Reaction Processing (LRP), which is a chemical vapor deposition technique. These analysis techniques were also used to study dislocation formation during annealing of material grown by both LRP and by molecular beam epitaxy (MBE). The thickness at which misfit dislocations first appear in as-grown material was similar for both growth techniques. The thermal stability of capped and uncapped films was also investigated after rapid thermal annealing in the range of 625 to 1000°C. Significantly fewer misfit dislocations were observed in samples containing an epitaxial silicon cap. Some differences in the number of misfit dislocation generated in CVD and MBE films were observed after annealing uncapped layers at temperatures between 625 and 825°C.

## INTRODUCTION

The recent fabrication of near-ideal heterojunction bipolar transistors [1] has increased the need to understand the structural stability of $Si_{1-x}Ge_x$ alloy thin films grown on silicon. It is important to prevent strain relaxation in these pseudomorphic device structures to preserve the desired band gap and avoid recombination currents associated with closely spaced misfit dislocations [2]. Measurements to date of the generation of misfit dislocations during growth of $Si_{1-x}Ge_x$ have focussed on epitaxial layers grown by MBE [3,4]. Some of these studies have used Rutherford backscattering (RBS) and x-ray diffraction to characterize the onset of strain relaxation. The sensitivity of these techniques is not sufficient to study the early stages of misfit dislocation generation. A combination of techniques has been used to characterize individual dislocations in as-grown MBE samples [5,6]. However, little experimental data has been presented concerning the onset of dislocation generation induced by post-growth annealing and no comparison has been made between $Si_{1-x}Ge_x$ formed by MBE and that formed by chemical vapor deposition.

In the present report, results are presented on material grown by limited reaction processing (LRP), a CVD technique that relies on the rapid heating and cooling of the substrate to initiate and terminate growth [7]. These $Si_{1-x}Ge_x$ films have been found to contain approximately $10^{20}$ oxygen atoms/$cm^3$. Previous reports [4,5] have shown that as-grown, MBE $Si_{1-x}Ge_x$ films are metastable with dislocation spacings larger than predicted by equilibrium theory. In the present study the thermal stability of both uncapped and epitaxial-silicon capped structures were characterized by x-ray topography and transmission electron microscopy. The observed stability was compared to the stability of the film predicted by equilibrium theory.

## EXPERIMENTAL DETAILS

The epitaxial layers grown by LRP had a germanium mole fraction x=0.21 as measured by Rutherford backscattering (RBS). These films were grown at 625°C using $GeH_4$ and $SiH_2Cl_2$ in an apparatus that has been described elsewhere [7]. Thicknesses were found by RBS to be 50 to 200 nm. The capped samples were prepared by growing approximately 200 nm of silicon at 840°C at a steady-state growth rate of 0.8 nm/sec on top of the $Si_{.79}Ge_{.21}$ layers. Epitaxial structures were also grown by MBE with a measured Ge mole fraction x=0.15 at a growth temperature of 520°C. Both sets of films were grown on (001)-oriented silicon substrates.

Annealing studies were performed *in-situ* for some of the LRP samples and in a commercially available RTA unit for both LRP and MBE samples. Times and temperatures were chosen to study the early stages of misfit dislocation generation.

Misfit dislocation spacings were measured directly using x-ray topography and plan-view transmission electron microscopy. In combination, these techniques have the required sensitivity to study individual misfit dislocations of arbitrary spacing. When average spacings less than 50 µm are mentioned in this report the spacings were determined by TEM; when greater, they are estimated from topographs. The x-ray topographs were taken in transmission using a rotating-anode x-ray generator and a Lang camera with Cu$K_{\alpha 1}$ radiation. The (220) and (400) type substrate reflections were employed. This arrangement enabled misfit dislocations to be studied with epitaxial films as thin as 50 nm. A JEOL 200CX TEM was used to image the dislocations in plan-view. Care was taken to analyze the same wafer area with all techniques employed to minimize the effects of spatial variations in alloy composition and thickness. Diffraction imaging techniques were chosen instead of EBIC [6] to avoid ambiguity introduced by possible differences in the electrical activity of the dislocations.

## RESULTS

### As-Grown Dislocation Spacing

Uncapped LRP samples were dislocation-free at thicknesses up to 55 nm. Misfit dislocations were observed at a thickness of 60 nm, and the spacing decreased with increasing layer thickness as shown in Fig. 1. The dislocation configuration and type consisted of an orthogonal array of dislocations and appeared similar to earlier descriptions of MBE samples [4,5]. An example of Hagen-Strunk multiplication [8], which was evident in samples with numerous dislocation intersections, is illustrated in Fig. 2. The majority of the dislocations were straight lengths of misfit dislocations extending to the edge of the specimen. 55 nm thick SiGe samples with approximately 200 nm-thick epitaxial-silicon caps grown at 840°C had a misfit dislocation spacing of about 6 µm while samples with caps grown at 750°C had a spacing greater than 20 µm. As will be discussed below, we believe that these defects are introduced during the temperature ramp or early stages of cap growth.

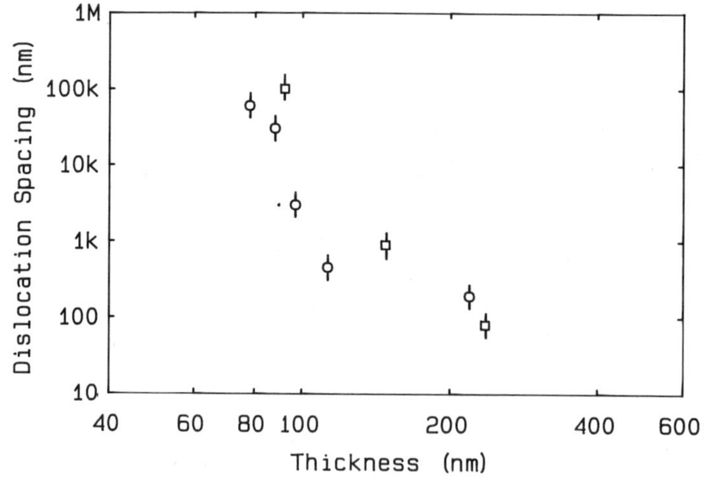


Figure 1: Mean misfit dislocation spacing in LRP $Si_{.8}Ge_{.2}$ films. O denote undoped films, □ denote boron doped films, [B] = $10^{19}$ cm$^{-3}$.

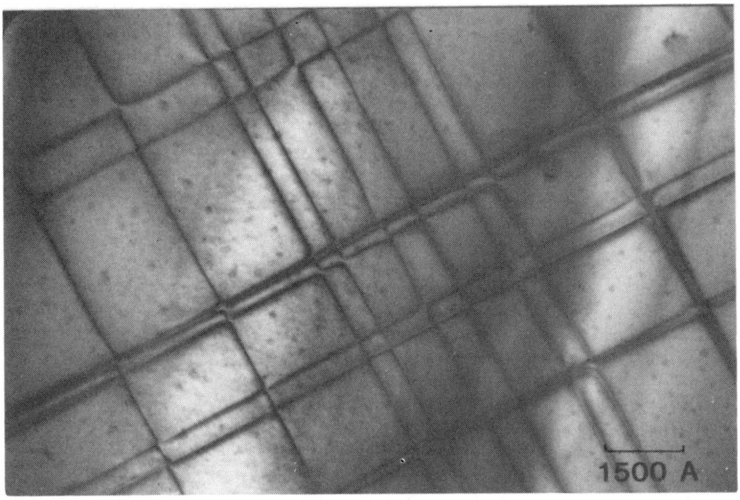


Figure 2: Transmission electron micrograph showing Hagen-Strunk dislocation multiplication in a 120 nm thick uncapped $Si_{.8}Ge_{.2}$ LRP film.

The uncapped 55 nm thick MBE $Si_{.85}Ge_{.15}$ layers were also free of misfit dislocations even though they contained threading dislocations with a density of about $40/cm^2$ which exhibited strong diffraction contrast in the x-ray topographs. Another sample was capped with about 200 nm of silicon at a growth temperature of 520°C. This capped sample was also free of misfit dislocations as-grown.

Effects of Post-Growth Annealing

The mean dislocation spacing in uncapped LRP and MBE films decreased as the annealing temperature was increased from 625°C to 825°C. Data presented in Fig. 3 show the temperature dependence of the mean misfit-dislocation spacing. The LRP samples annealed at 625°C were free of misfit dislocations while the MBE samples contained misfit dislocations with an average spacing of about 9 µm. The dislocation morphology was the same in annealed and as-grown material for samples with equal dislocation spacing. Islanding was observed for *in-situ* anneals of LRP samples above about 800°C. LRP films which were annealed in an RTA unit following exposure to air were less likely to island than those annealed *in-situ* after film growth.

A silicon cap was effective in reducing the rate of misfit-dislocation generation during annealing. Although the capped LRP samples contained misfit dislocations with an average spacing of about 6 µm as-grown, there was no significant increase in the dislocation density after annealing for one hour at temperatures up to 900°C. The capped MBE samples were annealed for four minutes at temperatures up to 1000°C and remained free of long misfit dislocations of the type observed in uncapped films. However, short (100-300 µm-long) dislocations were present with a density less than $15/cm^2$ in the MBE samples annealed above 900°C. These dislocations grew from an average length of 100 µm after four minutes at 900°C to 400 µm after four minutes at 1000°C. The very slow velocity compared to dislocation glide in uncapped samples suggests a different formation mechanism than that for the long misfit dislocations in uncapped films. Some of these dislocations appeared connected to a threading defect at one end. Plan-view TEM confirmed the absence of long misfit dislocations in the sample annealed at 1000°C.

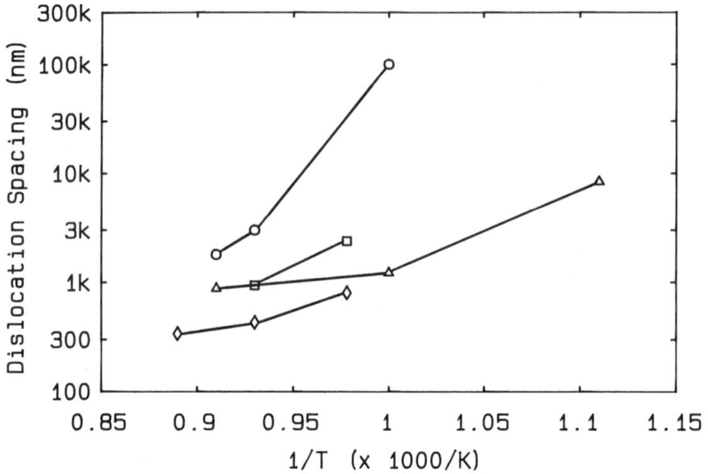


Figure 3: Dislocation spacings measured after annealing $Si_{1-x}Ge_x$ films. MBE films had a composition of x=0.15; LRP films, x=0.21. ○ denote 4 min. anneals of 55 nm thick LRP films,□ denote 1 hr. anneals of 41 nm thick LRP films,△ denote 4 min. anneals of 55 nm thick MBE films, and ◊ denote 1 hr. anneals of 63 nm thick LRP films.

DISCUSSION

The microstructural observations presented in this work on uncapped films are consistent with the strain-relaxation process described by Kvam [5]. This process involves the expansion and interaction of 60$^{\circ}$ mixed dislocation half-loops which nucleate at the epilayer surface in order to lower the system energy by relieving misfit strain at the expense of increasing dislocation line length. Quantification of the as-grown misfit-dislocation density in LRP samples formed at 625$^{\circ}$C showed the thickness for the appearance of such defects to be about five times the thickness predicted by equilibrium theory. Literature reports [6] on similar MBE samples grown at 510$^{\circ}$C cite the thickness for the generation of misfit dislocations during growth to be almost the same as that observed in this study. This suggests that growth technique and oxygen content do not dominate the onset of misfit-dislocation generation during growth.

Differences were, however, observed in the behavior of uncapped LRP and MBE samples on annealing. Despite the six-percent-lower Ge mole fraction in the MBE samples, more rapid misfit-dislocation generation was observed at temperatures up to 825$^{\circ}$C. One possible explanation for greater misfit-dislocation generation in the MBE samples is an increase in nucleation rate provided by the higher threading-defect density, although detailed observations have not clearly confirmed this mechanism. Observations of dislocation interactions with precipitates in LRP samples after annealing for sixty hours at 900$^{\circ}$C suggest that oxygen may also be involved. The increase in dislocation-generation rate with temperature in both kinds of material, however, suggests that kinetic limitations dominate the dislocation-generation process.

An important observation was the significant reduction in misfit dislocation generation provided by an epitaxial silicon capping layer. No long, misfit dislocation segments were generated in capped MBE films annealed at temperatures up to 1000$^{\circ}$C for four minutes. Capped LRP films which contained dislocations after growth showed no appreciable increase in dislocation density after annealing for one hour at 900$^{\circ}$C. The minimum capping-layer thickness for a significant reduction in misfit-dislocation generation was estimated by comparing capped and uncapped films with similar misfit-dislocation densities after annealing. An uncapped sample annealed for

15 sec at 850$^{o}$C and a sample with a cap grown at 850$^{o}$C for 2.7 minutes had approximately equal dislocation densities. Assuming the dislocations were introduced during the first 15 sec of growth, the minimum cap thickness for a reduction in dislocation generation is estimated to be about 20 nm using the growth rate of SiGe at 850$^{o}$C of 1.2 nm/sec.

Equilibrium theory [9] predicts a critical thickness for the growth of a misfit dislocation segment from a threading defect in a single strained $Si_{1-x}Ge_x$ layer on silicon to be about 24 nm with x=0.15 and 14 nm with x=0.21. The capped structure discussed above has a stress-free cap that is four times as thick as the strained layer. Additional energy is required to expand a half loop in this structure due to the additional dislocation line length required at the second interface between the strained layer and cap. The equilibrium critical thickness of the capped structure is estimated to be twice as large as that of the uncapped sample; *ie* 48 nm for the x=0.15 MBE samples and 28 nm for the x=0.21 LRP samples. Therefore, the MBE samples were slightly above, and the LRP samples were almost double, the equilibrium critical thickness. Because no long misfit dislocations were generated during the annealing of either type of sample, kinetic limitations on the nucleation and/or multiplication of misfit dislocations appear to dominate. Because the top surface of the silicon capping layer is stress free, there is no driving force to nucleate a half loop at that surface. It is also unlikely that a dislocation half loop of sufficient radius to relieve stress in the strained layer would spontaneously nucleate during annealing. This reasoning leads us to believe that the mechanism for stress relief in capped films might be different from that likely to operate in uncapped samples.

## CONCLUSION

The growth of a silicon cap on a coherently strained $Si_{1-x}Ge_x$ layer reduces misfit-dislocation generation during annealing at temperatures up to 1000$^{o}$C. Such films can withstand moderate device processing temperatures without structural degradation. The temperature dependence of misfit-dislocation generation for LRP and MBE films suggests that kinetic limitations dominate in both materials. Differences in the number of misfit dislocations between LRP and MBE samples after annealing were most pronounced at lower temperatures. This observation suggests that additional impediments to dislocation generation in LRP samples or sources of dislocation generation in MBE material may exist.

## ACKNOWLEDGEMENTS

The authors would like to acknowledge Kang Wang and Jin Park of UCLA for providing the MBE samples used in this study and Bob Smith and Mark Juanitas for expert structural characterization. The authors from Stanford University are grateful for the support of the Advanced Research Projects Agency.

## REFERENCES

1. C.A. King, J.L. Hoyt, C.M. Gronet, J.F. Gibbons, M.P.Scott and J. Turner, accepted for publication in IEEE Electron Device Letters

2. C.A. King, J.L. Hoyt, D.B. Noble, C.M. Gronet, J.F. Gibbons, M.P. Scott, T.I. Kamins, and S.S. Laderman, submitted to IEEE Electron Device Letters

3. J.C. Bean, L.C. Feldman, A.T. Fiory, S. Nakahara, and I.K. Robinson, J. Vac. Sci. Technol.A, 2, 436 (1984)

4. E. Kasper and H.J. Herzog, Thin Solid Films 44, 357 (1977)

5. E.P. Kvam, D.J. Eaglesham, D.M. Maher, C.J. Humphreys, J.C. Bean, G.S. Green, and B.K. Tanner, Materials Research Society Symposium Proceedings Volume 104, 623 (1988)

6. Y. Kohama, Y. Fukuda, and M. Seki, Appl. Phys. Lett. 52, 380 (1988)

7. J.F. Gibbons, C.M. Gronet, and K.E. Williams, Appl. Phys. Lett. 47, 721 (1985)

8. W. Hagen and H. Strunk, Appl. Phys. 17, 85 (1978)

9. J.W. Matthews and A.E. Blakeslee, J. Cryst.Growth 27, 181 (1974)

# MISMATCH DISLOCATION NUCLEATION AND PROPAGATION IN $Si/Si_{1-x}Ge_x$ HETEROSTRUCTURES

C G TUPPEN, C J GIBBINGS AND M HOCKLY
British Telecom Research Laboratories, Martlesham Heath, Ipswich, UK IP5 7RE

## ABSTRACT

Mismatch dislocation nucleation mechanisms have been studied in the $Si/Si_{1-x}Ge_x$ materials system using the simple application of a selective, defect revealing chemical etch, followed by standard Nomarski optical microscopy. The onset of mismatch dislocation formation was found to occur at a point just above the equilibrium critical thickness limit set by Matthews and Blakeslee ($h_e$). However, the bulk of the material remained coherently strained until it reached a metastable critical thickness ($h_m$), where $h_m >> h_e$. Nomarski micrographs clearly show the initiation of the mismatch dislocation network and the major dislocation multiplication processes that occur at $h_m$. At low levels of relaxation the individual mismatch dislocation half loops can be identified, and we show, for the first time, the evolution and propagation of these dislocations from nucleation sites in the epilayer. A dramatic decrease in $h_m$ has been observed for epilayers that contain a large number of intrinsic dislocations.

The mismatch dislocations were found to nucleate at extended defects and further relaxation occurred by dislocation multiplication at these sites. This gave rise to a non-uniform distribution of mismatch dislocations, with patches of highly relaxed material surrounded by material that was still coherently strained.

## INTRODUCTION

The concept of a "critical thickness" in describing the onset of strain relief in strained layer systems has received considerable discussion. A number of different definitions for the "critical thickness" have been proposed and we shall use the following notations: $h_e$ will represent the calculated thickness at which it is energetically favourable for mismatch dislocations to form, ie. the equilibrium critical thickness as originally defined by Matthews and Blakeslee[1]. $h_m$ will represent the metastable critical thickness, ie. the thickness at which the extent of relaxation exceeds ~1%. This corresponds to the critical thickness measured by techniques such as RBS[2] and X-ray diffraction[3] and will depend on a number of factors, including the growth temperature[4], growth rate and other parameters that control the thermal history of the layer. Finally, $h_F$ will represent the thickness at which the first mismatch dislocation is formed. To detect the $h_F$ point Fritz[5] showed that very sensitive analytical techniques are required such as photoluminescence microscopy[6] or electron beam induced current (EBIC)[7].

A number of theoretical models have been proposed to describe the relaxation process and these have been reviewed fully elsewhere[8]. Basically they can be broken down into the energy or force balance models such as the well known one proposed by Matthews and Blakeslee[1], models based on the homogeneous nucleation of dislocation half loops at the free surface[9,10] and those based on an excess stress approach[11] followed by the application of standard plastic deformation theory[12]. To date these models have primarily been supported by one set of measurements for $h_m$ in the $Si/Si_{1-x}Ge_x$ system[2] and have used a variety of mismatch nucleation mechanisms based on a minimum of experimental evidence.

Here we report on the use of a selective chemical etch to delineate mismatch dislocations, and show it to be a very sensitive technique. The relaxation process in $Si_{1-x}Ge_x$ alloys is studied from its onset at $h_F$, through the metastable critical thickness point at $h_m$ and on to near 100% relaxation. The nucleation points of individual mismatch dislocations are also examined.

## EXPERIMENTAL

Epitaxial layers were produced on a VG Semicon V80 MBE system fitted with Airco Temescal electron beam evaporators and a Sentinel flux control system. Wacker n+ (001) substrates were introduced to the vacuum system with no special pre-cleaning stage. The native oxide was reduced at 850°C for 5 mins, using a low silicon flux, equivalent to a growth rate of $0.5Ås^{-1}$. This process has been described in detail elsewhere[13] and reliably produces Si homoepitaxial layers with dislocation densities in the region of $10^3 cm^{-2}$. Following the surface cleaning stage an intermediate buffer layer of Si was deposited during which time the deposition temperature was lowered to 550°C. $Si_{1-x}Ge_x$ alloy structures were then deposited at a growth rate of $10Ås^{-1}$, with the Ge content controlled by the Sentinel flux control system.

X-ray rocking curves were recorded using Cu $K\alpha_1$ radiation near the 004 reflection with an InP first crystal. $1cm^2$ samples of the layers were also defect revealed using a diluted Schimmel etch (4 parts 48% HF : 5 parts 0.3M $CrO_3$) for 30-60s at room temperature. This reveals individual mismatch lines at low levels of relaxation and can also be used to measure threading dislocation densities. All the micrographs presented in this paper are of etched samples taken using Nomarski interference contrast.

## RESULTS AND DISCUSSION

A high magnification Nomarski micrograph of a 2.4μm thick $Si_{0.95}Ge_{0.05}$ layer is shown in fig.1a. This micrograph shows two intercepting slip planes, each containing a number of mismatch dislocation half loops. In the vertical direction the slip plane contains just two half loops. These are identified by four etch pits due to the threading arms of the dislocation half loops and a faint etch line resulting from the interfacial components of the dislocations. The 45° angle subtended by the etch pits with respect to the interfacial component of the mismatch dislocation implies that the threading arms are oriented along <110> directions and lie on a {111} plane as shown in fig.1b. In the horizontal direction a large number of individual half loops are observed on the same {111} slip plane and the series extends for a considerable distance beyond the edge of the micrograph.

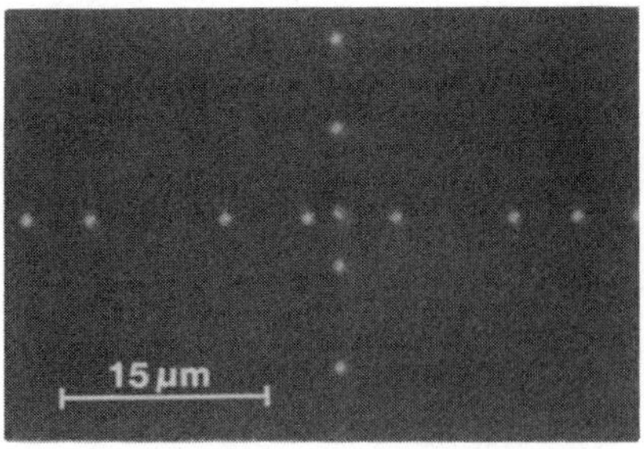


(a)

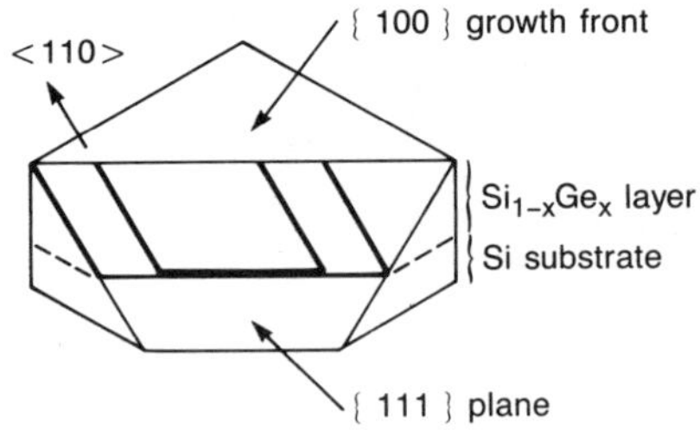


(b)

Fig.1 (a) Micrograph of an etched 2.4μm thick layer of $Si_{0.95}Ge_{0.05}$ showing individual mismatch half loops and the nucleation point. (b) Schematic diagram showing two half loops on a {111} plane.

At the intersection point of the two orthogonal slip planes is an etch pit that is similar in size and shape to the etch pits produced by the threading components of the misfit half loops. In nearly all cases we find that this point corresponds to the midpoint of the two intersecting misfit lines and that the individual half loops are arranged symmetrically about it. Based on this observation we conclude that this point represents the nucleation point for the misfit dislocations. However, it is difficult to identify the exact nature of the nucleation source from micrographs of etched layers.

MBE material generally has two sources of extended defects. One is due to an incomplete cleaning process at the substrate/epilayer interface and gives rise to threading dislocations in the layer[13]. The other type of defect results from Si particulates incorporated in the layer during growth. These are generally known as particulate induced defects or PID's. Although the shape and size of the central etch pit shown in fig.1a is characteristic of a threading dislocation it is not possible unambiguously to identify it as such. However, the central etch pit in fig.2 is typical of a PID and has clearly nucleated a vast number of misfit half loops which are located on multiple slip planes. Over an area of $1cm^2$ the density of PID nucleation sources was low and, on average, we found the type of nucleation point shown in fig.1a to dominate.

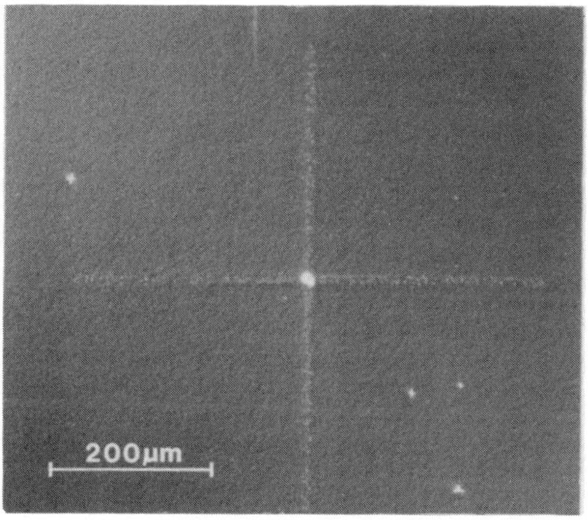


Fig.2 Micrograph of an etched, 2.4$\mu$m thick layer of $Si_{0.95}Ge_{0.05}$ showing large numbers of misfit half loops, sitting on multiple slip planes and nucleated by a PID.

To study the relaxation process in more detail a series of $Si_{0.875}Ge_{0.125}$ layers were grown with thicknesses ranging from $\sim h_e$ to $\sim 2\times h_m$ (ie. 25nm to 2$\mu$m). To provide sufficient material for etching in samples where the alloy layer thickness was below 0.1$\mu$m, a 1$\mu$m thick cap of Si was deposited. As the level of relaxation in these samples was minimal (ie. $h \ll h_m$) the Si cap remained effectively unstrained. A zero strain component in the Si means there is a reduction in the driving force for the continued glide of any mismatch dislocations formed during the deposition of the alloy layer. Thus the further movement of misfit dislocations is prohibited and they are essentially "frozen in".

A 1cm × 1cm sample of the 25nm thick layer of $Si_{0.875}Ge_{0.125}$ was etched and examined by optical Nomarski microscopy. No evidence for the formation of mismatch dislocations was found. However, a 50nm thick layer had ~5 mismatch dislocations $cm^{-2}$ where the lengths of the individual lines were <5$\mu$m. This implies that the relaxation process had not begun at $h \sim h_e$ ($h_e$ = 22nm for $Si_{0.875}Ge_{0.125}$) but had started in a small but finite way by $h \sim 2\times h_e$. Thus, in terms of the critical thickness definition proposed by Fritz[5], $h_F \sim h_e$ within experimental error.

Increasing the layer thickness from 50nm to 500nm resulted in an increase in both the density and length of the mismatch dislocations. At this stage of the relaxation process the lengths of the interfacial components of the mismatch dislocations were generally in the range 300-600nm with a crossing point density of ~400 $cm^{-2}$. The misfit lines in the 500nm thick sample remained in single pairs of orthogonal lines with no evidence for dislocation multiplication. However, for a 700nm thick $Si_{0.875}Ge_{0.125}$ layer the original dislocation lines had begun to multiply, clearly shown by a degree of branching away from the original point source in fig.3. Even so, there are still large areas of the sample free from any misfit dislocations.

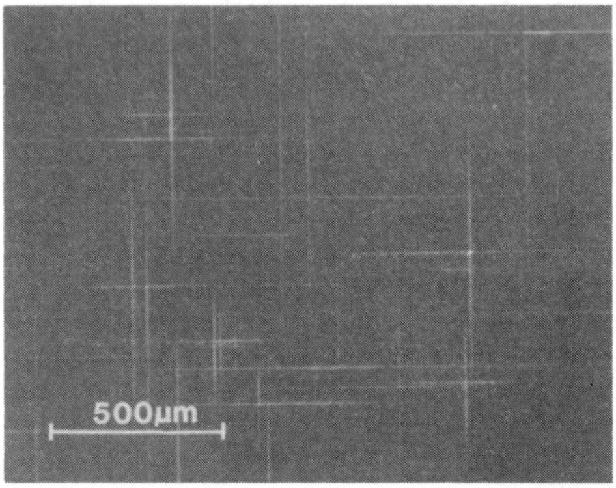


Fig.3 Micrograph of an etched 700nm thick layer of $Si_{0.875}Ge_{0.125}$

A 1$\mu$m thick $Si_{0.875}Ge_{0.125}$ layer, fig. 4, continues the general trend, with misfit dislocation multiplication centered about the the original defect nucleation point. This gives rise to patches of highly dislocated material surrounded by areas with a very low dislocation density. X-ray diffraction measurements on this series of layers, fig. 5, shows that the width of the 004 alloy layer diffraction peak begins to broaden at a layer thickness of 1$\mu$m. This corresponds to the definition of metastable critical thickness ($h_m$) and a value of $h_m$ = 1$\mu$m is in good agreement with the original work of Fiory et al[2]. Fig.6 shows the effect of increasing the alloy layer thickness to 1.5$\mu$m. The relaxation process has now progressed to a stage where the network of misfit dislocations covers the entire surface, although the original nucleation points can still be identified as areas of extended relaxation.

From the above data it is not possible to identify the precise nature of the misfit dislocation nucleation source. From fig.2 it is obvious that PID's can act as sources but, in general, the etch pit at the nucleation point was more representative of a threading dislocation. To assess the effect of the intrinsic threading dislocation density on the relaxation process a 1$\mu$m thick layer of $Si_{0.875}Ge_{0.125}$ was deposited on a specially prepared substrate with a threading dislocation density of ~$10^6 cm^{-2}$. Such high dislocation levels were achieved in a controlled fashion using solid phase epitaxial growth of Si, a technique which is reported in more detail elsewhere[8]. The 1$\mu$m thick alloy layer deposited on to this highly dislocated substrate showed greatly enhanced relaxation and no evidence of patches was observed by Nomarski microscopy which showed a uniformly dense network of misfit dislocations. The x-ray rocking curve for this sample gave a peak width equivalent to a 1.8$\mu$m thick layer grown on a standard substrate (see fig.5).

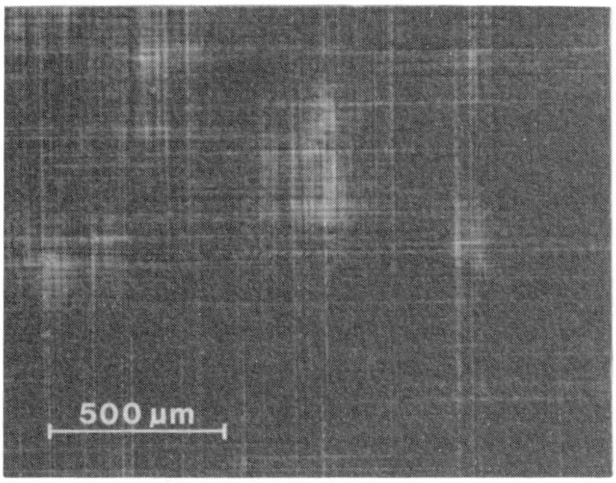


Fig.4 Micrograph of an etched 1$\mu$m thick layer of $Si_{0.875}Ge_{0.125}$.

Fig.5 (004) X-ray diffraction peak widths for the series of $Si_{0.875}Ge_{0.125}$ layers grown on substrates with an intrinsic dislocation density of $10^3$cm$^{-2}$ (□) and a 1$\mu$m thick layer of $Si_{0.875}Ge_{0.125}$ grown on a substrate with an intrinsic dislocation density of $10^6$cm$^{-2}$ (+).

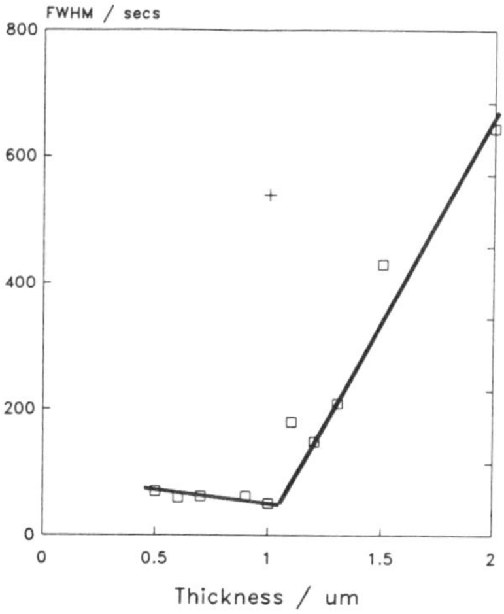


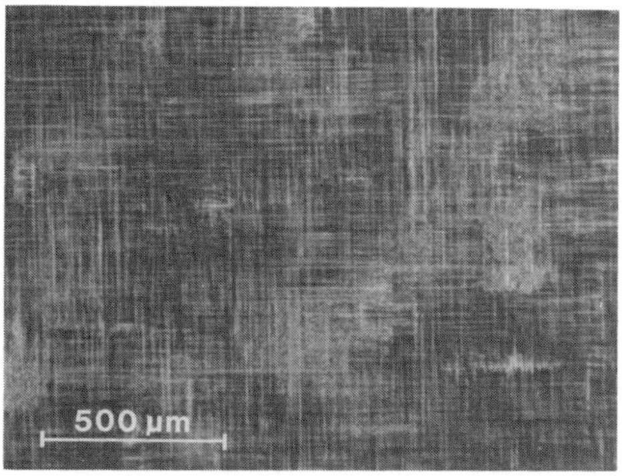


Fig.6 Micrograph of an etched 1.5$\mu$m thick layer of $Si_{0.875}Ge_{0.125}$.

## CONCLUSIONS

Schimmel etching in combination with Nomarski interference contrast microscopy has been used to examine the nucleation and propagation of mismatch dislocations in strained $Si_{1-x}Ge_x$ layers deposited on (001) Si substrates. Two types of nucleation sites have been identified. Those due to particulates incorporated in the layer during growth (PID's), act as very efficient sources for misfit dislocations which are present on multiple slip planes. The second type of nucleation source was dominant, and appeared to be related to intrinsic threading dislocations originating in the underlying epitaxial Si with a typical areal density of $\sim 10^3 cm^{-2}$. From the position and orientation of the etch pits due to the threading component of the misfit dislocations we conclude that the majority of dislocations take the form of the classical half loop configuration where the two arms of the half loop glide out symmetrically from the central nucleation point.

A detailed examination of $Si_{0.875}Ge_{0.125}$ layers with thicknesses ranging from $\sim h_e$ to $\sim 2 \times h_m$ showed that the first misfit dislocations were formed when the thickness was just greater than $h_e$, ie. $h_F \sim h_e$. As the thickness was increased it became apparent that the relaxation was occurring in patches which gradually increased in size until they covered the entire surface. At the metastable critical thickness point ($h_m$) the patches were ~ 200$\mu$m in diameter and consisted of highly relaxed material. Growth of a 1$\mu$m $Si_{0.875}Ge_{0.125}$ layer on a highly dislocated Si substrate showed a substantial reduction in the value of $h_m$, being equivalent to a 1.8$\mu$m layer grown under standard conditions. This supports the hypothesis that intrinsic threading dislocations can act as nucleation points for misfit dislocations.

## ACKNOWLEDGEMENTS

The authors would like to thank S T Davey, M A G Halliwell, M H Lyons and P J Skevington for analytical support, G J Davies for useful discussions and the Director of Research and Technology, British Telecom for permission to publish.

## REFERENCES

1. J W Matthews, J. Vac. Sci. Technol. 12, 126 (1975).
2. A T Fiory, J C Bean, L C Feldman and I K Robinson, J. Appl. Phys. 56 1227 (1984).
3. M A G Halliwell, M H Lyons, S T Davey, M Hockly, C G Tuppen and C J Gibbings, to be published in Semicon. Sci. & Technol.
4. R H Miles, T C McGill, P P Chow, D C Johnson, R J Hauenstein, C W Nieh and M D Strathman, Appl. Phys. Lett. 52, 916 (1988).
5. I J Fritz, Appl. Phys. Lett. 51, 1080 (1987).
6. P L Gourley, I J Fritz and L R Dawson, Appl. Phys. Lett. 52, 377 (1988).
7. Y Kohama, Y Fukuda and M Seki, Appl. Phys. Lett. 52, 380 (1988).
8. C G Tuppen, C J Gibbings and M Hockly, to be published in J. Cryst. Growth.
9. P M J Marée, J C Barbour, J F van der Veen, K L Kavanagh, C W T Bulle-Lieuwma and M P A Viegers, J. Appl. Phys. 62 4413 (1987).
10. R People and J C Bean, Appl. Phys. Lett. 47 322 (1985).
11. J Y Tsao, B W Dodson, S T Picraux and D M Cornelison, Phys. Rev. Lett. 59, 2455 (1987).
12. B W Dodson and J Y Tsao, Appl. Phys. Lett., 51, 1325 (1987).
13. C J Gibbings and C G Tuppen in Epitaxy of Semiconductor Layered Structures, edited by R Tung, L R Dawson and R L Gunshor (Mater. Res. Soc. Symp. Proc. 102, Pittsburgh, PA 1988) pp. 311-314.

# CRITICAL THICKNESS FOR THREE-DIMENSIONAL EPITAXIAL ISLAND GROWTH

K. JAGANNADHAM AND J. NARAYAN
Materials Science and Engineering, North Carolina State University, Raleigh, North Carolina 27695

## ABSTRACT

The generation of misfit dislocation loops in three-dimensional epitaxial islands grown on thick substrates is analyzed. The coherent strain in the island is described by virtual interfacial dislocation loops situated in the interface. The traction free surface boundary conditions are satisfied by the surface dislocation loops situated on the surface of the island. A misfit dislocation loop is formed and the changes in the energy of the configuration used to determine if the total energy is lowered. The numerical analysis is carried out for hemispherical islands of GaAs grown on (100) silicon with a misfit dislocation of Burgers vector 3.84 Å. It has been found energetically favorable to nucleate a misfit dislocation loop at a distance of 3 Å from the interface when the radius of the hemispherical island is equal to or greater than 40 Å. In addition, a misfit dislocation loop could be nucleated at a larger distance from the interface when the size of the island is larger.

## INTRODUCTION

The transformation of an island from a coherent pseudomorphic state to an incoherent equilibrium state has been analyzed quite extensively, both experimentally [1-3] and using analytical modelling [4-7]. The growth of an epilayer has been classified into three categories [8]. In the first, a two-dimensional island on a substrate is attributed to low interfacial energy associated with the epilayer/substrate interface. The second category (Volmer-Weber) of growth of epitaxial layers consists of the islands grown on a substrate as a result of high interfacial energy. Finally, a combined planar and island growth (Stranski-Krastanov) is also observed in certain systems with an initial two-dimensional growth. In this paper, growth of epitaxial islands is considered in order to determine the critical size for nucleation of lattice misfit dislocations. Specifically, the analysis is applied to GaAs growth on (100) silicon surfaces which tend to form as hemispherical islands.

## A HEMISPHERICAL ISLAND

A hemispherical island grown on an infinitely thick substrate is considered. Therefore, we assume the strain in the substrate to be vanishing. The lattice mismatch with respect to the (100) silicon in the GaAs island in the x and y directions is described by virtual interface dislocation loops with mutually orthogonal Burgers vectors and located in the interface (x-y plane). Figure 1 is a schematic illustration of the virtual interfacial

Mat. Res. Soc. Symp. Proc. Vol. 130. 

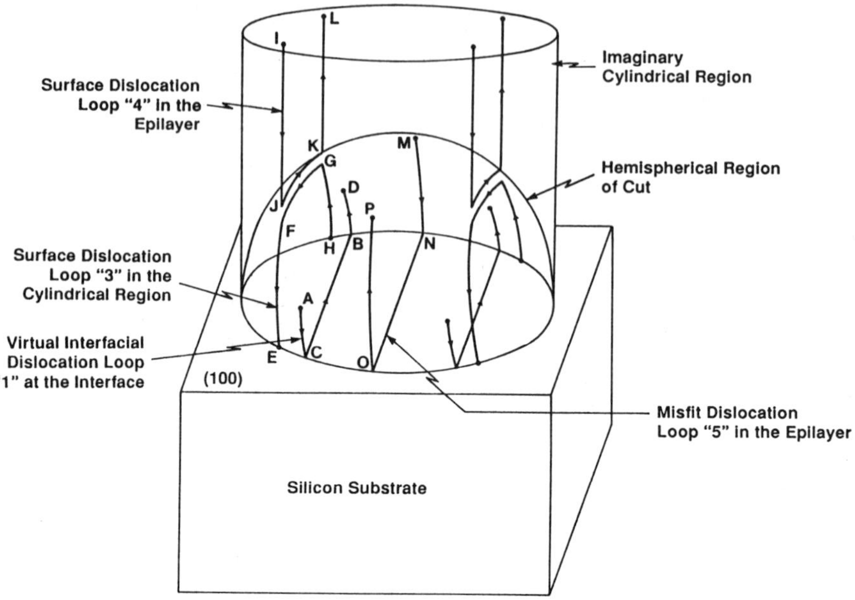


Figure 1. A schematic illustration of the hemispherical island grown on a substrate. The virtual interfacial dislocation loops describe the lattice mismatch, the surface dislocation loops satisfy the free surface boundary conditions and the misfit dislocation loop represents the relaxation of the lattice strain.

dislocation loop "1" with the segment CD situated at the interface. The dislocation with Burgers vectors in the x direction terminates at the free surface at points A and B. Similar dislocation loops with Burgers vector in the y direction describe the mismatch in the perpendicular direction. Additionally, the tractions should vanish at the free surface , i.e. $\sigma_{rr}=0$, $\sigma_{r\theta}=0$, and $\sigma_{r\phi}=0$ in the spherical coordinate system $(r,\theta,\phi)$. These free surface boundary conditions have been satisfied in the past [10,11] by a continuous distribution of surface dislocations with mutually orthogonal Burgers vectors. The surface dislocation loops "3" and "4" shown in figure 1 are situated in the island and the imaginary cylindrical region of GaAs respectively. The dislocation loop "3" (HGFE) in GaAs hemispherical island terminates at the free surface at points E and H and the dislocation loop "4" (IJKL) in the imaginary cylindrical region also terminates at the free surface but at points I and L. Similar dislocation loops are present symmetrically in the hemispherical island. The sense vector of the dislocation loops "3" and "4" near the hemispherical surface is opposite so that one can consider this configuration to be a dislocation segment starting at point I(or L) and ending at point E(or H). The Burgers vector of these dislocation loops is in the x direction and similarly dislocation loops can be constructed with Burgers vectors in the y and z directions. The distribution of such dislocation loops can be determined in order to satisfy the free surface boundary conditions. Next a cut can be made along the hemispherical surface without disturbing the free surface boundary conditions and thus separate the cylindrical region from the island. It has been further illustrated [12] that the traction free boundary conditions also correspond to the minimization of the energy of the dislocation configuration so that the positions and the Burgers vectors of the surface dislocations can be determined from the lowest energy configuration. It is necessary to use a discrete dislocation model [12] to arrive at a numerical solution to this complex problem. The nucleation of a misfit dislocation loop "5" (MNOP)in the hemispherical region of GaAs is considered to take place in the center of the island from the surface [9], as shown in figure 1.

The determination of self and interaction energy terms associated with the dislocation configuration representing the generation of a misfit dislocation, as shown in figure 1, is difficult, particularly, since the stress field expressions associated with the prismatic dislocation loops in a two-phase medium are not readily available. However, in the past, where tensile cracks can not easily be analyzed, an analogous shear type crack is used to represent the situation and arrive at an equivalent configuration. While the quantitative results may, in fact, differ by certain magnitude [13], the conceptual nature of the problem could be understood very clearly with the results providing valuable information. An equivalent dislocation configuration, although, not describing the same distortion, but with equivalent self and interaction energy terms is shown in figure 2. The dislocation loops are not prismatic in nature. Instead, glide loops with Burgers vectors in the x and y directions are used. The stress fields and the interaction energy terms associated with the glide dislocation loops parallel to the interface in a two-phase medium can be derived [14].

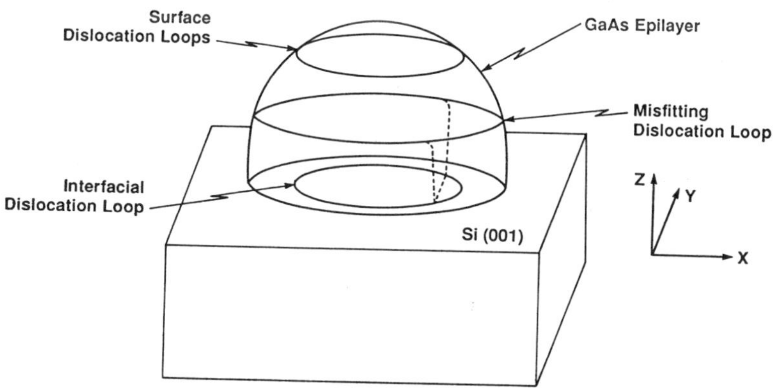


Figure 2. An equivalent dislocation configuration to that shown in figure 1. A lattice misfit dislocation with Burgers vector in the x or y direction only is shown. Glide of a portion of the misfit dislocation loop towards the interface is shown by the dotted line.

**NUMERICAL ANALYSIS**

The numerical analysis of the energy terms associated with dislocation configuration shown in figure 3 is carried out using four interfacial dislocation loops, with Burgers vectors in the x and y directions. The magnitude of the Burgers vectors associated with each loop is based on the lattice mismatch between GaAs and silicon. The energy associated with the dislocation configuration in the absence of the misfit dislocation is given by

$$E_{coherent}=E_C=2[E_S+E_I(S)+E_I(I,S)] \quad (1)$$

where, $E_S$ is the self energy associated with the surface dislocation loops, $E_I(S)$ the interaction energy between the surface dislocation loops and $E_I(I,S)$ the interaction energy between the surface and the interface dislocation loops. The cross terms of interaction energy, i.e, between dislocation loops with Burgers vectors in the x and y directions are neglected in comparison to the interaction energy between dislocation loops with parallel Burgers vectors. Similarly, the energy of the configuration in the presence of the misfit dislocation with Burgers vector in the x direction only is given by

$$E_{misfit}=E_m=2[E_S+E_I(S)+E_I(I,S)] +E_S(M)+E_I(S,M)+E_I(I,M)+E_F+E_\gamma \quad (2)$$

where in addition to the terms defined in equation (4), $E_S(M)$ is the self energy associated with the misfit dislocation, $E_I(S,M)$ the interaction energy between the surface dislocation loops and the misfit dislocation, $E_I(I,M)$ the interaction energy between the virtual interface dislocation loops and the misfit dislocation, $E_F$ the frictional energy spent in creation of the misfit dislocation loop, and $E_\gamma$ the free surface energy associated with the ledge step created at the surface of the hemispherical island. The frictional energy is given by

$$E_F=\pi b_m r_m^2 \tau_F \quad (3)$$

where $\tau_F$ is the lattice frictional stress opposing the formation of the misfit dislocation loop of radius $r_m$ and Burgers vector $b_m$. The lattice frictional stress $\tau_F$ is a strong function of the temperature so that formation of the misfit dislocation loop is favorable at a higher temperature. The surface energy $E_\gamma$ created by the ledge step is neglected in this analysis. The determination of self and interaction energy terms from the stress field expressions follows the standard procedure [15]. The condition that $E_m<E_C$ is used for nucleation of the misfit dislocation loop. In the present numerical analysis, the Burgers vector of the misfit dislocation loop is assumed to be that of a $90^o$ dislocation so that $b_m$=3.84 Å. The magnitude of the shear modulus of GaAs is assumed [15,16] to be 32.5 Gpascals whereas that of silicon to be 68.1 Gpascals. Also, the Poisson's ratio of GaAs, 0.20 is used whereas that of silicon, 0.218 employed for numerical purposes. The magnitude of lattice frictional stress of GaAs at 0.01XShear Modulus of GaAs is also used in the numerical

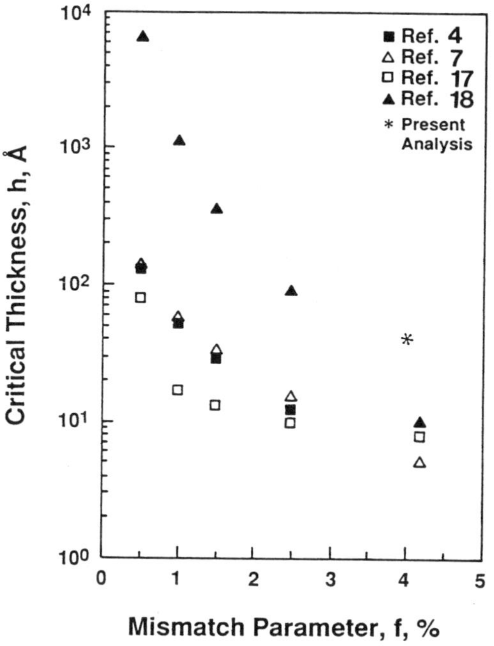


Figure 3. Comparison of critical thickness predictions with the misfit parameter obtained from different theories.

Table I. The coherent energy, $E_c$ and the energy of the misfit dislocation configuration, $E_m$ shown for different values of the sizes of the hemispherical epilayer and for different values of the height, z at which a misfit dislocation is nucleated. The magnitudes of the energy terms are expressed in $10^{-10}$ ergs.

| R (Å) | 25 | 30 | 40 | 50 | 75 | 100 | 200 |
|---|---|---|---|---|---|---|---|
| $E_{coh}$ | -0.080 | -0.163 | -0.574 | -1.584 | -8.895 | -26.477 | -340.250 |
| $E_{mis}$ | | | | | | | |
| z=3 (Å) | 0.694 | 0.065 | -3.686 | -13.410 | .... | .... | .... |
| z=5 (Å) | 0.989 | 0.899 | 0.375 | - 7.514 | -41.540 | -114.200 | -1671.00 |
| z=10(Å) | 1.071 | 1.911 | 1.772 | 1.251 | - 8.573 | - 47.580 | -1238.10 |
| z=15(Å) | 0.882 | .... | .... | .... | -10.133 | - 34.700 | .... |
| z=20(Å) | .... | .... | 1.493 | 0.998 | -10.523 | - 47.478 | - 785.38 |
| z=30(Å) | .... | .... | .... | .... | -10.726 | - 47.550 | .... |
| z=50(Å) | .... | .... | .... | .... | .... | .... | -649.11 |

evaluation of the energy terms. The free surface boundary conditions are satisfied by six surface dislocation loops, also, with Burgers vectors in the x and y directions. The positions and Burgers vectors of these surface dislocation loops should be determined by minimization of energy of the configuration. However, evaluation of the integrals to determine the energy terms has been found to require extensive computational time so that it is found economical to fix the positions and only vary the Burgers vectors. Specifically, since the magnitudes of the Burgers vectors remain as a coefficient in the expression for the interaction energy terms, it is possible to tabulate the energy terms between dislocation loops for unit Burgers vectors and a given set of radii and subsequently use these values in the minimization with respect to the Burgers vectors only. Through this procedure, the minimization can be completed. In order to compare the energy terms associated with the two dislocation configurations, one with no misfit dislocation ($E_{coherent}$) and the second with the misfit dislocation present ($E_{misfit}$) in the epilayer, the self and interaction energy terms associated with the virtual interfacial dislocation loops among themselves need not be included since these are common to both the configurations. This result is valid only in the absence of any structural changes taking place at the interface when the misfit dislocation loop is created, an assumption in our analysis.

**RESULTS OF NUMERICAL ANALYSIS**

The two energy terms associated with the dislocation configurations are presented in the table I for different sizes of the hemispherical islands and different distances from the interface at which a misfit dislocation is nucleated. The coherent energy, $E_c$ remains the same but the energy of the misfit dislocation configuration '$E_m$' changes with the distance z, from the interface as shown in table 1. The lowering of the energy in the presence of the misfit dislocation is not present for smaller sizes of the hemispherical islands. The self energy of the misfit dislocation loop $E_s(M)$ is larger than the negative interaction energy with the virtual interface dislocation loops, presumably because of the smaller loop size. Therefore, the hemispherical island tends to remain coherent for all positions of the loop from the interface. However, when the size of the hemispherical island is above a critical value, i.e. 40Å for GaAs on (100) Si, the negatve interaction energy dominates, so as to lower $E_{misfit}$ compared to $E_{coherent}$, only for smaller values of the distance of formation of the loop from the interface, as shown in table I. The results shown in table 1 also indicate that the lowering of the total energy is not present when the dislocation loop is nucleated at larger distances from the interface.

When the size of the hemispherical island increases, the magnitude of $E_m$ remains smaller than $E_c$ for larger values of the distance of nucleation so that it becomes energitically favorable to generate the dislocation. However, it is recognized from the results presented in table I that a larger decrease in $E_m$ occurs when the loop formation takes place at the interface rather than away from the interface. A step on the surface of the hemispherical island which is inevitably present in the formation of all crystalline epilayers also aids the formation of the misfit dislocation loop away from the interface. It is favorable to form the dislocation loop away from the interface so that interfacial ledge steps are avoided, particularly, in the presence of an increase in the interfacial energy due to formation of a incoherent interface. Since the interfacial energy is expected to be high for island growth to occur, this result also points to favorable nucleation of the dislocation loop away from the interface. The results of critical thickness of the epilayer obtained in the present analysis are compared with that of other models [4,7,17,18] in figure 3. The larger values of the critical size obtained in the present analysis compared to the other values may be explained in terms of the smaller coherent energy present in the hemispherical islands so that dislocation loops can not be nucleated easily.

The elastic strain energy associated with a hemispherical nucleus with inhomogeneous radial strains presented by Cabrera [19] along with the energy of a cross-grid of dislocations and that associated with the interface were used [20] to minimize the energy of the configuration. The critical radius of the hemispherical island for a misfit strain near 4% such as in platinum/gold or palladium/gold systems was found to be 82 Å and 74 Å respectively. These values are higher than the critical radius of 40 Å obtained in the present analysis for a misfit of 4% in GaAs/Si system. Firstly, the self energy associated with the cross-grid of dislocations in an infinite crystal is used in the previous analysis [20]. Secondly, the interfacial energy term is also included in the total energy assuming that the misfit dislocations are formed at the interface. On the other hand, a single misfit dislocation loop is formed with no changes in interfacial energy in the present analysis.

If the dislocation loop forms during growth at a higher temperature by a climb process in the presence of excess vacancy concentration and their high mobility, we can consider the formation of a dislocation loop away from the interface. Further, the loop can move towards the interface and combine with the virtual interface dislocations of opposite sign in the presence of the driving force associated with the negative interactive energy. Similarly, the movement of the misfit dislocation loop towards the interface by glide is schematically illustrated in figure 2 where the dotted line illustrates the loop motion. If prismatic dislocation loops are used to describe the lattice mismatch in the epilayer, as illustrated in figure 2, glide dislocation loops with Burgers vectors oriented at $60^\circ$ to the line vector will form on the surface and move along the glide plane towards the interface. These mechanisms of nucleation will become part of our subsequent analysis with the stress field terms determined for the two-phase medium.

**ACKNOWLEDGEMENTS**

The authors wish to acknowledge the support of this research by the Kopin Corporation, Inc. (Dr. J. C. C. Fan). The authors wish to thank Professors W. A. Jesser and J. H. Van der Merwe for valuable discussions.

**REFERENCES**

1. R. Hull, A. Fischer-Colbrie, S. J. Rosner, S. M. Koch, and J. S. Harris, Jr., M. R. S. Society, Symposium, 94, 25, 1987.
2. E. R. Kvam, D. J. Eaglesham, D. M. Maher, C. J. Humphreys, J. C. Bean, G. S. Green and B. K. Tanner, M. R. S. Society, Symposium, 104, 623, 1987.
3. N. Otsuka, C. Choi, Y. Nakamura, S. Nagakura, R. Fischer, C. K. Peng, and H. Morkoc, Appl. Phys. Lett., 49, 277, 1986.
4. J. W. Matthews, in "Dislocations in Solids", Ed. F. R. N. Nabarro, North Holland, Amsterdam, New York, 1975, p. 559.
5. W. A. Jesser and D. Kuhlmann-Wilsdorf, Phys. Stat. Solidi., 19, 95, 1967.
6. J. H. Van der Merwe, J. Woltersdorf, and W. A. Jesser, Mat. Sci. Engng., 81, 1, 1986.

7. S. Sharan, J. Narayan, and K. Jagannadham, in "Dislocations and Interfaces in Semiconductors", Eds. K. Rajan, J. Narayan and D. Ast, A publication of AIME, 420 Commonwealth Drive, Warrendale, PA. 15086, also J. Narayan, S. Sharan, A. Srivatsa, and A. S. Nandedkar, Mat. Sci. Engng., Bl, 105, 1985.
8. D. W. Pashley, J. Mat. Education., 7, 729, 1985.
9. J. Narayan, B. C. Larson, and W. H. Christie, in "Laser Solid Interactions and Laser Processing", Eds. S. D. Ferris, H. J. Leamy and J. M. Poate, M. R. S. Society, Boston 1978, p. 440.
10. K. Jagannadham, in "Structure and Deformation of Boundaries" Ed. K. N. Subramanian and M. A. Imani, The Metallurgical Society, Warrendale, PA 15086, 1986, p. 51.
11. K. Jagannadham and J. Narayan, J. Appl. Phys., 62, 1698, 1987
12. K. Jagannadham and M. J. Marcinkowski, in "Unified Theory of Fracture", Trans. Tech SA, Switzerland, 1983.
13. M. J. Marcinkowski and R. W. Armstrong, J. Appl. Phys., 43, 2548, 1972.
14. N. J. Salamon and J. Dundurs, J. Elasticity, 1, 153, 1971.
15. J. P. Hirth and J. LOthe, Theory of Dislocations, McGraw Hill, 1968.
16. S. Atachi, J. Appl. Phys., 58, Rl, 1985.
17. J. H. Van der Merwe, J. Appl. Phys., 34, 123, 1962.
18. R. People and J. C. Bean, Appl. Phys. Lett., 49, 229, 1986.
19. N. Cabrera, Surf. Sci., 2, 320, 1964.
20. W. A. Jesser and D. Kuhlmann-Wilsdorf, Phys. Stat. Solidi., 19, 95, 1967.

# SUBSTRATE STEP INDUCED STRAIN IN HETEROEPITAXIAL GROWTH

W. R. L. LAMBRECHT*, B. SEGALL*, and P. PIROUZ †

* Department of Physics, Case Western Reserve University, Cleveland, Ohio 44106

† Department of Materials Science and Engineering, Case Western Reserve University, Cleveland, Ohio 44106

## ABSTRACT

During epitaxial growth of lattice-mismatched materials, substrate surface steps induce *vertical misfit* between substrate and epilayer. It is shown that the resulting strain is accomodated by an interface dislocation, which is of the Read-Shockley type for a complete step and supplemetary displacement type for a demistep, using R.C. Pond's classification. Alternative models for the strain accomodation are considered. A localized "displacement boundary" is shown to be unfavorable irrespective of the film-thickness. Residual shear strain in the film requires debonding from the substrate, which is shown to be unfavorable with respect to the dislocation model. The distribution of strain between film and substrate is shown to depend weakly on film-thickness but strongly on the material's elastic constants.

## INTRODUCTION

Recently, there has been considerable interest in the role of substrate surface steps in semiconductor heteroepitaxy [1–7]. In the case of a sphalerite semiconductor grown on a Si (001) substrate, monoatomic steps induce *Inversion Domain Boundaries* (IDB) between neighboring domains which have grown independently on different terraces and coalesced. A slight tilt of the substrate surface off the low index plane, which favors diatomic steps, was therefore found to be beneficial because it eliminates the IDB's [2]. The energetics and structure of single and double steps on Si surfaces have been studied by means of tight-binding atomistic calculations by Chadi [5]. An electronic structure study of the (110) IDB in SiC is presented in another contribution to this Meeting [8].

In a recent paper, Pirouz et al. [7] pointed out that, in addition to the "chemical bond" mismatch, the step also induces a vertical displacement between domains growing on neighboring terraces if the materials are lattice-mismatched. This should lead to stress in the epitaxial film and possibly also in the substrate. This effect also exists for even-numbered steps in which case there is no IDB. In fact, the vertical misfit increases with the step size. In general the vertical misfit is given by:

$$v = nd_s - md_e, \tag{1}$$

where $n$ is the order of the substrate step and $m$ the number of layers deposited, and $d_s$ and $d_e$ are the interplanar spacings in substrate and epilayer respectively for the set of planes parallel to the film-substrate interface. For small steps and reasonable lattice-mismatch $n = m$.

The defect introduced by the vertical displacement was tentatively called a *displacement boundary* (DB) [7], and the question was raised at which critical thickness of the film it would become favorable to form a dislocation in order to release the strain energy associated with the DB. The question was posed in a similar context as the problem of lateral lattice-mismatch. In that case, as is well known, the growth is pseudomorphic up to a certain thickness, after which it becomes favorable to relieve the mismatch by means of misfit dislocations at the boundary [9–11]. The critical thickness in pseudomorphic growth is in fact not only governed by energetics but also by kinetic considerations [12], but this is outside the scope of the present study.

Here, a simple energetic analysis, is presented for the additional vertical misfit introduced by the presence of steps. Using linear continuum elastic theory, it is argued that in this case there is *no* critial film thickness associated with the problem in the sense discussed above. The very concept of a localized DB is unrealistic because it assumes complete rigidity of the material outside the DB-deformed region. This means that either the strain will spread over the whole domain, leading to a situation of homogeneous shear strain or an interfacial dislocation will appear at the step edge. The former alternative requires debonding the film from the substrate and will thus cost adhesion energy.

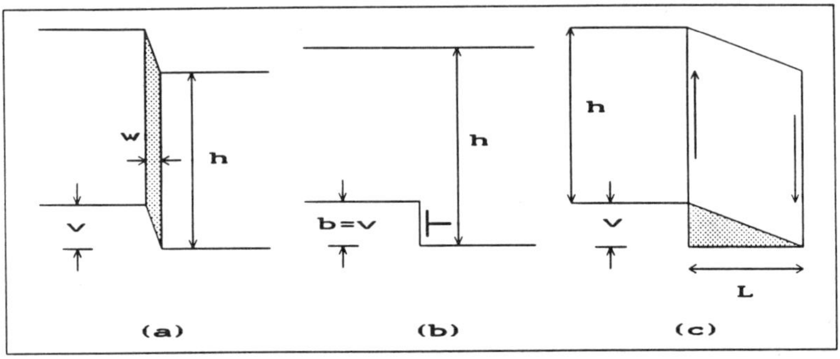


*Fig. 1: Schematic representation of limiting cases for accomodating the vertical misfit between substrate and epilayer: (a) localized displacement boundary plane, (b) interface dislocation, and (c) residual shear in film through debonding from substrate.*

These three alternatives, as shown schematically in Fig. 1, are compared and the dislocation model is found to be the favorable one for realistic assumptions about the adhesion energy. The type of dislocation and the distribution of strain between substrate and epilayer will be discussed.

THEORY

First consider the situation depicted in Fig. 1 (a). Let the vertical displacement $v$ be accomplished over a distance $w$, perpendicular to the DB. Clearly, this corresponds to a pure shear strain $v/w$ and the total energy in the sheared region per unit area of the DB plane is given by $E_{\text{shear}} = \mu v^2/2w$ where $\mu$ is the shear modulus. For a film of thickness $h$, the total energy per unit length of the step-edge is given by:

$$E_{\text{DB}} = \frac{\mu v^2}{2w} h. \tag{2}$$

A localized DB assumes that $w$ has an atomic dimension. As the strain energy is inversely proportional to $w$ the strain will try to spread as much as possible and therefore this assumption is unrealistic unless some other force keeps the distortion localized. In the present context, it is the adhesion between substrate and film which forces the step to be accomplished over a distance of an atomic dimension. This assumption will be discussed in what follows. In fact, the localized DB-model assumes the two domains to be completely rigid outside the sheared region. If, on the other hand, one allows the system to relax elastically, it is clear that the planar defect will be reduced to a line defect along the step edge, i.e. an interface dislocation. Interface linear defects are most easily discussed in terms of a Volterra-construction, (see e.g. Ref.13). For the continuum-model presently under discussion, this immediately shows that the step leads to a pure edge dislocation with Burgers vector perpendicular to the substrate-film interface and equal to the vertical displacement $b = v$.

The crystallography of interface defects has been discussed extensively by Pond [13]. The following discussion about the crystallography of the interface dislocation is restricted to epitaxy of sphalerite crystals on a (001) substrate of the diamond cubic structure. According to Pond [13], an interface dislocation corresponds to a Burgers vector

$$\vec{b} = \vec{t_1} - \mathbf{T}\vec{t_2}, \tag{3}$$

where $\vec{t_1}$ and $\vec{t_2}$ are lattice translation vectors in either of the two half-crystals or linear combinations of a translation vector and a supplementary displacement. For a nonsymmorphic crystal, the latter is the non-primitive translation corresponding to certain spacegroup operations, i.e. $\frac{1}{4}a\langle 111\rangle$ for the cubic diamond structure. For a symmorphic but nonholosymmetric crystal such as the sphalerite crystal, the supplementary displacement is associated with the exchange operator. The latter is not a spacegroup element but the lack of the inversion for the sphalerite structure as compared to the diamond cubic structure gives rise to a multiplicity of two possible domains which can be related by an atom exchange operation. This generalized symmetry operation must be taken into account in order to find all possible interface defects. $\mathbf{T}$ is a global transformation of the coordinate system of one half-crystal with respect to the other, here assumed to equal the identity operator because we assume parallel epitaxy. This type of dislocations has been called a *Read-Shockley dislocation* by Pond [13] if only lattice translation vectors occur. This is the case for complete steps, i.e. even numbered atomic steps and the smallest possible Burgers vector (for $n = m = 2$ in Eq. (1)) is then given by $\vec{b} = \frac{1}{2}(a_s - a_e)\langle 011\rangle$. For demisteps, i.e. monoatomic steps or odd multiples thereoff one is dealing with a *supplementary displacement interface dislocation* in Pond's classification [13] and the smallest possible Burgers vector (for $n = m = 1$ in Eq. (1)) is given by $\vec{b} = \frac{1}{4}(a_s - a_e)\langle 111\rangle$. Note, that in addition to the dislocation, an IDB is associated with the demistep because of the exchange operator involved. This assumes a preferred bonding of one of the sublattices of the sphalerite to the substrate. In general the Burgers vector is seen to have a lateral component and the dislocation can thus actually contribute to the release of lateral strain at the cost of introducing some vertical strain [6]. If the surface is a vicinal surface with regularly spaced steps, these dislocations form a dislocation network. The dislocation-line lies along the step-edge and thus can have many possible orientations in the interface plane. A more realistic model would include relaxations from the ideal geometry, e.g. the two half-crystals can be translated with respect to each other, both perpendicular and parallel to the interface, thus introducing a non trivial transformation operation $\mathbf{T}$ in Eq. (3) above. In the following, we do not further discuss the crystallography but focus on the continuum elastic model.

The energy of an edge dislocation at a distance $h$ below a free surface is given by

$$E_{\text{edge dislocation}} = \frac{\mu b^2}{4\pi(1-\nu)} \ln\frac{h}{r}, \tag{4}$$

where, $\nu$ is Poisson's ratio and $r$ is the core radius of the dislocation, which is of the order of magnitude of the Burgers vector $b$. For the interface dislocation under study, the expression will be more complicated, involving the elastic constants of both materials, but Eq.(4) is sufficient for the present purpose. In fact, it is clear that with $w \approx r$, the localized DB energy (Eq. (2)) scales linearly with the film-thickness $h$ while the dislocation energy (Eq. (4)) scales only logarithmically. Therefore, the localized DB-model can be ruled out for all film thicknesses for which it makes sense to consider the question. For a monoatomic layer, the two models are indistinguishable, because one can not speak of a planar boundary in that case. It is intuitively obvious that even for thicknesses of a few atomic layers, the notion of a region of atomic width of constant shear strain, which is characteristic of the DB-model, is untenable.

The next question to be considered is whether the adhesion energy is in fact sufficiently large to produce an abrupt step between the neighboring domains at the interface. As already mentioned, an alternative mechanism to spread the strain would be to shear the film homogeneously over the whole domain. In that case the width $w$ becomes the terrace width $L$ and the residual shear energy can be made arbitrarily small with respect to the dislocation energy. However, as shown in Fig. 1 (c), this requires a debonding of the film from the substrate, in order to accomplish the vertical displacement gradually over the whole domain. This necessitates the formation of a "wedge" between substrate and film as indicated schematically in Fig. 1 (c). The energy cost involved in making this wedge will now be estimated.

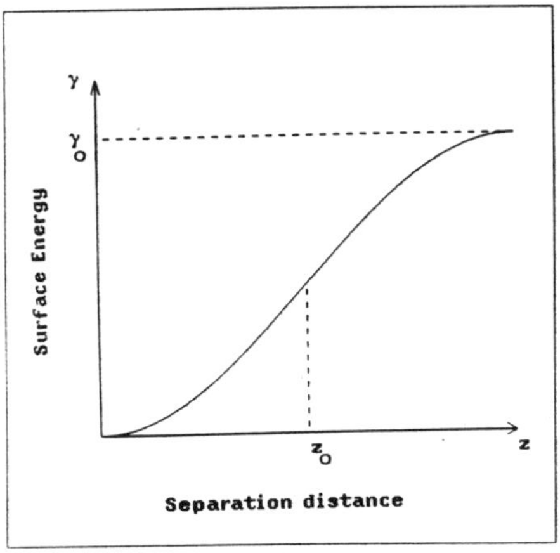


*Fig. 2: Schematic representation of the adhesion energy as a function of surface separation*

This requires a model for the adhesion energy as a function of separation of the two materials: $\gamma(z)$. The details of this function are not important for the estimate here. The present choice:

$$\gamma(z) = \begin{cases} \frac{\gamma_0}{2}[1 - \cos\frac{\pi z}{2z_0}], & \text{if } z \leq 2z_0; \\ \gamma_0, & \text{if } z > 2z_0, \end{cases} \tag{5}$$

schematically shown in Fig. 2 has a turn-over point at $z_0$, which basically determines the length-scale over which the separation takes place. The adhesion energy per unit area at large separation is given by $\gamma_0$. The energy required to produce the wedge is then given by:

$$E_{\text{wedge}} = \int_0^L \gamma(vx/L)\, dx. \tag{6}$$

Using the specific model of Eq. (5), one obtains for $v \leq 2z_0$:

$$E_{\text{wedge}} = \frac{\gamma_0 L}{2}[1 - \frac{2z_0}{\pi v}\sin\frac{\pi v}{2z_0}]. \tag{7}$$

As $v$ should be comparable in magnitude to $z_0$, the factor between square brackets is of order one and the essential result is thus that the energy required to debond the substrate and the film sufficiently to accomodate the vertical displacement $v$ by residual shear in the film is proportional to the terrace width over which the debonding has to take place and to the adhesion energy per unit surface area.

This energy must now be compared with the elastic energy of the dislocation model as given in Eq. (4), i.e the quantities to be compared are: $\gamma_0 L$ and $\mu b^2$. A simple order of magnitude estimate can be made for SiC/Si (001) adhesion energy assuming Si-C bonding. Each Si has two dangling bonds on the (001)-surface and a SiC-bond energy is approximately half the cohesive energy of SiC per atom, i.e. 3.17 eV. Thus, one has 6.34 eV per 2D-unit cell of the Si surface, i.e. 14.85 Å$^2$ or an adhesion energy of $\sim 7$ Jm$^{-2}$. This is an upper limit because the SiC bonds will actually be stretched and are thus weaker than perfect SiC-bonds and the mismatch of $\sim 20$ % will cost considerable misfit dislocation strain energy. Nevertheless, the result is of the expected order of magnitude for adhesion energies of ceramic materials [14]. A typical terrace width is 1000 Å thus giving $\gamma_0 L \approx 10^{-7}$ Jm$^{-1}$. The shear moduli for a very stiff material such as SiC is of order $10^{11}$ Jm$^{-3}$ and the vertical misfit is a fraction of an interatomic spacing i.e. of the order $10^{-10}$ m, so $\mu b^2 \approx 10^{-9}$ Jm$^{-1}$. Thus, under the usual circumstances, the debonding energy is much larger than the energy involved in the dislocation model. In case of very weak adhesion and small terrace widths, i.e. essentially a rough surface, it might be favorable to accommodate some part of the vertical displacement through this mechanism of debonding.

Next, the question of how the dislocation strength and thus stress is distributed between substrate and epilayer will be adressed. In principle the interface dislocation strain field should be calculated, taking into account the boundary conditions at the interface between the two materials with different elastic constants. In order to make a qualitative estimate, we replace the interface dislocation by two partial dislocations with Burgers vectors $b_s$ and $b_e$ in the substrate and epilayer respectively, such that their sum equals the total vertical misfit: $b_s + b_e = v$. Minimizing the sum of the two partial dislocation energies with respect to $b_e$ leads to:

$$\frac{\mu_e}{(1-\nu_e)} b_e \ln\frac{h}{b_e} = \frac{\mu_s}{(1-\nu_s)} b_s \ln\frac{R}{b_s}. \tag{8}$$

Here $R$ is an upper cut-off for the radius of the strained region in the substrate, which is usually assumed to equal the average distance between dislocations. Also, $b_e/h$ and $b_s/R$ are taken to be small. In practice, the presence of lateral misfit dislocations will determine the cut-off $R$. This equation implies a weak dependence on film-thickness. For $R \approx h$, the dislocation strengths are shared in inverse proportion to the shear moduli:

$$\frac{b_e}{b_s} = \frac{\mu_s/(1-\nu_s)}{\mu_e/(1-\nu_e)} \approx \frac{\mu_s}{\mu_e}. \tag{9}$$

CONCLUSIONS

In summary, the main conclusions from the present analysis are:

(1) localized displacement boundary planar defects are not favorable under any circumstances;

(2) vertical misfit, induced by steps on the substrate surface, will be accomodated by an interfacial dislocation of the Read-Shockley or supplemetary displacement type depending on whether the step is a complete one or a demistep using the classification of Pond [13];

(3) residual shear strain in the film is unfavorable because of the large energy cost of reducing the adhesion between substrate and epilayer over the whole size of the terrace;

(4) the distribution of strain between substrate and epilayer depends weakly on the film-thickness but is essentially inversely proportional to their elastic constants.

Clearly, a more detailed study of the atomic relaxations and strain fields around interface dislocations would be very interesting but because the associated Burgers vectors are quite small, it may be anticipated that atomistic models will be required.

Stimulating discussions with R. Mullen, R. Ballarini and H.-J. Möller are gratefully acknowledged. This work was supported by DARPA-ONR Contract No. N000-13-86-K-0773.

## REFERENCES

[1] D. E. Aspnes and J. Ihm, *Phys. Rev. Lett.* **57**, 3054 (1987)

[2] T. Nakayama, Y. Tanashiro, and K. Takayanagi, *Jpn. J. Appl. Phys. Pt. 2* **26**, L280 (1987)

[3] R. J. Fisher, N. C. Chand, W. F. Knopp, H. Morcoc, L. P. Ericson, and R. Youngman, *Appl. Phys. Lett.* **47**, 397 (1985)

[4] S. Sakamoto and G. Hashiguchi, *Jpn. J. Appl. Phys. Pt. 2* **25**, L78 (1986)

[5] D. J. Chadi, *Phys. Rev. Lett.* **59**, 1691 (1987)

[6] J. H. van der Merwe, *Phys. Rev.* **B 37**, 2892 (1988); *S.-Afr. Tydskr. Fis.* **9**, 55 (1986); J. H. van der Merwe and H. Kunert, *Phys. Rev.* **B 37**, 2902 (1988)

[7] P. Pirouz, C. M. Chorey, and J. A. Powell, *Appl. Phys. Lett.* **50**, 221 (1987)

[8] W. R. L. Lambrecht and B. Segall, *Mat. Res. Soc. Proc.* MRS Fall Meeting 1988, Symposium Q, submitted

[9] J. H. van der Merwe, *J. Appl. Phys.* **34**, 123 (1962)

[10] J. W. Matthews and A. E. Blakeslee, *J. Crystal Growth* **27**, 118 (1974)

[11] R. People and J. C. Bean, *Appl. Phys. Lett.* **47**, 322 (1985)

[12] B. W. Dodson and J. Y. Tsao, *Appl. Phys. Lett.* **51**, 1325 (1987)

[13] R. C. Pond, in *Dislocations and Properties of Real Materials*, (London: Inst. Metals 1987), p. 71

[14] W. D. Kingery, H. K. Bowen, and D. R. Uhlmann, *Introduction to Ceramics*, Second Edition (John Wiley, New York 1976), p. 177

# X-RAY DIFFRACTION STUDIES OF ANNEALED SiGe/Si SLS

WALTER P. LOWE, R. A. MACHARRIE AND J.C. BEAN
AT&T Bell Laboratories, 600 Mountain Ave., Murray Hill, NJ 07974

## ABSTRACT

We have done x-ray diffraction to study the affect of annealing on a SiGe/Si strained-layer superlattice (SLS). The post anneal spatial distribution of the Ge and Si atoms within the alloy layer is of particular interest. X-ray data taken from an annealed sample along the superlattice direction (100) show two sharp and well separated superlattice envelope functions. In contrast to the unannealed sample, and what would be a typical scattering spectra for a superlattice, the envelopes bisect the position of the unannealed sample's envelope function. The position of the envelope function is related to the tetragonal distortion $\varepsilon_T$ in the SiGe alloy layers. The two separate envelopes ($\alpha$ and $\beta$) show that the SiGe alloy (100) planes have divided themselves into two groups with different interplanar spacings after the anneal. Each group with a tetragonal distortion such that $\varepsilon_T^\alpha < \varepsilon_T$ , $\varepsilon_T^\beta > \varepsilon_T$ and $\varepsilon_T^\alpha = - \varepsilon_T^\beta$ where $\varepsilon_T$ is the tetragonal distortion in the unannealed sample. This is strong evidence for coherent phase separation perpendicular to the growth direction, along the SiGe alloy layer. In another investigation a search of reciprocal space for the 2 X 2 {. . . SiSiGeGe . . .} (111) chemically ordered state showed no evidence for the existence of such an ordered state.

## INTRODUCTION

The strain-layer superlattice because of its highly strained state is studied with respect to how the strain changes in response to external stimuli. One such important and interesting stimulus is prolonged elevated temperatures. The dependence of the systems behavior on temperature is paramount in that device fabrication requires that the superlattice undergo anneals. Since the electronic energy bands are very sensitive to strain, any slight change in the strain away from the intended profile could prove harmful to device characteristics. The relationship between strain, energy gap and band offset will not be discussed here, but rather this discussion will be limited to the subtle changes in strain that can occur during annealing. Several theories have been presented recently on what happens when an SLS relaxes from its highly strained state [1-4]. The theories that address the issue structurally allow the interface to become incommensurate resulting in a non-pseudomorphic structure [4]. Some of them describe a phase separated state [3,4] which lends support to our interpretation of x-ray data from this experiment.

## DATA AND ANALYSIS

This particular sample of $Si_{.5}Ge_{.5}/Si$ was previously reported to transform to a chemically ordered state after being annealed [6]. The ordered state was described as consecutive pairing of pure Si and pure Ge planes along the (111) direction in the fashion {. . . SiSiGeGe . . .}. Half integral reflections at 1/2(111), 1/2(311), 1/2(331), etc., consistent with this ordering scheme were observed using TEM. Using the synchrotron source the annealed sample was examined for the 1/2(911) and $1/2(9\overline{1}\overline{1})$ reflections. A series of transverse scans (*h*=constant) followed by a series of orthogonal scans (*l*=constant) was done to form a grid centered at 1/2(911) and $1/2(9\overline{1}\overline{1})$. No evidence of half-order superlattice reflections was found using the grid scans at any of the half-order positions examined. It should be noted that although this the same sample used in the TEM study by Ourmazd et. al. [6], the actual sample parameters are different than those previously reported. However the concentration of 50% Ge would be more consistent with a 2 X 2 ordering than the previously reported 40% Ge.

Further scattering data was collected along the (00*l*) direction about $l = 4$. The scattered intensity here is modulated by the form factor and strain profile between the Si and SiGe layers. This modulation produces a set of satellites with their intensities governed by an envelope function. The position and intensity of these satellites in reciprocal space directly reflect the attributes of the superlattice. Shown in figure 1 is the scattering data collected from the sample in this geometry. Upon annealing this sample, (shown in fig. 1b) each of the satellites split to form doublets. The doublets are centered about the positions of the unannealed satellites. The doublets can be divided into two sets, those at the large $l$ values (on the right) and those at the lower $l$ values (on the left). Each set can be encompassed by separate envelope functions. The overall intensity of the doublets is approximately equal to the intensity of the unannealed satellites. The presence of two envelopes indicate two composition waves, with different strain profiles, but with equal periods. This is accomplished by increasing the strain in one area of the alloy layer while decreasing the strain in an adjacent area by the same amount. Therefore the SiGe alloy layer has separated into two regions - one with less strain and more intense satellites ($\alpha$) and the other with more strain and less intense satellites ($\beta$).

The most reasonable situation would be that a strain wave (**k** ) purely along a direction orthogonal to (00*l*) must exist in the alloy layer. Using a scattering model that assumes an abrupt interface it is possible to obtain $\varepsilon_T$ for each of the satellite envelopes. The difference in $\varepsilon_T$ for the $\alpha$ and $\beta$ regions is the amplitude of **k** . Relating this difference to the strained SiGe cell in the c-direction results in a .006 Å difference between the $\alpha$ regions and the $\beta$ regions. This would correspond to a composition variation of approximately 1% change in the Ge concentration between the $\alpha$ and $\beta$ regions.

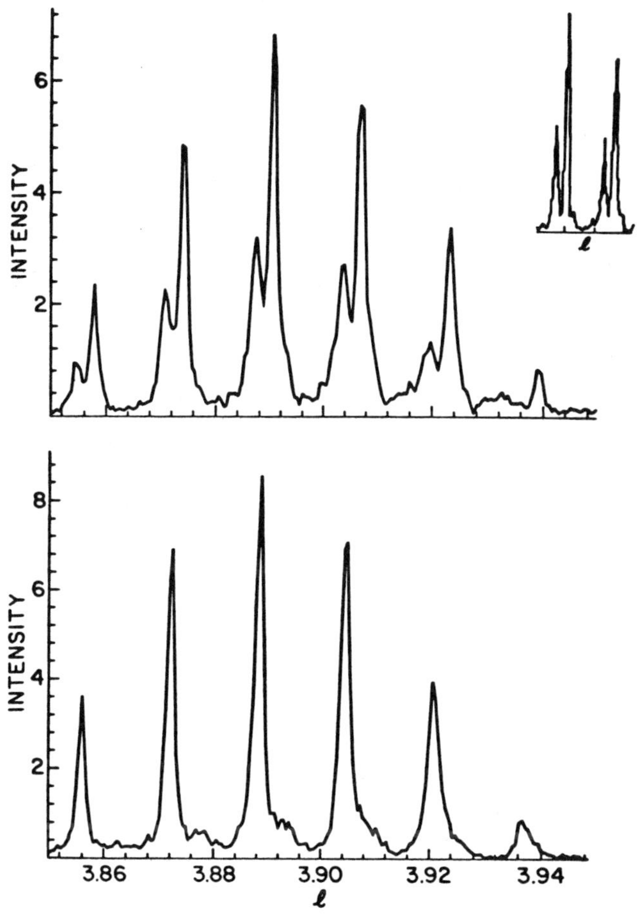


Figure 1. The x-ray diffraction satellites from the $Si_{.5}Ge_{.5}/Si$ SLS before (fig. 1a) and after annealing (fig. 1b). Two set of doublets recorded (fig. 1b inset) using the synchrotron source highlight the discrete doublet separation.

DISCUSSION

Since the variation in Ge concentration is small the interface between $\alpha$ and $\beta$ regions may not be observable in high resolution lattice imaging. However in TEM micrographs of several periods, stress fields indicative of the schematic in figure 2 are visible[7]. A potentially similar effect has also been seen in TEM studies of SiGe/Si SLS in other studies [8]. Superlattices grown with compositions very near their critical thickness exhibit an in-plane strain modulation in TEM cross-sections. This "pagoda-defect" [8] is shown to be coherent with the surrounding matrix in many cases. Spinodal decomposition is suggested for these types of modulated systems, [9] however no time dependence of the amplitude of **k** has been measured in this sample.

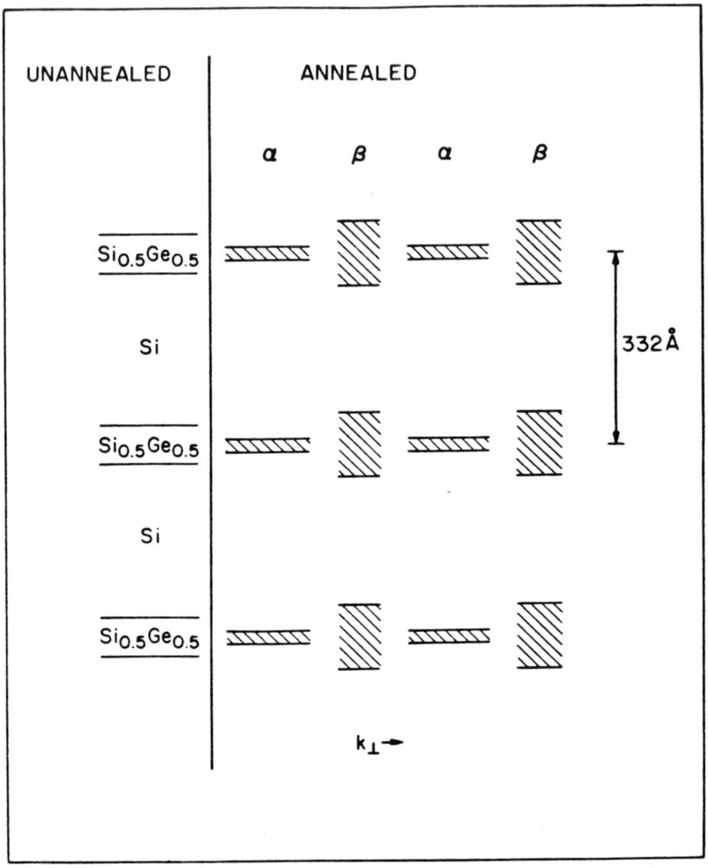


Figure 2. A schematic representation of layer strain demonstrating that the α and β regions are correlated along the (00*l*) direction.

SUMMARY

In summary we have observed a transition in an annealed $Si_{.5}Ge_{.5}/Si$ superlattice in which the $Si_{.5}Ge_{.5}$ alloy layer divides into two discrete regions with respect to the tetragonal distortion. The tetragonal distortion is reduced for one region and increased by the same amount for the other region. The regions do not appear to be equal in sample volume, with the larger being where $e_T$ is decreased. Since the regions extend throughout the entire 20 periods of the superlattice the nucleation of the two regions must be coordinated over a large sample volume. As for the order-disorder transition the half-order reflections observed in this sample using TEM are not observed in the x-ray data. One possible explanation for this is that the ordered state does not exist in a substantial fraction of the whole samples volume. Since the TEM data is taken from very small areas it may not be representative of the entire sample.

## ACKNOWLEDGEMENT

We wish to thank A. Ourmazd for the annealed $Si_{.5}Ge_{.5}/Si$ superlattice sample. Some of the studies presented here were done at NSLS at Brookhaven National Labs.

## REFERENCES

[1] J. L. Martins and A. Zunger, Phys. Rev. Lett. **56**, 1400 (1986).

[2] C. P. Flynn, Phys. Rev. Lett. **57**, 599 (1986)

[3] S. Ciraci and I. P. Batra, Phys. Rev. Lett. **58**, 2114 (1987)

[4] J. Y. Tsao, B. W. Dodson, S. T. Picraux and D. M. Cornelison, Phys. Rev. Lett. **59**, 2445 (1987).

[5] J. C. Bean and E. A. Sadowski, J. Vac. Sci. Technol. **20**, 137 (1982).

[6] A. Ourmazd and J. C. Bean, Phys. Rev. Lett. **55**, 765 (1985).

[7] R. Hull, J. C. Bean, J. M. Gibson, K. J. Marcantonio, A. T. Fiory and S. Nakahara, in *Layered Structures, Epitaxy, and Interfaces*, edited by J. M. Gibson and L. R. Dawson, MRS Symposia Proceedings No. 37 (Materials Research Society, Pittsburgh; PA, 1985), p. 261.

[8] D. D. Perovic, G. C. Weatherly and D. C. Houghton, to appear in the proceedings of the NATO Advanced Research Workshop on " Advanced Characterization of Semiconductor Materials" held in Bristol, England Sept. 1988.

[9] F. Spaepen, in *Layered Structures, Epitaxy, and Interfaces*, edited by J. M. Gibson and L. R. Dawson, MRS Symposia Proceedings No. 37 (Materials Research Society, Pittsburgh; PA, 1985), p. 295.

PART III

# Fracture and Related Failure Mechanisms

# THE CRACKING AND DECOHESION OF THIN BRITTLE FILMS

MU-SAN HU
Materials Department, College of Engineering, University of California, Santa Barbara, CA 93106

## ABSTRACT

Film cracking and decohesion are two typical failure modes observed in thin coatings. In this investigation, failures were initiated in brittle Cr films deposited on both ductile and brittle substrates to study these phenomena. Fracture mechanics models were proposed to interpret the experimental observations. Excellent agreement between theory and experiment has been demonstrated.

## INTRODUCTION

Thin films are widely used in applications, such as electronic devices, magnetic recording media, cutting tools and optical lenses. Besides the specific material properties needed for each technical application, the mechanical integrity of the film is the key to a high manufacturing yield and reliable service. In this article, film cracking (splitting) and film decohesion, the two typical failure modes when films are stressed in tension, will be described and analyzed. To observe the sequence of failure events, brittle Cr films were deposited onto various substrate materials by electron-beam evaporation. Cracks were then initiated and monitored through bending tests [1] and by controlled stress corrosion cracking [2]. The results are used to validate the analytical predictions.

## FILM CRACKING

### Theoretical Background

For a steady state, long crack in an elastic system (Fig. 1), non-dimensional analysis shows that the mode I stress intensity factor at the crack front, $K_I$, is independent on the crack length, and can be expressed as [1,2,3]:

$$K_I = \Omega_c(\Sigma)\,\sigma\sqrt{h} \qquad (1)$$

where $\sigma$ is the film stress, h is the film thickness, $\Sigma$ is the Young's modulus ratio of the film to the substrate, ie. $E_f/E_s$, and $\Omega_c(\Sigma)$ is a dimensionless quantity.

When ductile substrates are used, the stress concentration at the interface crack tip can cause substrate plasticity and blunt the tip (Fig. 5). In this case, the function $\Omega_c$ is also dependent on the ratio of the film stress to the substrate yield strength,

$$K_I = \Omega_c(\Sigma, \sigma/Y)\,\sigma\sqrt{h} \qquad (2)$$

By rearranging and imposing the critical condition, a critical film thickness $h_c$ can be derived as:

$$h_c = \left(\frac{K_{Ic}}{\sigma}\right)^2 \frac{1}{\Omega_c^2(\Sigma, \sigma/Y)} \qquad (3)$$

In this expression, $K_{Ic}$ is the mode I fracture toughness of the film, which is a material constant. Below the critical thickness, $h_c$, film cracking is completely suppressed. Furthermore, since the film stress $\sigma$ can be controlled or measured in experiments, the only unknown needed to calculate $h_c$ is $\Omega_c$. Therefore, the following discussion will concentrate on $\Omega_c$ and how it changes with respect to changes in material properties. Note that a higher $\Omega_c$ results in lower $h_c$, ie. the film is more likely to crack.

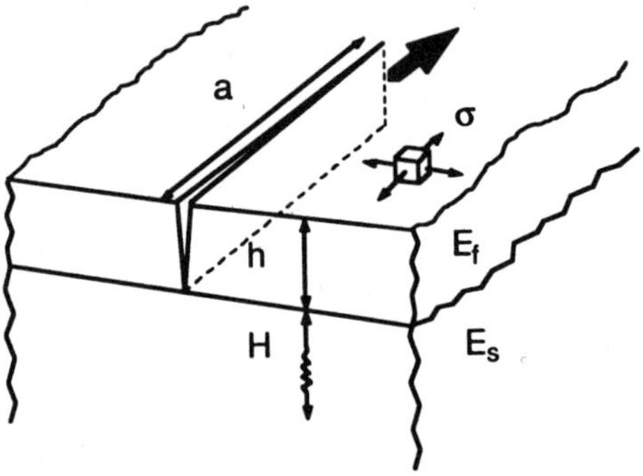


Figure 1 Schematic illustration of a steady state film crack in an elastic system

Calculation of $\Omega_c$ based on a shear lag model [1] (Fig. 2) reveals the enhancement of film cracking with a low yield strength substrate. This can be visualized as follows: as Y decreases the substrate plasticity tends to increase the crack tip opening at the interface. This, in turn, encourages the film crack to propagate further. An increase in $\Omega_c$ with respect to higher $\Sigma$ is also observed.

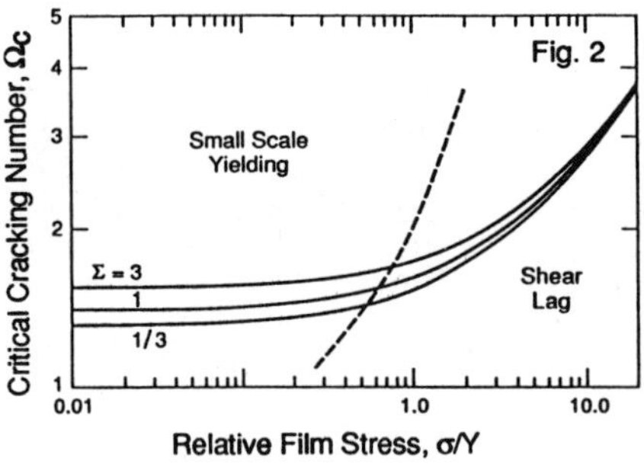


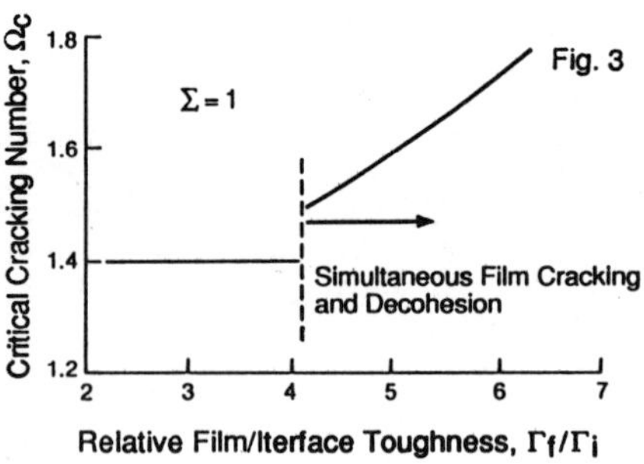


Figure 2 $\Omega_c$ vs. $\sigma/Y$ at different modulus ratios, $\Sigma$

Figure 3 Trends in $\Omega_c$ with $\Gamma_f/\Gamma_i$ at different modulus ratios, $\Sigma$

If decohesion initiates after film cracking, the two processes are independent and the previous film cracking results are still applicable. However, if the two precesses happen simultaneously, then $\Omega_c$ will also be dependent on the fracture toughness ratio of the film to interface, $\Gamma_f/\Gamma_i$. This effect is shown in Fig. 3 for an elastically homogeneous system with steady state interface decohesion. Specifically, when $\Gamma_f/\Gamma_i$ is greater than about 4, simultaneous film cracking and decohesion occur, which results in a higher $\Omega_c$. This trend is expected because decohesion relaxes the constraint at the interface and encourages the film crack to propagate.

## Experiment

To evaluate the above predictions, brittle Cr films were deposited onto Al (6061-T6) and stainless steel (440-C) substrates by electron beam evaporation. Ductile substrate are used to prevent cracks from penetrating into the substrate. The Cr films, with thicknesses ranging from 400 nm to 1 µm, have tensile residual film stresses on the order of 800 to 900 MPa. Specimens were loaded in four-point bending with successively increasing loads.

The morphology of the film suface was recorded by optical microscopy with Normarski differential interference. At the critical loads, film cracks were observed between the inner loading points. (Fig. 4) The cracks are perpendicular to the direction of the applied stress and parallel to each other. In both systems, the tendency of thicker films to crack at lower critical stresses is observed.

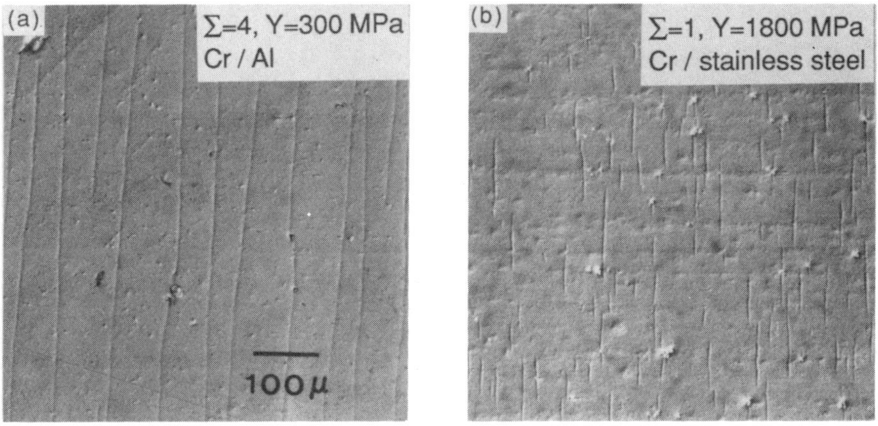


Figure 4 Optical micrograph showing film cracks in (a) Cr/Al (b) Cr/stainless steel systems

The Al substrate has a low yield strength, therefore, substrate plasticity near the crack tip is suspected. A specimen is prepared for a cross sectional SEM study to examine this effect near the interface. (Fig. 5) Note that film crack is stopped at the interface. Also, the crack surfaces are almost parallel throughout the film thickness, indicating that the tip is blunted at the interface.

Conversely, the high yield strength of the steel causes the near-tip stress concentration to persist and may even encourage the nucleation of an interface crack. When film cracks first appeared, no decohesion could be observed at this load. (Fig. 4b) However, as the load was further increased, the short cracks not only grew longer but also exhibited stable interfacial decohesion. (Fig. 6) The debond length increased with load, and from this information the ratio $\Gamma_f/\Gamma_i$ was calculated to be 2.8. [1] Consequently, decohesion and film cracking are independent at the critical load for film cracking, as confirmed by experimental observation.

The difference in substrate properties explains why film cracks on Al are generally longer than those on stainless steel at the critical stresses. (Fig. 4) Specifically, the higher $\Sigma$ and lower Y of the Cr/Al system results in higher $\Omega_c$, ie. easier crack propagation.

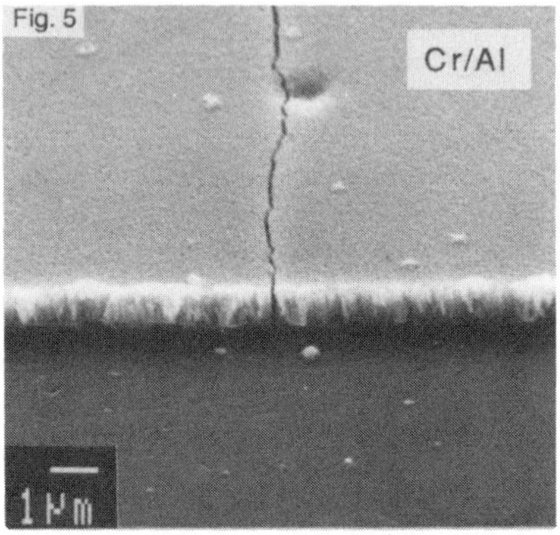


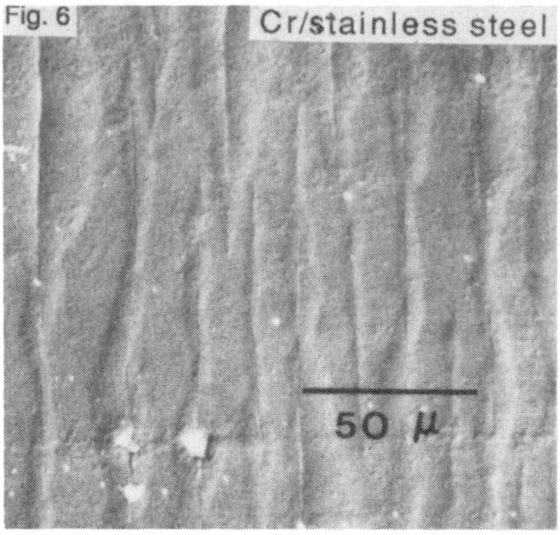


Figure 5 Scanning electron micrograph showing cross sectional view of a film crack running toward film edge in a Cr/Al system

Figure 6 Optical micrograph showing interfacial decohesion in Cr/stainless steel systems

## FILM DECOHESION BY SUBSTRATE CRACKING

### Theoretical Background

Film decohesion can occur either along the interface or into the substrate. Therefore, it is important to know not only *when* but also *where* the crack grows. For decohesion *along the interface*, a recent study [4] has shown that, $\Gamma_s/\Gamma_i > 4$ is a necessary condition, with $\Gamma_s$ being the fracture toughness of the substrate. This means that interface decohesion is possible only with a very tough substrate or weak interface, as demonstrated by the Cr/ stainless steel system.

When brittle substrates are used, cracks are more likely to propagate into the substrate parallel to the interface with a characteristic crack depth [2,5,6]. There are several postulates on the governing criterion of this steady state crack depth. The one used here is that the steady state crack trajectory follows $K_{II} = 0$ [7,8]. Thus, the steady state stress intensity factor for such a long substrate crack can be expressed as,

$$K_I = \Omega_c(\Sigma, \lambda)\,\sigma\sqrt{h} \tag{4}$$

where $\Omega_c$ is now a function of $\Sigma$ and $\lambda$, with $\lambda h$ being the steady state crack depth beneath the interface. Note that under critical conditions, $K_{Ic}$ will now be the fracture toughness of the *substrate*. Note that $\Omega_c$ is implicitly a function of $\Sigma$ only (Fig. 7), since the relation between $\Sigma$ and $\lambda$ is pre-determined by $K_{II} = 0$.

## Experiment

In this series of experiments, Cr films were evaporated onto glass substrates. Glass was chosen as the substrate because its stress corrosion cracking property is well studied [8,9], and its transparency facilitates crack observation. Stress corrosion was used to control cracking by changing the moisture content of the environment.

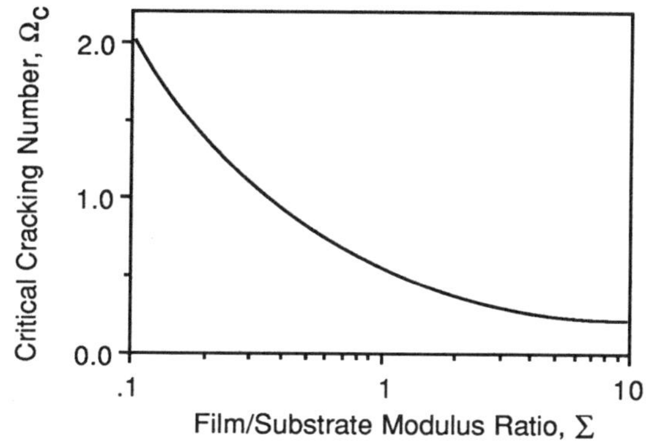


Figure 7 Trends in $\Omega_c$ with different modulus ratio $\Sigma$ for substrate decohesion

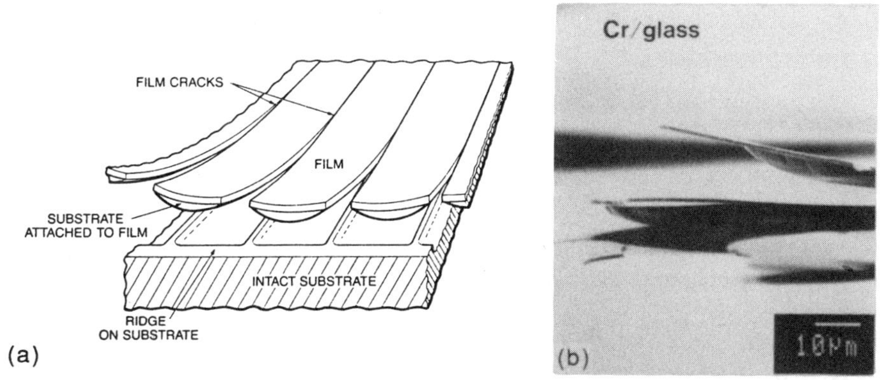


Figure 8 (a) Schematic illustration of the fracture in Cr/glass system (b) Scanning electron micrograph showing substrate crack parallel to the interface in a Cr/glass system

The failure in this system develops in three stages (Fig. 7) and can occur both from the film edges or from defects within the film. When specimens exceeded 650 nm in thickness and were in a 100% relative humidity environment, the fracture process started by film splitting. Some bifurcation of these cracks was also observed occasionally. This was followed by decohesion from the edges, initially along the interface (for about one film thickness) and then into the substrate, gradually reaching a stable crack depth of about 4 times the film thickness (Fig. 8).

The substrate decohesion velocities were determined by measuring the decohesion length at different times with optical microscopy. Experimental results show that a higher relative humidity leads to a higher crack velocity, confirming that decohesion in the glass substrate is controlled by stress corrosion cracking. Once the substrate crack reaches the steady state crack depth, decohesion can be analyzed by the model previously described.

The mode I stress intensity factor at the substrate crack tip can be determined by two independent processes. It can be calculated from Equation (4) for different thicknesses of films, with $\Omega_c$ ($\approx$ 0.3) from Fig. 7. The results can be compared with another set of $K_I$ extrapolated from known experimental data [8] with the measured crack velocities. The agreement is excellent with $K_I$ ranging between 0.30 to 0.45 MPam$^{1/2}$. $K_{II}$ can also be calculated using the steady state crack depth, $\lambda$, measured by scanning electron microscopy. It is essentially zero - ranging from 0.02 to 0.06 MPam$^{1/2}$ - consistent with the $K_{II}$ = 0 fracture criterion.

Results in Fig. 7 suggest that this type of substrate cracking is more likely to happen in low modulus films. Therefore, when systems incorporate polymer or metal films on ceramic substrates, this failure mechanism is preferred.

## CONCLUDING REMARKS

This article has demonstrated the important factors concerning thin film failure. Specifically, stress, interfacial fracture energy, film thickness, elastic modulus ratio, the substrate yield strength, and the loading can impact the failure mode. Valuable information can be derived from quantitative analysis, such as the critical film thickness needed to completely suppress film cracking/ decohesion. The fracture toughness of thin brittle coatings can also be estimated in some cases.

## ACKNOWLEDGEMENT

This work was supported under ONR Contract Number N00014-86-K-0753. The author would like to thank Dr. A.G. Evans for his helpful suggestion in this work.

## REFERENCES

[1] M. S. Hu and A. G. Evans, Acta Metall., in press.
[2] M. S. Hu, M. D. Thouless and A. G. Evans, Acta Metall., 36, p. 1333, (1988).
[3] G. Gille, Thin Solid Films, 110, p. 219, (1984).
[4] M. Y. He and J. W. Hutchinson, to be published.
[5] M. D. Drory, M. D. Thouless and A. G. Evans, Acta Metall., 36, p 2019, (1988).
[6] M. D. Thouless, A. G. Evans, M. F. Ashby and J. W. Hutchinson, Acta Metall., 35, p. 1333-1341, (1987).
[7] S. M. Wiederhorn, Frac. Mech. of Ceramics, 2, p. 613-644, (1974).
[8] S. W. Freiman, G. S. White and E. R. Fuller, Jr., J. Am. Ceram. Soc., 68, p. 108-112, (1985).

# TAILORED MICROSTRUCTURES TO CONTROL INTERFACIAL TOUGHNESS

T. S. OH, R. M. CANNON, and R. O. RITICHIE
Center For Advanced Materials, Lawrence Berkeley Laboratory, and Materials Science and Engineering Dept., University of California, Berkeley, CA, 94720

## ABSTRACT

Failure of ceramic-metal interfaces in response to residual or applied stress is frequently effectively brittle owing to the propensity for interfacial cracking caused by weaker than average chemical bonds and elastic or plastic discontinuities at the interface. Stronger atomic bonding derived from improved interfacial chemistry can enhance fracture resistance to a degree which may be limited by diversion of the crack into the brittle ceramic.

Alternatively, methods under study promote interfacial fracture toughness in glass-Cu bonds via near-interfacial microstructures that encourage greater energy dissipation in a region near the interface fracture. Particular success has obtained from microvoid toughening wherein placement of controlled arrays of microcrack-like-voids in the ductile metal, suitably near the interface, can markedly enhance interfacial fracture energy, e.g. by as much as two orders of magnitude and to much greater toughnesses than for glass. The toughness develops with extension of the crack owing to formation of a bridging zone behind the crack wherein the crack flanks are spanned by ligaments of plastically deforming metal film.

## INTRODUCTION

Stresses due to service environment, thermal cycling or processing (e.g. thin film deposition) can cause interfacial fracture. As depicted in Fig. 1, decohesion proceeds by expansion of a preexisting or newly created crack, the driving force for which can be exacerbated by discontinuities in mechanical properties at the interface as well as geometrical stress concentrations (1). Crack extension is resisted by the fracture energy, (i.e. toughness) which reflects the energy for bond rupture as well as energy dissipation associated with several irreversible processes (2,3). Thus, it is pertinent to assess interfacial adherence in terms of the fracture energy, $G_c$, and moreover, to enhance the interfacial fracture energy by chemical and microstructural means.

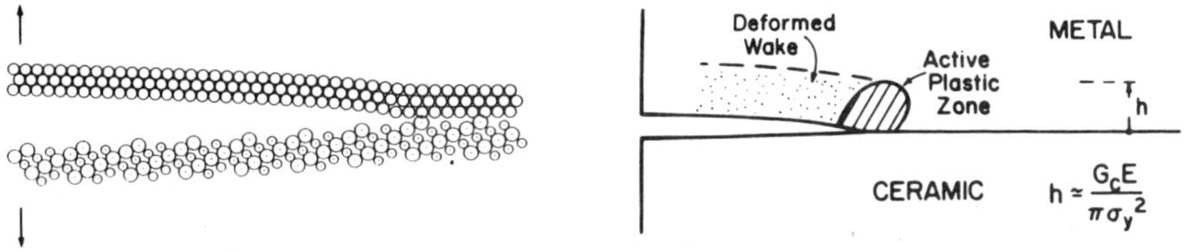


**Fig. 1.** Sketches of ceramic-metal interfacial fracture emphasizing a) atomic bond rupture and interfacial structure for a perfectly brittle (elastic) crack and b) the development of a dissipative plastic zone in the ductile metal.

The interfacial fracture energy embodies contributions from interfacial bond scission and the unavoidable crack-tip plasticity which dictate the inherent interfacial fracture resistance, $G_0$, Fig. 1. Additional toughening can also derive from crack-microstructure interactions, which often depend upon crack-tip shielding mechanisms, i.e. interactions which reduce the crack extension force experienced locally near a crack-tip (2-5). Such mechanisms can produce much greater toughness than is achievable solely by control of

Mat. Res. Soc. Symp. Proc. Vol. 130. © 1989 Materials Research Society local interface chemistry to enhance $G_0$. This paper illustrates features and limitations of various mechanisms that use local microstructure and microchemistry to control interfacial toughness

## EXPERIMENTAL PROCEDURES

The method used to measure the fracture resistance, described previously (4,5), entails splitting a ceramic-metal-ceramic sandwich along one interface using a double cantilever beam (DCB) configuation. Samples are formed by vapor depositing thin (0.3-2 micron) metal films onto two ceramic substrates and then diffusion bonding the metals together; pore-free, glass-Cu interfaces obtain with relatively low bonding temperatures. Fracture is confined to one interface by a very thin (10-20 nm) Cr bond layer deposited at the other. This geometry minimizes interference from thermal expansion mismatch stresses and from interfacial shear stresses induced by elastic and plastic compliance discontinuities.

To isolate and study specific crack-microstructure interactions, we have developed novel methods to make samples with prescribed microstructural and interfacial characteristics. These entail photolithographic, deposition, and diffusion bonding processes described elsewhere (4-7). In particular, glass-copper interfaces with periodic arrays of microcrack-like-voids, located suitably near the interfaces have been produced, and combined, if desired, with inclined steps in the interface to be fractured, Fig. 2.

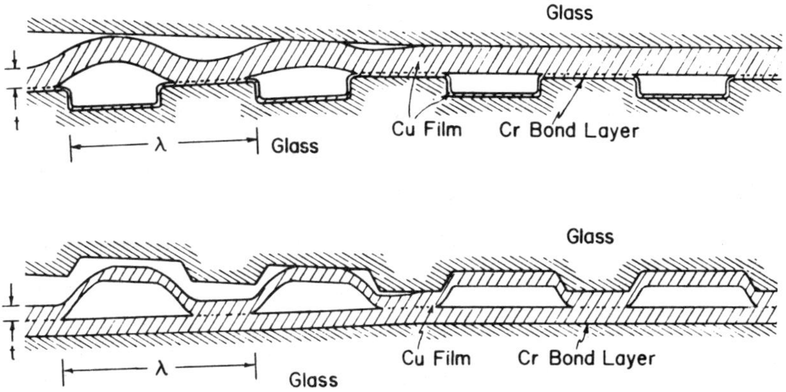


**Fig. 2.** Tailored microstructures that toughen glass-Cu interfaces using out-of-plane microcracks (to promote crack bridging) and interface tilts (that cause crack-tip deflection) are made by etching patterns into the glass, depositing metal and diffusion bonding. Switching the interface designated to fracture (dictated by emplacing a thin Cr layer on the other) yields specimens that experience: a) only ligament bridging with crack extension along a flat interface, or b) bridging plus deflection via fracture on a patterned glass-Cu interface.

## RESULTS AND DISCUSSION

### Chemical Effects

Fracture energies for these silicate glass-Cu interfaces increase markedly and diminish in variability with increasing Cu bonding temperature, Fig. 3. Presumably degrading impurities, remaining after cleaning and deposition, diffuse away from the interface during heating. Annealing at 500°C or hotter yields interfacial fracture energies comparable to or exceeding those of glass (∿10 J/m²); for such samples, cracks often travel partially or completely within the glass.

**Fig. 3.** Results showing the increase in fracture energy for glass-Cu or silica-Cu interfaces with the temperature for diffusion bonding the copper. Dispersal of interfacial impurity left from preparation is regarded as a major cause of the toughness increase with heating; differing amounts of such impurity is considered to cause the variability among samples, although the role of the cation modifiers within the glass is not yet known. At higher temperatures, the variability reflects differing degrees of fracture within the glass and resulting nonplanar cracks (6).

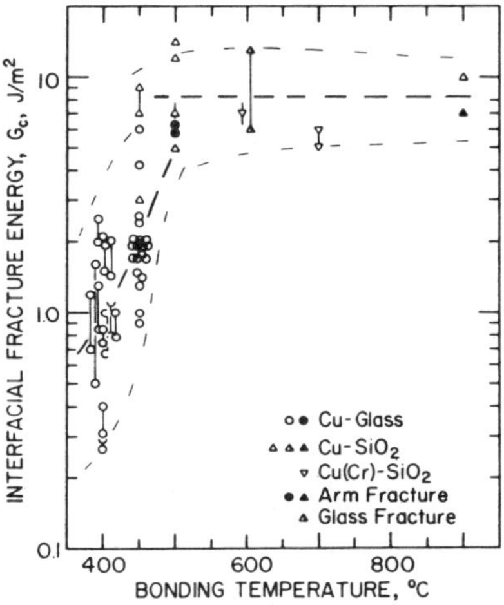


Auger spectroscopy analyses from in situ fracture studies permit identification of degrading impurities (8). Preliminary results indicate that Cu-glass samples with $G_c \sim 2$ J/m² typically have more than a monolayer of interfacial impurity, e.g. C, Cl, and sometimes S and N. Moreover, some of this impurity desorbs (as CO and $CH_4$) during fracture and more can be desorbed by the electron beam, indicating many of the segregated species are rather weakly bound, consistent with their deleterious effect on fracture resistance.

## Microcrack Induced Bridging

The presence of periodic arrays of microcrack-like-voids in the Cu near the interface significantly increases the energy for interfacial fracture, as illustrated in Figs. 4-6. During crack extension, interfacial decohesion of the regions of Cu adjacent to the microcracks occurs more slowly than does fracture elsewhere along the interface. As a result, ligaments of Cu film form that bridge across the crack flanks, and are stretched as the crack advances, as in Figs. 2 and 4. The toughness increase largely derives from plastic deformation in the stretching ligaments (4-7). This, explains the observation that this toughening mechanism is potent when the fracturing interface is perfectly flat, e.g. Figs. 4 and 6; the bridging is triggered by the out-of-plane, preexistent microcracks. Order of magnitude increases in fracture energy have resulted for Cu-glass interfaces with $G_0$ = 1-3 J/m².

As shown in Fig. 4a, the fracture toughness increases with crack extension as the crack passes the microcrack arrays, until an asymptotic value obtains. This slowly rising fracture energy (R-curve behavior) results because the toughening derives from the development of a zone of partially severed, bridging ligaments behind the advancing crack tip.

The attainable toughening depends upon the inherent interfacial toughness, $G_0$, and the geometry of the bridging ligaments through their effect on the toughness increment due to ligament stretching, which is given by (4):

$$\Delta G_c = \sigma_y \varepsilon_p t X \qquad (1)$$

where $\sigma_y$ is the flow stress, $\varepsilon_p$ the plastic stain from ligament stretching, t, the ligament thickness, and X the area fraction of such ligaments. Each increment of crack extension and opening entails a competition between further stretching of the bridging ligaments and further severing of the ligaments by interfacial fracture, the balance of which depends upon the hardening behavior of the Cu and on $G_o$. As a result of this competition, the toughening increment from this crack-tip shielding mechanism increases almost linearly with $G_o$, Fig. 5, until $G_o$ becomes high enough to permit fracture within the glass. With greater ligament thickness, less plastic work is done prior to complete interfacial fracture leading to the behavior in Fig. 6. However, for sufficiently thin ligaments, there is not enough ductility to cause complete interfacial fracture; instead, tearing of the ligaments occurs as in Fig. 7.

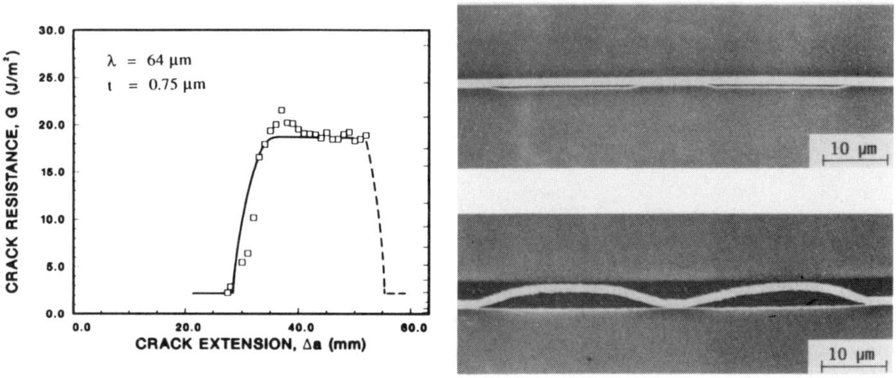


**Fig. 4.** a) Resistance curve showing the development of interfacial fracture energy during crack extension as a crack approached and then traversed a region containing an array of microcrack-voids within the Cu and b) scanning electron micrographs of sample cross-sections showing the "microcrack-void" geometry before (above) and after (below) crack extension.

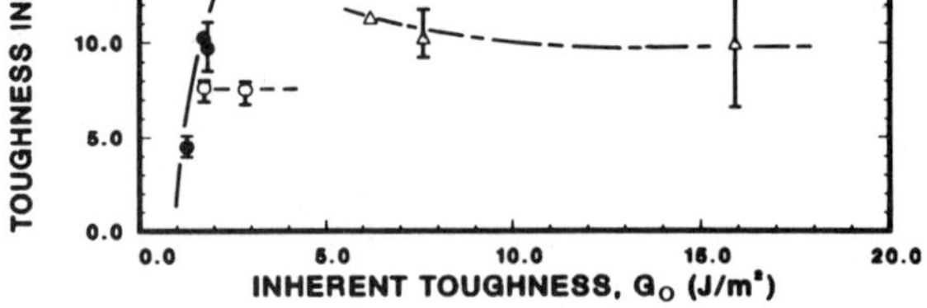


**Fig. 5.** Variation of the toughness increment from microcrack/bridging with the inherent toughness of the plain glass-Cu interface.

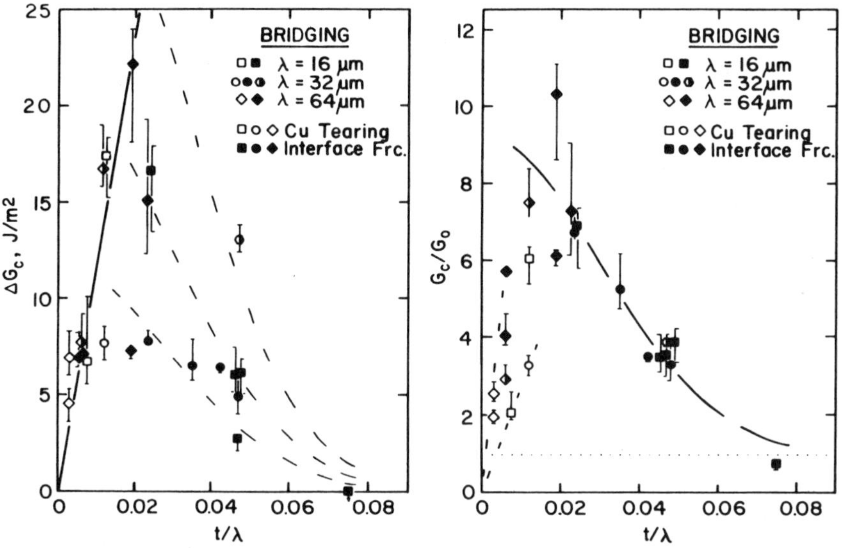


**Fig. 6.** Experimental results (for samples with randomly varying $G_0$ values) illustrate how toughening from microcrack induced bridging varies with the thickness, t, and periodic spacing, λ, of the bridging ligaments. When fracture is entirely along the interface (Fig. 2), the toughening increment diminishes with ligament thickness but is proportional to the inherent interfacial toughness, $G_0$, as illustrated by the normalization in the right figure.

**Fig. 7.** For sufficiently thin metal ligaments, fracture entails tearing of the ligaments. The toughening increment does not directly depend upon $G_0$ as can be seen in Fig. 6, but increases with film thickness.

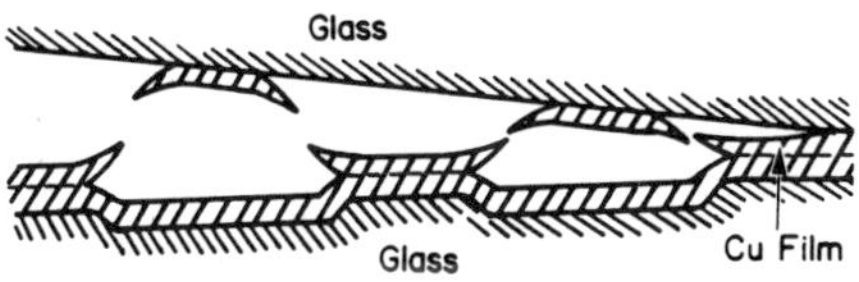


## Crack tip Deflection

Crack tip deflection caused by irregular interfaces, Fig. 2b, should lead to higher toughness owing to the reduced crack tip stresses at the deflected tip (9). Further increases are expected if the higher local shear stresses induce greater deformation in the metal. Finally, if some far field shear loading exists, the ensuing crack face contact behind the tip may markedly reduce the crack tip stresses (10). Although such toughening is expected, few definitive experiments exist that illustrate solely deflection effects. Recent DCB measurements on sandwich samples of Cu and as-fired commercial polycrystalline $Al_2O_3$ yield toughness values exceeding 20 $J/m^2$, double that expected for the tensile loading, likely reflecting crack deflection effects owing to the roughness of the as-fired surface (11).

Combining interface deflection with microcrack toughening may yield substantial benefits. The best microcrack toughened glass-Cu samples to date gave 160 $J/m^2$, far tougher than the glass (5). These values may have resulted because deflection effects from the nonplanar interface enhanced ligament stretching. However, refinement is required; in other instances, interface steps lead to diminished ligament toughening, perhaps because geometrical irregularities reduce the ductility of the bridging Cu films.

## CONCLUDING REMARKS

When pore-free Cu-glass interfaces are processed to eliminate deleterious interfacial impurities, the fracture energies approach or exceed those of the glass. Similar toughnesses can obtain by the use of bonding layers of more chemically reactive metals such as Cr. However, in both instances, the attainable toughness can be limited after achieving good enough bonding to cause cracks to pass from the interface into the brittle ceramic (12).

Much higher (one to two orders of magnitude) interfacial fracture energies can be obtained by emplacing or inducing microcrack-like-voids in the metal adjacent to the interface. The greater toughness primarily results from extra plastic deformation that occurs in the Cu between the interface and the microcracks. This microstructural toughening can be applied in instances when the inherent interface toughness is low enough to preclude cracks passing into the brittle glass. The greater energy dissipation occurs in a larger region around the crack than for a simple crack. As a result, crack tip stresses do not reach levels required to induce the crack to enter the brittle ceramic. An adverse aspect is that full toughness is not realized until the cracks are long enough to develop a bridging zone. This will limit the strengthening which can be attained with such a mechanism and the protection against small flaws in microscale devices. Continuing effort is being devoted to refining the mechanism to enhance the rate of toughness increase. Recent results show this mechanism is effective under subcritical and fatigue induced fracture.

Crack tip deflection, induced by a nonplanar interface can yield appreciable toughening, but which is less dramatic than that from microcrack induced bridging. However, it can operate at much smaller crack sizes. It can be especially effective in conjunction with bridging mechanisms.

## ACKNOWLEDGEMENTS

This work was supported by the Office of Energy Research, Office of Basic Energy Sciences, Materials Sciences Division of the U.S. Department of Energy under Contract No. DE-AC03-76SF00098. Gratitude is expressed to G. Chan for experimental assistance.

## REFERENCES

1 B.J. Dalgleish, M.C. Lu, and A.G. Evans, Acta Metall, **36** 2029-36 (1988).
2. R.M. Cannon, V. Jayaram, B.J. Dalgleish, R.M. Fisher, Mat. Res. Soc. Symp. Proc. **72**, 121-26 (1986).
3. R.M. Cannon, V. Jayaram, B.J. Dalgleish, and R.M. Fisher, in: Ceramic Microstructures '86: Role of Interfaces, J. A. Pask and A. G. Evans, eds., Plenum, New York, pp. 959-68 (1987).
4. T.S. Oh, et al, Acta. Metall. **36**, 2083 (1988).
5. R.M. Cannon, T.S. Oh, J. Rödel, A.M. Glaeser, and R.O. Ritchie, J. Am. Ceram. Soc. Comm., in review.
6. T.S. Oh, et al, in Interfaces in Polymer, Ceramic, and Metal Matrix Composites, H. Ishida, ed. Elsevier Publishing Co. pp. 567-81 (1988).
7. T.S. Oh, R.M. Cannon, and R.O. Ritchie, in Proc. Mat. Res. Soc. Symp. on Joints of Metals and Ceramics, Tokyo, June 1988, in press.
8. G. Blackman, et al, to be published Rev. Sci. Instr.
9. B. Cotterell and J.R. Rice, Int. J. Fract. **16**, 155-69 (1980).
10. A.G. Evans and J.W. Hutchinson, to be published.
11. B.J. Dalgleish, V. Jayram, and R.M. Cannon, to be published.
12. R.M. Cannon, R.M. Fisher, and A.G. Evans, Mat. Res. Soc. Symp. Proc. **54**, 799-804 (1986).

# THERMAL STRESS-INDUCED GRAIN BOUNDARY VOIDING AND NOTCHING IN NARROW Al-BASED METALLIZATIONS

Che-Yu Li, R.D. Black, and W.R. LaFontaine
Department of Materials Science and Engineering
Cornell University, Ithaca, NY 14853

## ABSTRACT

Grain boundary voiding has been identified as a diffusional creep mechanism that produces failure of narrow Al-based metallizations during thermal aging. It is considered to be a reliability concern for sub-micron metallizations because the resulting failure rate has been observed to be strongly line width dependent. This paper presents a theoretical model for stress-induced grain boundary voiding. The proposed model is shown to account for the experimentally observed temperature and time dependence of thermal aging-induced line failure data reported in the literature.

## INTRODUCTION

During a constant temperature hold the relaxation of thermal stresses in narrow, thin, Al-based alloy metallizations deposited on a Si substrate can be accomplished via dislocation creep processes or diffusional creep processes, or both. If the nucleation of voids, such as those which nucleate on grain boundaries as a result of grain boundary sliding (GBS), occurs readily, then the existence of void surfaces will provide additional sources of atoms which can participate in diffusional creep. This type of void growth and the resulting failure of the narrow metallization has been extensively reported in the recent literature [1,2,3]. According to Hinode and co-workers, it can be considered to be as important a reliability concern as electromigration [1].

This paper addresses, as a function of thermal history, the role of thermal stress in stress-induced grain boundary cavitation; the kinetics of the line failure process is also considered. The theoretical model proposed by Li, Black, and LaFontaine [4] is shown to accurately predict failure data of narrow Al-based alloy metallizations which have been reported in the literature.

## PHENOMENOLOGY

Nucleation and growth of cavities during creep under an applied load is a common failure mechanism of bulk materials aged at elevated ($>0.4T_m$) temperatures [5]. In this case cavities nucleate on grain boundaries at stress concentration sites, such as steps and second phase particles, as a result of GBS [6]. For the case of voiding in thin, narrow, metallizations nucleation occurs by the same mechanism, except that the stress causing the GBS and subsequent nucleation results not from an applied load, but from the thermal expansion mismatch between the Al-based alloy metallization and the Si substrate. When the metallizations are narrow ($\sim 1\mu m$) this thermal stress is essentially uniaxial. The traction-free void surface provides a source of atoms and both the grain boundaries and the edges of edge dislocations oriented at a finite angle to the stress axis provide sinks for the atoms. The driving force for cavity growth is the difference in chemical potential of Al atoms, $\Delta\mu$, between the source and the sinks. For dislocations oriented normal to the stress axis a virtual work argument can be used to give $\Delta\mu = \sigma\Omega$, where $\sigma$ is the tensile stress in the line and $\Omega$ is the atomic volume of an

Al atom.

The thermal history which has been found to result in grain boundary voiding typically involved annealing at 450°C, a cool-down to room temperature, and then thermal aging at an elevated temperature (~200-300°C) [1]. Since the temperature during the anneal corresponds to approximately $0.6T_m$ of Al, the build-up of thermal stress in the metallization during this process is small. After the cool-down, provided there is good adhesion between the metallization and the Si substrate, the difference in thermal expansion coefficients will produce a shear displacement in the metallization which results in a corresponding uniaxial tensile stress.

The stress in the metallization varies with temperature such that a stress-temperature hysteresis loop occurs during a temperature cycle. Such a loop can be calculated by integrating the governing equation

$$\dot{\sigma} = E(\Delta\alpha\dot{T} - \dot{\epsilon}) \quad (1)$$

where $\dot{\sigma}$ is the time rate of change of the stress, $\dot{T}$ is the time rate of change of the temperature, $\Delta\alpha$ is the difference in thermal expansion coefficient between the Si substrate and the Al-based alloy metallization, $\dot{\epsilon}$ is the inelastic strain rate of the Al-based alloy metallization, and E is the elastic modulus of the Al-based alloy metallization. The rate of inelastic deformation is determined based on experimentally determined constitutive equations between $\sigma$ and $\dot{\epsilon}$. Fig. 1 shows a typical stress-temperature loop ($\dot{T}$ = 0.1°C/s) calculated for an Al metallization from constitutive equations based on experimental data from Flinn and co-workers [7] and Doerner and co-workers [8]. The stress in the loop during annealing at 450°C is assumed to be approximately zero. The top curve represents the cool-down from the aging temperature and the bottom curve represents a subsequent reheating. During reheating the stress decreases elastically due to a decrease in thermal expansion mismatch and eventually becomes compressive. As the temperature continues to increase the film finally begins to deform inelastically. During thermal aging at a constant temperature the thermal stress will relax as a function of time according to equation (1) with $\dot{T} = 0$.

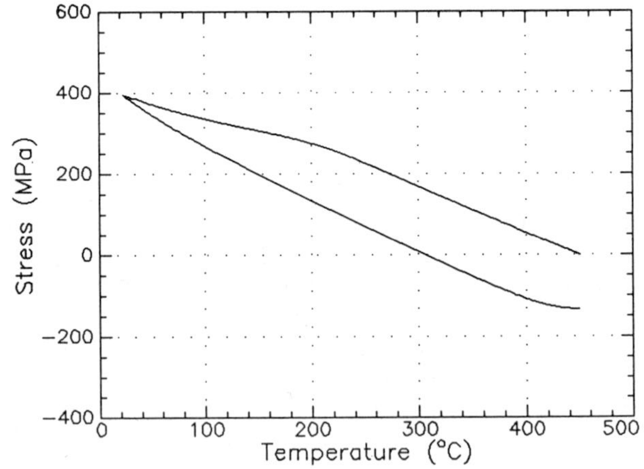


**Figure 1. Stress versus temperature plot for an Al metallization deposited on a Si substrate.**

VOID GROWTH KINETICS

For the case where growth of a grain boundary void is controlled by grain boundary diffusion the simple mass balance shown in equation (2) can be used to calculate the rate of increase in the cavity radius with time.

$$dV/dt = 4\pi r^2\, dr/dt = J_{gb}(2\pi r\delta)\Omega \qquad (2)$$

In the above equation $J_{gb}$ is the flux of atoms out along the grain boundary, $\delta$ is the boundary width, $\Omega$ is the volume of an Al atom, and r is the cavity radius. The balance equates the change in volume of the cavity, assumed to be spherical, to the flux of atoms out along the grain boundary. Since at quasi steady-state $J_{gb} = (D_{gb}/\Omega k_BT)(\sigma\Omega/r)$ [ref. 9], the rate of increase in the void radius is proportional to $1/r^2$.

The time to one percent failure can be estimated by calculating the time, $t_f$, needed for a cavity, located in the center of a grain boundary, to grow and sever the line. This time is given by integrating equation (2) from an initial radius, which is approximately zero, to w/2, where w is the width of the line (equation (3)).

$$\int_0^{w/2} 4\pi r^2\, dr = \int_0^{tf} J_{gb}\Omega(2\pi r\delta)\, dt \qquad (3)$$

The resulting cubic line width dependence of the failure rate, assuming that the stress remains constant, is close to the observed $w^{2.7}$ dependence reported by Hinode for Al-2%Si metallized lines [1].

The relationship between the rate of line failure, the place on the grain boundary where the void nucleates, and the rate of void growth can be conveniently illustrated, as depicted in Fig. 2, by considering a metal line of a circular cross-section. The radius of the metallization is $r_o$. If a void is nucleated on the circumference of an inner circle of radius $r = (r_f-r_o)$, it must grow to a radius of $r_f$ to cause failure. The area of this inner circle is $\pi(r_f-r_o)^2$ so that the fraction of line failures, f, corresponding to a failure radius $r_f$, is given by the ratio of areas in Fig. 2. That is

$$f = \{\pi(r_f-r_o)^2)\}/\pi r_o^2. \qquad (4)$$

Upon differentiation with respect to time equation (4) yields:

$$df/dt = 2\pi\ (r_f-r_o)/(\pi r_o^2) \bullet dr_f/dt. \qquad (5)$$

Since $dr_f/dt$ is proportional to the stress in the metallization (equation (1)), it is possible to calculate the fraction of line failure as a function of time by using equation (4).

Fig. 3 shows stress versus time curves, at 250°C and 295°C, calculated from (a) equation (1) with $\dot{T}=0$, (b) constitutive equations based on experimental data from Flinn and co-workers [8] and Doerner and co-workers [9], and (c) the initial stresses shown in Fig 1. By using these profiles according to equation (2), and using a modified form of equation (4) which takes into account the observed slit-like shape of the voids [1], the percentage failure data reported by Hinode and co-workers [1] can be simulated. These simulations, along with the percentage failure data from Hinode and co-workers, are shown in Fig. 4. For these calculations the percentage of line failures due to processing defects were subtracted from the experimental data [3].

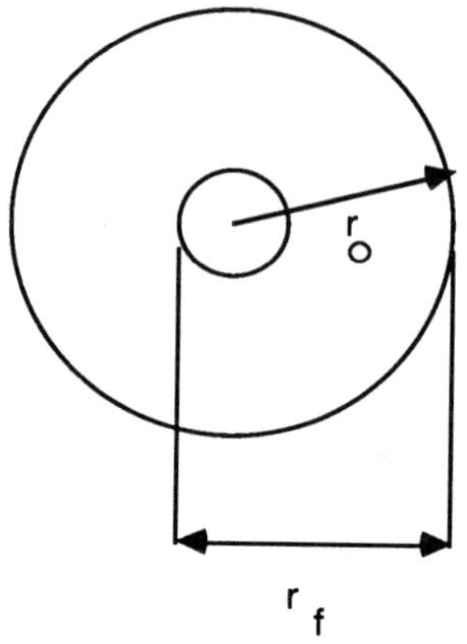


Figure 2. Schematic of metallization used to illustrate relationship between rate of void growth and failure rate.

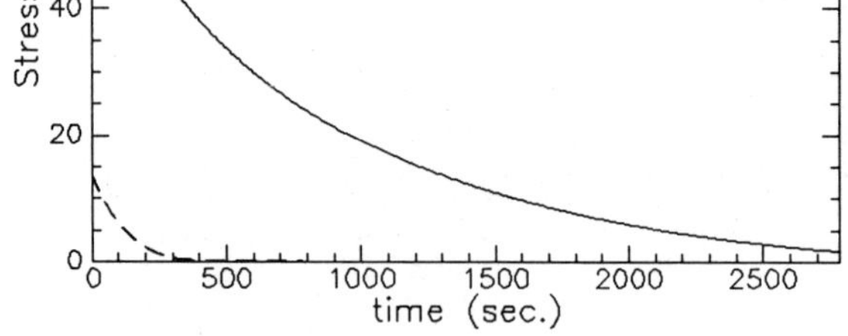


Figure 3. Stress vs. time curves at 250°C and 295°C.

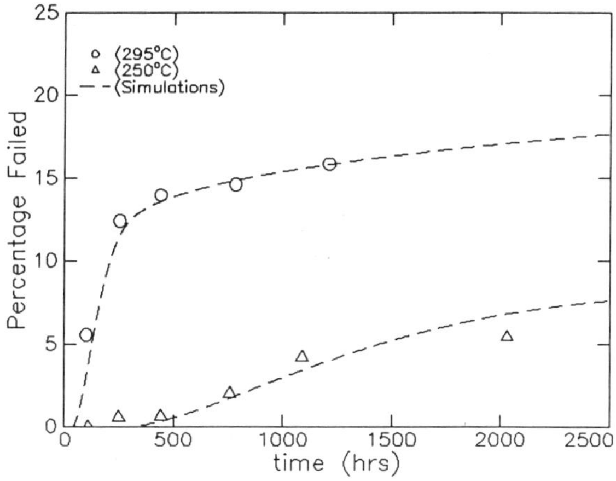


Figure 4. Simulations of percentage failure data.

The complexity of the stress-induced grain boundary failure process in aluminum-based alloy metallizations is clearly demonstrated by the change in shape of the percentage line failure data as a function of time and temperature. According to equation (2) the kinetics of cavity growth depend only on the grain boundary diffusivity and the stress in the metallization. The rate of growth, however, also depends upon the rate of surface diffusion along the void wall. Thus the width of the slit-like void depends upon both the grain boundary and void surface diffusivities, which are temperature dependent, and the stress, which is temperature and time dependent. For the present theoretical simulations time independent slit widths of 0.36$\mu$m and 0.11$\mu$m were used to simulate the 250°C and the 295°C data, respectively. With these values the activation energy for grain boundary diffusion was determined to be ~0.88eV. This value is close to that reported in the literature for Al [10].

## CONCLUSION

Examples of numerical simulations, based on a theoretical model for thermal stress-induced voiding in narrow Al-based alloy metallizations, were reported. The simulations were found to describe well line failure data reported in the literature.

REFERENCES

1. K. Hinode, N. Owada, T. Nishida, and K. Mukai, J. Vac. Sci. Technol. B 5, 518 (1987).
2. J.W. McPherson and C.F. Dunn, J. Vac. Sci. Technol. B 5, 1321 (1987).
3. J. Klema, R. Pyle, and E. Domangue, Proc. of the IEEE Trans. on Reliability Physics, 1 (1984).
4. C.-Y. Li, R.D. Black, and W.R. LaFontaine, Appl. Phys. Lett. 4, 31 (1988).
5. D. Hull and D.E. Rimmer, Philos. Mag. 4, 673 (1959).
6. I.-W. Chen and A.S. Argon, Acta Metall. 29, 1759 (1981).
7. P.A. Flinn, D.S. Gardner, and W.D. Nix, IEEE Trans. Electron Devices ED-34 689 (1987).
8. M.F. Doerner, D.S. Gardner, and W.D. Nix, J. Mater. Res. 1, 845 (1986).
9. M.E. Fine, Phase Transformations in Condensed Systems, (MacMillan, New York, 1964), p. 56.
10. H.J. Frost and M.F. Ashby, Deformation Mechanisms, (Permagon Press Oxford, 1982), p. 21.

# DELAMINATION OF LPD TUNGSTEN FILMS BY RESIDUAL STRESS

A. BORER*, B. DERBY* AND I. MAY**
*Department of Metallurgy and Science of Materials, Oxford University, Parks Road, Oxford OX1 3PH, U.K.
**GEC Research Limited, Hirst Research Centre, East Lane, Wembley, Middlesex HA9 7PP, U.K.

ABSTRACT

Thin tungsten films have been deposited on sapphire and glass substrates using laser photolytic deposition from a tungsten hexa-fluoride precursor. Failure of the films is caused by the presence of large residual stresses (1-5 GPa). Compressive and tensile stresses are found at different stages of film growth and either may be responsible for delamination. Early compression is caused by oxidation and impurity influenced intrinsic stress. Tensile stress occurs with microstructural changes but is not maintained with film growth. Large stress gradients, indicative of up to 0.75-1.2% lattice strain difference across the film, are found through both compressive and tensile films and are not explained.

## Introduction

Tungsten combines the advantages of a refractory film with low electrical resistivity and high resistance to electromigration. Consequently its use is of great interest in very large scale integration processing. Laser photolytic deposition can be used for direct writing of very fine features at lower temperatures than conventional chemical vapour deposition. In theory this should enable minimisation of thermal stresses and reduce the risk of process induced damage.

Tungsten films have been deposited from a tungsten hexafluoride precursor onto sapphire substrates up to a depth of 2000nm. Deposition temperatures ranged between 80 and 450°C. The adhesive failure of these films is often found and has been attributed to the presence of large residual stresses in the films. The mode of failure is neither uniform across a film nor consistent between films. Initially thin films are observed to suffer from a compressive stress. As the film grows a tensile stress is induced which is in turn replaced by compressive stress in the thickest films.

## Failure of Thin Tungsten Films

A compressive stress is apparent in the thinnest films through surface buckling. This occurs in films 20 to 40nm deep (Dektak profilometer measurements) and may be either disc or ridge bucking (Figs 1,2). The biaxial compressive stress residual in the disc buckled film is calculated by measuring the expansion in radius of disc from the height of the buckle. Using elementary elasticity theory

$$\sigma_o = \frac{\delta a . E}{(1-\nu)a} \qquad (1)$$

where a is the radius of the film prior to buckling, $\delta a$ is the expansion in radius of the buckled film, E the film elastic modulus and $\nu$ the film Poisson ratio. The magnitude of the compressive stress present to cause buckling appears to be thickness related (as seen in fig.1). Stress values of $\sigma_o$ = 3-5GPa are found for most of this film. However in the

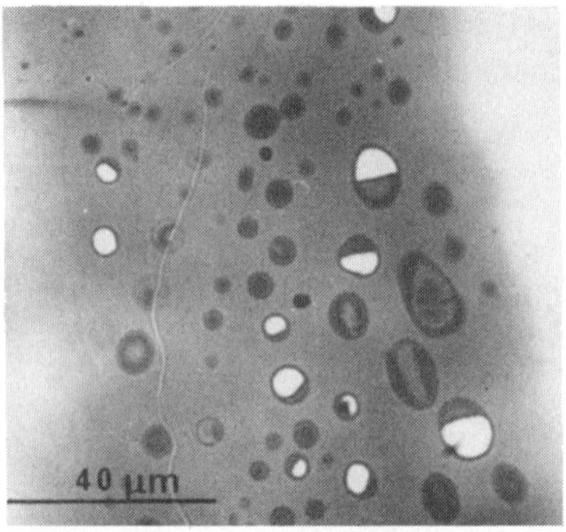


Fig.1 Disc buckling of 20-30nm amorphous film (optical).

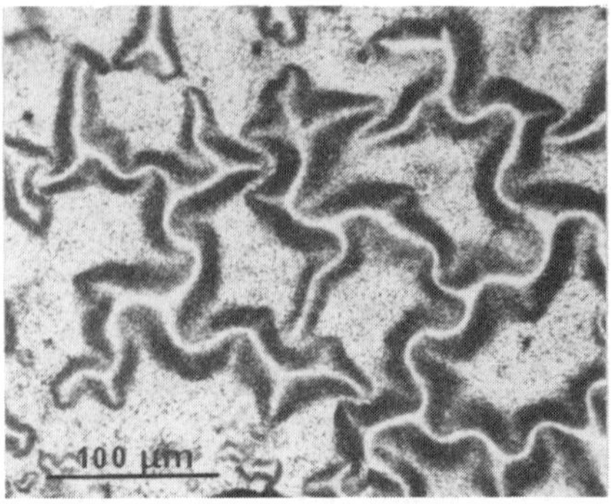


Fig.2 Ridge buckling of 40-50nm amorphous film (optical).

thicker areas the residual stress is found to be reduced, buckle heights are smaller and stresses are 0-1 GPa. Very thin areas 20nm do not show buckling. This is expected considering the residual stress required for buckling calculated from Marshall and Evans 1984[1]

$$\sigma_R^2 = \frac{G_c \cdot E}{(1-\nu)(1-\alpha)t} + \sigma_C^2 \tag{2}$$

where $G_c$ is the initial strain energy release rate for film decohesion and $\alpha$ is the relative compliance of the buckled film ($\alpha$ = 0.404 calculated from Evans and Hutchinson 1984)[2], t is the film thickness and $\sigma_C$ the critical stress for buckling a decohered film. Where $\sigma_C$ is given by

$$\sigma_C = \left[\frac{kE}{12(1-\nu^2)}\right] \left(\frac{t}{a}\right)^2 \tag{3}$$

and k a constant = 14.68[1]. Hence a 20nm film is calculated to require in excess of 6.7GPa of residual stress to buckle.

Some disc buckles crack around the base leaving the delaminated flakes free on the surface. The flakes curl up (in opposite direction to the buckle bend) indicating the bottom surface is in greater compression than the top. This stress differential is quickly calculated using simple beam theory.

$$\sigma_d = \frac{Et}{r} \tag{4}$$

For a 20-30nm film $\sigma_d$ = 1.5-2.5GPa indicating a lattice strain in the order of 0.5%. Lattice strains of 0.75% have been found in even thinner films. For example the stress differential of the film in Fig.3 is in the range 1.5-3.1GPa, given an estimated film depth as little as 10nm.

Buckling of the film into ridges is also found (Fig.2). The calculated strain here due to film expansion is again 1% (c.4GPa). Cracking along the peaks of ridges allows curling or in thinner areas rolling of the film. The stress differential between top and bottom surfaces is found to range between 1.1-2.4GPa. This decreases as film thickness increases so that in thicker areas the stress is not great enough to cause cracking of the buckle. The cracking of buckled films occurs around the base of disc buckles, but along the peaks of ridge buckles. This is expected from shortest edge energy considerations.

## Mechanisms of Internal Stresses

There is no simple explanation for the presence of stresses of such magnitude. TEM shows these thin films tend to be amorphous. Hence epitaxial and morphological considerations can be ruled out. The ridge buckled case does show a change in morphology with $\alpha$ phase tungsten grains growing within the initial amorphous layer, but this would not be expected to cause a compressive stress. $\alpha$-W grains are found to grow defect free and in random orientation, which also generates no epitaxial strain. It has been reported that tungsten films grown on sapphire by chemical vapour deposition[3] and by electron beam evaporation[4] do exhibit epitaxial growth, but these depositions were at higher substrate temperatures than here.

Another possible mechanism is the effect of differences in the chemical purity of the two surfaces causing large enough differences in surface tension to induce the observed strains. Equating the observed bending moment of the film with the bending moment due to a difference in surface tension using simple beam theory gives.

$$(\Delta\gamma)t = \frac{Et^3}{12r} \qquad (5)$$

Measured radii of curl of $r = 1\text{-}10\mu$ for film thicknesses of $t = 0.01\text{-}0.05\mu$ give values for $\Delta\gamma$ of $3\text{-}9\,Nm^{-1}$. Given the surface energy of tungsten is c.$3\,Jm^{-2}$ these differences in surface tension seem too large to be physically reasonable.

An identifiable source of compressive stress is thermal expansion mismatch arising after deposition at high temperatures. Tungsten has a lower expansion coefficient than the sapphire or glass substrates used.

$$\sigma_\alpha = \frac{\Delta\alpha . \Delta T . E}{(1-\nu)} \qquad (6)$$

where $\Delta_\alpha$ is the difference in expansion coefficients of tungsten and substrate over the temperature range $\Delta T$. Values for compressive stress induced via this mechanism range from 0.03GPa for low deposition temperatures on sapphire, up to 0.91GPa for 400°C deposition on glass. However these still do not match the c.4GPa observed compression.

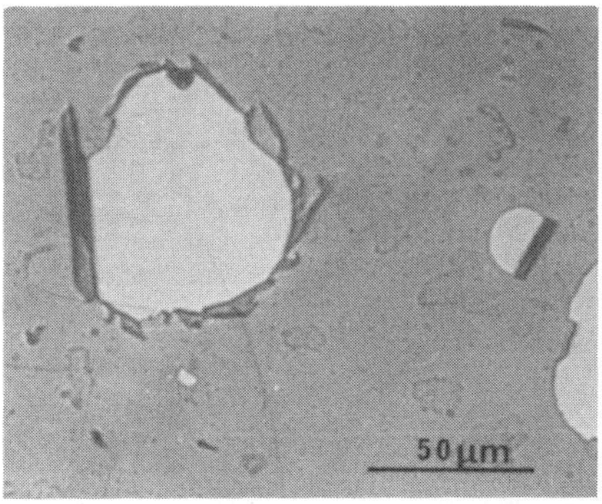


Fig.3 Compressive failure showing rolled, delaminated film (optical).

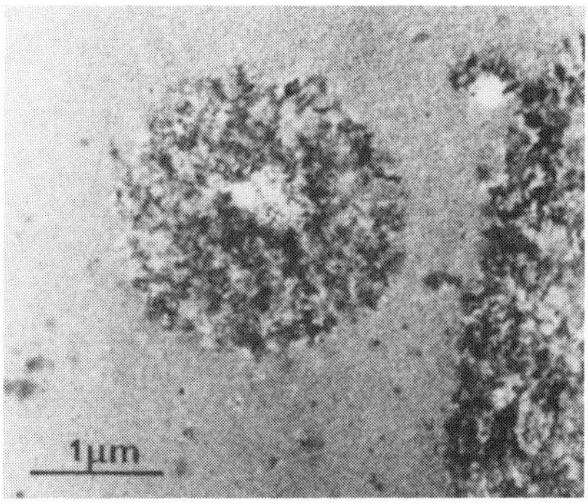


Fig.4 $\alpha$-W phase recrystallising from $\beta$-W film (TEM).

A further cause of compression is thought to be oxidation induced volume expansion. The surfaces of these amorphous films appear highly coloured orange to blue consistent with native tungsten oxides and have been analysed by Auger as $WO_{2.7 \rightarrow 3.0}$. Auger depth profiling and XPS surface analysis both show heavy oxygen and fluorine contamination through the film. The presence of such impurities is known to reduce atomic mobility and thereby reduce the mean grain size. Aberman and Koch, 1986[5] report that this will increase the compressive stress found at the end of coalescence. This is caused when isolated clusters of depositing metal grow with a slightly reduced lattice spacing due to surface tension effects. The presence of impurities is very much diminished in later crystalline films which are found not to exhibit such stresses. β-W films (which are known to be stabilised by impurities such as oxygen[6]) show higher oxygen and fluorine levels than α-W films.

## Effect of Film Thickness

It is found that the extent of buckling and curling of the films is reduced as thickness increases. In the first instance this may be the result of a corresponding increase in film stiffness with depth. In the second instance it could be caused by factors which reduce either the differential stress through the film or the compressive stress of the film. These factors are large enough to cause a change in the overall film stress to a tensile stress.

The most important feature here is recrystallisation. This is found to occur with a lengthening of the deposition time such that thicker amorphous films are seen to be crystallising into α-W phase. Raising the deposition temperature has an accelerating effect so that even very thin films (t = 20nm) are fully crystalline by the end of depositon. This is presumably due to mobility considerations. In such films the meta-stable β-W phase (stable up to 650°C[7]) is found to crystallise first. The α-W phase subsequently recrystallises from the β-W phase. This is seen in Fig.4, patches of α-W are growing into the β-W film. Again this recrystallisation is favoured by the combination of longer deposition time and higher substrate temperature. The recrystallisation process can be followed through stages from the rough edged α-W grains growing in β-W, to rough α-W with β-W at grain boundaries to all rough-edge α-W and finally to smooth edged α-W grains (Fig.5). That recrystallisation should cause a tensile stress is widely reported for the features which exist here. The crystalline phase is more dense than the amorphous, the α-W phase is also 1.4% denser than the β-W phase. A large reduction in grain boundaries occurs from β-W mean grain diameter 5-10nm to α-W phase mean grain diameter 200-300nm.

Stresses in these recrystallised films have been measured using an X-Ray back reflection technique[8] with monochromated CuKα radiation. 30-40nm films (the minimum thickness to give results) showed tensile stresses in the range 0.85 → 2.1GPa. The lowest value here is also a β-W film, i.e. recrystallisation not complete, greater tensile stress yet to be induced.

TABLE 1 MEASURED STRESS WITH FILM THICKNESS

| Film Depth nm | 30 | 35 | 40 | 90 | 250 | 420 | 2000 | 2500 |
|---|---|---|---|---|---|---|---|---|
| Mean Stress GPa<br>Compresson/Tension | 0.85<br>T | 2.1<br>T | 1.5<br>T | 2.5<br>T | 1.1<br>T | 0.3<br>T | -0.27<br>C | -0.24<br>C |

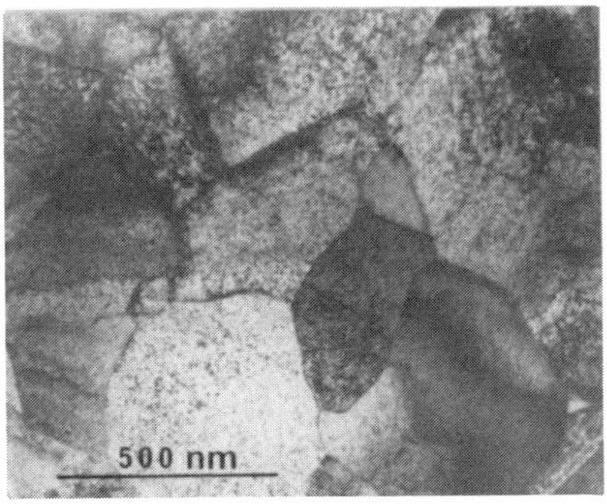


Fig.5 Recrystallised α-W grains (TEM).

Fig.6 Tensile failure showing cracked and rolled film (Optical).

It is found that the stress values of the thin α-W phase films do not increase further as film thickness continues to increase. Indeed a down-turn is soon reached and the films go on to suffer overall compression again. This indicates the tensile stress is generated solely by the recrystallisation process, and is not maintained during film growth. Epitaxial consideration can again be neglected since the grains are of random orientation. Cracking of these tensile films is found to occur along grain boundaries suggesting some embrittlement exists. This could be caused by the diffusion of impurities such as oxygen. In areas of widespread cracking a contraction of up to 0.5% can be seen in the film.

Large stress gradients are also found across recrystallised films. The differential is in the same direction as in the amorphous film, i.e. the top surface is in greater tension and spontaneous delamination results in complete rolling of the film (Figure 6). Stress differentials in the range 0.8-5.2GPa (i.e. up to 1.2% difference in lattice strain) have been measured for tensile films. In this case however, the thicker films despite having lower overall tensile stress now show the larger differential stress. Indeed in partially recrystallised (β-W phase) and very thin α-W phase films ( 30nm) no curling of the film occurred even when delaminated.

## Discussion

The initial thin amorphous films were found to contain unexpectedly large compressive stress together with an equally large residual stress differential. The major cause of these stresses has not been identified and since summing the calculated possible stresses does not yield the observed figure, we must presume that intrinsic growth stress coupled with a large degree of oxygen contamination to be the major factors responsible.

The transition of films to tensile stress is quite clearly explained by recrystallisation to one, or both of the crystalline phases. Of particular note is the large reduction in oxide or other contamination in these films. As film thickness increases the gradual reduction in tensile stress shows that the film is going down stress free. Growth is found to be columnar with mean grain diameter maintained through the depth of film. In the thickest films the overall stress is once again compressive. The possibility of large scale diffusion of oxygen down the grain boundaries is unlikely since film resistivity (7.5μΩcm) is approaching bulk values (5.65μΩcm). However the slight compressive stress found is attributable to thermal mis-match. The observed stress is of the same order as the mean film stress expected due to the compression of thermal expansion overcoming the recrystallisation tensile stress. Little light is cast on the large differential stresses found throughout the range of films. TEM shows no sign of plastic deformation, i.e. films have very low dislocation density and so these stress gradients have not been satisfactorily explained.

## References


1. D.B. Marshall, A.G. Evans, J. Appl. Phys. 56, 2632 (1984).
2. A.G. Evans and J.W. Hutchinson, Int. J. Solids Structures, 20 (5), 455 (1984).
3. A. Miller, H.M. Manasevit, D.H. Forbes and I.B. Cadoff,, J. Appl. Phys. 37, 2921 (1966).
4. J.H. Souk, J.F. O'Hanlon and J. Angilello, J. Vac. Sci. Technol. A. 3, 2289 (1985).
5. R. Aberman and R. Koch, Thin Solid Films 142, 67 (1986).
6. C.C. Tang and D.W. Hess, Appl. Phys. Lett. 45 (6), 633 (1984).
7. A.F. Wells, Structural Inorganic Chemistry, (Clarendon 1984), p. 1283.
8. C.S. Barrett, T.B.Massalski, Structure of Metals, 3rd rev.edn. (Pergamon 1980) p. 472.

# INTERFACIAL SHEAR STRENGTH OF THIN POLYMERIC COATINGS ON GLASS

J. E. Ritter, L. Rosenfeld, M. R. Lin, and T. J. Lardner
Mechanical Engineering Department, University of Massachusetts, Amherst, MA 01003

## ABSTRACT

The interfacial adhesive shear strengths of epoxy and acrylate coatings on glass substrates were measured by the indentation and lap shear tests. The lap shear strengths were about an order of magnitude less and exhibited considerably more variability than those measured by indentation. It is believed that the lap shear strength is controlled by large processing flaws (pores in this study); whereas, the indentation test measures the "intrinsic" strength of the coating.

## INTRODUCTION

The adhesion of thin polymer coatings is often critical to their successful performance in service. However, despite its importance, adhesion is difficult to quantify in a physically meaningful way [1-5]. The microindentation test has recently been developed to measure the interfacial shear strength of thin coatings [6,7] and this test is believed to solve many of the problems associated with thin coating adhesion measurements. The test basically consists of loading a coated surface with an indenter (ball or Vickers). The coating underneath the indenter is deformed and displaced laterally. This lateral movement of the coating results in a shear stress across the interface that initiates an interfacial crack at sufficiently high indenter loads. The indenter load at which debonding initiates is related to the interfacial shear strength.

The purpose of this research is to compare the adhesive shear strength measured by indentation to that measured in a conventional lap shear test [4]. To this end, the adhesive strengths of epoxy and acrylate coatings in soda-lime glass was measured.

## EXPERIMENTAL

The polymer coatings utilized in this study were epoxy* and acrylate** and the substrates were soda-lime glass plates. Prior to coating, the glass slides were cleaned in a methanol ultrasonic bath. All coatings were deposited by the doctor blade technique and the epoxy coatings were cured in ambient air and the acrylate coatings were cured under ultraviolet radiation. The exact curing details followed manufacturer's recommendations. Thickness of the coatings was measured by a micrometer and varied from 25 to 200 μm.

The polymer surfaces were indented with either a Vickers indenter or a ball indenter (4 mm diameter). The indenter was attached to a load cell that is bolted to the bottom of the crosshead of an Instron testing machine.

---

* DGEBA (diglycidyl ether-bisphenol A) epoxy resin, with polyamide hardener, Devcon Corp.

** DeSoto Corp. (Desolite 950-008).

Specimens are placed coating side up on the stage of an inverted microscope that is positioned directly underneath the indenter. This permitted in situ viewing of the indentation-induced debonding through the backside of the substrate. A filar eyepiece attached to the microscope allowed direct measurements of the indentation-contact dimensions. A video camera and a 35 mm camera were also attached to the microscope to record the entire indentation debonding process. For all the indentation tests, the Instron crosshead speed was fixed at 0.5 mm/min.

The single lap shear specimens consisted of two aluminum holders that were mounted to an Instron testing machine. Bonded to the aluminum holders were soda lime glass microscope slides that were in turn bonded to each other by the epoxy or acrylate coatings. Air bubbles within the coatings were minimized by shearing and establishing bonding from one edge. Care was taken to align the two aluminum holders parallel to the tensile axis of the specimen. The thickness of the epoxy coating was 60 μm and that of the acrylic coatings 50 and 120 μm. The interfacial lap shear area was 2.5 x 0.6 cm. The Instron crosshead speed was 0.5 mm/min.

## RESULTS AND DISCUSSION

With the ball indenter debonding occurred with the deformations in the coating being essentially elastic. Under these conditions, the interfacial shear strength $\tau_c$ is given by [6,7]

$$\tau_c = G[\frac{\alpha}{2} K_1(\frac{\phi a_c}{h}) + \frac{\nu a_c}{(1-2\nu)R} - \frac{\beta}{2} I_1 (\frac{\gamma a_c}{h})] \tag{1}$$

where $a_c$ is the contact radius at debonding, G and $\nu$ are the coating shear modulus and Poisson's ratio, R is the ball indenter radius, h is the coating thickness and where

$$\phi = \sqrt{\frac{6(1-\nu)}{4+\nu}}$$

$$\gamma = \sqrt{\frac{3(1-2\nu)}{2(1-\nu)}}$$

$$\beta = \frac{[\frac{\phi a}{h} K_1' (\frac{\phi a}{h}) - K_1(\frac{\phi a}{h})](\frac{h}{2R})(1-6\nu)}{[\gamma K_1(\frac{\phi a}{h}) I_1'(\frac{\gamma a}{h}) - \phi I_1(\frac{\gamma a}{h}) K_1'(\frac{\phi a}{h})](1-2\nu)}$$

$$\alpha = \frac{-4}{(4+\nu)\phi K_1'(\frac{\phi a}{h})} [\frac{(1-6\nu)h}{2R(1-2\nu)} + \beta\gamma I_1'(\frac{\gamma a}{h})]$$

and where $I_1(x)$ and $K_1(x)$ are the first order modified Bessel functions of the first and second kind and $I_1'(x)$ and $K_1'(x)$ are their derivatives. The indenter load can be related to the contact radius by [6,7]

$$P = \pi a^2 G[\frac{2\nu h(1-6\nu)}{3R(1-2\nu)^2} + \frac{4\nu\beta h}{3a(1-2\nu)} I_1(\frac{\gamma a}{h}) + \frac{2(1-\nu)}{1-2\nu} (\frac{a^2}{4Rh} - \delta)] \tag{2}$$

where

$$\delta = - \frac{\nu(4+\nu)h\alpha}{12a(1-\nu)} [\frac{\phi a}{h} K_1'(\frac{\phi a}{h}) + K_1(\frac{\phi a}{h})]$$

By measuring the load at debonding $P_c$, the contact radius $a_c$ was calculated from Eq. (2) and the interfacial shear strength was then determined from Eq. (1). For both the epoxy and acrylate coatings $\nu$ was taken to be 0.38 as given by the manufacturer and G was measured in situ by fitting Eq. (2) to data of the contact radius as a function of indenter load, e.g. see Fig. 1. In this way G for the acrylate coatings was determined to be 614 MPa and for the epoxy coatings 1180 MPa.

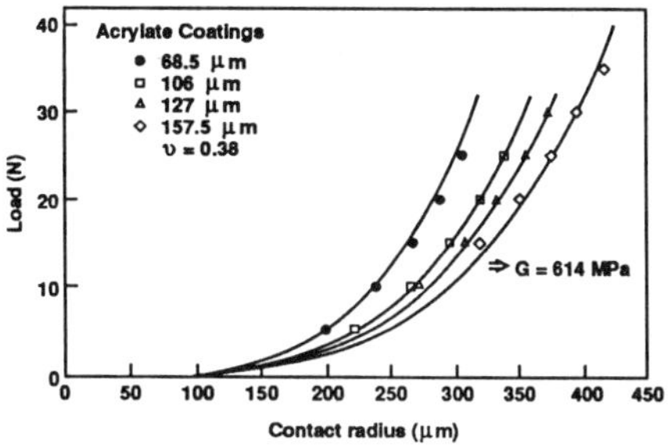


Figure 1. Contact radius as a function of indenter load for acrylate coatings on glass. Data fit to Eq. (2) with G = 614 MPa.

For Vickers indentation debonding occurred with the deformations in the coating underneath the indenter being predominantly plastic but with no penetration of the substrate. In this case the interfacial shear strength is given by [6,7,8]

$$\tau_c = \frac{-0.56\, H_c}{K_1'(\phi b_c/h)/\phi K_1'(\phi b_c/h) + \nu\, h/b_c\, \phi^2} \qquad (3)$$

where $H_c = P_c/2b_c^2$ and $b_c$ is half the diagonal length of the coating/indenter contact area.

Figure 2 summarizes the interfacial shear strength data for the epoxy and acrylate coatings. The overall agreement between the interfacial shear strengths, as measured by either ball or Vickers indentation, of the epoxy and acrylate coatings as a function of thickness, clearly demonstrates the self-consistency of the indentation technique. It can further be seen that the interfacial shear strength of the epoxy coatings is greater then that of the acrylate coatings, whether measured by indentation or lap shear. However, the interfacial shear strengths measured by indentation are almost an order of magnitude greater than those measured by lap shear.

With both the indentation and lap shear tests, interfacial failure occurred at the polymer/glass interface and was characterized by a brittle fracture. Consequently, it would be expected that failure was initiated from a preexisting flaw at the interface and that the strengths between the two tests would scale with the area under stress since the probability of having a large strength-controlling flaw increases with the area under stress. Based on Weibull failure statistics [9,10], the interfacial strength distribution for a uniformly applied shear stress is:

$$\mathrm{LnLn}\, \frac{1}{1-F} = m\, \mathrm{Ln}\, \tau_c + \mathrm{Ln}\, A/\tau_o^{\,m} \qquad (4)$$

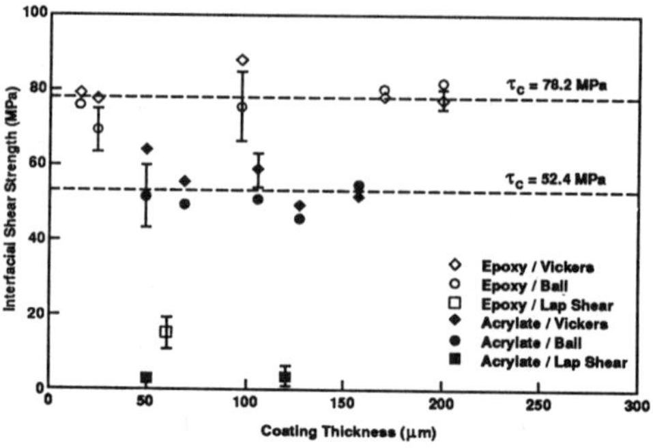


Figure 2. Interfacial shear strength data for epoxy and acrylate coatings on glass as measured by ball and Vickers indentation and by the lap shear test.

where F is the cumulative failure probability, A is the interfacial area under shear, and m and $\tau_o$ are constants. Assuming that the strength-controlling flaw populations of the indentation and lap shear specimens are the same, then from Eq. (4) the interfacial shear strength ratio of the lap shear and indentation tests at equal levels of failure probability should scale with area as

$$\frac{\tau_{c\ lap}}{\tau_{c\ ind}} = \left(\frac{A_{ind}}{A_{lap}}\right)^{1/m} \tag{5}$$

The area under shear in the lap shear test is simply 1.50 $cm^2$. The equivalent area under shear in the indentation test is difficult to estimate since underneath the indenter large compressive stress exist that inhibit debonding in this region [6]. Thus, debonding initiates near the edge of the contact zone where the compressive stress is small and the effective shear stress is a maximum [6]. Assuming that failure initiates in the narrow annular region of $\pm$ 5% $a_c$, then the area of the indentation test is 0.20 $\pi\, a_c^2$. Based on the average indentation interfacial strength, the corresponding area of the indentation test is 1.6 x $10^{-4}$ $cm^2$ for the epoxy coatings and 3 x $10^{-4}$ $cm^2$ for the acrylate coatings. Figure 3 compares the predicted lap shear strengths based on the indentation strength data to the measured lap shear data for the epxoy and acrylate coatings. Note that for these comparisons, since the interfacial shear strengths are independent of thickness (Fig. 2), all the indentation strength data were combined, independent of thickness. It is evident from this figure that the predicted lap shear strength is considerably greater then the measured and that the lap shear strength distribution has greater variability then the indentation strength distribution.

The obvious reason for the discrepancies in the lap shear and indentation strengths is that the strength-controlling flaw populations between the two tests are different. Fractographic analysis of the interfacial fracture surface of the lap shear tests indicates that relatively large pores at the interface of both epoxy and acrylate coatings act as the source of fracture, see for example Fig. 4. These large pores can explain the low interfacial shear strengths of the lap shear test in comparison to the indentation shear strengths, as well as the larger

# SIMPLE MODELS FOR SURFACE CRACKING

PAUL MEAKIN

Central Reseach and Development Department, E. I. du Pont de Nemours and Company, P. O. Box 80356, Wilmington, DE l9880-0356

## ABSTRACT

A simple model for crack growth in thin films deposited onto a substrate has been developed. This model results in the generation of cracking patterns which have a remarkably realistic appearance. In the simulations a slow crack initiation period is followed by a period of rapid crack growth in which quite linear cracks are formed. Later on the crack growth process becomes slower and less regular as the cracking process relieves much of the stress in the surface layer. In this model pre-existing defects have a large effect on the crack growth kinetics and morphology.

## INTRODUCTION

Thin film deposits or coatings are used in a broad range of important applications to improve optical, electronic and tribological properties, corrosion resistance, wear resistance, appearance and other properties. Familiar examples include paint films, ceramic coated metal and optical coatings. Very often the performance of these systems is compromised by cracking of the thin surface layer which is a consequence of large stresses in the surface film brought about by processes such as differential contraction or expansion, phase transitions or physico-chemical changes such as the drying of paint films or oxidation of metals. Very often surface cracking processes lead to the growth of cracking patterns which resemble mud cracks in a dried up lake bed. Here a simple model for this type of material failure process is described [1] and some results obtained with this model are presented.

### Computer Models

In the model for crack growth in thin films the thin film is represented by a network of bonds and nodes which, in the most simple case, form a triangular network or lattice at the start of the simulation. The bonds are considered to consist of Hookean (central force) spring and the elastic energy of the surface· layer is given by

$$E = 1/2 \sum_{ij} k_{ij}(\ell_{ij}-\ell_0)^2 \qquad (1)$$

where $\ell_{ij}$ is the length of the bond joining the ith and jth nodes and $k_{ij}$ is the force constant associated with that bond. If the bond is broken, $k_{ij} = 0$. At the start of a simulation the network forms a regular triangular lattice with periodic boundary conditions and the length of the bonds in the network may be larger than or smaller than the equilibrium bond length ($\ell_0$). Periodic boundary conditions were used in all of the simulations. At the start of a simulation each node in the network is attached to the substrate at its equilibrium position by a "weak" bond (Figure 1). The force exerted by this bond on the ith node

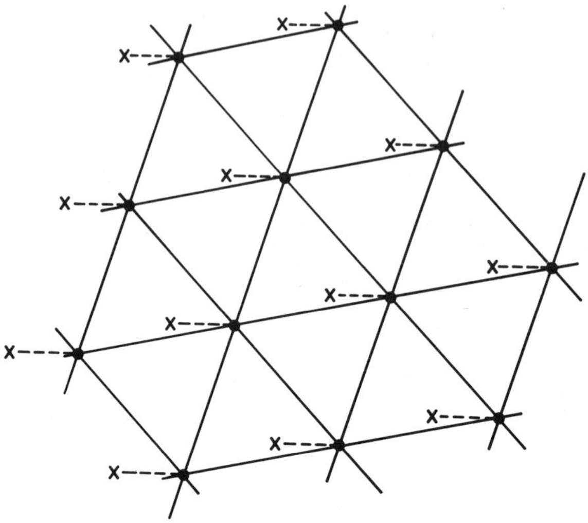

Figure I

A schematic representation of a model for elastic fracture in thin films. The nodes (large dots) are connected by strong bonds to form a triangular lattice. Each node is joined to the underlying structure by a weak bond (---) at its original position at the start of the simulation. Throughout the simulation the distance from the nodes to the underlying substrate is constant (only horizontal motion is allowed). This figure shows the original configuration in which each node is associated with six bonds in the surface network.

is given by

$$\mathbf{f} = k_2 \ (\mathbf{r}_i^o - \mathbf{r}_i) \tag{2}$$

where $\mathbf{r}_i$ is the position of the node and $\mathbf{r}_i^o$ is the position of attachment to the underlying substrate. The crack initiation and growth process is represented by a sequence of thermally activated bond breaking events [2]. It is generally assumed that bond breaking rate constants for all of the bonds in the network are given by

$$k_r = we^{-E_a^o/k_B T} \tag{3}$$

where $E_a^o$ is the bond breaking activation energy, $k_B$ is the Boltzmann constant and T is temperature. It is also assumed that the "weak" bonds connecting the surface layer to the substrate do not break. For a strained film the bond activation energy is given by

$$E_a = E_a^o - f\,(\delta) \tag{4}$$

where $\delta$ is the bond strain

$$\delta_{ij} = \ell_{ij} - \ell_o \tag{5}$$

If we further assumed that the bond breaking activation energy is reduced by the strain energy associated with that bond and use reduced units in which $k_B T = 1$ the probability that the bond joining the ith and jth nodes will be the next to break is given by

$$P_{ij} \sim e^{(k\delta_{ij}^2)/2)} \qquad (6)$$

The simulation then consists of a sequence of steps in which a bond is randomly broken (with the bond breaking probabilities given by equation (6)) followed by a relaxation [3] of the network to a new mechanical equilibrium. The bond breaking and relaxation processes are repeated many times until a large number (typically several thousand) of bonds

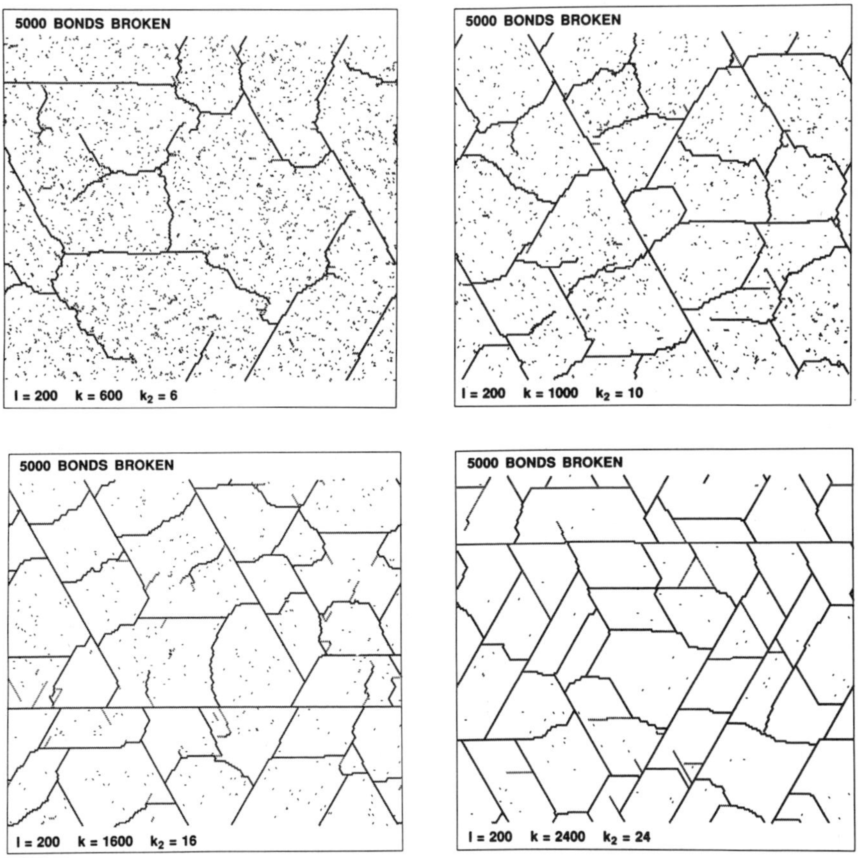


Figure 2
Simulations carried out using the surface cracking model with $k/k_2$ = 100 and four different values for k (600, 1000, 1600 and 2400 for Figures 2a, 2b, 2c and 2d, respectively). These simulations were carried out using a 200x200 node triangular lattice with $\ell_0$ = 0.90909 and an initial bond length of 1.0.

have been broken.

Figure 2 shows the results of simulations carried out using a 200x200 array of nodes with 120,000 connecting bonds. In these simulations the initial bond extension is 10% (at the start of the simulation $\ell$ = 1.0 and the equilibrium bond length $\ell_o$ is 0.90909..). Figures 2a-2d show the cracking patterns generated with four different values for the bond force constant k (k = 600, 1000, 1600 and 2400 in units in which $k_Bt$ = 1) and with a constant ratio of 0.01 for $k_2/k$. In this figure only the locations of the broken bonds in the original (undistorted) network are shown. For large values of k the first cracks to be formed are quite linear but as the cracking process proceeds the new cracks become less and less regular.

These simulations can easily be made time-dependent by selecting bonds at random and breaking them with a probability of $P_{ij}/P_{max}$ where the quantities $P_{ij}$ are given by equation (6) and $P_{max}$ is the maximum value of $P_{ij}$ for any bond in the system. Each time a bond is selected (whether or not it is actually broken), the time is incremented by $1/(NP_{max})$ where N is the total number of unbroken bonds. Time dependent simulations similar to those shown in Figure 2 suggest the following scenario for large values of k and $k/k_2$>>1. Initially isolated defects are formed and the rate of bond breaking is relatively slow. However, the stresses associated with bonds near these defects are larger than those further from defects and they have a relatively high probability of breaking. As the defects grow larger, the stress concentration effects become more important and once a crack has developed it grows very quickly. The fast growth of the first few cracks reduces the stress in the surface film and the rate of bond breaking becomes slower and slower as more of the elastic energy in the film is released. As the stress is progressively released, the cracks become less linear because of the reduced effects of stress concentration at the growing crack tips.

For polymers, a large body of empirical evidence suggests that the reduction of the activation energy is linearly dependent on the stress (see reference 4, for example). Consequently, simulations have also been carried out using the expression

$$P_{ij} \sim e^{v\delta_{ij}} \qquad (7)$$

in place of equation (6). For large values of v ($v >> \delta_o^{-1}$ where $\delta_o$ is the initial bond strain) this model gives results which are qualitatively similar to those shown in Figure 2. The results shown in Figure 2 were obtained from a model in which all of the bonds in the network are equivalent. In many cases it would be more realistic to have more than one type of bond or a continuous distribution of bond activation energies and force constants. Such simulations have been carried out and, in general, the early stage cracks are less linear than those shown in Figure 2. It is also interesting to simulate the effects of pre-existing "damage" in the surface film. Such damage might be caused by radiation, defects, impurities, etc. Figure 3 shows the results obtained from simulations carried out on a 300x300 triangular lattice array of nodes with k = 1600 and k = 16. Figure 3c shows the results of a simulation similar to that shown in Figure 2c. To obtain the results shown in Figure 3b, 333 nodes and their associated bonds were randomly selected and removed at the beginning of the simulation.

Figure 3c shows the results of a similar simulation in which 667 nodes were randomly selected and removed. It is apparent from Figures 3a-3b that the initial "damage" leads to a less regular cracking pattern. By observing the evolution of the cracking patterns on a computer terminal, it can be seen that the growing crack tips are occasionally diverted towards a defect (removed node) at a position which they would not ordinarily intersect.

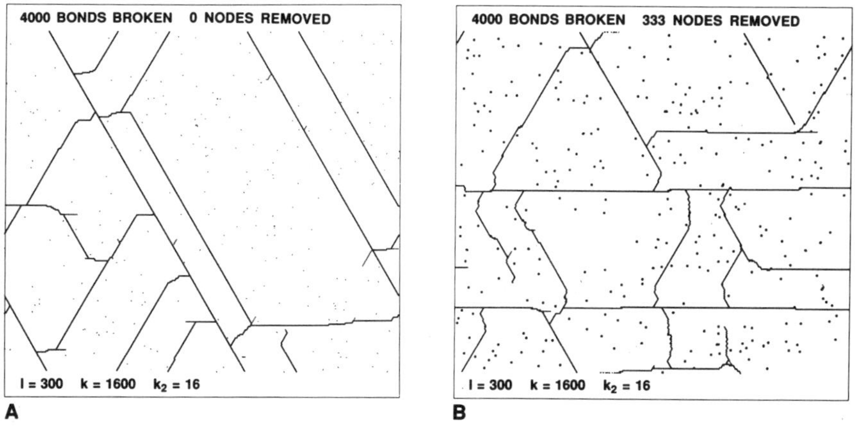


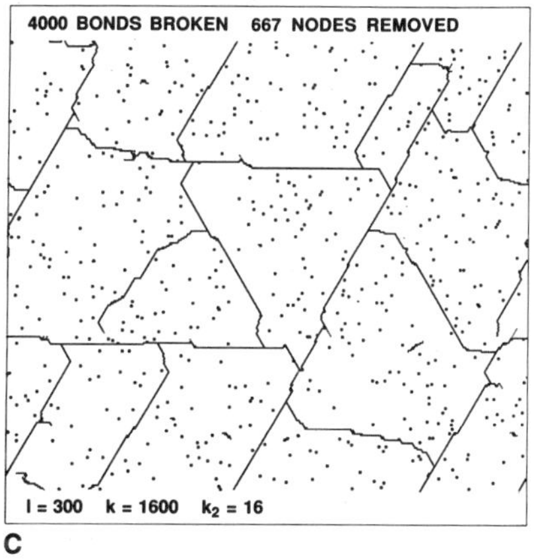


Figure 3
Simulations carried out using a 300x300 array of nodes with 270,000 bonds with k = 1600 and $k_2$ = 16. Figure 3a shows the results of a simulation similar to that shown in Figure 2c. The results shown in Figure 3b were obtained with 333 randomly removed nodes and the results shown in Figure 3c were obtained from a simulation in which 667 nodes were randomly removed at the start of the simulations.

Figure 4 shows the time dependence of the number of broken bonds obtained from the simulations shown in Figure 3. It is apparent that the presence of the defects (removed nodes) results in a much more rapid crack growth process.

## SUMMARY

The simple model described here results in the formation of quite realistic looking cracking patterns. The essential features required to produce the cracking patterns seen in these simulations seem to be (1) Attachment of a surface layer to a substrate so that the first few cracks do not relieve all the strain, (2) A bond breaking probability which grows rapidly with increasing strain, (3) A value for $k_2$ that is substantially smaller than k so that the stress/strain field is not too strongly localized.

Very similar cracking patterns have been produced on an analog experiment [5] using a model consisting of a monolayer of 3.4 μm (±1%) sulfonated polystyrene microspheres. The microspheres were

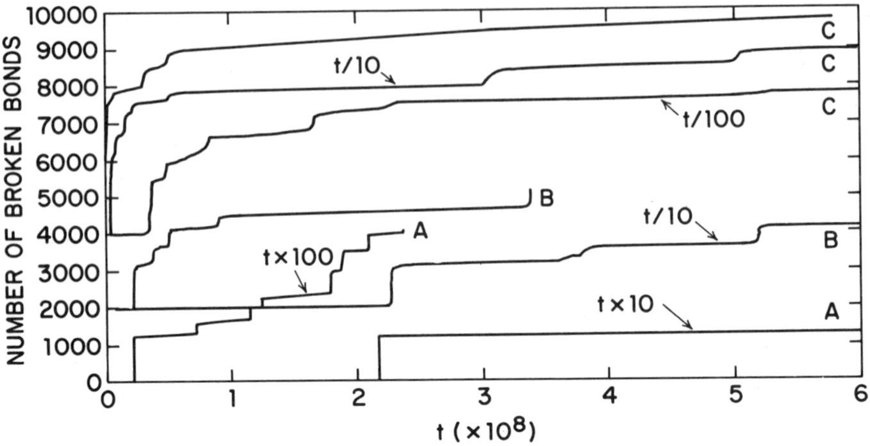


Figure 4

Time dependence of the number of broken bonds obtained from the simulations shown in Figure 3. The curves marked tx10 and tx100 are displayed on a time scale which is 10 times and 100 times longer than that given along the abscissa. Similarly, the curves marked t/10 and t/100 are displayed on time scales which are 10 and 100 times shorter.

dispersed in water to produce the monolayer and a strained film was produced by allowing the monolayer to slowly dry. During the drying process, the sphere diameter is reduced to 2.7 μm. At the first stages of failure, small (localized) defects are formed at a relatively slow rate. This stage is followed by the rapid growth of more or less linear cracks. In the later stages, the crack growth process becomes slower and the cracks become less and less linear as the stress in the monolayer is released by the cracking process. A computer model closely related to that described here has been used to simulate this process [6].

REFERENCES

1. P. Meakin, Thin Solid Films 151, 165 (1987).
2. H. Eyring, J. Chem. Phys. 3, 107 (1953).
3. D. M. de G Allen, "Relaxation Methods", McGraw Hill, New York (1954).
4. A. J. Kinloch and R. J. Young, Failure Behavior of Polymers, Applied Science, London (1983).
5. A. T. Skjeltorp, preprint (1988).
6. A. T. Skjeltorp and P. Meakin, Nature 335, 424 (1988).

# STRESS GRADIENTS AND ANISOTROPY IN THIN FILMS

**ROBERT M. FISHER, J. Z. DUAN, and ALAN G. FOX**
Center for Advanced Materials, Lawrence Berkeley Laboratory, University of California, Berkeley, CA 94720

## ABSTRACT

Indications that steep through-thickness strain gradients occur in vapour-deposited chromium films stemming from previous observations of film curling during spontaneous delamination from substrates, have been substantiated by analysis and simulation of Bragg X-Ray diffraction peaks. The presence of large through-thickness compressive strains, that increase quadratically with distance from the substrate from about zero at the interface to around 0.8% at the film surface, was deduced by empirical computer matching of diffraction peak shapes.

## INTRODUCTION

Quantitative interpretation of spontaneous, i.e. residual stress-driven, delamination of thin films from substrates requires precise knowledge of the internal stress state at the interface and throughout the film [1]. The direction and magnitude of stresses in films on substrates can be determined by measurements of substrate distortion. However, any preferred crystallographic texture, steep through-thickness stress gradients or large in-plane stress anisotropy, as well as the occurrence of any brittle, ductile or diffusive stress relaxation processes during or subsequent to film deposition can lead to an erroneous interpretation of substrate bending [2,3]. On the other hand, measurements of residual stresses in crystalline materials by x-ray diffraction are less influenced by localized film relaxation and particularly by cracking.

## STRAIN MEASUREMENTS BY X-RAY DIFFRACTION

Representative tracings of 110 Bragg peaks from several thin Cr films vapor-deposited on glass substrates are presented in Figure 1 along with the peak from a reference sample of annealed Cr powder. The large shift in peak position reveals that lattice planes parallel to the surface are in compression with a mean strain of about 0.5%, i.e. the films are in tension. The considerable broadening of the lines is indicative of a substantial range of elastic strains in the films (the pronounced asymmetry of the peaks derives from non-uniformities within the films as discussed below). The observed broadening is a composite of effects due to the fine grain size and non-uniform strain in the films, as well as instrument broadening, is given by [4],

$$B_{obs}^2 = B_I^2 + B_{GS}^2 + B_{\Delta\varepsilon}^2 \qquad (1)$$

The grain size broadening $B_{GS}$ and non-uniform strain broadening $B(\Delta\varepsilon)$ are given by,

$$B_{GS} = \frac{k \bullet \lambda}{t \bullet \mathrm{Cos}\theta_B} \qquad (2) \qquad \text{and } B(\Delta\varepsilon) = -2\left(\frac{\Delta d}{d}\right)\tan\theta_B \qquad (3)$$

Subtracting instrument broadening, established with a well-annealed, coarse grain standard, and rearranging terms gives,

Mat. Res. Soc. Symp. Proc. Vol. 130. 

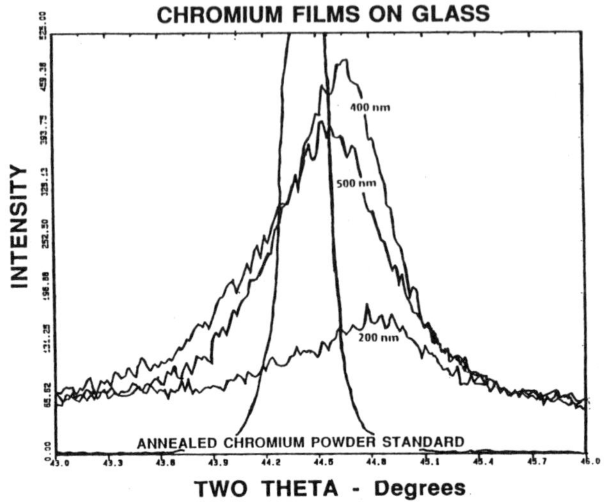


Fig. 1. X-Ray diffraction peaks (110 bcc - Cu K alpha) for thin Cr films vapour-deposited on glass - superimposed peak from annealed Cr powder.

$$\frac{B^2 \cos^2 \theta_B}{\lambda^2} = \frac{k^2}{t^2} + 4\left(\frac{\Delta d}{d}\right)^2 \frac{\sin^2 \theta}{\lambda^2} \tag{4}$$

Here $B^2 = B_{obs}^2 - B_I^2$, t is the diameter of the grain size, and $k^2$ is the constant depending on the shape of the grain. It may be seen that the slope and the intercept of a plot of equation (4) using experimental values for $B^2$,$\cos^2\theta$, $\sin^2\theta$ are directly related to the non-uniform strain and the grain size respectively. The representative results for grain size and strain range as tabulated below in Table I are in the right range but are about a factor of 2 too large. This will be discussed below in relation to the unusual asymmetry of the x-ray peaks.

TABLE I. Results of grain sizes and non-uniform strains from equation (4)

| film thickness | intercept | slope | GS Å | $\Delta e$ % |
|---|---|---|---|---|
| 300 nm | $0.35 \times 10^{-4}$ | $5.856 \times 10^{-4}$ | 169 | 1.21 |
| 600 nm | $0.20 \times 10^{-4}$ | $10.858 \times 10^{-4}$ | 223 | 1.65 |

A more fundamental approach is to fit the full intensity distribution of both first and second order reflections with a Fourier series (Warren-Averbach) method [5]. Data analysis using proprietary software on a Siemens diffractometer indicated grain sizes in the films of about 150 - 200 Å, in fair agreement with the Cullity method, and gave non-uniform strain values around 0.5%, but with persistent inconsistencies in results. It became clear that for both the Cullity and Warren-Averbach methods, very low diffracted intensities from the films made it impossible to obtain good data from the high order reflections that are needed for reliable analyses. Furthermore the marked asymmetry of line shape, indicative of the presence of contributions to line broadening that are not randomly distributed, is not contemplated in either formalism. In fact, diffraction peaks obtained using more highly absorbed Fe K-alpha varied significantly in shape and peak position from the Cu K-alpha results, indicating a change in strain with depth in the film. Thus it is evident that conventional x-ray methods are not adequate when steep, through-thickness strain gradients are present.

## COMPUTER SIMULATION OF X-RAY PEAKS

A more straightforward, albeit empirical, approach to interpreting the x-ray observations is to compute matching intensity distributions based on information about grain size and strain distribution obtained from electron microscopy studies that are summarized below.

When the adhesion of a Cr film to the substrate is poor, it may peel away from the substrate as illustrated in Figure 2. As mentioned previously, the occurrence of a strain gradient is suggested from this scanning electron micrograph which shows that the film has curled into a very tight roll after delamination from the substrate with the outer surface of the film in tension. Peak surface strains greater than 2 % are suggested by the small radius of curvature of the curls. Oblique deposition results in the formation of narrow hoops (by film cracking and followed by delamination) indicating a large stress anisotropy between directions parallel and perpendicular to the line of sight to the source [6]. Transmission electron microscopy of cross-sections of a number of typical films revealed that the grain size was not uniform through the film but increased from the substrate to the top by a factor of about 5 for the thickest films.

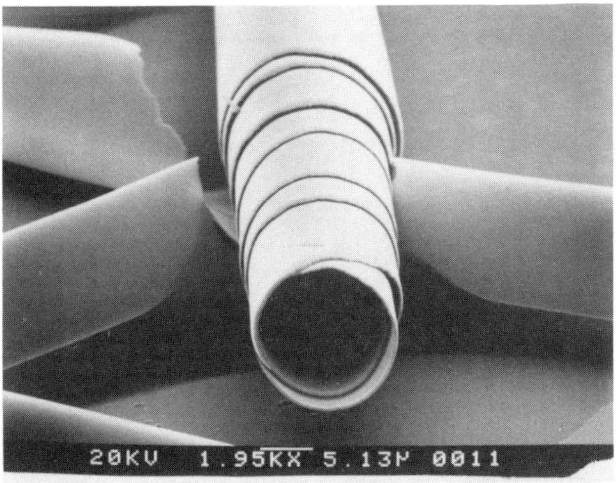


Fig. 2. Scanning electron micrograph showing curling of delaminated Cr film.

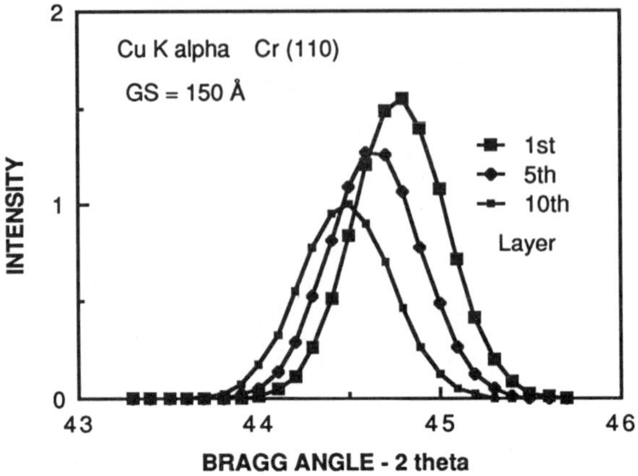


Fig. 3. Illustration of method of calculating intensities from individual layers in thin films. Effect of absorption in middle and bottom layers may be seen.

In computing x-ray diffraction peaks, each film was considered to be a stack of 10 layers with different (110) d spacings as shown in Figure 3. The intensity diffracted from each layer was corrected for absorption and for grain size and instrument broadening and summed to generate simulated peaks.

Simulated (110) peaks for 3 different variations of d spacing with distances from the substrate are shown in Figure 4. Although the fit is not particularly good for

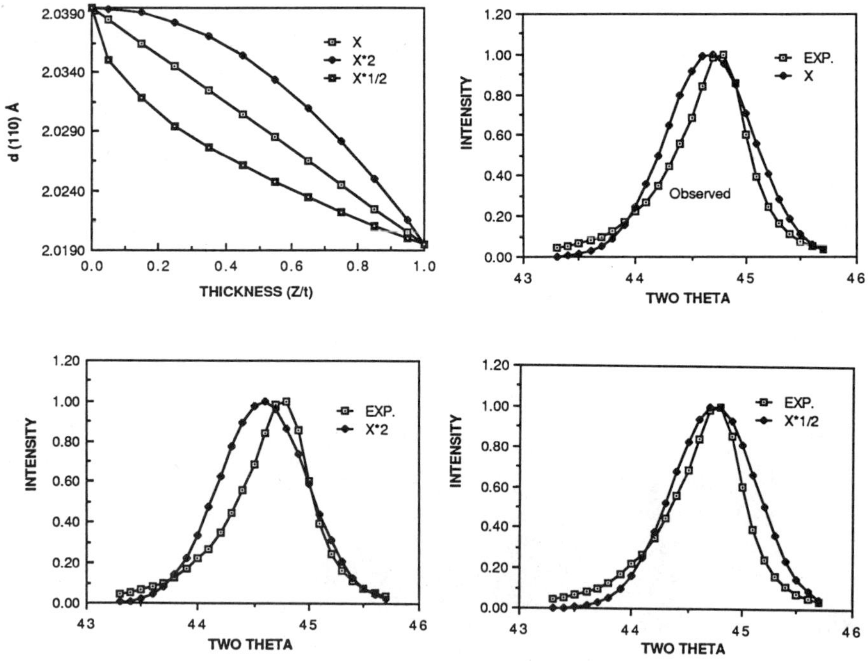


Fig. 4. Comparison of observed and simulated X-ray diffraction peaks for linear and non-linear lattice parameter gradients in a 3000 Å Cr film. Depth correction for absorption included – grain size assumed to be 100 Å.

any of the assumed strain distributions, the general peak shape is about right. Much better fits were obtained by using the actual grain size values determined by electron microscopy as illustrated in Figure 5. Very good fits were also achieved for Fe k-alpha diffraction peaks from the same films. As anticipated, the need to fit experimental peaks obtained with 2 different radiations narrowed the possible range of strain distributions considerably.

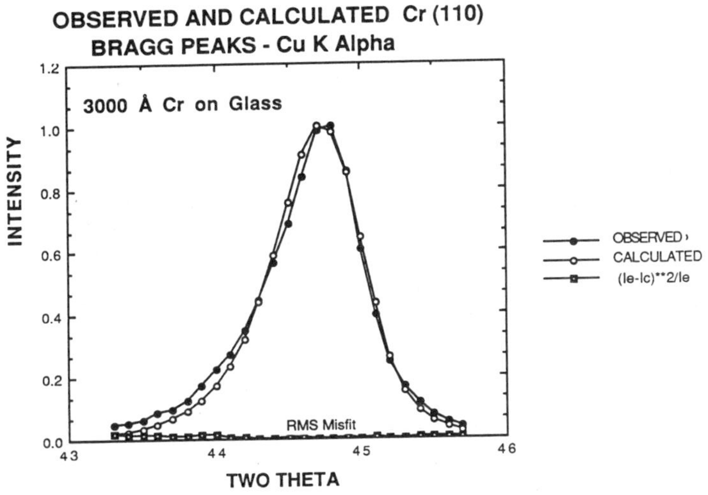


Fig. 5. Observed and simulated X-ray diffraction peaks for 3000 Å of Cr on glass. A linear grain size gradient and a non-linear strain gradient assumed - corrected for absorption.

## RESULTS AND DISCUSSION

The elastic strain distributions in two Cr films derived from computer matching of (110) Bragg peaks are shown in Figure 6. The plots illustrate a general feature about strain vs thickness that was noted throughout this study. In particular the strain at the substrate and the rate of increase with distance from the substrate tend to be smaller in the greater final film thickness. Increasing the substrate temperature is known to result in lower internal stresses in Cr films and so it is reasonable that the slight temperature rise that occurs during deposition will also result in lower strain values for thicker films.

Interpretations of film cracking and delamination based on the newly established knowledge of strain gradients in Cr films will be reported separately.

## ACKNOWLEDGEMENTS

This work was supported by the Director, Office of Energy Research, Office of Basic Energy Sciences, Materials Science Division of the U.S. Department of Energy under Contract No. DE-AC03-76-SF00098. We are indebted for helpful discussions with Rowland Cannon and collaboration with C.J.D. Hetherington.

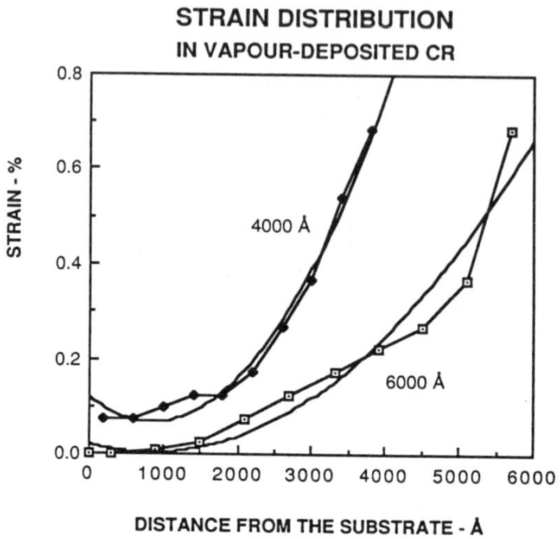


Fig. 6. Strain distribution in Cr films determined by simulation of X-ray peaks.

## REFERENCES

[1] R.M. Fisher, J.Z. Duan and A.G. Fox, Proc. of the Materials Research Society Spring 1988 Meeting, Reno, to be published.
[2] R.M. Fisher, EMSA proc. (1987), p. 236.
[3] R.M. Fisher, J.Z. Duan and A.G. Fox, J. of Mat. Sci. and Eng. (in press).
[4] B.D. Cullity, Elements of X-Ray Diffraction (Addison Wesley Pub. Co., 1956).
[5] B.E. Warren and B.S. Averbach, J. Appl. Phys. **21**, 595 (1950).
[6] J.Z. Duan, C.J.D. Hetherington and R.M. Fisher, EMSA Proc. (1988), p. 560.

# EFFECT OF HYDROGEN ON THERMAL STRESS IN LAYERED STRUCTURE OF Ti AND Ti(Al) THIN FILMS.

C.-K. Hu, P.S. Ho, and D. Gupta, IBM Research Division, Thomas J. Watson Research Center, Yorktown Heights, N.Y. 10598. S.T. Chen, IBM General Technology Division, Hopewell Junction, N.Y. 12533

## ABSTRACT

The effect of hydrogen on stress in thin film structures has been investigated for Ti, Al, Ti(Al) and Al(Ti) films. The stress in the metal films was measured *in-situ* as a function of temperature and ambient (vacuum of $10^{-6}$Torr., forming gas and nitrogen) using a bending beam technique. The metal films were deposited on fused quartz or polyimide/fused quartz substrates using a dual e-gun evaporation method. The stress in pure Ti films was found to be strongly dependent on the annealing environment. Upon thermal cycling in a forming gas, an abrupt change of stress in the Ti film due to crack formation and delamination in the Ti film was observed. The possibility to eliminate the hydrogen effect by Al addition has been investigated. The hydrogen effect can be reduced by an addition of 5 at.% Al, but stress induced crack formation similar to the pure Ti film was still observed in this alloy film. An increase to 20 at.% Al in the alloy films eliminated the hydrogen effect, however, such an alloy film was brittle as cracks and delamination were observed after one thermal cycle from room temperature to 350 °C. A similar behavior in stress relaxation was observed in an Al-riched film with 20 at.% Ti.

## INTRODUCTION

Thin film of Ti and its alloys are commonly used in VLSI metallizations for interconnection and packaging applications. The Ti film was used as a diffusion barrier layer, for a reduction of interfacial contact resistance, and as an adhesion layer to the dielectric materials. During VLSI processing, thin film metallization often undergoes thermal cycling and annealing in different ambients which can cause reliability problems due to the mechanical failures. Consequently, understanding of the stress relaxation behavior in thin films under processing conditions is important. For this purpose, we have used *in-situ* measurements to study stress as a function of film thickness, temperature, and ambient (vacuum, forming gas and nitrogen). We have previously reported the temperature dependence of stresses in metallic thin films(Ti, Cr, Cu) [1,2], polymeric films [3], and metal/polyimide [4] on fused quartz substrates annealed in nitrogen and vacuum.

We report here an investigation of the effect of hydrogen on thermal stress in the Ti, Al, and Ti-Al alloy films. The exposure to hydrogen comes from annealing in forming gas (90% nitrogen and 10% hydrogen) which is often used in VLSI process steps to reduce metal oxidation. Annealing in forming gas brought marked changes in the mechanical behavior of Ti films, in particular, crack formation and delamination were observed. The possibility of eliminating the hydrogen effect by alloying with Al was investigated. The results will also be reported.

# EXPERIMENTAL PROCEDURES

The temperature dependence of stress was measured *in-situ* using a cantilever bending beam technique similar to that reported in Ref. 1. The system consisted of two fused quartz cantilevers. One, 75 $\mu$m thick, was used as a substrate and the other, 0.16 cm thick, as a reference quartz beam. The substrate and the reference beam were mounted in a graphite holder with a tungsten spring. The polyimide (PMDA-ODA) film was spun on the fused quartz substrate and fully cured in an oven with nitrogen ambient. The final curing step was to bake polyimide at 400°C for 30 mins. and resulted in a 1 $\mu$m thick film. Prior to the metal deposition, the substrates were baked in vacuum at 150 °C for 15 mins.. Pure Ti, pure Al, Ti(5-20 at.% Al), and Al(20 at.%Ti) alloy thin films was deposited by a dual e-gun evaporator directly on the fused quartz substrates or or a polyimide interlayer *In-situ* measurements were made either in a vacuum of $10^{-6}$ Torr., or in 1 bar pressure of nitrogen or forming gas at a temperature in the range of 25-400 °C. The surface morphology of these films were examined after thermal cycles using scanning electron microscopy (SEM) and an optical microscope.

The experimental procedure for *in-situ* stress measurements has been described in Ref. 1. The deflection ($\delta$) of the substrate along its length was measured using a microscope. The stress in a trilayer structure, e.g. metal/polyimide/$SiO_2$, has been analyzed using an elastic approximation [4]. The average film stress (S) in terms of the deflection of the beam at any point along its length (L) can be obtained from the equation given below:

$$\delta = 3(1 - \nu_s)L^2(t_m + t_p)S/(E_s t_s^2) \quad (1)$$

where $t_m$, $t_p$ and $t_s$ are the thicknesses of metal film, polyimide film and substrate, respectively. $E_s$ ($7.3 \times 10^{11}$ dynes/cm$^2$) and $\nu_s$ (0.16) are Young's modulus and Poisson's ratio for the fused quartz substrate, respectively. The average stress in the metal film can be calculated as follows:

$$S_m = S(t_m + t_p)/t_m \quad (2)$$

The calculation of the thermally induced stress $S_{th}$ of three elastic components is rather cumbersome [5]. For the metal/polyimide/$SiO_2$ trilayer structure under the conditions of $t_s >> t_m + t_p$ and $E_s$ or $E_m >> E_p$, the equation of the thermal-induced stress can be simplified as:

$$S_{th} = E_m \Delta T(\alpha_m - \alpha_s)/(1 - \nu_m) \quad (3)$$

where $E_m$ and $E_p$ are Young's moduli for the metal film and polyimide film, respectively. $\alpha_m$ and $\alpha_s$ are the coefficients of thermal expansion of metal and fused quartz, respectively. $\nu_m$ is Poisson's ratio for the metal film. The $\Delta T$ is the difference between two measured temperatures. Under these conditions, the thermal-induced stress can be evaluated independent of the polyimide thickness. This relationship is derived on the basis of elastic approximation. This ignores the plastic deformation of the polymer which has been found to be effective in relieving the stress in the metal film [4]. Further studies will be carried out to investigate this effect on stress. The thermal stress reported in this paper is calculated using Eq. (3).

# RESULTS AND DISCUSSION

The stress in as-deposited Ti films is strongly dependent on film thickness. This residual stress is compressive and become tensile for film thickness > 100 nm [1]. Figure 1 shows average stress in Ti (1 $\mu$m) on fused quartz measured during thermal cycling in a vacuum of $10^{-6}$ and a nitrogen ambient. The arrows show the heating and cooling sequence. No significant difference was observed in stress annealed in a vacuum or nitrogen which suggests that the reaction between Ti and nitrogen is negligible for temperatures below 350°C. During the first heating period, a thermal-induced compressive stress is superimposed on the residual stress upon deposition. During heating, the kinetics of stress relaxation can release only a part of total stress, leaving behind a residual tensile stress at 400°C. The behavior during the cooling period is different; as a higher tensile stress is generated at increasingly lower temperature, most of thermal-induced stress is not relaxed. The value of stress during the cooling cycle can be well accounted for by Eq. (3). Further cycling shows that the thermal stress becomes dominant as cycling hardening of the film reduces substantially the stress relaxation. The stress behavior is drastically different for the sample heated in a forming gas. The temperature dependence of stress in pure Ti(0.5 $\mu$m) on a Ti/$SiO_2$ bilayer film annealed in forming gas is shown in Fig. 2. The heating rate is about 10 °C/min and the samples remained at 350 °C for 30 mins. before the cooling cycling. The data of the first heating cycle are similar to the results of heating in vacuum. However, during cooling from 350°C to room temperature, an abrupt change of stress in the Ti film from tension to compression occurred around 200 °C. Cracks and delimitations were observed in this film using an SEM. This behavior was also found in the Ti(1 $\mu$m)/$SiO_2$ bilayer and Ti(0.5$\mu$m)/Polyimide(1 $\mu$m)/$SiO_2$ trilayer structures (not shown). The constitutional diagram of Ti-H system [6] shows two phases $\alpha$(HCP) + $\beta$(BCC) at temperature above 300 °C, and $\alpha + \gamma$ at temperature below 300 °C in the Ti-rich region.

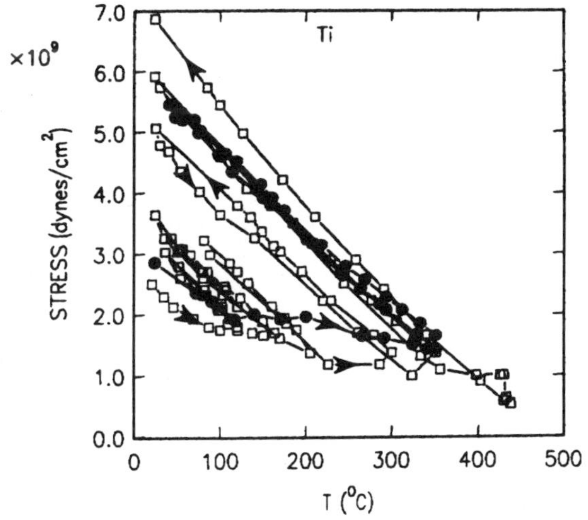


Figure 1. The In-situ stress as a function of temperature. The open circles are the sample annealed in vacuum and the closed circles are annealing in a nitrogen atmosphere. The lines have been added to guide the eye.

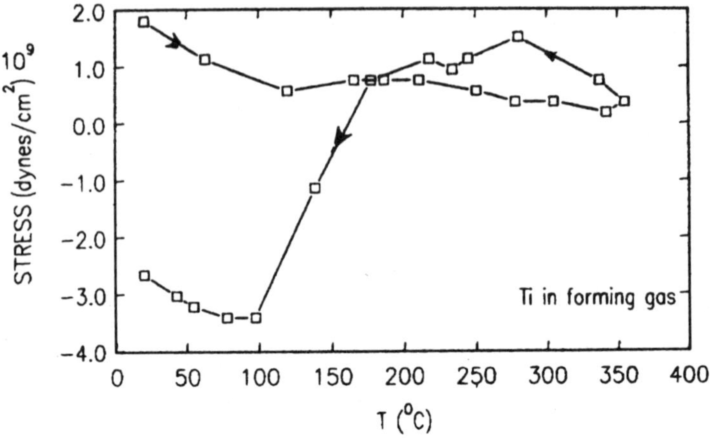


Figure 2. Stress of 0.5 µm thick Ti measured in forming gas ambient. An abrupt change in stress state was clearly shown after one thermal cycle.

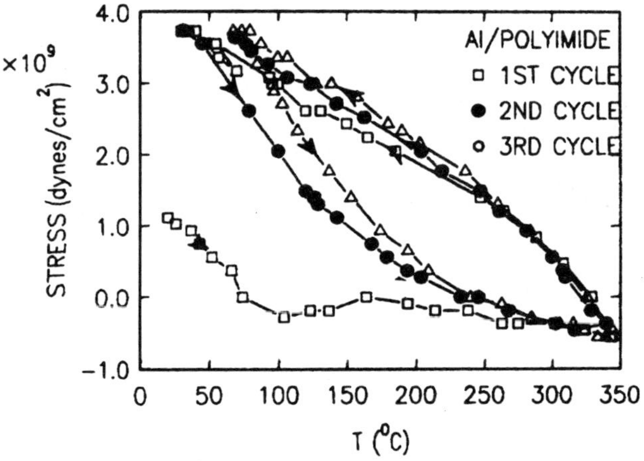


Figure 3. The stress of Al in Al/polyimide on a fused quartz substrate. No hydrogen effect on stress was observed.

In addition, the hydrogen atoms can rapidly diffuse into the Ti film at 350 °C [7]. This suggests that, upon cooling, hydrogen can be incorporated into the Ti film, resulting in a change in the microstructure and causing the abrupt change in the stress.

Fig. 3 shows the temperature dependence of stress in Al in the trilayer structure of Al(0.4 $\mu$m)/Polyimide(1$\mu$m)/$SiO_2$ measured in forming gas. Unlike the Ti film, the solubility of hydrogen in Al is extremely small, only about 3 ppm [8]. No abrupt change in the stress is seen in Fig. 3. The residual stress at room temperature increased after thermal cycles, but it is considerably smaller than that calculated from Eq. (3). This indicates that a substantial amount of the stress relaxation has taken place in Al, resulting in the loop type curves. This is consistent with the results of a study reported earlier [9]. The addition of Al in Ti is commonly used for stabilizing Ti against hydrogen embrittlement. The effect of hydrogen on stress was found to be reduced by an addition of 5 at.% Al, but the changes in the stress state similar to the pure Ti film were still observed in this alloy after four thermal cycles. An increase in the concentration of Al up to 20% eliminated the effect of hydrogen on stress. Fig. 4 shows the stress in Ti(20%Al) in Ti(20%Al)/Polyimide/$SiO_2$ annealed in forming gas during six thermal cycles. Only partial data are presented in Fig. 4. The behavior of this alloy film is similar to those shown in Fig. 1. After the first cycle the stress in Ti(20%Al) can be well predicted by Eq. (3), but shows a systematic drop after every thermal cycle. This suggests that a delamination between metal to polyimide has occurred. The residual stress in Ti(20%Al) is rather high

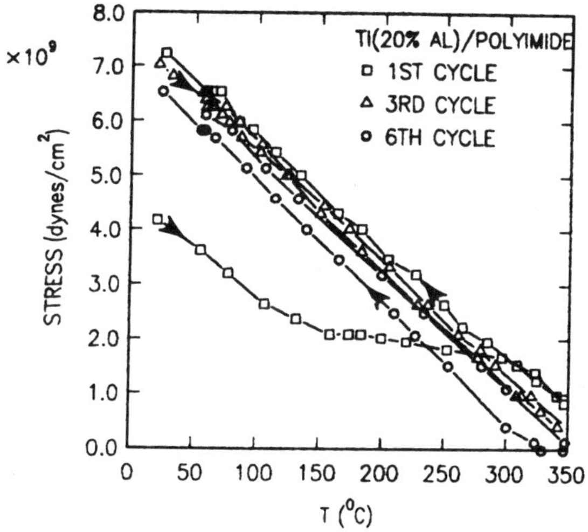


Figure 4. Stress of Ti(20 at.% Al) in Ti(20% Al)/polyimide/fused quartz structure during thermal cycling in forming gas.

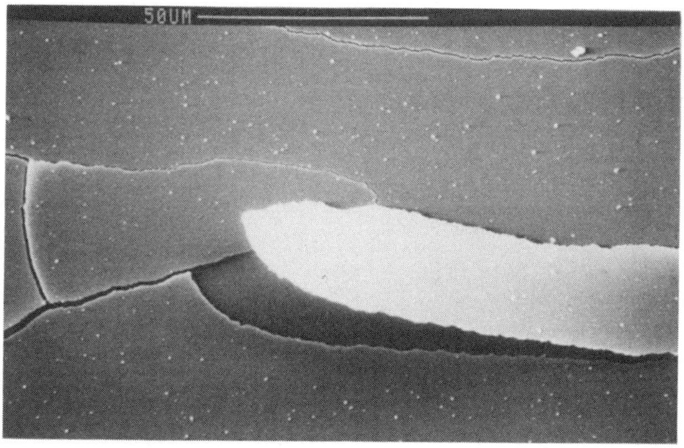


Figure 5. The SEM micrograph of the top surface of the Ti(20% Al)/polyimide/fused quartz structure after 6 thermal cycles in forming gas.

after the thermal cycling and results in cracking. The cracks and delamination are observed and shown in Fig. 5. In the case of Al(20%Ti) (not shown), the stress behavior under thermal cycling are similar to those shown in Fig. 4. The crack formation was found after the first thermal cycle due to the Ti alloying which greatly reduces stress relaxation in Al.

# CONCLUSIONS

We have carried out a study of stress relaxation during thermal cycling in Ti, Al, Ti-Al and Al-Ti thin films using the cantilevered bending beam technique. The stress behaviors among the various thin films were different. The effect on stress due to annealing in forming gas was investigated. The presence of hydrogen was found to have a strong effect on the state of stress in Ti but not in Al, Ti(20% Al) and Al(20%Ti) films. Alloying the Ti with Al can retard the hydrogen embrittlement. Increasing the Al concentration to 20 at.% eliminated the effect of hydrogen on stress, but resulted in crack formation in the film due to a high level of residual stresses after thermal cycles. The stress in pure Al film is relieved during thermal cycles, but Al(20% Ti) did not exhibit such a stress relaxation, instead stress-induced cracking of the film was observed.

# ACKNOWLEDGMENTS

We express our gratitude to M.B. Small for many comments on the manuscript. We also want to thank IBM Center of Scientific Service at Yorktown Heights for metal evaporations.

# REFERENCES

1. C.-K. Hu, D. Gupta, and P.S. Ho, Proceedings of 2nd International VLSI Multilevel Interconnection Conf.,187 (1985)
2. C.-K. Hu, D. Gupta, J.T. Wetzel and P.S. Ho, Mat. Res. Soc. Sypm. Proc. 54, 153 (1986).
3. C.-K. Hu, H.M. Tong, C. Feger, and P.S. Ho, Proceedings of 2nd International VLSI Multilevel Interconnection Conf.,280 (1985)
4. S.T. Chen, C.Y. Yang, F. Faupel and P.S. Ho, to be published in J. of Appl. Phys.
5. B.S. Berry and W.C. Pritchet, IBM J. of Res. Develop. 32, 682, (1988).
6. *Binary Alloy Phase Diagrams* edited by T.B. Massalski (Amer. Soc. for Met. 1986)
7. J. Volkl and G. Alefeld in, *Diffusion in Solid-Recent Development,* edited by A.S. Nowick and J.J. Burton (Academic, NY 1974),Chap. 5.
8. *Constitution of Binary Alloys, First Supplement,* edited by R.P. Elliott (McGraw-Hill, 1965)
9. A.K. Sinha and T. T. Sheng, Thin Sol. Film, 48, 117 (1978).

# ADHERENCE OF DIAMOND FILMS PRODUCED BY MICROWAVE PLASMA DEPOSITION ON SiAlON TOOL INSERTS

ROBERT C. McCUNE, D.W. HOFFMAN, T.J. WHALEN AND C.O. McHUGH
Ford Motor Company, P.O. Box 2053, Dearborn, MI 48121-2053

## ABSTRACT

The adherence of diamond films produced by microwave-enhanced chemical vapor deposition (CVD) on SiAlON tool inserts was assessed by scratch, indentation and machining tests on aluminum alloys. While the films are nominally adherent during handling and mild cutting operations, the measured adherence values are appreciably lower than values associated with conventional CVD coatings on cobalt-tungsten carbide cutting inserts. An indentation type adhesion test using polished substrates was found to yield reproducible behavior for interfacial crack extension at loads ranging from 15 to 150 kg. The "critical load" in scratch testing was found to be less than 10 N.

## INTRODUCTION

Recent progress in the ability to form crystalline diamond particles and continuous films from various vapor-phase processes has suggested that cutting tools incorporating such coatings may be consequently developed. Angus and Hayman [1] have reviewed a number of the novel processes for deposition of diamond and diamondlike carbon films from the vapor phase. Because hard coating of metal-cutting tools has achieved widespread acceptance among end users of such products [2], it is reasonable to predict that coated cutting inserts will be among early applications of the new diamond technologies. Two particular areas of interest to manufacturers are: 1.) the prospects of developing low-cost, ultra-hard coatings for three-dimensional tool surfaces, and 2.) the possibility of exploiting new engineering materials such as ceramics and fiber-filled composites in applications that had previously been uneconomical due to the high cost of diamond tooling required for machining.

As with earlier cutting tool coatings, however, widespread acceptance of diamond-coated tools by the end user will be largely determined by the price/performance characteristics of such tools. Early reports on diamond-coated, cemented-carbide cutting tool inserts [3-5] suggest that growth and adherence of the coating on compatible insert materials will be a significant problem in establishing a family of robust metal-cutting tools.

This work reports observations of diamond film adherence on SiAlON cutting tool inserts using techniques which have been previously employed in assessment of film adhesion on cemented carbide inserts. In that regard, the scratch and indentation tests performed, as well as several tests of metal cutting capabilities, are preliminary in nature due to the extremes in physical properties of diamond relative to even the hardest films produced by current physical and chemical vapor deposition technologies. The spirit of the work was, however, to assess the adherence and performance of diamond films made by a repeatable process in terms of currently achievable cutting-tool coating behaviors.

EXPERIMENTAL

Materials

The substrate material chosen for the depositions and subsequent metal-cutting tests was a commercial grade of SiAlON in the SNG 432 geometry. In the first series of experiments the inserts were coated in the as-received condition; however, pronounced crack propagation of the coating along finishing ridges of the insert prompted use of a polishing sequence concluding with 3 $\mu$m diamond paste before application of the coating. Such polished inserts were used only for indentation testing. Prior to the coating process, all inserts were "seeded" by lightly polishing with 0.25 $\mu$m diamond paste followed by rinsing in methanol.

Diamond depositions were conducted through facilities at the Materials Research Laboratory of The Pennsylvania State University, using a Toshiba microwave plasma apparatus with quartz specimen tube configured perpendicular to the waveguide. The insert to be coated was supported in the densest region of the plasma using a boron nitride susceptor. The only heat supplied to the substrate during deposition was via coupling to the microwave power, which was adjusted to yield a specimen temperature at the growth surface of 1000 °C as measured by optical pyrometry through a window in the growth tube. The system pressure was maintained at 90 torr with total gas flow rate of 100 sccm. The feed gases were $CH_4/H_2$ mixtures at 1% or 2% methane. Growth times were typically 5 hours. Figure 1 shows a typical cutting edge surface for a 2% methane deposition on a polished insert.

Figure 1. Morphology of 2% $CH_4$ diamond-coated SiAlON tool cutting edge.

Adhesion Testing

There are a variety of adhesion tests available for hard coatings; many of which have been reviewed in the literature [6,7]. Scratch testing using a diamond indenter has been described by Perry [8], and several of the complicating factors have been discussed [9-11]. The particular instrumentation used in these investigations was the Revetest Automatic Scratch Tester manufactured by Centre Suisse d'Electronique et de Microtechnique (CSEM). This device provides for indentation by Rockwell 'C' diamond indenter having a cone angle 120° and tip radius of 0.2 mm, at a linearly increasing load of up to 100 N over a track length of 10 mm. An acoustic emission sensor allows assessment of the load at which the coating de-adheres by spallation from the surface of the substrate.

Chiang, Marshall and Evans [12] developed adhesion measurements for thin films predicated on crack propagation induced at the film/substrate interface as a consequence of indentation with a pyramidal diamond indenter. Jindal, et al [9] proposed use of a spherical indenter in a similar approach. In the indentation methods, the extent of the interfacial cracking is determined as a function of the applied load, leading to a measure of interfacial toughness. High intrinsic stresses within the film may provide additional driving force for crack extension [13].

In this work, a 'C' Brale indenter of radius 0.2mm was employed with a Rockwell hardness tester at loads of 15, 30, 45, 60, 100 and 150 kg. Diamond films grown on the polished SiAlON yielded the most readily observable cracking patterns. Use of an SEM generally did not enhance the ability to determine the extent of interfacial cracking for these films.

RESULTS and DISCUSSION

Scratch Testing

In general, scratch tests yielded spallation of the diamond films at critical loads of less than 10 Newtons. Figure 2 shows a typical scratch pattern, on which the point of de-adherence can be detected visually. Acoustic emission detection confirmed the occurrence of film break-up at these loads. Although direct comparisons between materials and films in scratch tests have been cautioned [11], it is worthwhile to point out that a typical range of critical loads for such coatings as CVD TiC on cobalt-cemented WC may be in the range of 30-70 Newtons as measured on the Revetest machine under similar conditions of loading rate, indenter speed and indenter wear.

Figure 2. Scratch behavior of diamond film on SiAlON showing decohesion.

Indentation Tests

Attempts to produce measurable interfacial cracks using the Vickers indenter method of Chiang, et al. [12] were unsuccessful on the inserts coated in the as-received condition. The first indications of film spallation, however, were observed at a load of 30 kg, and removal of plate-like regions was clearly observable at a load of 50 kg.

By pre-polishing the SiAlON insert, and through use of a diamond Brale indenter, it was possible to obtain reproducible measurements of crack diameter versus applied load. In these cases, the delamination of the diamond layer could be observed using optical microscopy with polarized light. Figure 3 shows a typical indent/crack pattern using a Brale indenter and 100 kg load. Figure 4 illustrates the crack diameter versus load behavior for a 1% methane/hydrogen film grown on polished and seeded SiAlON. The behavior for a CVD TiC film on a cobalt-cemented WC substrate [9] is shown for comparison.

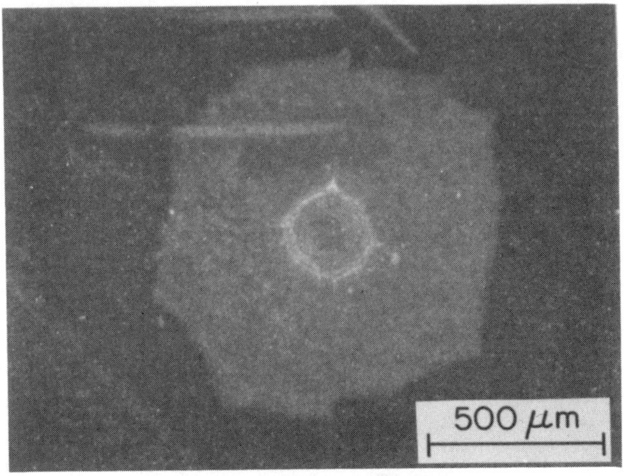


Figure 3. Optical micrograph of delamination crack resulting from 100 kg loaded diamond Brale indenter on diamond film/SiAlON substrate.

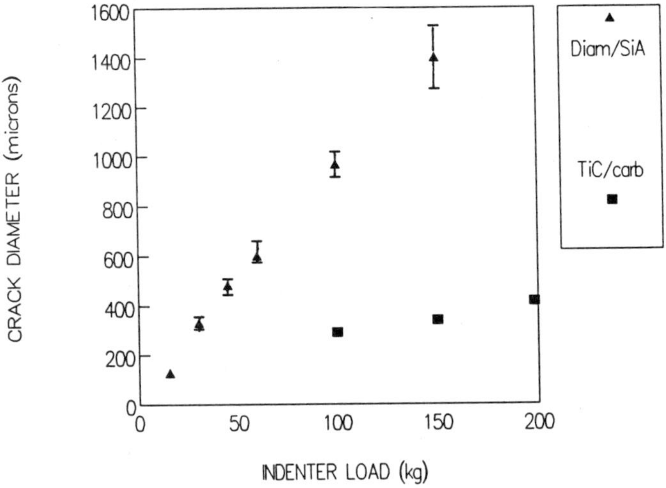


Figure 4. Crack extension vs. load behavior using the Brale indenter for a 1% $CH_4$ diamond film grown on polished SiAlON. Literature values for a CVD TiC film on cobalt-cemented tungsten carbide shown for comparison [9].

Machining Studies

Cutting tests on aluminum-silicon alloys were performed with several diamond-coated SiAlON inserts in both turning and milling operations. Such experiments are necessary to more completely reproduce the stress and temperature conditions experienced by cutting tools in actual practice. Turning studies were conducted on an 8" diameter "log" of AA 333 aluminum (8-10% Si, 3-4% Cu) at speeds of 1000 and 2000 sfpm, with feeds of 0.014 and 0.007 ipr respectively, and 0.040" depth of cut. Uncoated SiAlON was used as a reference material. Flank wear on both coated and uncoated tools was generally less than 0.003" and difficult to discern for both coated and uncoated tools due to the generally good machinability of the particular workpiece alloy. While the diamond-coated SiAlON showed generally larger dimensions to the wear scars at both speeds, the resulting surface finish from the coated tool was somewhat better (49-52 microinches) than for the uncoated reference (88-92 microinches) after thirty minutes of cutting at 2000 sfm.

Milling tests were conducted on a more difficult to machine alloy, AA 390 aluminum (nom. 16-18% Si, 4-5% Cu), at speeds of 544 and 1171 sfpm, with tool feed rate of 0.008 ipr and 0.060" depth of cut. The wear behavior of the coated and uncoated inserts was virtually indistinguishable in these tests. Figure 5 shows the progression of flank wear for coated and uncoated SiAlON; with tool wear data for a C2 cobalt-cemented tungsten carbide and compacted polycrystalline diamond (PCD) shown for comparison. Figure 6 shows a region of flank wear on the diamond-coated SiAlON where the spallation of the film appears to occur in advance of the wear scar. This would be consistent with the observation that the wear was occurring at the same rate on both the coated and the uncoated tool inserts.

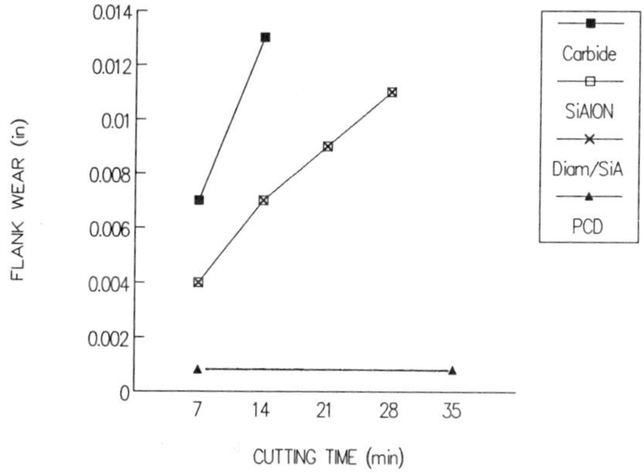


Figure 5. Flank wear vs. cutting time for carbide, SiAlON, diamond/SiAlON and compacted polycrystalline diamond (PCD) tools in milling of an AA390 aluminum alloy at 544 sfpm, .008 ipr feed and .060" doc.

CONCLUSIONS

Early versions of diamond-coated cutting tool inserts prepared by microwave-enhanced CVD using methane-hydrogen mixtures with SiAlON substrates suggest film adherence performance which is apparently less than for conventional CVD coatings on Co-WC materials as reported in the literature. Loss of the coating during turning of a type 333 alloy and milling of type 390 alloy permit the wear of the tool to progress at rates comparable to the uncoated inserts. The indentation type of adherence test appears to be useful with polished insert pieces.

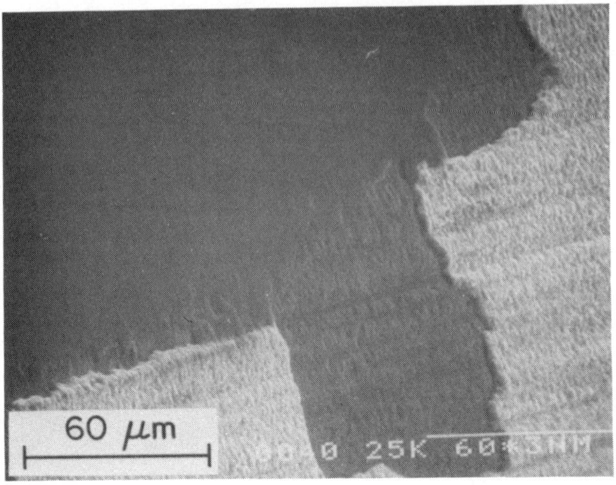


Figure 6. Diamond film spallation in advance of flank wear scar on a diamond-coated SiAlON insert used for milling AA 390 aluminum.

ACKNOWLEDGEMENTS

The authors would like to acknowledge the assistance of William R. Drawl of Pennsylvania State University who prepared the diamond coatings, and W. Ron Felcher of Ford Motor Company who performed the cutting tests.

REFERENCES

1. J. C. Angus and C. C. Hayman, Science 241, 913 (1988).
2. C. Wick, Manuf. Eng. 97, 26 (1986).
3. N. Kikuchi and H. Yoshimura, in New Diamond, (Japan New Diamond Forum, 1988) p. 42.
4. K. Shibuki, M. Yagi, K. Saijo and S. Takatsu, presented at the 15th. Int'l. Conf. on Metallurgical Coatings, San Diego, CA, 1988 (to be published in Thin Solid Films).
5. M. Murakawa, S. Takeuchi, H. Miyazawa and Y. Hirose, presented at the 15th. Int'l. Conf. on Metallurgical Coatings, San Diego, CA, 1988 (to be published in Thin Solid Films).
6. D. S. Rickerby, presented at the 15th. Int'l. Conf. on Metallurgi cal Coatings, San Diego, CA, 1988 (to be published in Thin Solid Films).
7. J. Valli, U. Makela and A. Matthews, Surface Engineering 2 [1], 49 (1986).
8. A. J. Perry, Thin Solid Films 107, 167 (1983).
9. P. C. Jindal, D. T. Quinto and G. J. Wolfe, Thin Solid Films 154, 361 (1987).
10. A. J. Perry, Surface Engineering 2 [3], 183 (1986).
11. P. A. Steinmann, Y. Tardy and H. E. Hintermann, Thin Solid Films 154, 333 (1987).
12. S. S. Chiang, D.B. Marshall and A. G. Evans, in Surfaces and Interfaces in Ceramic and Ceramic-Metal Systems - Materials Science Research v. 14 , edited by J. Pask and A. Evans (Plenum, New York, 1981), p. 603.
13. D. B. Marshall and A. G. Evans, J. Appl. Phys. 56 [10], 2632 (1984).

PART IV

# Mechanical Properties of Thin Films

# DEFORMATION IN THIN FILMS DEPOSITED ON RIGID SUBSTRATES

Masanori Murakami
IBM Research Division, T. J. Watson Research Center,Yorktown Heights, New York 10598

## ABSTRACT

Stresses and strains are always present in thin films deposited on rigid substrates. They are introduced into the films during the deposition and/or subsequent heat and chemical treatments. Since the origins of the strains (stresses) in thin films and their shapes are different from those of bulk materials, the distribution of strains supported elastically by the films is different from that of the bulk materials. The present paper reviews deformation in thin films caused by elastic strains, using illustrative data of Pb and Nb films that were deposited on Si wafers. Also, x-ray diffraction techniques to detect such local deformation are briefly reviewed.

## INTRODUCTION

The use of thin films has become very important in various components of electronic devices. Strains (or stresses) are always present in these thin film electrode materials and may deteriorate both the device performance and reliability. Two kinds of strains are present in thin films; "intrinsic" (or "grown-in") and "extrinsic" strains. The intrinsic strains are frequently observed in as-deposited "hard" films such as W, Fe, Mo, Cr, Mn, Nb, etc [1]. The origins of these strains have been discussed extensively by Doerner and Nix [2], Hoffman [3], and Chopra [4]. The extrinsic strain is primarily caused by mismatch of thermal coefficients between the films and the substrates. The large strains are observed in "soft" films such as In, Au, Al, Pb, etc. which have thermal expansion coefficients much larger than those of commonly used substrates like Si, Ge, GaAs, etc.

The universal problem caused by high strain is loss of adhesion between the electrode materials and the substrates. The problem is particulary serious for refractory materials used widely in VLSI devices as reviewed exhaustively by d'Heurle [5]. The problem has been frequently overcome by improving the substrate cleaning procedure, the deposition of thin adhesive layers, etc. Recently, to obtain the optimum device performance, both horizontal and vertical dimensions of the devices have been reduced significantly to almost the technological limit. The modern electron-beam lithography technique allows fabrication of devices with electrode width as small as $0.1\mu m$. Muliple-layer structures have been used in which extremely thin layers act as barriers between the two different electrode materials. For such devices, the microstructure changes in the electrode materials, as small as a few nanometers, caused shorts, leading to device failures. However, such small deformation is extremely difficult to observe directly even with the modern analytical techniques because of the complexity of the device structure. Therefore, it is very important to predict such deformation of thin film electrode materials induced by small strains (even within the elastic limit) either by theories or diagostic experiments in order to increase the reliability of these devices.

The mechanical properties of thin films were generally treated based on an assumption that the films and substrates were elastically isotropic. This assumption was valid for only a few films such as W. Other films have anisotropic elastic constants, and unique strain distributions were observed in thin films, which had never been observed in the films with isotropic elastic constants [6,7]. In particular, when the film thickness was comparable to the mean grain size, the effect of anisotropy in elastic constants became the primary cause for non-uniform strain distribution, and the distribution was observed to be strongly affected by the origins of the strain or stress [7-9]. In addition, one must consider deviations from truly planar geometry. The strains supported elastically by line or island-shaped films were different from those of the blanket films, and non-uniform strain distributions which depended on the film shapes were observed in films even with isotropic elastic constants [8,10-12].

In this paper we review deformation caused by elastic strains present in thin films which were deposited on rigid substrates. In order to show the effect of anisotropy in elastic constants on the deformation, we review theories and experiments carried out for Pb and Nb with large anisotropy in the elastic constants which were deposited on Si substrates. In these experiments, the stresses and strains were introduced extrinsically and intrinsically into the Pb and Nb films, respectively, and strain behavior of the two films was compared. The relationship between the origin of the stresses and strains and the deformation in the films was elucidated. There are several techniques to measure the strains in thin films as reviewed by Cambell [13]. Among them x-ray diffraction techniques provide information about distribution of such small strains [14] and some of these techniques are also reviewed briefly.

## DEFORMATION DUE TO ANISOTROPY IN ELASTIC CONSTANTS

### Strain and Stress Relation

For a (001)-oriented crystal, the principal axes $x_1, x_2$, and $x_3$ coincide with [100], [010], and [001] crystal directions, respectively. The strains $\varepsilon_{33}$ are related to the stresses $\sigma_{kl}$ using the matrix notation given by Nye [15] as

$$\varepsilon_{ij} = S_{ijkl}\sigma_{kl}$$

where $S_{ijkl}$ are the compliances which are calculated from the elastic constants $C_{ij}$'s and i, j, k, or l are integers of 1, 2, or 3. For an arbitrary $(hkl)$-oriented crystal in which two orthogonal axes $x'_1$ and $x'_2$ are in the plane parallel to the substrate surface and the third axis $x'_3$ is perpendicular to the substrate surface, a transformation of these axes is given by

$$x'_i = a_{ij}x_j$$

where $a_{ij}$ is the direction cosine of $x'_i$ with respect to $x_j$. The strain and stress are related by

$$\varepsilon'_{ij} = S'_{ijkl}\sigma'_{kl} \tag{1}$$

where the subscripts $ij$ denote the directions of orthogonal axes: 11 and 22 are directions parallel to the substrate surface, and 33 is the one normal to the substrate, and the transformed compliance $S'_{ijkl}$ is given by

$$S'_{ijkl} = a_{im}a_{jn}a_{ko}a_{lp}S_{mnop}.$$

Equation (1) consists of six equations and contains 12 unknown parameters (six $\varepsilon'_{ij}$ and six $\sigma'_{ij}$ values). Thus, to solve Eq. (1), six boundary conditions are required. These boundary conditions are deduced from the origins of strains and/or stresses acting on the films. Since most of films are biaxially strained or stressed, the strain and stress relations for these two cases are given below.

### Biaxial Strain Model

The biaxial strain is introduced into the films upon changing temperature from the deposition temperature ($T_o$) to an ambient temperature ($T_1$) due to mismatch of thermal coefficients between the films and substrates. Behavior of such the biaxially strained films was analyzed by Vook and Witt [16]. The strain in the planes parallel to the substrate surface is,

$$\varepsilon'_{11} = \varepsilon'_{22} = \varepsilon_{therm}. \tag{2}$$

In this equation the thermal strain, $\varepsilon_{therm}$, is given by,

$$\varepsilon_{therm} = \int_{T_1}^{T_o}(\alpha_{film} - \alpha_{sub})dT \tag{3}$$

where $\alpha_{film}$ and $\alpha_{sub}$ are thermal expansion coefficients of film and substrate, respectively. The strains acting on the films are applied at the interface between the film and the substrate, and, consequently, all tensile and shear stress components in directions normal to the film surfaces are zero,

$$\sigma'_{3i} = 0. \tag{4}$$

Because the film is assumed to adhere rigidly to the substrate, $\varepsilon'_{12}$ equals 0.

Using the above six boundary conditions, unknown strain and stress components are calculated by solving the set of linear equations (1). The strain components $\varepsilon'_{13}$, $\varepsilon'_{23}$, and $\varepsilon'_{33}$ are computed by,

$$\varepsilon'_{33} = S'_{3311}\sigma'_{11} + S'_{3322}\sigma'_{22} + (S'_{3312} + S'_{3321})\sigma'_{12} \tag{5a}$$

$$\varepsilon'_{23} = S'_{2311}\sigma'_{11} + S'_{2322}\sigma'_{22} + (S'_{2312} + S'_{2321})\sigma'_{12} \tag{5b}$$

$$\varepsilon'_{31} = S'_{3111}\sigma'_{11} + S'_{3122}\sigma'_{22} + (S'_{3112} + S'_{3121})\sigma'_{12}. \tag{5c}$$

**Biaxial Stress Model**

The boundary conditions imposed on this biaxially stressed film are

$$\sigma'_{11} = \sigma'_{22} = \sigma, \sigma'_{12} = 0,$$

where $\sigma$ is stress applied to the grains in the planar direction [9]. Also, the boundary conditions expressed by Eq. (4) are applied to the present biaxially stressed film. Thus, the remaining six unknown strain components $\varepsilon'_{ij}$ are calculated by

$$\varepsilon'_{ij} = (S'_{ij11} + S'_{ij22})\sigma. \tag{6}$$

**Biaxially Strained Pb Films Deposited on Si Substrate**

To illustrate the important effect of anisotropy in elastic constants on deformation of biaxially strained thin films, the values of strain $\varepsilon'_{33}$ normal to the film surface, strain ellipsoid, interfacial stress $\sigma'$, and elastic strain energy were calculated for Pb thin films that were deposited onto the Si substrate at 298K and then cooled to 4.2K [6-8]. Upon cooling, the strain $\varepsilon_{therm}$ = 0.704% was introduced biaxially into the film as calculated by Eq. (3). The elastic constants $C_{ij}$ of Pb at 4.2K used in the calculation were $C_{11} = 5.554$, $C_{12} = 4.452$, and $C_{44} = 1.942$ in units of $10^{11}$dyn/cm$^2$ [17].

Upon cooling the film/substrate composite, although the same amounts of the isotropic strains $\varepsilon'_{11}$ and $\varepsilon'_{22}$ (in the film plane) were introduced into each grain, the $\varepsilon'_{33}$ value (normal to the film surface) of each grain was different and dependent on the grain orientation (hkl). The strains $\varepsilon'_{33}$ was calculated by Eq. (5a) for Pb films with various crystal orientations using six boundary conditions given by Eqs. (2)-(4) [7,8] and are shown in Fig. 1, where the $\varepsilon'_{33}$ values were normalized by the $\varepsilon'_{11}$ value. In order to read off easily the $\varepsilon'_{33}$ values of any orientation, the values were plotted on the upper half-sphere of a (111) stereographic projection, where [111] direction is normal to the figure and the [$\bar{1}$10] is the vertical axis. Usually, the film surfaces of most grains are not parallel to specific crystal orientations such as (111), (100), etc. and are always several degrees off from these orientations. The $\varepsilon'_{33}$ values of grains with any crystal orientation can be read from this stereographic projection. From Fig. 1 it is seen that the ratio $\varepsilon'_{33}/\varepsilon'_{11}$ has a minimum value of 0.96 for (111) orientation and a maximum value of 1.635 for (100) orientation.

The validity of using the biaxial strain model to analyze the strain and stress components of thermally strained Pb films was evaluated experimentally. The $\varepsilon'_{33}$ values of grains with various crystal orientations were measured and compared with those calculated using the biaxial strain model. Pb and PbInAu films were cooled from 298 to 4.2K and the strains $\varepsilon'_{33}$ (normal to the film surface) were measured at 4.2K by the conventional Bragg-Brentano x-ray

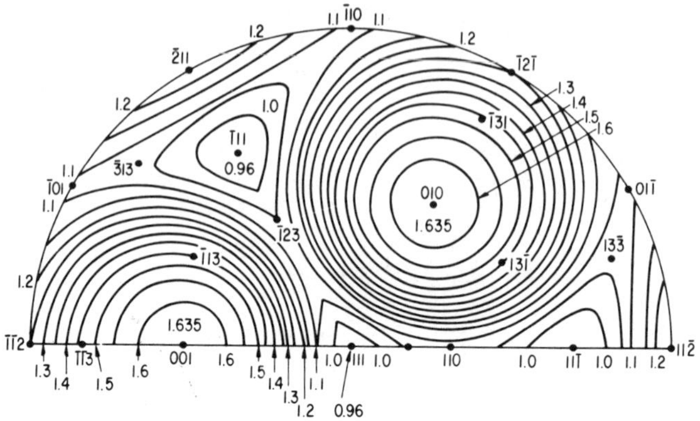

*Fig. 1. Stereographic projection of equicontour lines of the ratio $\varepsilon'_{33}/\varepsilon'_{11}$ for Pb films at 4.2K [7].*

diffraction technique [7,14,18]. For example, the strains $\varepsilon'_{33}$ measured for grains with various (hkl) orientations (shown in the upper horizontal axis) of Pb films are shown in Fig. 2. As expected, the $\varepsilon'_{33}$ values have a strong dependence on the crystal orientation. A large strain of ~1.1% was obtained for (311) oriented grains and a small strain of ~ 0.6% was obtained for (111) oriented grains. The dashed line shows the calculated strain values using a biaxial strain model with $\varepsilon'_{therm}$=0.704%. Note that the experimental strain values agree very well with the theoretical values, indicating that the films are biaxially strained due to thermal mismatch between the films and the substrates. This indicates that shrinkage of (100) oriented grains is much larger than that of (111) oriented grains and local deformation is expected at boundaries between grains with different crystal orientations.

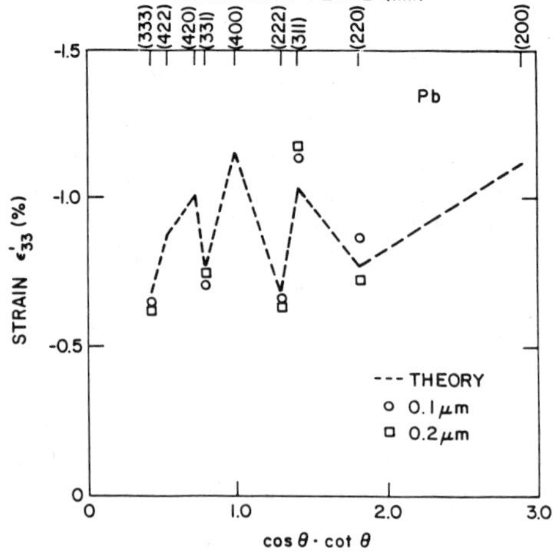


*Fig.2. Strain $\varepsilon'_{33}$ measured for Pb films at 4.2K. The dashed line is strain values calculated based on the biaxial strain model [8,11].*

The interspacings of (hkl) crystal planes of the biaxially strained films were found to be a function of the angle $\psi$ between the (hkl) plane and the interface [14]. As examples, the strain ellipsoids for the (001), (111), (011) and (112) interface planes, cut by a plane perpendicular to the interface, were calculated for Pb films at 4.2K [14] and are shown in Fig. 3 as dashed curves. The solid circles represent the unstrained interspacings. In order to emphasize the differences, a hypothetical interface strain of $\varepsilon = 0.20$ was chosen for the plots. The strain ellipsoids for the (001), (111) and (011) cases had rotational symmetries around the interface normal. However, the strain ellipsoid for the (112) interface lost the rotational symmetry and one of its major axes was tilted by an angle of 17.25° from the interface normal around the [110] axis, perpendicular to the plane of the cut. This indicates that the maximum deformation is not on the substrate plane for (112) oriented grains.

The interfacial stress $\sigma'$ defined by $[(\sigma'_{11})^2 + (\sigma'_{12})^2]^{1/2}$ on a given crystal plane were calculated for Pb thin films at 4.2K [6] and plotted in Fig. 4 for (001), (111), and (011), and (112) interface planes. Isotropic stresses were obtained on the (100) and (111) planes. The stress values on the (111) plane was about three times larger than that on the (100) plane even though the same amount of the strain was biaxially introduced into the films. Twofold symmetries were observed for the (011) and (112) planes. The $\sigma'$ value at vertical axis was about two thirds of that at the horizontal axis on these planes. The difference in the stresses gives non-uniform deformation on the substrate.

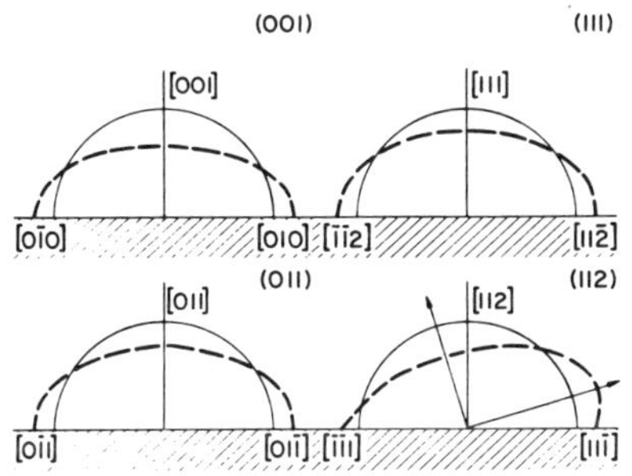


*Fig.3. Strain ellipsoids of Pb films at 4.2K calculated based on the biaxial strain model with interfacial strain of 20 % [14].*

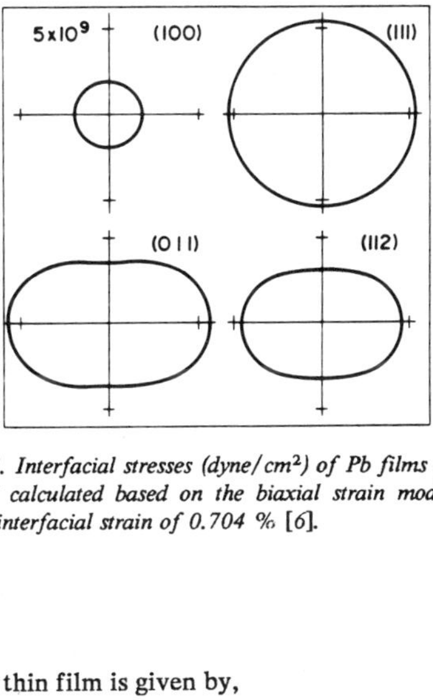


*Fig.4. Interfacial stresses (dyne/cm²) of Pb films at 4.2K calculated based on the biaxial strain model with interfacial strain of 0.704 % [6].*

The strain energy W (per unit volume) stored in a thin film is given by,

$$W = \frac{1}{2}\sigma'_{ij}S'_{ijlm}\sigma'_{lm}. \tag{7}$$

For a polycrystalline film in which grains have various surface crystal planes (hkl), the strain energy stored in each grain depended on the (hkl) plane. The W values stored in a Pb film upon cooling from 298K to 4.2K were calculated by using Eq.(7) [6]. The W values, normalized by $1/2(\varepsilon'_{11})^2$, were plotted in Fig. 5 on the (111) stereographic projection. Since Pb has strong anisotropy in the elastic constants, the W values were observed to be strongly dependent on crystal orientation. The maximum and minimum W values were observed in the (111) and (100) oriented planes, respectively, and the W value of the (111) oriented planes is about a factor of three larger than that of the (100) oriented planes.

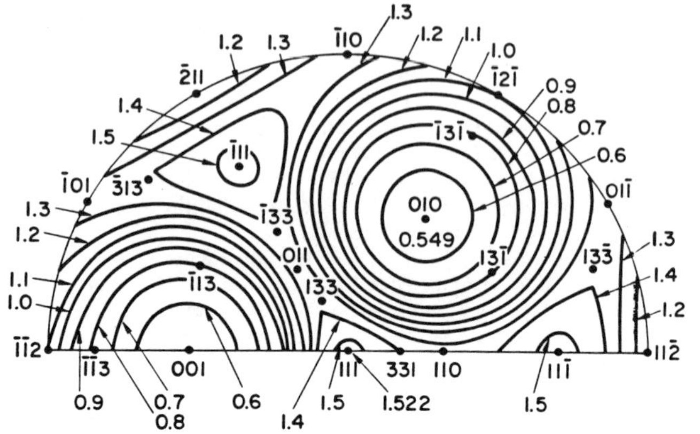


*Fig.5. Stereographic projection of equicontour lines of the elastic strain energy W ($10^{12}$dyne/cm$^2$) normalized by $\varepsilon'^2_{11}$ /2 for Pb films at 4.2K* [6].

From the above strain and stress calculations and experiments on biaxially strained films, it is learned that the strains supported by the film are strongly dependent on both the crystal planes of the films and the direction on a given plane. Therefore, non-uniform deformation, such as shrinkage difference, is expected upon cooling Pb films. Note that direct imaging of the deformation is extremely difficult.

**Biaxially Stresses Nb Films Deposited on Si Substrate**

The effect of anisotropy of elastic constants on deformation of biaxially stressed films has been studied for Nb films which were deposited on Si substrates [9]. The elastic constants of Nb at 298K used for calculations were $C_{11}$=2.456, $C_{12}$ = 1.387 , and $C_{44}$ = 0.293 in units of $10^{12}$dyne/cm$^2$ [19].

The strain $\varepsilon'_{33}$ values (normal to the film surface) were calculated using the biaxial stress model for Nb films which were hypothetically stressed by an amount of $\sigma$ = 1x$10^{10}$dyne/cm$^2$. These values for various crystal orientations are plotted on the (111) stereographic projection of Fig. 6. Note that the $\varepsilon'_{33}$ value has a maximum value of 1.01% for the (111) orientation and a minimum value of 0.497% for the (100) orientation. When the film was stressed in tension, about a factor of 2 difference in shrinkage was expected between neighboring grains with the (100) and (111) orientations, which causes local deformation at the grain boundaries.

The strains $\varepsilon'_{33}$ supported elastically in grains with various crystal orientations were measured by the x-ray diffraction method for as-deposited Nb films with 0.23$\mu$m thickness [9], and are plotted in Fig. 7. The theoretical strains read from Fig. 6 are shown by a dashed line, where $\sigma$ = 1x$10^{10}$dyne/cm$^2$ was used in the calculation. The theory and the experiment fit very well, indicating that the films were biaxially stressed by this amount. Note that the crystal orientation dependence calculated based on the the biaxial "strain" model did not fit the experimental strain values of the Nb films.

The interfacial strain defined by $\varepsilon' = [(\varepsilon'_{11})^2 + (\varepsilon'_{12})^2]^{1/2}$ was calculated using Eq.(6) for the biaxially stressed Nb films with the (001), (011), (111), and (112) orientations. The interspacings $d = d_0(1 + \varepsilon')$ are shown by dashed curves in Fig. 8 where $d_0$ = 1 represent the spacings of the undeformed spheres which are shown by solid circles. In this calculation, $\sigma$ = 1x$10^{11}$dyne/cm$^2$ was used to emphasize the difference. It is seen that although the same

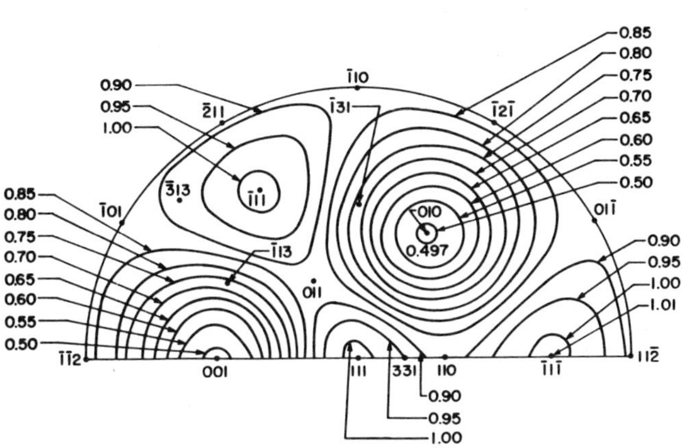

*Fig.6. Stereographic projection of equicontour lines of strain $\varepsilon'_{33}$ (%) for Nb films at 298K. The interfacial stress of $1x10^{10}$dyne/cm² was used for the calculation [9].*

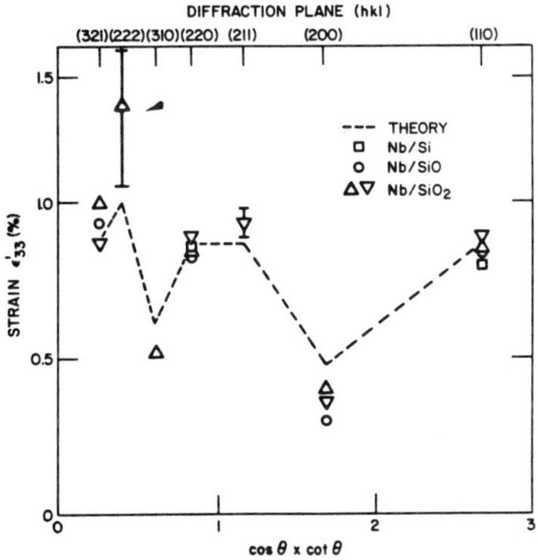


*Fig.7. Strains $\varepsilon'_{33}$ measured for Nb films. The dashed line is the strains calculated based on the biaxial stress model [9].*

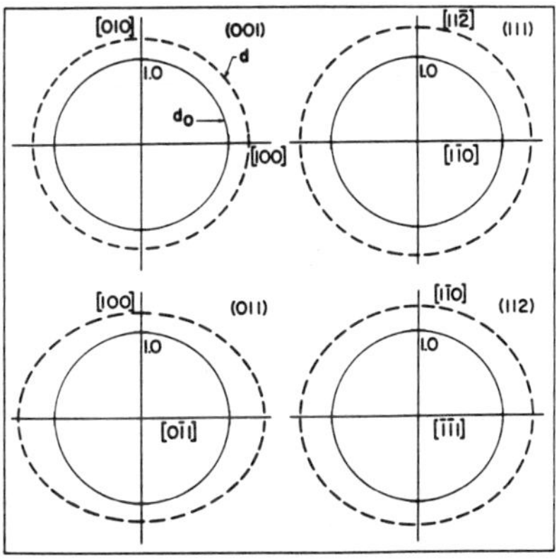


*Fig.8. Interfacial strains of Nb films calculated based on the biaxial stress model with the interfacial stress of $1x10^{11}$dyne/cm² [9].*

amounts of the stresses were applied biaxially to the film, the interfacial strains depended on the crystal orientations. Strains on the (001) and (111) planes are isotropic and the strain on the (111) plane is about a factor of 2 larger than that of the (001) plane. However, strains on the (011) and (112) planes showed twofold symmetries.

The interspacing of (hkl) planes of the biaxially stressed Nb films were also calculated as a function of the angle $\psi$ between the (hkl) planes and the interface. The interspacing values d of the deformed films with the film surface planes parallel to (001), (011), (111) and (112) are shown by dashed curves of Fig. 9, where a hypothetical interface stress of $\sigma = 1x10^{11}$dyne/cm² was again applied. The solid circles indicated by $d_0$ correspond to the interspacings of the undeformed films. The strain ellipsoids for (001), (011), and (111) planes had rotational sym-

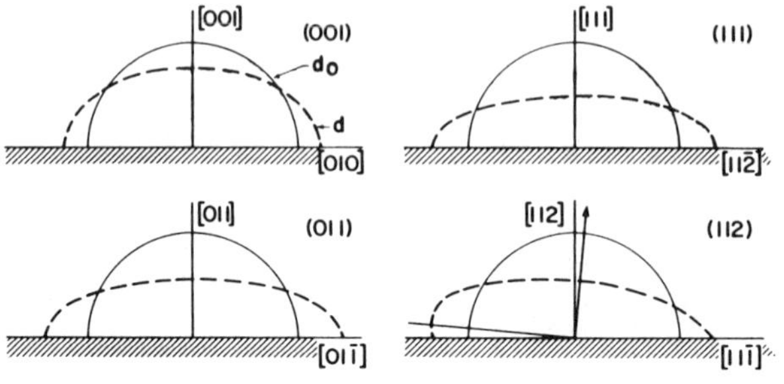


*Fig.9. Strain ellipsoids of Nb films calculated based on the biaxial stress model with the interfacial stress of* $1x10^{10}$*dyne/cm*$^2$ [9].

metries around the axis normal to the substrate. However, the ellipsoid for the (112) planes had lost the rotational symmetry as was seen in Fig. 3 and one of its major axes was tilted by an angle of about 5° from the normal axis.

Strain energies W stored in the biaxially stressed polycrystalline Nb film were calculated by Eq.(7). The constant contour curves of the W values are plotted on the (111) stereographic projection of Fig. 10 where $\sigma = 1x10^{10}$dyne/cm$^2$ was used to calculate the W values. In this figure a strong dependence of the W values on the crystal orientations was seen, and the maximum and minimum values were observed for the (111) and (001) oriented planes, respectively, which is similar to the biaxially strained Pb films of Fig. 5.

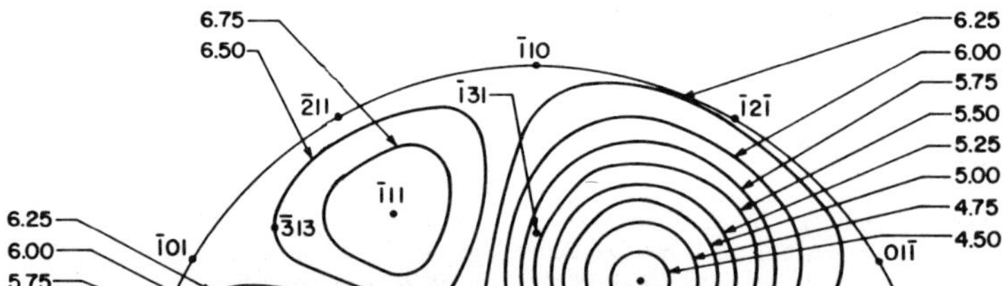


*Fig.10. Stereographic projection of equicontour lines of the strain energy W (* $10^8$*dyne/cm*$^2$*) for Nb films. The interfacial stress value of* $1x10^{10}$*dyne/cm*$^2$ *was used for the calculation* [9].

## DEFORMATION DUE TO GEOMETRY DEVIATIONS

Shapes of thin films are different from those of the bulk material and unique deformation for the thin films due to the specific shapes is expected for the films even with isotropic elastic constants. Strains and stresses in thin films are generally induced from the substrate and, therefore, non-uniform strain distribution was observed for blanket films in direction normal to the substrate surface. In addition, for line and island-shaped films, strain relaxation at the film edges was observed and non-uniform strain distributions in directions both normal and parallel to the substrate surface were observed in these films. Techniques to detect such non-uniform deformations using x-ray diffraction and the experiments carried out for Pb films [8,10,12] are reviewed below.

### Blanket Pb Films Deposited on Si Substrates

The large strain was observed close to the substrate and a method [8] developed to detect such strain gradient normal to the interface was reviewed. In thin films bonded to rigid substrates, strains were introduced from the film-substrate interface and they were partially relaxed close to the top film surface. Based on an assumed strain distribution normal to the interface, an expected shape of the x-ray intensity profile diffracted by crystal planes parallel to the interface were computed using the modified Houska's method [20]. The line broadening was due to the range of lattice spacings encountered by the incident beam traversing regions of varying spacings. For the analysis, it is convenient to subdivide the total film volume into small volume elements. The x-ray intensity profile $P_m$, diffracted from a volume element $\Delta V_m$ is given by

$$P_m(\theta) = I_0 Q_m \Delta V_m A_m p(\theta - \theta_m) \tag{8}$$

where $I_0$ is the intensity of the incident beam, $Q_m$ is the reflectivity per volume unit, slightly depending on the strain $\varepsilon_m$ inside $\Delta V_m$, $p(\theta - \theta_m)$ is an instrumental "standard" profile, normalized to enclose unit area and centered on the Bragg angle $\theta_m$ for the strained lattice spacing of the element $\Delta V_m$ , and $A_m$ is the x-ray absorption correction factor. Apparently a knowledge of the strain distribution suffices to synthesize the diffracted intensity profile $P(\theta)$ of the film obtained by adding up the profiles $P_m(\theta)$ over all elements $\Delta V_m$,

$$P(\theta) = \sum_m P_m(\theta) = I_0 \sum_m Q_m \Delta V_m A_m p(\theta - \theta_m). \tag{9}$$

This method was applied to simulate the elastic strain distribution of thin Pb films that were deposited at room temperature onto oxidized Si wafers and subsequently cooled to 77 K [8]. In this analysis, an empirical strain distribution given by,

$$\varepsilon_m = \varepsilon_0 \exp\left( - \frac{h - h_m}{\eta} \right) \tag{10}$$

was used, where $\varepsilon_m$ is the strain normal to the interface at a depth $h_m$ from the top film surface, $\varepsilon_0$ is the strain at the film substrate interface, h is the total film thickness, and $\eta$ is the distance from the interface at which the strain has relaxed to 1/e of its value at the interface. This exponential strain variation was theoretically predicted by van der Merwe [21]. Intuitively, as the metal film becomes thicker and thicker, the influence of the first strained film layer at the interface on the layers close to the film surface becomes weaker and weaker, and eventually the lattice constant of the film approaches a state characteristic of the bulk metal. The theory was developed based on the assumption that some strain relaxation toward the bulk lattice constant would occur from atomic layer to layer, and that the amount of relaxation is proportional to the deviation from the bulk lattice constant. This model should be applicable to strained polycrystaline films bonded to rigid substrates, if the grain sizes are much larger than the film thicknesses.

Figure 11 shows x-ray diffraction patterns of films with different strain distributions. The 1 μm thick Pb films were deposited onto the Si substrates at room temperature. The (333) diffraction profile of the film measured at 298 K shortly after deposition is shown by dots in Fig. 11(a). It is noted that the diffraction profile is narrow and symmetric around the dotted centroid. The intensity profile given by the dashed curve was calculated for a strain-free 1 μm thick film, using a linear combination of the Gaussian and Cauchy distribution for the standard profile p [22]. The diffraction profiles of the films cooled from room temperature to 200 K and to 77 K are shown in Fig. 11(b) and 11(c), respectively. The peak positions of the profiles shifted to higher Bragg angles, indicating the introduction of tensile strains into the film parallel to the interface. A larger average strain and a wider strain distribution was observed in the film cooled to the lower temperature. The diffraction profile of the film cooled to 77 K was observed to have a noticeable asymmetric line broadening as compared to that cooled to 200 K. Again, the dashed curves are diffraction profiles, computed using Eqs. (9) and (10). At both temperatures, the agreement between experimental and computed intensities was excellent. For the film cooled to 77 K, the strains at the film-substrate interface and at the free surface, determined by this method, had values of 0.18 and 0.04 %, respectively. Therefore, the large strain gradient normal to the interface existed throughout the film. The strains at the top and bottom of the 1 μm thick film differed by a factor of about four.

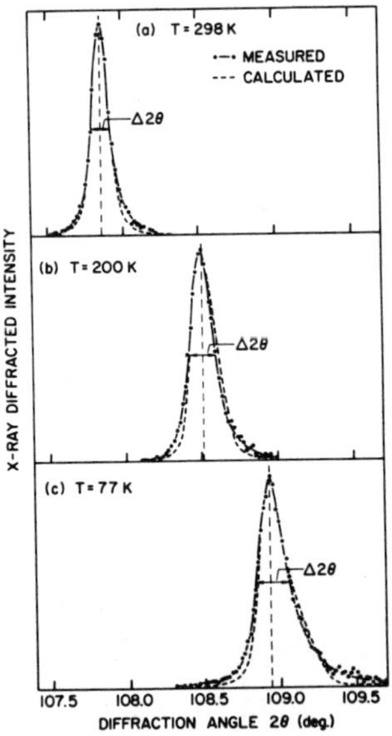


*Fig.11. X-ray (333) diffraction profiles of a Pb film measured at (a) 298K, (b) 200K, and (c) 77K [8].*

**Island-shaped Pb Films Deposited on Si Substrate**

The strain distribution was calculated for island-shaped Pb films in which no mechanical interaction existed between neighboring islands [10] The isolated island was approximated as a cylindrical shape. It was assumed that on a given plane cutting an island parallel to the substrate at a distance $h_m$ from the film surface the elastic strain was a maximum at the center of the plane and decreased exponentially with increasing distance r from the center point reaching zero at the island edge. The strain in the volume element $\Delta V_m$ located at a point $(r, h_m)$ was given by,

$$\varepsilon(r,h_m) = \varepsilon_m(h_m)\left\{1 - \exp\left(-\frac{0.5g - r}{\eta_1}\right)\right\} \tag{11}$$

where $\varepsilon_m(h_m)$ is the strain at the center point and is given by Eq.(10), g is the island diameter, and $1/\eta_1$ is a proportionality constant. The parameter $\eta_1$ is qualitatively indicative of the volume fraction (inside an island) which is influenced by "zero strain" at the island edge. The strain $\varepsilon_m(h_m)$ was found to increase exponentially with increasing $h_m$ as seen by Eq. (10).

Island-shaped (0.2 μm thick) Pb films with various island sizes g were prepared at 298K, and then cooled to 4.2K where the strain $\varepsilon'_{33}$ was measured [10]. If there was no strain relation occurring upon cooling to 4.2K, the strain $\varepsilon'_{33}$ of 0.68% should be supported by the film. It was seen the $\varepsilon'_{33}$ values decreased with decreasing g and a reasonable linearity was observed in the dependence of $\varepsilon'_{33}$ on 1/g. From the plots of $\varepsilon'_{33}$ vs 1/g, $\eta$ =1.54μm and $\eta_1$=2.74 x $10^{-2}$μm were determined [10].

Using these values the strain distributions inside the island were calculated by Eq. (11) for 0.2 μm thick films. The calculated constant strain contours at intervals of 0.1% are shown in Fig. 12 for a film with g = 0.2μm (a) and with g = 1.0μm (b). Note that the volume fraction in which $\varepsilon'_{33} > 0.6$ % was small for the film with small islands, and large for the film with large islands, corresponding to the smaller value measured for $\varepsilon'_{33}$ in films of small island size.

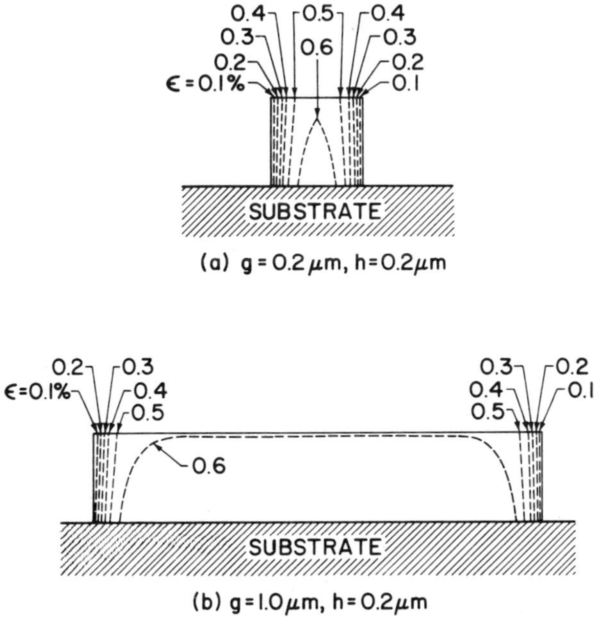


*Fig.12. Constant strain equicontours inside islands for 0.2 μm thick films with (a) g = 0.2 μm and (b) g = 1.0 μm [10].*

SUMMARY

The deformation in thin films caused by strains and stresses has been frequently treated by ignoring the importance of anisotropy in the elastic contants and also shapes of thin films. With such assumptions, only uniform deformation is expected over the films.

In the present paper we reviewed studies carried out local deformations in thin films with anisotropic elastic constants and with line or island shapes. The calculation and experimental results obtained for Pb and Nb films, which were deposited onto the Si substrates, were used to explain behaviors of extrinsic and intrisic strains, respectively. For the blanket films, an extremely large strain was observed close to the film/substrate interface and was relaxed toward the top film surface. For the island or line-shaped films, additional strain relaxation at the film edges was observed. In the films which had strong anisotropy in the elastic constants, the strain was found to have a strong crystal orientation dependence, and shrinkage (or expansion) of each grain was found to be different for various crystal orientations. Also, the possibility of revealing the origin of the strain introduced in thin films was demonstrated by comparing the strains measured for various crystal orientations with those calculated based on various models. These non-uniform strains caused local deformation in thin films, which were not noticed if the strain-stress analysis was carried out by ignoring the importance of anisotropy in the elastic constants and also shapes unique to the thin films. Finally, the non-destructive x-ray diffraction method was demonstrated to be the most powerful method to detect such local deformation. The non-uniform strain distributions both normal and parallel to the film surface were detected by analysing the diffracted x-ray profiles.

REFERENCES

1. E. Klokholm and B.S. Berry, J. Electrochem Soc. **115**, 823 (1968).
2. M.F. Doerner and W.D. Nix, CRC Critical Rev. in Sol. Stat. and Mat. Sci. **14**, 225 (1988).
3. R.W. Hoffman, "Physics of Nonmetallic Thin Films", eds. H.S. Dupuy and A. Cachard, Plenum Press, New York (1974).
4. K.L. Chopra, "Thin Film Phenomena", McGraw-Hill, New York (1969).
5. F.M. d'Heurle, Int. Met. Rev. (in press)
6. M. Murakami, CRC Critical Rev. in Sol. Stat. and Mat. Sci. **11**, 317 (1984).
7. M. Murakami and P. Chaudhari, Thin Solid Films **46**, 109 (1977).
8. M. Murakamami, Acta Metall. **26**, 175 (1978).
9. M. Murakami and T. Yogi, J. Appl. Phys. **57**, 211 (1985).
10. M. Murakami, Thin Solid Films **69**, 253 (1980)
11. M. Murakami, J. Appl. Phys. **53**, 3560 (1982).
12. M. Murakami, T. S. Kuan and I. A. Blech, Treat. Mat. Sci. Technol. **24**, 163 (1982).
13. D. S. Cambell, "Handbook of Thin Film Technology", eds. L. I. Maissel and R. Glang, McGraw-Hill, New York (1983).
14. A. Segmuller and M. Murakami, Treat. Mat. Sci. Technol. **27**, 143 (1988).
15. J.F. Nye, "Physical Properties of Crystals", University Press, Oxford (1972).
16. R.W. Vook, and F. Witt, J. Appl. Phys. **36**, 2169 (1965).
17. D.L. Waldord and G.A. Alers, J. Appl. Phys. **33**, 3266 (1962).
18. M. Murakami, J. Appl. Phys. **52**, 1309 (1981)
19. D.I. Bolef and M. Menes, Bull. Am. Phys. Soc. **5**, 40 (1960)
20. C.R. Houska, Treat. Mat. Sci. Technol. **19A**, 63 (1980)
21. J.H. Van der Herwe, Proc. Conf. Single Crystal Films, Pergomon Press, New York (1964).
22. B. E. Warren, "X-ray Diffraction", Addison-Wesley, Reading, (1968).

# LONG-RANGE COHERENCY STRAIN AND TILTED EPITAXY IN Ag-Fe-Ag SANDWICH STRUCTURES ON GaAs(001) SUBSTRATES.

R.F.C. Farrow*, V.S. Speriosu*, S.S.P. Parkin*, C. Chien+, J.C. Bravman+, R.F. Marks**, P.D. Kirchner**, G.A. Prinz***, B.T. Jonker***

* IBM Almaden Research Center, San Jose, Ca. 95120.
+Stanford University, Department of Materials Science and Engineering, Stanford, Ca.94305.
** IBM Thomas J. Watson Research Center, Yorktown Heights, N.Y. 10598.
*** Naval Research Laboratory, Washington, DC 20375.

## ABSTRACT.

We find that epitaxial Fe films sandwiched between epitaxial Ag films grown on GaAs (001) substrates possess residual coherency strain at a thickness of 2000Å. The [001] directions of the Fe and Ag films are tilted with respect to the GaAs [001] axis. The tilts are coplanar with the tilt of the substrate surface normal to the [001] axis of GaAs and are qualitatively consistent with a recently proposed model for tilted epitaxy.

## INTRODUCTION.

Recent studies of magnetism in epitaxial Fe/Ag structures have revealed a variety of interesting properties. For example , studies of the magnetic properties of epitaxial Fe films grown on Ag(001) surfaces [1,2] and Fe-Ag multilayers grown on Ag(001) films on GaAs(001) surfaces [3] show that the Fe has perpendicular anisotropy for thicknesses in the range 2-5 monolayers. In addition , transient paramagnetism has been detected [4] in Ag films on Fe(001) surfaces due to spin-polarization of conduction electrons crossing the Fe-Ag interface. Such structures have not been structurally characterized. In particular the state of strain of the films and the microstructure of the various interfaces have not been examined. In this paper, we report the results of an analysis of coherency strain in epitaxial Fe-Ag sandwich structures grown by MBE on GaAs(001) surfaces. Double-crystal X-ray diffractometry was used to determine coherency strain and TEM (transmission electron microscopy) was used to probe the microstructure of the GaAs-Fe and Ag-Fe interfaces.

## FILM GROWTH AND IN SITU CHARACTERIZATION

The substrates for Fe and Ag epitaxy were undoped films of homoepitaxial GaAs , grown by molecular beam epitaxy at IBM T.J. Watson Research Center. The surfaces of the substrate wafers were inclined at about 2 ° to the (001) plane , the tilt being along a <100> direction. Following GaAs epitaxy the surface of the sample was protected by an overlayer (cap) of arsenic which was deposited at ~250K. After a brief exposure to air the sample was transferred in a vacuum flask to Almaden Research Center and loaded into a VG80-M, MBE machine ( VG Semicon Ltd.). The arsenic cap was desorbed by heating the sample to ~450°C for a few minutes. The sample surface exhibited either a (1x1) or (2x4) reconstruction , both of which were indicative of an As-stabilized surface. The results of this study did not depend on which of these reconstructions existed prior to epitaxy. Auger electron spectroscopy indicated ~1-3% monolayer of carbon as an impurity. No other surface impurities were detected.

Fe and Ag were deposited from e-gun and effusion sources , respectively, at a rate of ~0.5Å/s onto the substrate , held at room temperature. The substrate was rotated continuously at ~1rev/s during deposition. RHEED patterns recorded after Fe and Ag epitaxy were similar to those reported earlier [3]. In order to grow single orientation Ag films on GaAs it was necessary to first deposit at least 6ml of Fe. Subsequent Ag films grew with their (001) plane parallel to the (001) planes of Fe and GaAs(001) but with the Ag lattice rotated by 45° about the [001] axis. In this setting the lattice misfit between Fe and Ag is 0.8%.

Figure 1 shows a cross-section TEM micrograph of the interface between the GaAs and the Ag film. The interface is viewed along the [110] direction of the GaAs. The (200) planes of the Ag lattice are resolved. It is difficult to distinguish the 6ml Fe film from the Ag lattice but the dark band near the interface probably represents the Fe film. From this and other micrographs , in addition to in situ RHEED studies , it is clear that the Ag film is locked-in to a single orientation by the Fe film. For thinner or no initial predeposition of Fe we found that the Ag grew with a mixture of two different orientations: Ag(110) parallel to GaAs(001) or Ag(001) parallel to GaAs(001). This finding is in agreement with prior studies [5] of this interface. Details of our TEM studies of the Ag-Fe and GaAs-Fe interfaces will be presented elsewhere.

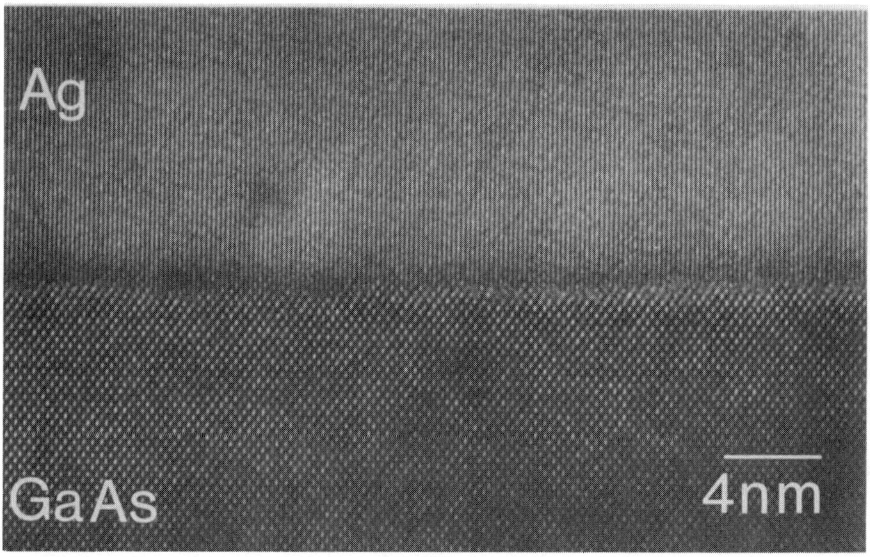


**Figure 1** - Cross-section transmission electron micrograph of the interface region of a structure comprising: 500Å Ag / 6ml Fe / GaAs(001). The view is along GaAs[110]. The fringes in the Ag film represent the (200) planes.

DOUBLE CRYSTAL DIFFRACTOMETRY OF Fe-Ag-Fe-GaAs STRUCTURES.

Two types of structures were examined in this study. In the first (type 1) , a 500 or 2000Å film of Ag was grown onto the 6ml Fe film. This was followed by ~2000Å of Fe and a final protective film of 100Å Ag. In the second type (type 2) of structure a 2000Å film of Fe was grown onto the GaAs. This was followed by a 2000Å Ag film.

The sample crystal was probed by CuK$\alpha_1$ radiation from the (004) reflection of a Si(001) monochromator. Both theta two-theta scans and rocking curves were recorded in symmetric and asymmetric settings [6]. Figure 2 shows theta two-theta scans for a type 1 structure comprising a 2000Å Fe film grown on a 2000Å Ag film. The scans were made in the symmetric setting for two different azimuths of the sample . As the sample was rotated about its surface normal the angular separations between Ag(002) and GaAs(004) peaks and between Fe(002) and GaAs(004) peaks varied cyclically with the cosine of the azimuthal angle. This is seen in Figure 2 where the Fe and GaAs peaks are nearly coincident for an azimuth of 12° but are separated by about 0.8 ° at an azimuth of 282° This behavior indicates that the [001] axes in the GaAs, Ag and Fe lattices are not coincident but point in different directions in space. We also found that the GaAs surface normal is not coincident with the GaAs [001] axis. Similar studies of the type 2 structures also revealed tilts of the various axes. Figure 3 illustrates the relative orientations of the axes in space for the type 1 and type 2 structures. The tilts are coplanar with the tilt of the substrate surface with respect to the GaAs(001) plane.

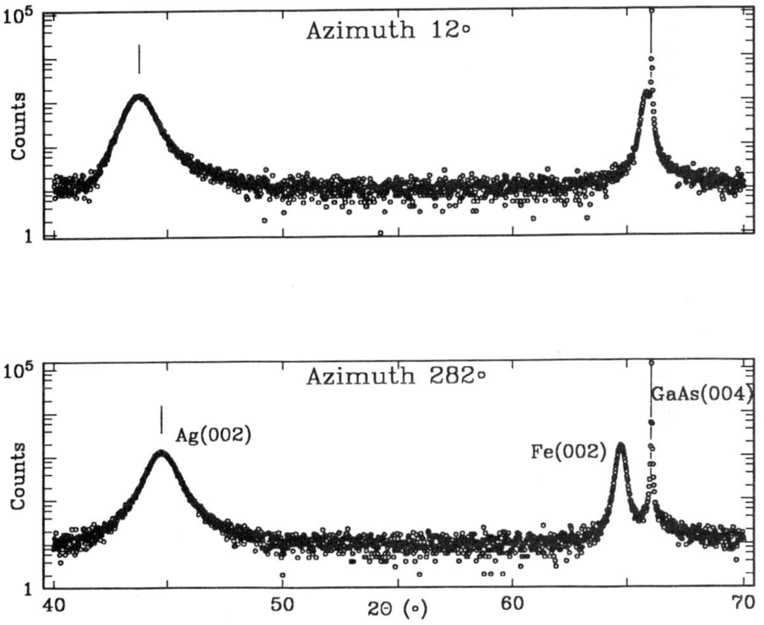


**Figure 2** - Theta two-theta X-ray scans for epitaxial sandwich structure comprising : 100Å Ag / 2000Å Fe / 2000ÅAg / 6ml Fe/ GaAs(001). CuK$\alpha_1$ radiation. The scans are for two different azimuths of the beam in the (001) plane of the GaAs substrate. Note that as the sample is rotated in its own plane , the Bragg peaks of Ag and Fe move relative to the GaAs(004) peak. This indicates tilts of the Ag[001] and Fe[001] axes away from GaAs[001].

Perpendicular strain ( $\varepsilon^{\perp}$ )values of the Fe films in type 1 structures were derived from the mean angular separation of Bragg peaks in rocking curves of the Fe(002) and GaAs(004) in the symmetric setting. Both the perpendicular and parallel strains ( $\varepsilon^{||}$) were derived from the mean angular separation of Bragg peaks in rocking curves of Fe(022)and GaAs (044) in the asymmetric setting [6]. For symmetric reflections the angular separation between the Fe and and GaAs peaks is

$$\Delta\omega = -\varepsilon^{\perp} \tan\theta_B + \phi_m \cos\alpha \qquad 1$$

where $\theta_B$ is the substrate Bragg angle, $\phi_m$ is the tilt of the Fe [001] axis from the GaAs [001] axis and $\alpha$ is the azimuth.

For asymmetric reflections $\Delta\omega$ is related to both $\varepsilon^{\perp}$ and $\varepsilon^{||}$ by:

$$\Delta\omega = -(\varepsilon^{\perp}\cos^2\phi + \varepsilon^{||}\sin^2\phi)\tan\theta_B \pm (\varepsilon^{\perp} - \varepsilon^{||})\sin\phi\cos\phi + \phi_m\cos\alpha \qquad 2$$

where the $\pm$ sign refers to the two possible paths for the X-ray beam in in the asymmetric setting and $\phi$ is the angle between the Bragg planes and the sample surface. Tables 1 and 2 summarize the X-ray and elastic strains for Fe films derived from rocking curves for several type 1 structures and a type 2 structure. Strain measurements for the Ag films confirmed , that in both types of structures , the Ag films were strain-free and fully relaxed to the bulk lattice constant for Ag.

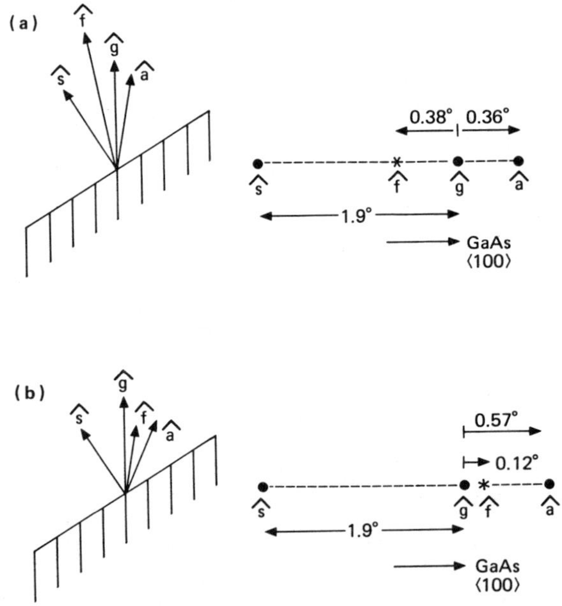


**Figure 3** - Schematic diagram illustrating the directions of the axes in epitaxial Ag / Fe structures grown on GaAs(001).
**a , f and g** are the [001]axes of Ag , Fe and GaAs respectively. **s** is the GaAs surface normal.

a) Sample 138 : 100Å Ag / 2000Å Fe / 2000Å Ag / 6ml Fe / GaAs(001).
b) Sample 139 : 2000Å Ag / 2000Å Fe / GaAs(001).

TABLE 1

X-RAY STRAIN ($\varepsilon^{\perp}$, $\varepsilon^{||}$) AND ELASTIC STRAIN ($e^{\perp}$, $e^{||}$) FOR EPITAXIAL FILMS OF Fe IN Ag / Fe / Ag / GaAs(001) SANDWICH STRUCTURES.

| SAMPLE | $\varepsilon^{\perp}$(%) | $\varepsilon^{||}$<100> | $\varepsilon^{||}$<010> | $\phi m^{(o)}$ | $e^{\perp}$(%) | $e^{||}$(%) |
|---|---|---|---|---|---|---|
| MBE 138 | 1.02 | 1.69 | 1.70 | 0.38 | -0.37 | 0.30 |
| MBE 66 | 0.86 | 1.84 | 1.89 | 0.54 | -0.52 | 0.45 |
| MBE 67 | 0.86 | 1.81 | 1.81 | 0.60 | -0.52 | 0.42 |

TABLE 2

X-RAY STRAIN ($\varepsilon^{\perp}$, $\varepsilon^{||}$) AND ELASTIC STRAIN ($e^{\perp}$, $e^{||}$) FOR AN EPITAXIAL FILM OF Fe IN Ag / Fe / GaAs(001) SANDWICH STRUCTURE.

| SAMPLE | $\varepsilon^{\perp}$(%) | $\varepsilon^{||}$<100> | $\varepsilon^{||}$<010> | $\phi m^{(o)}$ | $e^{\perp}$(%) | $e^{||}$(%) |
|---|---|---|---|---|---|---|
| MBE 139 | 1.63 | 1.14 | 1.13 | 0.12 | 0.24 | -0.25 |

Notes:

1. Samples MBE 66 and 67 comprised 100Å Ag / 1875Å Fe / 500Å Ag / 6ml Fe / GaAs. Sample MBE 138 comprised 100Å Ag / 2000Å Fe / 2000Å Ag / 6ml Fe / GaAs. Sample MBE 139 comprised 2000Å Ag / 2000Å Fe / GaAs(001). All samples were grown and measured at room temperature (20°C).

2. $\varepsilon^{\perp}$ and $\varepsilon^{||}$ are the X-ray strains ($\Delta d/d$) between the doubled Fe unit cell and GaAs measured perpendicular and parallel to the (001) film plane.

3. $e^{\perp}$ and $e^{||}$ are the perpendicular and parallel elastic strains in the Fe film with respect to the relaxed bulk unit cell of Fe.

4. $\phi_m$ is the measured tilt of the Fe [001] axis with respect to the GaAs [001] axis. The tilt is along a <100> direction in the GaAs.

In the case of the asymmetric setting $\varepsilon^{||}$ and $\varepsilon^{\perp}$ were determined from two pairs of reflections : one pair with
$\alpha = 0$, beam incident at $\theta_B + \phi$ ; $\alpha = 180°$, beam incident at $\theta_B - \phi$
the other pair with
$\alpha = 90°$, beam incident at $\theta_B + \phi$ ; $\alpha = 270°$, beam incident at $\theta_B - \phi$
The former pair gave $\varepsilon^{||}$ along the <100> direction in GaAs , the second pair gave $\varepsilon^{||}$ along <010>. Tables 1 and 2 show that the in-plane strain is isotropic since the values of $\varepsilon^{||}$ are in good agreement ( within 5% ). The values of $\varepsilon^{\perp}$ for the symmetric and asymmetric settings were also in good agreement. Elastic strains for the Fe films were calculated from the X-ray strain values using room temperature (bulk) lattice constants for

Fe and Ag of 2.8662Å and 4.08626Å respectively. These values are summarized in the last two columns of each table. Values of Poisson's ratio , derived from the data in tables 1 and 2 , were 0.39 and 0.32 respectively. The literature value [7] is 0.42.

In the case of type 1 structures the elastic strains show that the films are dilated in-plane and compressed along GaAs[001]. This is consistent with residual coherency strain arising from misfit between the Fe film and its Ag film substrate. The full values of coherency strain in this case would be $e^{||}$ = 0.8% and $e^{\perp}$ = -1.16%. In the case of the type 2 structure the measured elastic strains indicate residual coherency strain arising from the misfit between Fe and the GaAs substrate. In this case the full value of coherency strain would be $\varepsilon^{||}$ = -1.39% and $\varepsilon^{\perp}$ = 2.01%.

DISCUSSION.

The measured values of tilts in both type 1 and type 2 structures are qualitatively consistent with a recent model [8] for tilted epitaxy proposed by Bai et al. In this model the substrate surface is inclined at an angle $\phi_s$ to a low index plane . This breaks the surface symmetry and leads to a terraced growth surface with step edges along one direction. The total energy of the bicrystal system can then be lowered by a tilt of the film growth axis with respect to the surface normal. If the film lattice parameter in the growth direction is greater than the substrate parameter ( $\varepsilon^{\perp} > 0$ ) then a tilt of the film axis away from the surface normal leads to matching of atoms in the plane of the interface along a direction normal to the step edges. Similarly, if $\varepsilon^{\perp} < 0$ then matching occurs when the film axis tilts towards the surface normal. The matching condition is satisfied when $\delta = \varepsilon^{\perp} \tan \phi_s$ , where $\phi_s$ is the tilt of the substrate surface from the low index plane. In the present case the low index plane is (001) and the maximum tilt of the 6ml Fe prelayer should occur for a coherent interface for which $\varepsilon^{\perp}$ = 3.4%. The tilt would then be away from the surface normal by 0.065°. The maximum tilt for Ag on Fe is for a coherent interface for which $\varepsilon^{\perp}$ = 1.5%. In this case the Ag would tilt away from the surface normal by 0.029°. The total tilt of the Ag away from the GaAs surface normal would then be ~0.065 + 0.029 = 0.094°. We indeed measure a tilt of the Ag film away from the surface normal, coplanar with the tilt of the substrate surface , but by an amount (see Figure 3 a ) much greater than this. Similarly , for the 2000Å Fe film on Ag we expect a maximum tilt for the case where the Fe is coherent with the Ag and $\varepsilon^{\perp}$ = -2.0%. In this case , Fe tilts towards the surface normal by 0.03°. Again we measure a tilt in this sense but an order of magnitude greater. A similar discrepancy occurs for the type 2 structures. Furthermore , since we observe misfit dislocations at each of the interfaces the predicted tilts are less than the values estimated above. At this stage the reason for the tilt enhancement is not clear but we expect that full analysis of the high resolution images of the interfaces will suggest a mechanism. In this context it should be noted that recent studies by Schowalter et al [9] , of the epitaxial growth of GaAs films on misoriented Si (001) substrates , also suggest that the model of Bai et al is not applicable to all epitaxial systems. Schowalter et al find that the measured tilts do not agree in either direction or magnitude with those predicted by the model of Bai et al. In particular , they find that the tilts in this system are introduced by a particular type of dislocation , with its Burgers vector normal to the interface, which causes tilting of epitaxial islands of GaAs at the pre-coalescence stage of film growth. Thus the origin of tilt may depend on the growth mode and is clearly more complex than envisaged in the model of Bai et al.

The residual coherency strain of Fe films in both types of structure shows that strain relaxation to the thermal equilibrium value by nucleation and propagation of dislocations is inhibited by kinetics at room temperature. The complete relaxation of the Ag films may reflect the fact that Ag is mechanically softer than Fe and that dislocations may nucleate and propagate more readily. This phenomenon of residual coherency strain , at distances far from a lattice-mismatched interface , has also been observed [7] in the case of MBE-grown films of $CoSi_2$ on Si (111) surfaces. Such structures are metastable with respect to the nucleation of misfit dilocations.

## CONCLUSIONS.

We find that epitaxial Fe films , grown at room temperature, on either GaAs or Ag film surfaces possess residual coherency strain at thicknesses of 2000Å. The [001] directions of the Fe and Ag films are tilted with respect to the GaAs [001] axis. The directions of these tilts are in agreement with a recently proposed model for tilted epitaxy on misoriented substrates but the magnitudes of the measured tilts are much greater than predicted. Clearly , tilted epitaxy is a general phenomenon in growth on misoriented substrates and it must be included in any quantitative strain analysis of epitaxial structures.

## ACKNOWLEDGMENTS

This work was partially supported by the office of Naval Research: Contract number N00014-87-C-0339. The assistance of C.A Shen and K. Roche in the double-crystal diffractometry experiments is acknowledged with thanks.

## REFERENCES.

1. B.Heinrich , K.B. Urquhart , A.S. Arrott , J.F. Cochran , K. Myrtle , S.T. Purcell. Phys. Rev. Lett. **59**, 1756 , (1987)

2. M. Stampanoni , A. Vaterlaus , M. Aeschlimann , F. Meir. Phys. Rev. Lett. **59**, 2483 , (1987).

3. N.C. Koon , B.T. Jonker , F.A. Volkening , J.J. Krebs , G.A. Prinz. Phys. rev. Lett. **59**, 2463, (1987).

4. H. Hurdequint , G. Dunifer ( personal communication)

5. R. Ludeke , T.C. Chiang , D.E. Eastman. J. Vac. Sci. and Technol. **21**, 599, (1982).

6. R.F.C. Farrow , S.S.P. Parkin, V.S. Speriosu. J.Appl. Phys. **64**, 5315 (1988).

7. J.P. Hirth and J. Lothe , " Theory of Dislocations" , McGraw-Hill , (1976).

8. G. Bai , D.N. Jamieson , M.A. Nicolet , T. Vreeland Jr. , Mat. Res. Soc. Symp. Proc. **102**, 259, (1988).

9. L.J. Schowalter , E.L. Hall , N. Lewis , S. Hashimoto. Paper presented in Symposium C , MRS Fall Meeting 1988 , this volume.

# THE EFFECTS OF INTRINSIC IN-PLANE STRESS ON THE LOCAL ATOMIC STRUCTURE OF THERMALLY GROWN $SIO_2$

J.T. FITCH, C.H. BJORKMAN, J.J SUMAKERIS, AND G. LUCOVSKY
Departments of Physics and Materials Science and Engineering, North Carolina State University, Raleigh, NC 27695-8202

## ABSTRACT

The effect of thermal history on the local atomic structure of thermally grown and annealed $SiO_2$ films is discussed. Thin films of $SiO_2$ were prepared for study by (1) thermal oxidation of silicon at 850°C and 1050°C in a dry oxygen ambient and, (2) by subsequent rapid thermal annealing at temperatures between 850°C and 1200°C. After thermal processing, the film thickness was reduced incrementally by a wet chemical etch-back technique to within 50 Å of the growth interface. Relief of intrinsic growth stress by viscous flow was investigated, at each thickness increment, on a local atomic scale by infrared spectroscopy and on a macroscopic scale by beam-bending type stress measurements. Results were analyzed in terms of a Maxwell model. Changes in film thickness and refractive index resulting from viscous flow during annealing were shown to obey a Clausius-Mossotti relationship. Local atomic strain values, calculated from shifts in the IR bond-stretching frequency, showed a linear correlation with measured stress values giving a value for a plane-strain Young's modulus of $4.7 \times 10^{11}$ dynes/cm$^2$. This compares favorably with the literature value of $8.8 \times 10^{11}$ dynes/cm$^2$.

## INTRODUCTION

As the basis for Si metal-oxide-semiconductor technology, $SiO_2$ is the single most important electronic insulator. High quality interfaces are possible in thermally grown $SiO_2$ despite the 30% larger Si-Si distance in a-$SiO_2$ as compared to the single crystal Si substrate. In previous work [1-6] we have shown via infrared spectroscopy (IR), ellipsometry, and beam-bending stress measurements that the local atomic structure of thermally grown $SiO_2$ is quite sensitive to thermal history. Systematic variations in these measurements were examined within the framework of a model for local atomic structure. Growth stress generated at the Si/$SiO_2$ interface is relieved through changes in the bond angle, $2\theta$, at the oxygen atom sites. This paper is a continuation of previous work with emphasis on the non-homogeneous thermal history aspect of oxidation with respect to thickness which is inherent in the growth process. We examine thermal history effects using (1) a wet chemical etching technique to section the films, (2) the Clausius-Mossotti relationship [1,7] between density and refractive index, and (3) an analysis of elastic properties of $SiO_2$ films using a combination of stress and IR data.

## EXPERIMENTAL PROCEDURE AND RESULTS

Silicon dioxide films for this study were grown on "device quality" silicon wafers at 850°C and 1050°C using standard procedures which are described in references 1-5. Rapid thermal annealing (RTA) was performed at temperatures between 850°C and 1200°C for 100 sec in an argon ambient with an AG Heatpulse model 210M. After thermal processing was completed, the $SiO_2$ films were wet chemically etched back to within 50 Å of the growth interface in small increments using a P-etch solution [8]. Stress and IR measurements were taken at each thickness interval.

Samples for stress measurements were made using 75 μm thick silicon strips oxidized along with the other thicker samples. After oxidation, the 75 μm thick samples had the oxide stripped from one side to induce curvature. The radius of curvature of the strips, resulting from the stress in the oxide layer, was measured using a parallel laser beam reflection technique developed by Kobeda and Irene [9]. The total stress in the film can be calculated using a standard beam bending equation [9]. The stress arising from the mismatch in the coefficients of thermal expansion between Si and $SiO_2$ has been subtracted from the total stress leaving only the intrinsic growth stress remaining. A more detailed account of stress measurement procedures can be found in references 6 and 9. The stress values reported in this paper are an average of at least three measurements made on each of three samples. The multiplicity of stress measurements has enabled us to measure intrinsic growth stress to an average uncertainty of $\pm 1.0 \times 10^9$ dynes/cm$^2$.

Film thickness and refractive index were measured with a precision automatic ellipsometer at a wavelength of 632.8 nm. All thickness and refractive index values given are an average of between three and five readings. The average uncertainties of thickness and refractive index values are ±3 Å and ±0.001 Å, respectively. The IR measurements reported here were obtained as described in references 1-6. Ten spectra were taken on each sample and electronically averaged to achieve a better signal to noise ratio. Measurements on the very thin oxides involved an average of several different samples. Although $SiO_2$ has three vibrational modes only the location of the center of the bond stretching frequency, ν, is reported here. For the thinnest films, 60 Å, the experimental uncertainty is ±1 cm$^{-1}$ compared to an uncertainty of ±0.6 cm$^{-1}$ for thicker films.

Measurements from films incrementally etched back by the procedure described above are plotted in Figures 1-4. In Figures 1 and 2 the intrinsic stress levels near the interface are quite high compared to the bulk. Also, stress levels in the annealed films were lower than those of the grown films, and, films grown at 1050°C had a lower stress than films grown at 850°C. Variation of the bond stretching frequency, ν, shown in Figures 3 and 4 is opposite to that of the stress data. The frequency, ν, falls off near the interface. In general, films which have higher processing temperatures had higher values of ν. In Figure 5, changes in film thickness and refractive index as a result of annealing are inversely related to one another.

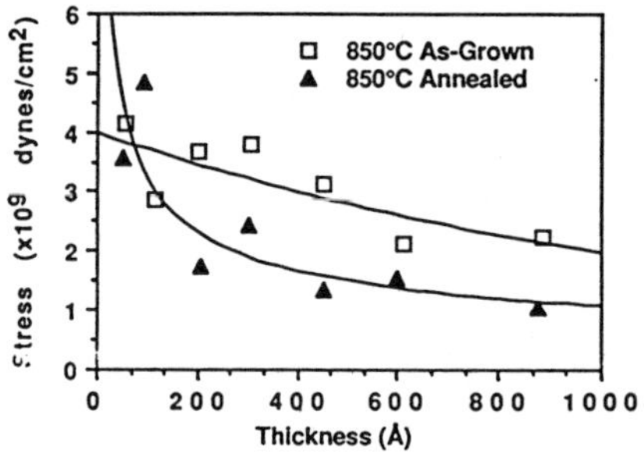


Fig. 1 Intrinsic growth stress of 850°C dry oxide as a function of film thickness after etch-back. Annealing was performed at 1100°C for 100 sec.

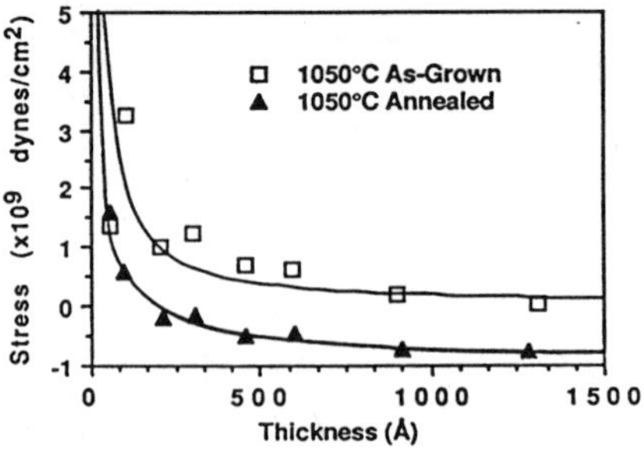


Fig. 2 Intrinsic growth stress of 1050°C dry oxide as a function of film thickness after etch-back. Annealing was perfomed at 1100°C for 100 sec.

## DISCUSSION

It is well documented that $SiO_2$ exhibits viscous flow in the range of temperatures used for conventional thermal oxidation of silicon [10]. Viscoelastic relaxation in thin films of $SiO_2$ has been successfully analyzed in terms of a Maxwell model by several authors [6,10]. Compressive stress, inherent in the oxidation process because the $SiO_2$ occupies a 120% larger molar volume than the silicon consumed to grow it, provides a driving force for viscous flow. If there was no viscous relaxation we would expect a constant value of stress across the thickness of the films. From inspection of Figures 1 and 2 this is clearly not the case. It is apparent that relatively high levels of stress generated at the growth interface relax as one moves further away from the interface. There is a non-homogeneous thermal history aspect to all thermally grown $SiO_2$ films: in a thick (say 1000Å) oxide film, the topmost increment of oxide was formed first and was being annealed while the underlying oxide was being formed. The increment of oxide nearest the interface has spent very little time at the growth temperature and is therefore expected to show the least amount of relaxation. Data in Figures 1 and 2 support this and it will be shown that the IR data in Figures 3 and 4 supports the same scenario. The familiar solution for stress relaxation within the context of the Maxwell model [10] is

$$\sigma = \sigma_0 * \exp(-t/\tau) \qquad (1)$$

where $\sigma_0$ is the interfacial stress, t is time, and $\tau$ is the relaxation time constant. A similar equation can be written for the IR bond stretching frequency of the form:

$$\nu_{max} - \nu = (\nu_{max} - \nu_{min}) * \exp(-t/\tau) \qquad (2)$$

where $\nu_{max}$ is determined from a high temperature oxide (1150°C, 1300Å thick) which is almost fully relaxed and is set equal to 1078.5 $cm^{-1}$ [2], and $\nu_{min}$ is assigned a value of 1062 $cm^{-1}$ from Figure 3. The IR bond stretching frequency, $\nu$, is significant in the following context [5]: (1) $\nu$ provides a measure of the bond angle at the oxygen bonding site between corner connected $SiO_4$ tetrahedra; (2) the bond angle can be used to calculate a silicon-silicon distance; and (3) the change in the silicon-silicon distance can be used as a measure of local atomic strain. The development of this argument begins with the introduction of a central force model for the bond stretching frequency [11,12]

$$\nu = \nu_0 \sin\theta; \quad \nu_0 = 1134 \text{ cm}^{-1} \qquad (3)$$

where $\theta$ is half of the Si-O-Si bond angle, and $\nu_0$ is determined empirically assuming a fully relaxed high temperature (1150°C) oxide has a bond angle $2\theta = 144°$ [13] which corresponds to a frequency of 1078.5 $cm^{-1}$. By simple geometry, the silicon-silicon distance is given in terms of the bond angle by

$$d_{Si\text{-}Si} = 2r_0 \sin\theta \qquad (4)$$

where $r_0$ is the Si-O bond length. Strictly speaking $r_0$ is a weak function of the Si-O-Si bond angle [14], but, the variation is sufficiently small so that it will be neglected here. The silicon-silicon distance can be converted into a relative strain if we write

$$\varepsilon = \Delta d/d = (2r_0 \sin\theta_1 - 2r_0 \sin\theta_2)/2r_0 \sin\theta_1 \qquad (5)$$

where $\theta_1$ is half of the bond angle of a fully relaxed oxide (72°) and $\theta_2$ is some oxide bond angle less than $\theta_1$. Cancelling factors of $2r_0$ and substituting $\nu/\nu_0$ for $\sin\theta$ we obtain a simplified expression for strain:

$$\varepsilon = (\nu_1 - \nu_2)/\nu_1 \qquad (6)$$

where $\nu_1$ is the 1078.5 $cm^{-1}$ value for an 1150°C oxide and $\nu_2$ the value of the oxide in question. It must be emphasized that eqn. (6) represents an expression for strain relative to an 1150°C oxide.

Differences in the thickness dependent properties of oxide films reflect the quantitative differences between the time scales required for oxide growth, and for viscoelastic relaxation at the growth temperature [6] (see Figure 6). One of the curves is derived from the viscoelastic relaxation times, $\tau$, calculated using eqn. (2), IR data from this paper, and data from references 2 and 6. Values of t, which represent a weighted average time at the growth temperature [6], were calculated using the Grove Deal model for silicon oxidation [15]. Figure 6 can best be explained in terms of two limiting cases: (1) at 700°C little or no relaxation of growth stress is expected and (2) at 1150°C the film should be completely relaxed. These assertions are independently supported by stress data from references 4 and 6.

In previous work [2-6] we have shown that $SiO_2$ films with high compressive stress levels are densified relative to relaxed films. The Clausius-Mossotti [7] equation given by

$$(\varepsilon - 1)/(\varepsilon + 2) = A/V; \ \varepsilon = n^2 \qquad (7)$$

gives the correct dependence of the optical frequency dielectric constant, $\varepsilon$, on the density of the material, where n is the refractive index, A is a normalized sum of silicon and oxygen atomic polarizabilities, and V is the molar volume. Relative values of the molar volume can be calculated from film thickness changes which result from rapid thermal annealing. If relaxation of strain is accompanied by an increase in the Si-O-Si bond angle, and if we assume that the molar volume is roughly proportional to $d_{Si-Si}{}^3$, then there must be a corresponding increase in the molar volume. Experimental data from Figure 5 has been analyzed in this context (see Figure 7). The relationship is almost perfectly linear with the exception of one sample which inadvertently oxidized slightly during annealing at 1200°C. The Clausius-Mossotti relationship quantitatively verifies that compressive growth stress in the plane of the film is relieved by squeezing material upwards much like a tube of toothpaste.

The elastic properties of $SiO_2$ can be investigated if the stress values in Figures 1 and 2 are plotted against strain values calculated from the data in Figures 3 and 4 using eqn. (6). Figure 8 is highly significant because the constant of proportionality between stress and strain is just a plane-strain Young's modulus:

$$\sigma = E/(1 - P)*\varepsilon \qquad (8)$$

where E and P are the Young's modulus and Poisson's ratio for $SiO_2$, respectively. A linear regression of the experimental data gives a plane strain modulus of 4.7 x $10^{11}$ dynes/$cm^2$. This compares favorably with the literature value of 8.8 x $10^{11}$ dynes/$cm^2$ [16]. Thus, the calculated microscopic strain values are accurate and correlate well with measured stress values giving a fundamental elastic constant of $SiO_2$.

## CONCLUSION

High interfacial growth stress is closely related to film densification. SiO2 films relieve growth stress by viscous flow, opening up the Si-O-Si bond angle and expanding in a direction normal to the plane of the film. The Clausius-Mossotti relationship has been verified to a high degree of accuracy so that changes in refractive index can be used to accurately predict dimensional changes. During relaxation, the calculated local atomic strain scales linearly with the intrinsic stress, the constant of proportionality being a plane strain Young's modulus which is very close to the literature value.

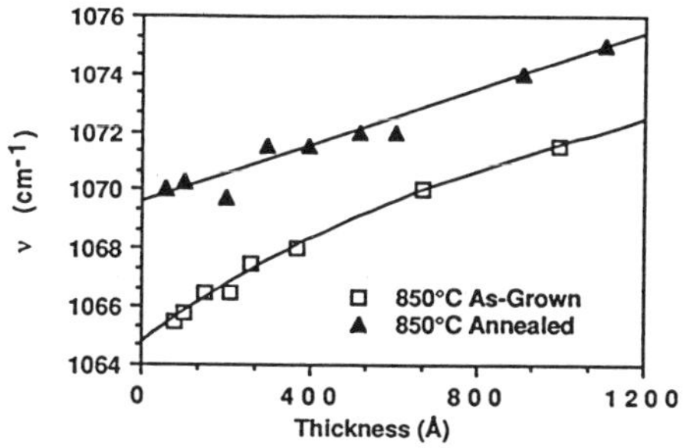


Fig. 3 IR Si-O bond stretching frequency of 850°C dry oxide as a function of film thickness after etch-back. Annealing was performed at 1100°C for 100 sec.

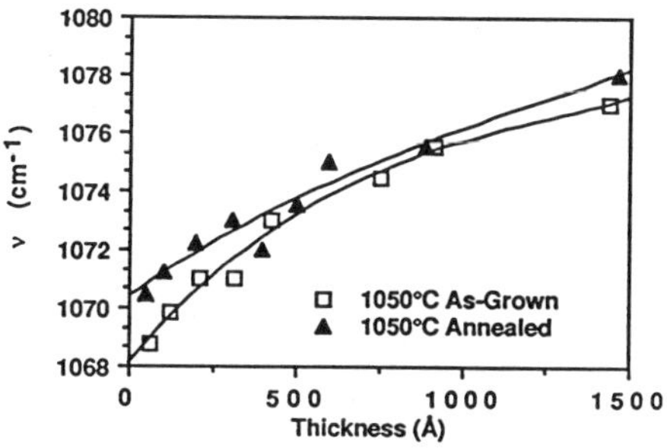


Fig. 4 IR Si-O bond stretching frequency of 1050°C dry oxide as a function of film thickness after etch-back. Anealing was performed at 1100°C for 100 sec.

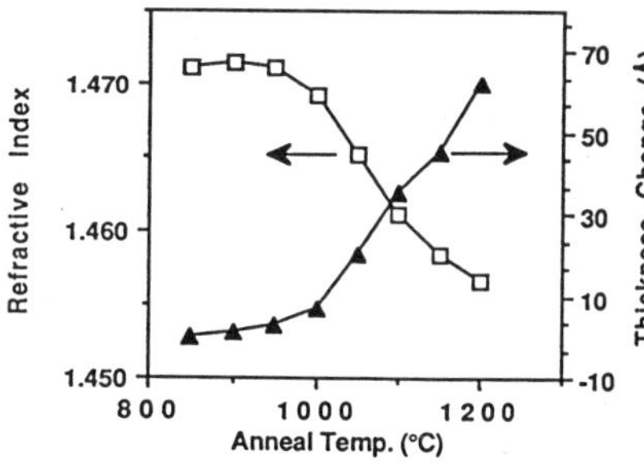


Fig. 5 Refractive index and film thickness change as a function of annealing temperature (100 sec at each temp.). Starting material was grown at 850°C in dry oxygen to a thickness of approx. 1300 Å.

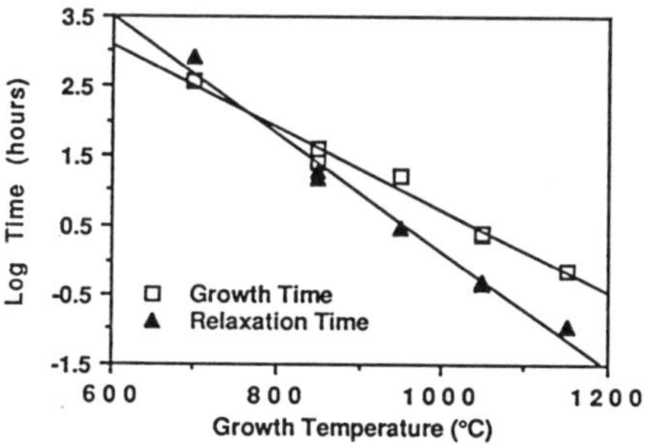


Fig. 6 Relative time scales for growth and viscous relaxation calculated from eqn. (2) are shown as a function of growth temperature.

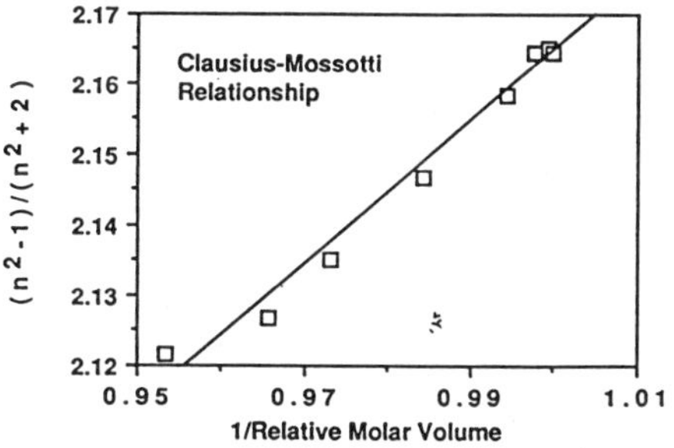


Fig. 7 Data from Fig. 5 analyzed via the Clausius-Mossotti relationship. The left hand side of eqn. (7) is plotted as a function of the reciprocal of the relative molar volume calculated from changes in film thickness.

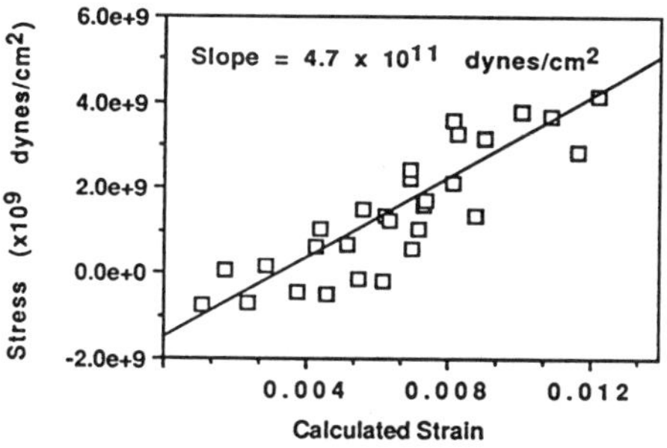


Fig. 8 Measured intrinsic growth stress vs. strain calculated from eqn. (6).

REFERENCES

1. G. Lucovsky, M.J. Mantini, J.K. Srivastava, and E.A. Irene, J. Vac. Sci. Technol. B5, 530 (1987).
2. J.T. Fitch and G. Lucovsky, MRS Symp. Proc. 92, 89 (1987).
3. J.T. Fitch and G. Lucovsky, MRS Symp. Proc. 105, 151 (1988).
4. J.T. Fitch and G. Lucovsky, AIP Conf. Proc. 167, AVS Sub-Series, ed. by G.W. Rubloff (1988) (in press).
5. G. Lucovsky, J.T. Fitch, E. Kobeda, and E.A. Irene, J. Electrochem. Soc. (1988) (in press).
6. J.T. Fitch, E. Kobeda, G. Lucovsky, and E.A. Irene, submitted to J. Vac. Sci. Technol.
7. W.K.H. Panofsky and M. Phillips, Classical Electricity and Magnetism, (Addison-Wesley, Reading Mass., 1955) p. 34.
8. W. A. Pliskin and R.P. Gnall, J. Electrochem. Soc. 111, 872 (1964).
9. E. Kobeda and E.A. Irene, J. Vac. Sci. Technol. B4, 720 (1986).
10. E.A. Irene, E. Tierney, and J. Angilello, J. Electrochem. Soc. 129, 2594 (1982).
11. F.L. Galeener and P.N. Sen, Phys. Rev. B17, 1928 (1978).
12. A.E. Geissberger and F.L. Galeener, Phys. Rev. B28, 3266 (1983).
13. R.L. Mozzi and B.E. Warren, J. Appl. Cryst. 2, 164 (1969).
14. P.J. Grunthaner and F.J. Grunthaner, Materials Science Reports 1, 65 (1986).
15. S.K. Ghandi, VLSI Fabrication Principles (John Wiley and Sons, New York, 1983) p. 380.
16. R.J. Jaccodine and W.A. Schlegel, J. Appl. Phys. 37, 2429 (1966).

# NANOMECHANICS OF THIN FILMS: EMPHASIS: TENSILE PROPERTIES

R. W. HOFFMAN
Department of Physics, Case Western Reserve University, Cleveland, OH 44106

## ABSTRACT

Tensile properties of thin films may be interpreted as a structure sensitive plastic region superposed on an elastic background in a manner similar to bulk specimen tensile testing. However, the limitations of both the material and tensile instrument have not usually been tested in detail.

We report our experience with aluminum and alumina films some 100 nm thick prepared by evaporation of Al followed by anodization for the alumina film. Self-supporting films are glued to glass "jaws" of the nanotensilometer and force-elongation data recorded. Mounting thickness, glue slippage, instrument calibration, and other possible artifacts will be examined in detail. A typical Al stress-strain curve has an initial small curved region interpreted as a mounting artifact, followed by a primarily elastic (near linear) region and increasing plastic deformation until failure occurs. Alumina films fail in a brittle manner. Characterization techniques include TEM, RBS, and other surface spectroscopies; selected examples will be reported. Strain rate and preliminary annealing data are presented with a microscopic interpretation.

In general, thin metal films are less ductile than their bulk counterparts, grain sizes are much smaller, and they may possess large stresses and unexpected impurities, but have mechanical properties that can be modelled.

## INTRODUCTION

The previous papers in this volume assess the role of growth stresses in thin films. It is obvious that the material of the thin film responds to the total local stress and its spatial distribution. Hence, one approach to thin film mechanics is to apply the same treatments that have been so successful in the bulk case. We in no way wish to reduce the importance of stresses in thin films. Indeed, the sum of the growth stresses and the stresses whose origin lies in differential thermal expansion plus their control in magnitude and location is the key to their utilization in critical applications. This has been the objective since the early papers and carries through to the present volume. With the application of finite element analyses to thin film structures came the realization that film material parameters are not so readily available. In some cases major changes in the composition and microstructure may be responsible. In general thin films have a smaller grain size and are more brittle than their bulk counterparts. It is the purpose of this paper to both review the literature and describe techniques for thin film mechanical property measurements and critique the results to date for the case of free standing aluminum films.

It is well known that both elastic and plastic properties of bulk materials are anisotropic. For films essentially the same thing is true, even for polycrystalline films a preferred, perhaps tilted, fiber texture is possible. The use of Si wafers for substrates and the mechanical anisotropies in $E/1-\nu$ have been known for over two decades.[1] Any other anisotropic property may be calculated[2] knowing the crystal symmetry. A property of interest for this paper is the thermal expansion coefficient, $\alpha$. In the cubic system it can be shown that the thermal expansion is isotropic even though the strain energy is not. Under the simplifying assumptions for an assumed isotropic film deposited on an isotropic substrate the following conditions hold. Except near the edges a planar biaxial stress generated either by deposition processes or differential thermal expansion exists. Oversimplified[3], the stress system is composed of $\sigma_1 = \sigma_2 = \sigma$ and $\sigma_3 = 0$. It is these simplifying assumptions that are commonly used even though we recognize that in detail complications may occur.

**THERMAL EXPANSION TENSILE DEVICES**

Assuming the film remains rigidly attached to the substrate and the film - substrate system is in mechanical equilibrium a convenient tensile machine exists because of the film deposition.[4] In cantilever plate or wafer geometry this is a commonly used device for stress measurement as well. Assuming a thermal expansion and temperature difference in the biaxial stress system the simplified equations may be solved for $d\sigma/dT = E/1-\nu \times \delta\alpha$. Although this $d\sigma/dT$ technique was first proposed for obtaining the elastic modulus or thermal expansion constants under conditions that the "same" film is deposited on two separate substrates, the same concept can be used in many ways from measurements of the stress relaxation modulus to small amplitude oscillations for creep studies. While this technique is well suited to high temperature relaxation studies, it is difficult to do detailed TEM characterization since the films are not easily detached from the substrates. Specialized silicon etch techniques forming a thin window have been published.[5]

**MECHANICAL PROPERTIES FROM CONSTRAINED FILMS**

From the preceding section we can see that values for the biaxial modulus and hence Young's modulus as well as some information about the plastic behavior may be obtained from such $d\sigma/dT$ curves. As far as the elastic modulus is concerned these data are summarized in Table I.[6-8] Since the elastic moduli are related to the curvature of the interatomic potential, we anticipate no thickness dependence nor important structure sensitive behavior in the modulus of thin films. The thickness dependence is borne out by reference [6] as well as our own uniaxial tensile data to be discussed later. However, the plastic deformation is highly structure sensitive for macroscopic samples and thin films are no exception. Doerner[8a] et al., indicate that these constrained substrate experiments give a yield strength which is inversely dependent on thickness but with a faster dependence on thickness than a Hall-Petch relation.

It is to be noted that a wide variation in elastic moduli is found between research groups and even particular samples whereas reproducibility with a given sample or technique generally is about 20%. Elastic moduli measured by most techniques appear to be smaller rather than larger compared to the corresponding values for macroscopic samples. While considerable information may be obtained about thin film mechanics by this technique we shall not consider it further but proceed to uniaxial tensile testers.

TABLE I Film Modulus and Grain Size Summary

| Film thickness | Grain size | Modulus Measurement Technique | | Value | Reference |
|---|---|---|---|---|---|
| 20-200 nm | N.A. μm | Cantilever | (C) | 72 GPa | 6 |
| 235 | N.A. | wafer | (C) | 39 | 7 |
| 210-1090 | 0.3-1.4 | wafer | (C) | N.A. | 8a |
| Al-1% Si | | wafer | (C) | N.A. | 8b |
| 35-75 HV | 0.05-0.12 | tension | (F) | 25-40 | This paper |
| 80 UHV | 0.16 | | | | |

(C) Film constrained on substrate
(F) Free standing film

Simple models for weaker bonding (springs) at the grain boundaries do not seem to be a significant effect even for grains less than 50 nm in diameter. However, the grain boundary potentials are not known for highly contaminated systems and the displacement continuity at boundaries is not known in detail. We suggest that the modulus per atom or per bond would be an improved way to count for an unknown low density film.

## FREE STANDING FILMS

### Film Deposition and Characterization

We restrict our discussion to HV and UHV deposited aluminum films although the technique has been applied to SiO amorphous and alumina formed by anodization.

We have found the best method for defining the specimen width is to evaporate a water soluble parting agent on the freshly cracked edge of a thin substrate with the film deposited on this parting layer (see Fig. 1). Using this technique no edge cracks or other artifacts are observed as indicated in Fig. 2. Grain diameters were determined by the line-intercept method with no three-dimensional correction under the assumption that the films are only angle grain thick. The data are in Table I and confirm an inverse thickness dependence. At the same thickness UHV prepared specimens appear to have approximately one third larger grain sizes than HV deposited aluminum specimens. AES and RBS as shown in Fig. 3 indicate the expected oxygen and carbon contamination at the film surfaces. It has been pointed out that[9] the determination of the cross sectional area of a thin film specimen is not an easy task.

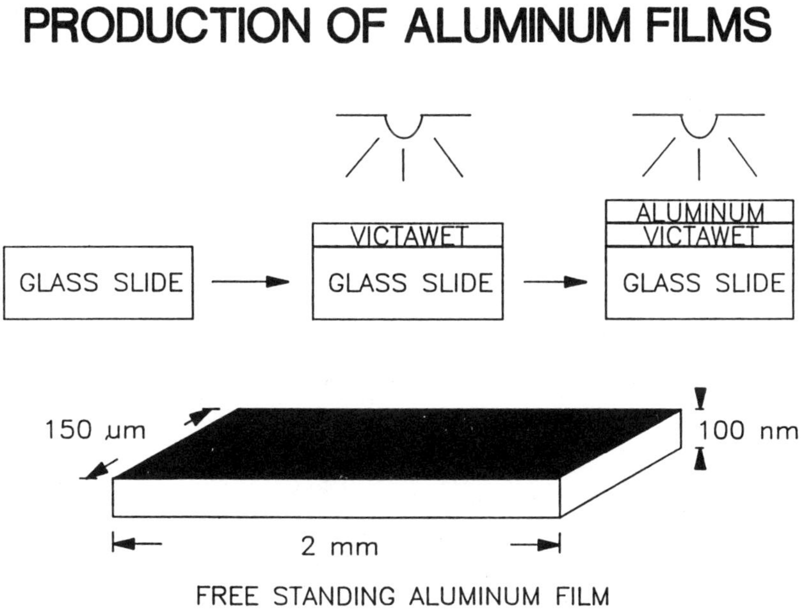


Figure 1 HV Film evaporation geometry

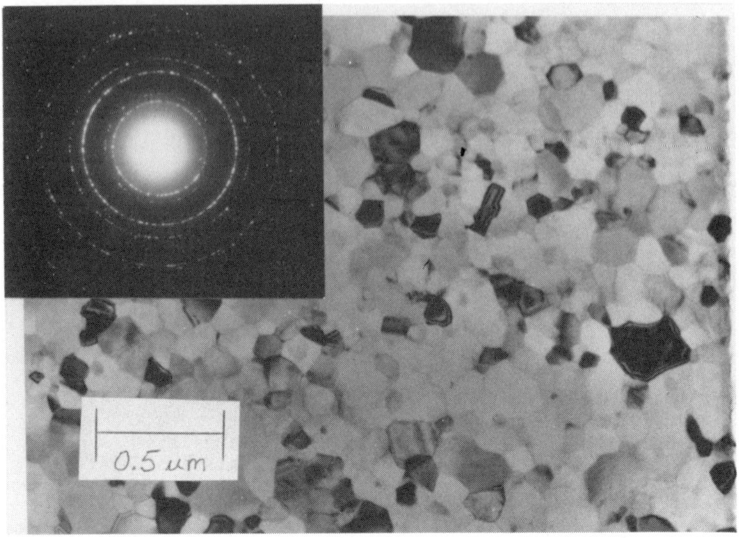


Figure 2 TEM of 75 nm Al film. Note well-defined sample free edge at right

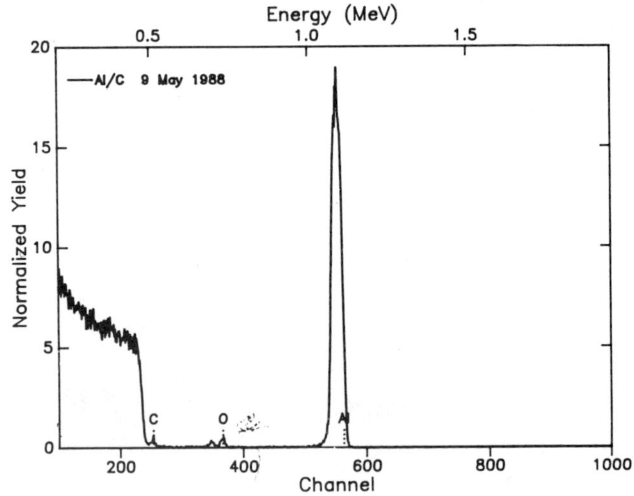


Figure 3 Representative RBS Spectrum showing oxygen and carbon at both film surfaces

**Thin Film Tensilometers**

Some of the basic TEM and uniaxial tensile data were obtained with instruments developed decades ago.[1] Since that time a search for materials of increased strength and understanding of the various contributions to flow stresses expected in thin film systems has emerged. Problems of mounting and premature specimen failure have always plagued the thin film tensilometers and the present design is no exception. In the following sections we illustrate some of the successes and difficulties associated with measuring the uniaxial tensile properties of thin films.

The CWRU nanotensilometer was designed and constructed for the purpose of obtaining force elongation data on 10 nm thick single crystal polymer specimens.[10] For thin films and fibers it is force limited to approximately 0.1 N which in turn limits the cross sectional area to some 10 $\mu m^2$ if fracture is desired.

Following the resurgence of interest some four decades ago in the mechanical properties of deposited thin films a number of tensile test instruments designed for small specimens were produced, and many of the experimental concepts obtained. The following decades demonstrated that single and polycrystalline films were similar[11], that the flow stress in thin films could be represented as a sum of contributions[12], and that film deformation maps could be obtained following detailed study of the deformation mechanisms.[13]

For the present study we have used free standing films, approximately 100 nm thick, 100 μm wide, and have having an initial guage length of some 0.5 - 1.5 mm. The nanotensilometer has the advantage of CCTV monitoring at optical magnifications. For most of the measurements, the nanotenilometer was operated in a "hard" mode in which the force necessary to elongate the specimen at a predetermined rate was measured as a function of the capacitively measured elongation. The block diagram is given in Fig. 4. We now describe some of the experiences to date.

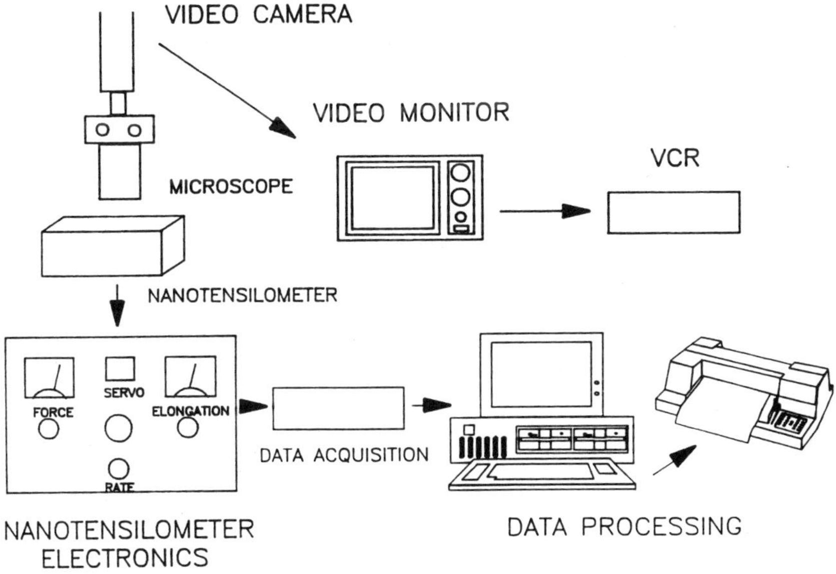


Figure 4 NANOTENSILOMETER BLOCK DIAGRAM

**Calibrations**

Optical magnifications were calibrated by direct observation of etched graticules on the high resolutions video screen. Overall instrument calibrations were obtained by using small diameter gold and aluminum wires. The same glues (Duco and epoxy) were used to fasten both calibration wires and thin film specimens to the nanotenilometer microscope slide "jaws". Representative calibration stress strain data for Al wire is shown in Fig. 5. Such calibration samples yield a long elastic region and a modulus equal to that of handbook data. Small diameter wires of gold, aluminum, and tungsten have been used.

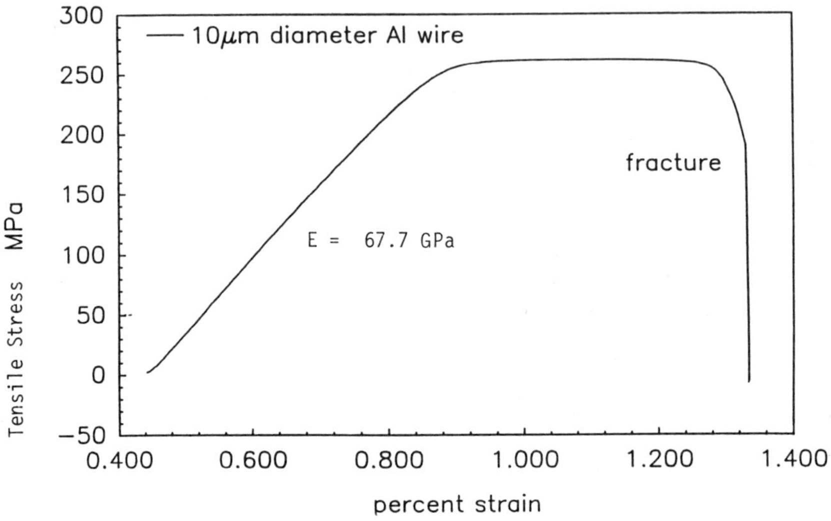


Figure 5 Aluminum wire Nanotensilometer calibration curve

**RESULTS**

In keeping with earlier tradition in tensile property measurements we translate the observed force elongation data into nominal stress strain data. We estimate the errors in sample dimensions to be about 10%. Because the elongations in the elastic range are comparable to optical resolutions the region at which the film specimen was glued to the nanotenilometer jaws was examined in special detail. Data for which the occasional slippage or a pullout of the specimen was observed were discarded. Typical stress strain curves for Al and amorphous $Al_2O_3$ films are shown in Figs. 6, 7, and 8 and look much like their bulk counterpart.

Because of the nanotensilometer construction no correction need be made to the force data, but there is a small servo error which may be electronically corrected for specimens whose material and dimensions make them stiff. This error has the effect of introducing a small mechanical hysteresis; for the data in this paper no such corrections were made so the reader may examine such effects directly. On a nanoscopic scale these nonlinearities make it difficult to decide whether a given film has both an elastic and plastic strain for small elongations.

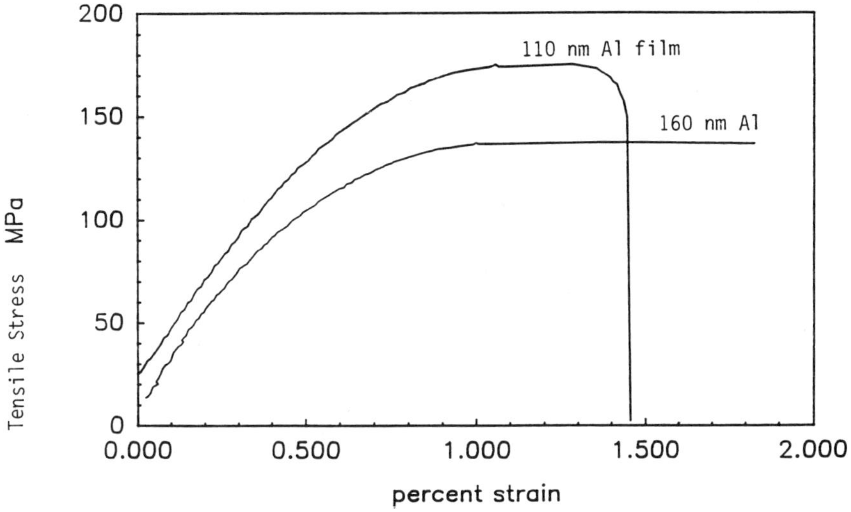


Figure 6 Nanotensilometer data for HV $N^5$ aluminum films

For films an artifact which we believe is related to mounting occurs upon initial loading. If we assume the effective cross sectional area of the specimen increases proportional to the extension a parabolic region is found in the initial portion of the stress strain curve and is seen in Figs. 6 and 8. This is commonly observed up to extensions of a few hundred nanometers but we feel of no practical consequence unless questions of

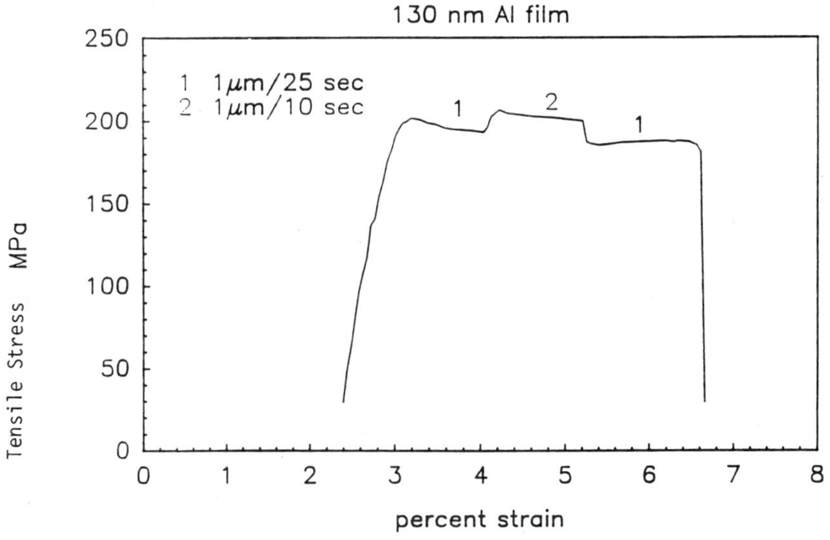


Figure 7 Nanotensilometer strain rate data for HV $N^5$ film

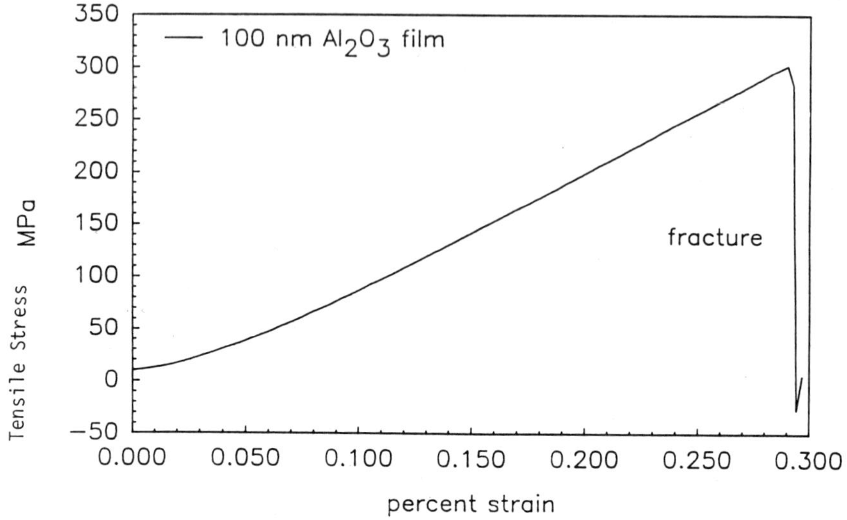


Figure 8 Tensile data for anodic alumina film

TABLE II Free Standing Al and $Al_2O_3$ Film Data

| Sample | Thickness | Modulus | Fracture Strain | UTS |
|---|---|---|---|---|
| 1 | 110 nm | 22 GPa | N.A. % | 88 MPa |
| 2 | 110 | 25 | N.A. | 124 |
| 3 | 110 | 25 | 1.1 | 173 |
| 4 | 110 | 22 | 7.0 | 145 |
| 5 | 110 | 20 | 2.8 | 175 |
| 6 | 110 | 22 | 1.4 | 174 |
| 7 | 110 | 8.2 | N.A. | 34.8 |
| 8 | 160 | 26 | 6.2 | 143 |
| 9 | 160 | 16 | 1.0 | 99 |
| 10 | 160 | 75 | 2.1 | 132 |
| 11 | 160 | 13 | 12.7 | 150 |
| 12 | 130 | 33.6 | 4.8 | 204 |
| 13 | 130 | 35.7 | 2.5 | 222 |
| 14 | 130 | 33.0 | 1.1 | 179 |
| 15 | 130 | 38.2 | 4.1 | 167 |
| 16 | 125 | N.A. | 1.1 | 277 |
| 17 | 125 | N.A. | 1.3 | 314 |
| 18 | 650 | 38.4 | 55.0 | 97 |
| | | Anodized $Al_2O_3$ films | | |
| 1 | 100 | 100 | 0.7 | 285 |
| 2 | 100 | 95 | 0.4 | 195 |
| 3 | 100 | 105 | 0.55 | 380 |

Notes: Samples 1-11 HV thermal evap. from m5N
Samples 12-18 HV thermal evap. from m6N

onset of plastic behavior at vanishingly small strains are considered. As a result we would estimate the values of mechanical parameters to be in error by no more than 20%. Table II summarizes the data for substantial number of films. Values for the elastic modulus, total strain at fracture and UTS are also listed. These data represent values from approximately 100 separate force elongation regions on some 25 separate specimens obtained from about ten successful depositions. Assuming constant thickness for all samples from a given deposition the reproducibility in the "modulus" is about 10%. Substantial differences are found between separate depositions and in isolated cases. These differences far exceed our estimates of errors in the specimen geometry, or any known impurity or grain size effects.

Approximately one half of these specimens mounted yielded useful data. Again mounting failures or specimens which fracture at the "jaw" edge are discarded. A substantial variation in the fracture strain is noted but in general large ductile regions are not found. Even those specimens which have substantial plastic deformation and show fracture in an unpredetermined position appear to fracture between one frame and the next (1/30th of a second) so we are unable to obtain crack propagation speeds. Brittle materials such as amorphous SiO and amorphous $Al_2O_3$ have stress-strain curves which are linear up to the point of fracture as one might expect (see Fig. 8). In addition to the tensile data directly obtained, the nanotensilometer is able to hold the elongation constant and plot the stress as a function of time at constant elongation. (see Fig. 9) Such stress relaxation data following a given strain are useful for creep applications. However, to date any relaxation in the glue has not been examined in sufficient detail to report quantitative results. The total stress appears to fall approximately exponentially at strains of approximately one half the fracture strain. In the nominal elastic region of the stress strain curve a strain reversal gives rise to a somewhat steeper unloading slope than upon initial straining. This has been interpreted in the past as an indication of softness or plastic deformation in the tensile test instrument.[11]

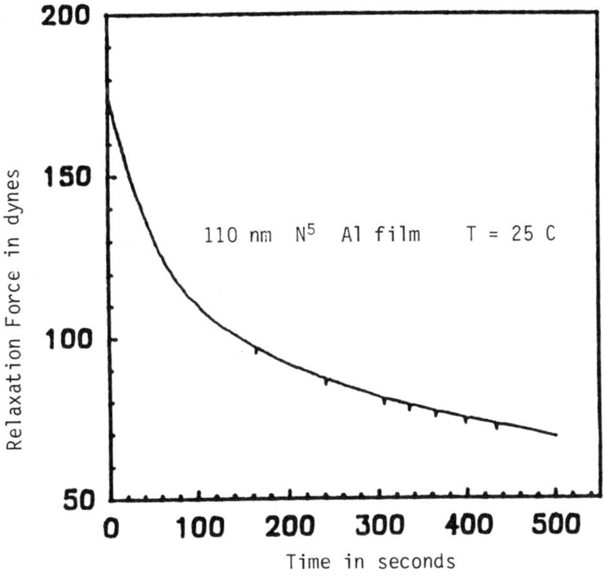


Figure 9 Room temperature relaxation data

**FINITE ELEMENT ANALYSES**

If the slope of the observed force elongation curve is recalculated as a uniaxial stress strain then the data indicates the following: First, small diameter wires yield a modulus equal to Young's modulus as one would expect, whereas films indicate an elastic modulus approximately one half of the Young's modulus literature values for bulk polycrystalline materials. In order to understand this difference we have started a series of finite element analyses using the program FEP2, a two-dimensional finite element program developed at CWRU.

Thin film fiber textures are commonly found and could lead to mechanical anisotropies in the modulus. However, for aluminum we would expect this effect not to exceed about 20% and hence not be large enough to explain the observed difference. Using the finite element code in two-dimensions and assuming an increasing displacement at one end of the film and a fixed center portion, the effective "experimental" slope may be calculated. The results of this calculation indicate that the specimen length to width ratio does not affect the observed modulus substantially as shown in Fig. 10. As a practical matter the specimens used in our experimental study are commonly 10:1 length:width ratio and no correction was made. From this finite element calculation one can see that it would be less than a 5% effect for 'square' samples. For our films the correction is insignificant, and even then in the wrong direction.

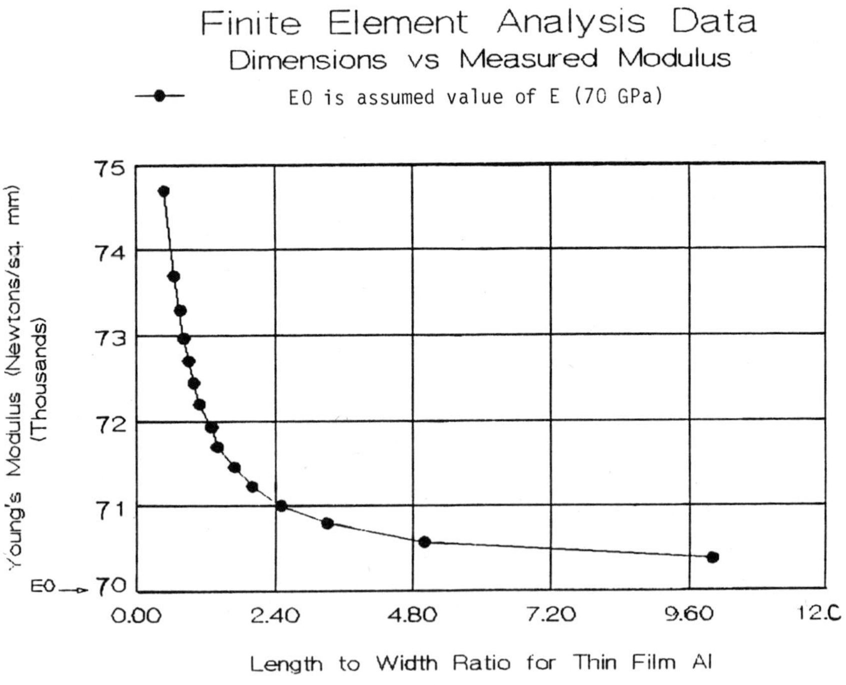


Figure 10 Geometric dependence of tensile modulus

One can also model the adhesive to holding the film as a region approximately 1000 times thicker than the film and with a shear modulus appropriate to polymers. An elastic continuum calculation for this case indicates that the correction to the film elongation would be negligible and no correction was attempted. Not unexpectedly a top view of the film shows that the width decreases substantially for a large extension. The shape crudely appears to be a constant but large radius and no modelling was attempted. None of the approaches to date has yielded a physically sensible explanation for the observed modulus. As a result, although we would predict that the observed stress strain slope should, for a thin film, be close to that of Young's modulus, the experimental results for both constrained and free standing films appears to indicate otherwise and presently remains unexplained.

Finally, a noncommercial video has been produced to demonstrate some of the effects and could be made available upon request.

## ACKNOWLEDGEMENTS

Even though a single author is listed at the beginning of this article I acknowledge with great thanks the contributions of G. Mearini, J. Eldridge, and G. DeRose for the sample preparation, R. Haaland and G. Michal for the TEM characterization, L.R. Doolittle for providing the RBS data analysis program, K. Chaffee and D. Jayne for the nanotensilometry, D. Krus and R. Mullen for the finite element analyses and R. Hoffman, Jr., for his insight in developing the technique and physical interpretation. Finally, we all acknowledge the partial support from DARPA under Grant N000141-80K-0723.

## REFERENCES

1. R.W. Hoffman, Physics of Thin films, Vol. 3, Hass and Thun, eds., Academic Press, (1966); NATO Advanced Topic Summer School, Corsica (1974). Published in NATO Advanced Study Institute Series B Vol. 14, Plenum Press (1976).

2. C.S. Smith in Solid State Physics, Vol. 6, Seitz and Turnbull eds. Academic Press (1956).

3. R.W. Hoffman, Mat. Res. Soc. Symp. Proc. Vol. 69, 95 (1986).

4. R.W. Hoffman, Mechanical Properties of Thin Films, AVS Short Course, 1st Edition (1978); 2nd Edition (1979); 3rd Edition (1983). Published under the auspices of the Education Committee, The American Vacuum Society (1983).

5. T-S. Kuan and M. Murakami, Metall. Trans. 13A, 383 (1982).

6. M. Laugier, Thin Solid Films 75, L17 (1981).

7. M. Janda, Thin Solid Films 412, 219 (1984).

8a. M.F. Doerner, D.J. Gardner, and W.D. Nix, J. Mater. Res. 1, 845 (1986).

8b. D.A. Flinn, D.S. Gardner, and W.D. Nix, IEEE Trans. on Electron Devices, Vol. ED-34, 689 (March, 1987).

9. R.W. Hoffman, R.W. Hoffman, Jr., and T.E. Mitchell, Thin Solid Films 154, 149 (1987).

10. R.W. Hoffman, Advances in Chemistry (J.L. Koenig, ed.) 174, 25 (1979).

11. J.M. Blakely, Journ. Appl. Phys. 35, 1756 (1964).

12. M. Ronay and C.F. Aliotta, Philos. Mag. A42, 161 (1980).

13. M. Murakami, T-S. Kuan, and I.A. Blech, in Treatise on Materials Science Technology, K.N. Tu and R. Rosenberg, eds., Academic Press (1982).

# RELATIONSHIP BETWEEN INTERFACIAL STRAIN AND THE ELASTIC RESPONSE OF MULTILAYER METAL FILMS

BRUCE M. CLEMENS AND GARY L. EESLEY
Physics Department, General Motors Research Laboratories, Warren, Michigan 48090-9055.

## 1 Abstract

We have used x-ray diffraction and picosecond transient piezoreflectance to investigate the structural and elastic properties of three distinct multilayer systems: Mo/Ni, Pt/Ni and Ti/Ni. We demonstrate that the commonly observed lattice expansion perpendicular to the film plane is not a bulk effect, but is localized at the interface between the contacting metals. Incorporating the measured interfacial expansion into a universal binding relation, we show that the measured elastic softening versus compositional repeat distance can result from the interfacial strain.

## 2 Introduction

There is considerable interest in the structural and mechanical properties of metal multilayer films[1]. These investigations utilize the opportunity to study the mechanical properties of extremely well controlled and fine spatial scale composite media. Interface mechanical properties can be isolated and investigated. It has been found that the film elastic properties can be affected when the thickness of the constituent layers approaches atomic scale dimensions. In most cases, the elastic properties of the multilayer films are found to soften as the bilayer thickness is decreased[2,3,4,5,6]. In rare instances, a hardening of elastic properties has also been observed and the so-called supermodulus effect has been a motivating factor in these studies [7,8]. The changes in elastic response have been attributed to electronic effects arising from the layering[9] as well as to the structural changes which accompany an observed lattice strain perpendicular to the film plane[10,11,12].

In an effort to understand these observations, we have investigated the structural and elastic properties of three distinct multilayer systems: Mo/Ni, Pt/Ni and Ti/Ni. We demonstrate that the observed atomic lattice expansion perpendicular to the film plane is not a bulk effect, but is localized at the interface between the contacting metal layers. Incorporating the measured interfacial expansion into a universal binding relation, we show that the measured elastic softening versus compositional repeat distance can result from the interfacial strain. Thus in spite of the rather dramatic differences in the interfacial structure of the metal pairs we have investigated, we show that interfacial strain is the underlying cause of the apparent changes in the bulk elastic response.

## 3 Sample Preparation and Characterization Technique

Multilayered samples were prepared in a cryopumped dual source magnetron deposition system as described previously[13]. The sputter rates were 0.3-0.5 nm/sec. and were set to produce an equal thickness of each constituent per layer. Samples with bilayer thicknesses between 1 and 20 nm, and total thickness between 200 and 300 nm were produced.

Our technique for measuring the elastic properties of these materials is an extension of conventional ultrasonic pulse-echo measurements into the picosecond time regime[14]. We use two synchronized lasers (excite and probe) which emit visible light pulses of approximately 4 psec duration. Local surface heating and expansion resulting from the excitation laser pulse result in a short duration (4 psec) acoustic pulse which propagates into the film. This acoustic pulse is partially reflected by the impedance mismatch at the film-substrate interface. Upon returning to the surface, the reflected pulse will induce a change in surface reflectivity due to the acoustic strain of the surface region of the crystal lattice (piezoreflectance). Thus, by using the second laser to monitor surface reflectivity as a function of time after the initial excitation pulse, we can determine when the reflected acoustic pulse returns to the surface, and thus determine the total time that the acoustic pulse spends on its trip from the surface to the interface and back.

Mat. Res. Soc. Symp. Proc. Vol. 130. 

Typical differential reflectivity vs time data for a Pt/Ni multilayer is shown in Fig. 1. The sharp rise at short times is the transient thermoreflectance due to sample heating by the excitation laser pulse. The sharp wiggles are the piezoreflectance signal. We can often observe two or more of these wiggles, allowing precise determination of the round trip time interval. This, in combination with the accurate determination of the sample thickness using small angle x-ray diffraction described below, gives a direct measurement of the longitudinal sound velocity in the growth direction. With this technique we can nondestructivly measure sound velocity in films of order 100 nm thick with a precision of better than a few percent.

# 4 Results

## 4.1 Structural Properties

X-ray diffraction was performed using $Cu_{k_\alpha}$ radiation in the conventional symmetric reflection geometry which confines the scattering vector to the growth direction. A composition modulation in the growth direction will produce a series of peaks in the low angle diffraction, and the position of these peaks can be used to determine the bilayer thickness($\Lambda$). The relative peak intensities yield the shape of the composition wave and an upper limit on the interface width can be estimated from the spatial scale corresponding to the highest-order peak resolved. We observe sharp, intense peaks in the low angle diffraction from all of the samples indicating the presence of a well defined composition modulation. Additionally we determine an upper limit interface width of 0.5 nm in the case of Pt/Ni and Mo/Ni, and 0.9 nm in the case of Ti/Ni.

High angle x-ray diffraction of compositionally modulated material is sensitive to the presence of disordered interfacial regions[15]. High angle superlattice lines persist only when fluctuations in the layer thickness are discrete, indicating crystalline interfaces. In the presence of a disordered interface region, fluctuations in the layer thickness will have a continuous distribution. Fluctuations of this type, with magnitudes as small as 0.1 nm, will broaden the high angle superlattice lines, leaving only diffraction characteristic of individual slabs of the constituent layers.

Looking at the diffraction data for a series of Pt/Ni multilayers (Fig. 2a), we see sharp high angle diffraction peaks for all bilayer thickness. This indicates that despite the 10% in-plane lattice mismatch there are crystalline interfaces for all bilayer thickness produced. Analysis of peak widths indicates that inhomogeneous strain (not particle size) is the dominant broadening mechanism, with structural coherence extending through several bilayer periods.

In the Ni/Mo system (Fig. 2b), there are sharp high angle superlattice peaks for samples with bilayer thickness greater than about 2.0 nm, indicating a crystalline registry between the Mo (110) and Ni (111) planes for these samples. The observation of long range coherence in the growth direction [3,16] may be related to the Nishiyama-Wassermann orientation which matches the spacing between rows of close packed atoms in the mating (110) BCC and (111) FCC planes[17]. In Mo/Ni the mismatch in row spacings is only 3%.

An interesting feature in the Mo/Ni diffraction data is the dramatic loss in long range order for the samples with bilayer thickness less than about 2.0 nm. This transition between the presence and absence of long range order is remarkable and abrupt. We find that in going from $\Lambda$=2 nm to 1.5 nm, the diffraction peak widths increase by an order-of-magnitude and the peak intensities decrease by two orders-of-magnitude. Similar behavior has been observed by other investigators[3], and has been attributed to the effect of strain which increases as the bilayer thickness decreases[11]. However, as we show below, the apparent increase in strain as the bilayer thickness decreases is only due to an increase of the density of interfaces with lattice spacing greater than the average of the constituents. The strain is localized to the interface, and thus it is not likely responsible for the loss of long range order. We feel that the dramatic change in structure as a function of bilayer thickness should be viewed as a *onset* of long range order as the layer thickness increases. We speculate during the initial deposition of Mo on Ni (or Ni on Mo) the deposited layer is highly disordered, and this disorder remains in the small bilayer thickness samples.

In the Ti/Ni system the (001) Ti planes and the (111) Ni planes have the same symmetry, but there is a 15% lattice mismatch. For $\Lambda > 4.6$ nm the diffraction pattern (Fig. 2b) can be described as a simple summation of the diffraction intensity from the Ni and Ti slabs

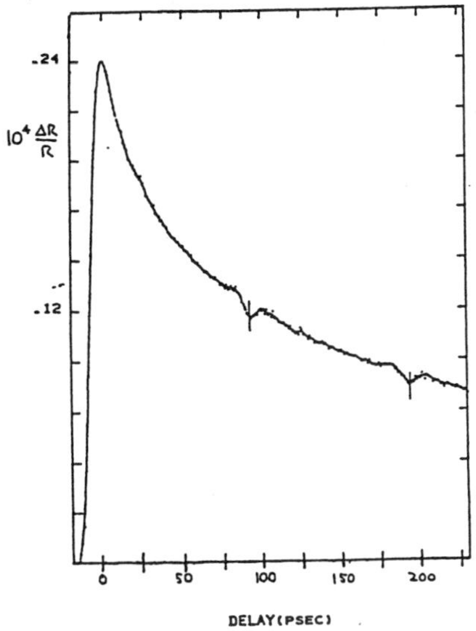


Figure 1: Transient reflectivity for a Pt/Ni multilayer sample

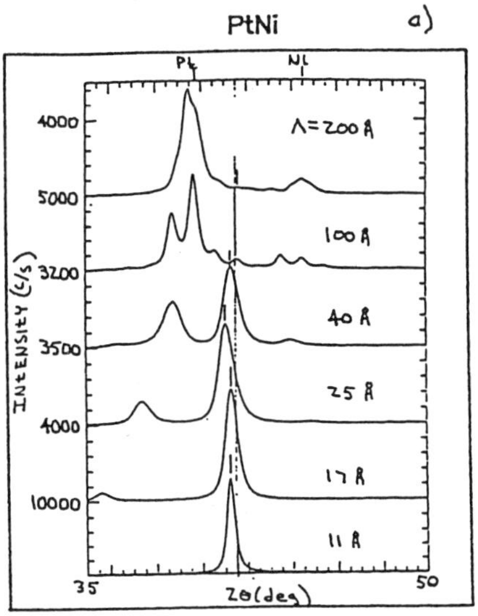


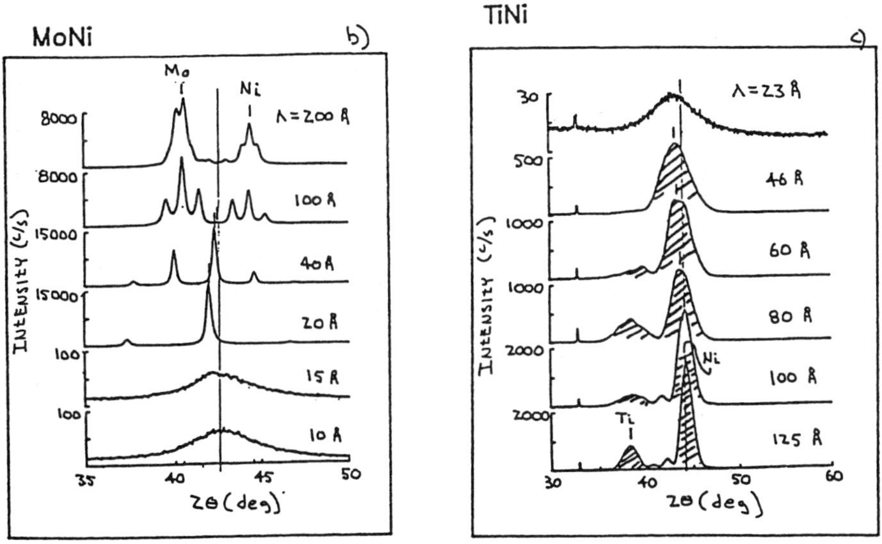


Figure 2: High angle x-ray diffraction for a series of a) Pt/Ni multilayers, b) Mo/Ni multilayers, and c) Ni/Ti multilayers.

comprising a bilayer. Thus, crystalline order exists within the elemental layers, but the absence of satellite peaks indicates that the interface region is disordered. In the range 2.3 nm $< \Lambda <$ 4.6 nm, we observe an abrupt transition from crystalline to disordered structure within the elemental layers.

Although these three metal-metal systems appear to exhibit radically different structural behavior as a function of $\Lambda$, they can be analyzed on common ground with respect to interfacial strain. When the interfaces are crystalline, there will be a diffraction peak which represents the average d spacing in growth direction $(\bar{d})$[3]. Thus in the case of the Mo/Ni and Pt/Ni systems, we can readily extract the average lattice spacing perpendicular to the film plane. However, the absence of satellite peaks in the diffraction from the Ti/Ni system allows a direct measurement of the lattice spacing of crystalline constituents. We can monitor the position of the Ni diffraction peak to determine an average $d$ spacing within the Ni layers. In Fig. 2, the peaks corresponding to the average $d$ spacing $\bar{d}$ for Pt/Ni and Mo/Ni and the $d$ spacing of Ni for Ti/Ni are identified with vertical lines. Note that in each case these peaks move to smaller angle (larger $d$ spacing) with decreasing bilayer thickness.

A simple one-dimensional model can be used to describe this behavior. If we assume that the atomic plane spacing at the interface $(d_i)$ differs from the average $d$ spacing of the constituents $(\bar{d}_o)$ by an amount $\delta_d$, then we find,

$$\Delta\bar{d} = \bar{d} - \bar{d}_o = 2\bar{d}_o\delta_d\frac{1}{\Lambda}. \tag{1}$$

This predicts that the observed expansion is linear in $1/\Lambda$. In Fig. 3 we plot the observed $\Delta\bar{d}$ versus $1/\Lambda$. Note that in the case of Ti/Ni, we have plotted the relative change in the average spacing of the Ni layers, $\Delta d_{Ni} = d_{Niexp} - d_{Nibulk}$. The observation that this expansion scales linearly with $1/\Lambda$, is a crucial point which cannot be accounted for by assuming that the interface spacing is an average of the constituent interplanar spacings.

We emphasize that this apparent change in bulk interplanar spacing is actually the consequence of a constant deviation at the interface which has no $\Lambda$ dependence at all. The linear relationship demonstrated in Fig. 3 results from the fact that the number of interfaces per unit volume scales as $1/\Lambda$. As the bilayer period is reduced, an increasing amount of interface material contributes to the x-ray diffraction from the multilayer film.

Although the Ti/Ni system exhibits highly disordered and extended interfaces relative to Mo/Ni and Pt/Ni, the same linear behavior is observed. This can be explained by the presence of Ni planes near the interface which have a larger $d$ spacing than the bulk of the Ni layer, leading to a similar expression for $\delta_d$ versus $1/\Lambda$. By fitting the linear regions of Fig. 3, we have obtained the interface expansion $\delta_d$ for each of the metal systems and the results are tabulated in Table 4.2.

Some insight into the nature of this interface expansion can be obtained by examining the deviation in $\bar{d}$ at small $\Lambda$. In Pt/Ni, dislocations are introduced at the interface when each layer exceeds the critical thickness[18], resulting in a relative expansion of the interface interplanar spacing. Below the critical thickness, interfaces will be dislocation free. We can estimate the critical thickness for Pt/Ni from the data in Fig. 3 to be about 1 nm, which is greater, but in fair agreement with the prediction of 0.2 nm from Van der Merwe[18].

In Mo/Ni the point at which the linear dependence on $1/\Lambda$ fails, corresponds to the $\Lambda$ where a sudden transition from crystalline to disorder in the diffraction from the bulk constituents. It is likely that the relaxation of the constraint of crystallinity within the layers allows the interplanar spacings to assume a value closer to the average of the bulk constituents.

In the case of Ti/Ni, we observe a rather large interface expansion. We note however, that this net expansion is actually distributed over a number of Ni layers and there may be some incorporation of larger Ti atoms into the Ni matrix.

## 4.2 Elastic Properties

The effect of the large interface interplaner spacing on the elastic properties can be seen as the number of interfaces increases ($\Lambda$ decreases). The square of our measurement of sound velocity is a direct measure of the change in elastic response. The results of our sound velocity measurements are shown in Fig. 4, where we have plotted the normalized velocity squared versus $\Lambda$.

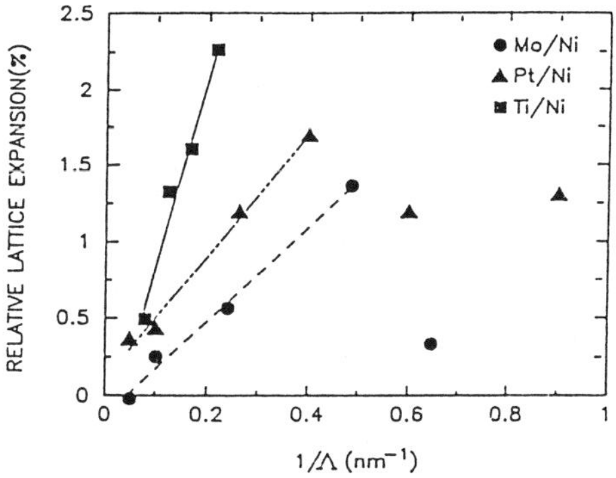


Figure 3: The relative change in average lattice parameter $(\Delta\overline{d}/\overline{d}_o)$ for Mo/Ni, Pt/Ni and the Ni lattice in Ti/Ni multilayers as determined from x-ray diffraction. The results are plotted versus the inverse bilayer thickness, and the straight lines are best fits to the regions of uniform expansion. as a function of $1/\Lambda$.

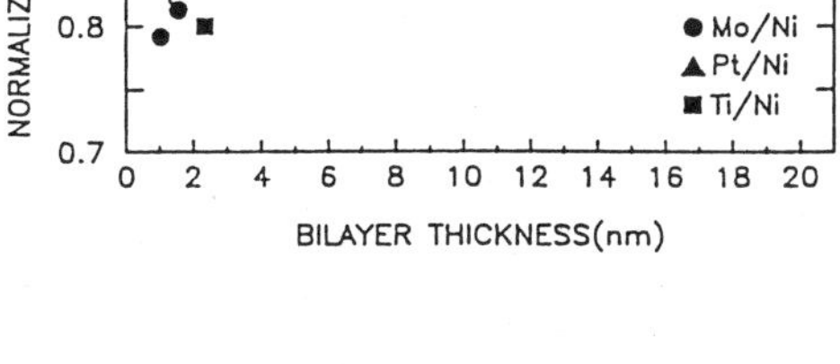


Figure 4: Experimentally measured sound velocity(squared) versus bilayer thickness, and normalized to the value(squared) measured at the maximum bilayer thickness: Mo/Ni($V_{max}$ = 5866 m/sec); Pt/Ni(4496 m/sec); TiNi(5800 m/sec). Error bars correspond to a maximum $\pm 2$ psec error in the measured pulse-echo time delay. The curves are a visual aid.

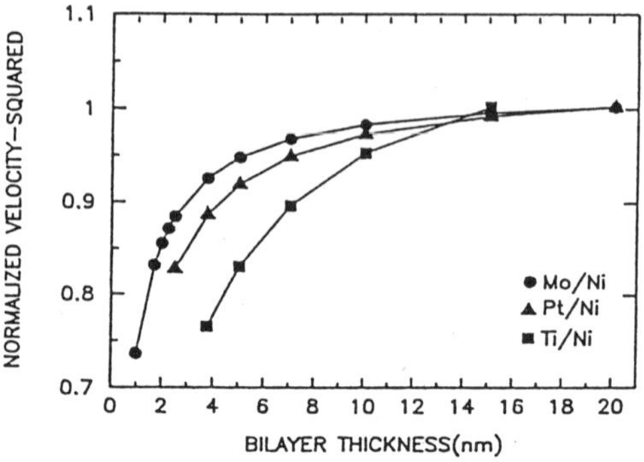


Figure 5: Calculated sound velocity (squared and normalized) versus $\Lambda$ for Mo/Ni, Pt/Ni and Ti/Ni multilayers.

| System | $\delta_d$(Å) | $l^{av}_{AB}$(Å) | $C_i/C_e$ |
|---|---|---|---|
| Mo/Ni | 0.15 | 0.54 | 0.54 |
| Pt/Ni | 0.20 | 0.54 | 0.44 |
| Ti/Ni | 0.32 | 0.64 | (0.5) |

Table 1: Observed interface expansion, universal scaling length and calculated relative interfacial elasticity. The value in parentheses for Ti/Ni is an assumed value(see text).

All three systems exhibit elastic softening(decreasing sound velocity) in the same region where the lattice expansion is observed to be proportional to $1/\Lambda$. We believe this softening is a direct consequence of the apparent lattice expansion which we have demonstrated to be the result of an expanded interfacial region. Further softening is observed in Mo/Ni and Ti/Ni as $\Lambda$ is decreased below the the point where long range order disappears. However, in the case of Pt/Ni which does not disorder at small $\Lambda$'s, the sound velocity is observed to increase. We believe this is associated with a reduction in the average lattice parameter which accompanies a reduction of the interfacial defect density. These results demonstrate in a consistent manner the effects of both interfacial and bulk structural properties on the macroscopic elastic response of multilayer metal films.

We can investigate our correlation between interface structure and elastic properties by noting that the binding energy versus distance relations for metallic systems exhibit universal behavior under a simple two- parameter scaling[19]. For adhesion between metals $A$ and $B$ the relation between the elastic stiffness constant perpendicular to the interface ($C_i$) and the equilibrium adhesion energy ($\sigma_{AB}$) is given by:

$$C_i = \frac{d_i \sigma_{AB}}{l^2_{AB}} \left[ \frac{d^2 E^*(a^*)}{da^{*2}} \right] \equiv C_e \left[ \frac{d^2 E^*(a^*)}{da^{*2}} \right], \qquad (2)$$

where $d_i$ is the interplanar spacing, $l_{AB}$ and $E^*(a^*)$ are the universal scaling length and energy function respectively, and $C_e$ is the equilibrium interface elasticity. The dimensionless parameter $a^*$ is a scaled length equal to $\delta_d/l_{AB}$. The universal scaling length for the adhesion region, $l_{AB}$, is estimated by averaging the scaling lengths appropriate for each constituent metal[19]. The universal energy function is well approximated by $E^*(a^*) = -(1+a^*)\exp(-a^*)$, and thus Eqn. 2 determines the reduction in the interfacial elastic response for a given $\delta_d$ [20].

The predicted interfacial softening can be used to determine the change in sound velocity of a multilayer film as a function of $\Lambda$. We have extended the formalism of Rytov[21] to include an interface of finite extent, with elastic properties different from the two constituent metals used to produce the multilayer film. The Rytov formalism applies to two-component films in limit that the acoustic wavelength is much larger than the bilayer repeat distance. As a consequence, there will be no dependence of the sound velocity on the $\Lambda$ assuming the material properties remain unchanged. However, by including a finite interface region (third component) with an elastic constant which differs from the effective constant of the two-component system, a dependence of the sound velocity on $\Lambda$ results.

Generalizing the Rytov formalism to include a third interfacial layer, we find the expression for the longitudinal sound velocity in the presence of perturbed interfaces ($V_i$) relative to that without ($V$) is given by:

$$\left(\frac{V_i}{V}\right)^2 = \left\{1 - \frac{2n_i d_i}{\Lambda}\left[1 - \frac{2}{C_i}\left(\frac{C_A C_B}{C_A + C_B}\right)\right]\right\}^{-1}, \tag{3}$$

where the velocity squared is indicative of the elastic response. In Eq. 3 the indices $A$, $B$ and $i$ represent metal $A$, metal $B$ and the interface region respectively. In deriving this expression we have assumed that the interface density is simply the average of the constituent metal densities [6,21,22] and that the interface is $n_i$ lattice spacings ($d_i$) thick. We see from this expression that the effective elastic constant of a lamination of $A$ and $B$ is $C_e = 2C_A C_B/(C_A + C_B)$, and if the interface elasticity ($C_i$) departs from this value then a $\Lambda$ dependence will occur.

Equations 2 and 3 provide us with a means of using the measured lattice expansion to calculate the interfacial softening and $\Lambda$ dependence of the sound velocity. The use of Eq. 2 to calculate the interfacial softening is justified in the case of Mo/Ni and Pt/Ni where the interfaces are relatively sharp on the atomic scale. The results of this calculation are tabulated in Table 1, including the measured interfacial expansion and the universal scaling length for the adhesion region.

In the case of Ti/Ni the presence of a relatively disordered and diffuse interface complicates a determination of the interface elastic response by a universal scaling. Previous measurements on TiNi alloys have yielded an elastic constant which is approximately 75% of the value we calculate for an unperturbed multilayer of crystalline Ni and Ti[23]. It is reasonable to expect the extensive interfacial disorder in our Ti/Ni samples will further reduce the interfacial elasticity. We will show that by using a value for $C_i = 0.5C_e$, we obtain a $\Lambda$ dependence comparable to that observed. In light of the rather dramatic softening found in the case of Mo/Ni and Pt/Ni, we believe this value of $C_i$ is physically plausible.

Combining the results in Table 1 with Eqn. 3, we have calculated the dependence of the relative velocity squared versus the bilayer thickness and the results are shown in Fig. 5. Comparison of the calculated results with the experimental results of Fig. 4) is rather striking, and several features are worth noting. First, we see that the qualitative shape of the measured data versus $\Lambda$ is reproduced by our model calculation. More importantly, we find that the relative degree of interfacial softening which is predicted by the universal binding model for the Mo/Ni and Pt/Ni is indeed that observed. That is, using our measured interfacial expansion, we predict that the Mo/Ni multilayer will show less apparent softening versus $\Lambda$ than the Pt/Ni multilayer.

We cannot be as forceful in our comparison with Ti/Ni. We note however, that we have attempted to realistically model the situation by distributing the interfacial expansion of the Ni lattice over four atomic planes ($n_i = 4$ and $di = d_{Ni} + \delta_d/4$ in Eqn. 3). The extent of this interface is in agreement with that determined by the x-ray diffraction and previous resistivity studies of the NiTi system[24]. Although we cannot use a simple universal scaling to determine the degree of interfacial softening, it undoubtedly occurs and is manifested as a decrease in sound velocity as the $\Lambda$ is reduced.

In conclusion, we have shown that the measured expansion in the average lattice spacing of metal multilayer films can be attributed to a local expansion at the interface between Ni and -Mo, -Pt and -Ti. The distinct differences in the interfacial structure for these three metal pairs leads us to believe that this phenomenon is not unique to the systems studied here. We have shown for the first time, that this interfacial expansion can explain the apparent softening of the bulk elastic response in multilayer metal films as the compositional modulation distance is decreased. Our measurements of the longitudinal sound velocity in these films are in substantial agreement with our model of the elastic softening.

Although we have not dealt with the possibility of elastic hardening, it is obvious that if the interface lattice spacing contracts then the universal binding theory will predict an increase in the interface elasticity. Likewise, our formulation for the sound velocity will predict an increased stiffness as the bilayer thickness is decreased. In a future publication we will provide a more detailed explanation of our model and application to other metal systems which have been studied.

## Acknowledgements

We would like to acknowledge the assistance of M. Devour in sample preparation and R. Waldo in electron beam microprobe determination of composition. We are also indebted to J. Smith for useful discussions on universal binding.

## References

1. I. K. Schuller, IEEE Ultrasonics Symposium **1985**, 1093 (1985).
2. A. Kueny, M. Grimsditch, K. Miyano, I. Banerjee, C. M. Falco and I. K. Schuller, Phys. Rev. Lett. **48**(3), 166 (1982).
3. M. R. Khan, C. S. L. Chun, G. P. Felcher, M. Grimsditch, A. Kueny, C. M. Falco, and I. K. Schuller, Phys. Rev. B **27**, 7186 (1983).
4. R. Danner, R. P. Huebener, C. S. L. Chun, M. Grimsditch, and I. K. Schuller, Phys. Rev. B **33**(6), 3696 (1986).
5. J. A. Bell, W. R. Bennett, R. Zanoni, G. I. Stegeman, C. M. Falco, and F. Nizzoli, Phys. Rev. B **35**(8), 4127 (1987).
6. P. Bisanti, M. B. Brodsky, G. P. Felcher, M. Grimsditch, and L. R. Sill, Phys. Rev. B **35**(15), 7813 (1987).
7. W. M. C. Yang, T. Tsakalakos and J. E. Hilliard, J. of Appl. Phys. **48**(3), 876 (1977).
8. U. Helmersson, S. Todorova, S. A. Barnett, and J.-E. Sundgren, J. of Appl. Phys. **62**(2), 481 (1987).
9. T. B. Wu, J. of Appl. Phys. **53**, 5265 (1982).
10. R. C. Cammarata, Scripta Met. **20**, 479 (1986).
11. I. K. Schuller and A. Rahman, Phys. Rev. Lett. **50**(18), 1377 (1983).
12. A. Banerjea and J. R. Smith, Phys. Rev. B **35**(11), 5413 (1987).
13. B. M. Clemens and G. L. Eesley, Phys. Rev. Lett. **61**, 2356 (1988).
14. G. L. Eesley, B. M. Clemens and C. A. Paddock, Appl. Phys. Lett. **50**(12), 717 (1987).
15. B. M. Clemens and J. G. Gay, Phys. Rev. B **35**, 9337 (1987) Rapid Communication.
16. I. K. Schuller, Phys. Rev. Lett. **44**, 1597 (1980).
17. E. Bauer and Jan. H. van der Merwe, Phys. Rev. B **33**(6), 3657 (1986).
18. J. H. Van der Merwe, J. of Appl. Phys. **34**, 123 (1962).
19. J. H. Rose, J. R. Smith, and J. Ferrante, Phys. Rev. B **28**(4), 1835 (1983).
20. P. Vinet, J. Ferrante, J. H. Rose and J. R. Smith, J. of Geophys. Res. **92**, 9319 (1987).
21. S. M. Rytov, Sov. Phys. Acc. **2**, 68 (1956).
22. If the interface density is different from the average of the two metals, an additional factor of $\left[1 - \frac{2n_i d_i}{\Lambda}\left(1 - \frac{\rho_i}{\rho_A V}\right)\right]^{-1}$ multiplies the right hand side of Eq. 2.
23. C.M. Jackson, H. J. Wagner and R. J. Wasilewski, *55-Nitinol- The Alloy with a Memory: Physical Metallurgy , Properties and Applications*, Volume NASA-SP5110 (Nat. Aero. and Space Adm., Washington,D. C., 1972).
24. B. M. Clemens, Phys. Rev. B **33**(11), 7615 (1986).

# THIN HgCdTe LPE LAYERS GROWN ON CdTe AND CdZnTe SUBSTRATES

KOSAKU YAMAMOTO, TOHRU MAEKAWA and HIROSHI TAKIGAWA
*FUJITSU LABORATORIES LTD.*, Infrared Devices Laboratory
10-1 Morinosato-Wakamiya, Atsugi, 243-01, Japan

## ABSTRACT

The dependence of the growth pattern of thin HgCdTe LPE layers on lattice mismatch was investigated. Epilayers grown on substrates with a lattice mismatch of more than 0.2% show island-like growth, which are aligned with the network of misfit dislocation lines. Thin epilayers grown on substrates with a lattice mismatch of less than 0.05% have mirror-like flat surfaces. It was found that the surface defects for thick epilayers also have a lattice-mismatch dependency.

## INTRODUCTION

HgCdTe is a direct band-gap semiconductor with many advantages when used in infrared detectors. Among the crystal-growth techniques studied, liquid-phase epitaxy is potentially the best process for producing HgCdTe crystals with high crystallinity, uniformity, and purity. CdTe and CdZnTe are generally used for LPE substrates. CdTe has a 0.2% larger lattice constant than $Hg_{0.8}Cd_{0.2}Te$. The $Cd_{1-Y}Zn_YTe$ substrate holds promise for lattice matching with $Hg_{1-X}Cd_XTe$ epilayers [1].

This paper reports the results of our investigation of the dependency of the growth pattern of thin epilayers on lattice mismatch. The island-like growth pattern of thin epilayers is presented, and we propose a model interpreting this growth. We also show how surface defects of thick epilayers depend on lattice mismatch.

## EXPERIMENT

The substrates were (111)B CdTe and $Cd_{1-Y}Zn_YTe$. The wafers were cut from an ingot grown by the Bridgman method. The Zn content of $Cd_{1-Y}Zn_YTe$ was measured by atomic absorption spectroscopy to an accuracy of 3%. Dislocation etch pit density of the CdTe and $Cd_{1-Y}Zn_YTe$ substrates was estimated to be $10^4$-$10^5$ $cm^{-2}$ using the Nakagawa etch [2]. Before epitaxial growth, the substrate surface was etched with a solution of 5% Br in methanol to remove polishing damage. $Hg_{0.8}Cd_{0.2}Te$ epilayers were grown from Te-rich growth solutions using a tipping technique. The epilayers were grown at 480°C with a cooling rate of 0.05°C/min. Supersaturation of the growth solution was 2°C.

## RESULTS AND DISCUSSION

Figure 1 shows the patterns of thin epilayers grown on different CdTe substrates. The growth times range from 20 to 120 seconds. The epilayer for 120-second growth is about 3 μm thick. Many islands form, grow laterally, become clearly visible, and finally combine. Many dents which are the traces of island boundaries can be seen in Fig. 1d.

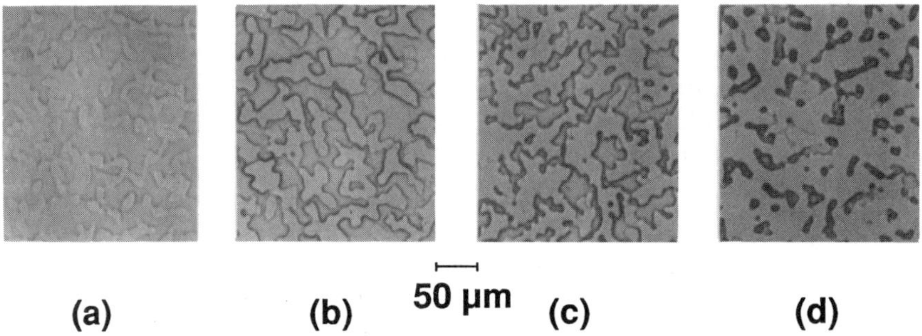


Figure 1. Photomicrographs of the surface pattern of thin HgCdTe epilayers grown on different CdTe substrates under the following growth times: (a) 20 s, (b) 40 s, (c) 60 s, (d) 120 s.

To investigate the cause of this growth, we examined a cross section of the epilayer by etch pitting. Figure 2 shows the cleaved, etched, (110) cross section of an epilayer grown on a CdTe substrate. Note that etch pits are distributed unevenly and that many accumulate under the dents, which are the traces of island boundaries. These etch pits seem to be misfit dislocations in the (111) heterojunction with a lattice mismatch of 0.2% between the epilayer and CdTe substrate [3]. The uneven distribution of etch pits was observed only on the cleaved ($\bar{1}10$) plane, and was not seen in the etched (111) surface of the epilayer.

The island-like growth pattern is thought to be produced from the initial growth mechanism shown in Fig. 3. First, micro-islands form and begin to grow laterally. Most of the lattice-mismatch stress is distributed at the edge of the islands and stress prevents lateral growth. As a result, island-like growth patterns become visible. When islands combine, their boundaries remain as dents, and many misfit dislocations are produced to accommodate the accumulating stress.

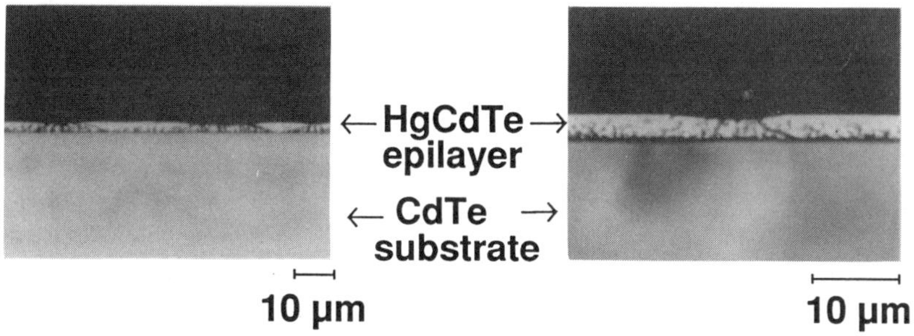


Figure 2. Photomicrographs of cleaved, etched ($\bar{1}10$) cross section of an HgCdTe epilayer grown on a CdTe substrate

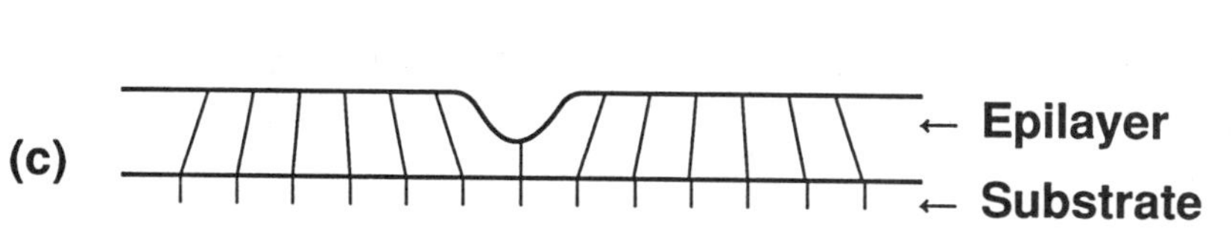


Figure 3. Schematic of the initial growth mechanism of a HgCdTe epilayer on a CdTe substrate.

Figure 4a shows the misfit dislocation lines in a (111) abrupt heterojunction [4]. The lines lie in the (111) plane in the $<\bar{1}10>$, $<01\bar{1}>$, and $<10\bar{1}>$ directions. As shown in Fig. 2, most dislocations are distributed under the dents. When dislocations under the dents are misfit dislocations, misfit dislocation lines are thought to be in bundles and to compose the network as shown Fig. 4b. From this, we consider that the growth pattern reflects the network pattern of misfit dislocation lines.

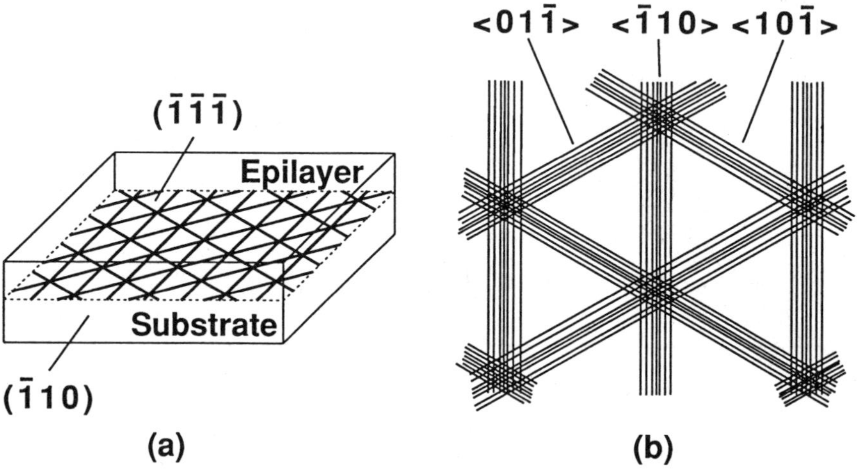


Figure 4. Schematic of misfit dislocation lines composing a planar network in a (111) heterojunction.

To confirm the influence of lattice mismatch on the growth pattern, we investigated the pattern of thin epilayers grown on lattice-matched CdZnTe substrates and lattice-mismatched CdZnTe substrates with excess Zn concentrations. Figure 5 shows the surface of 3 μm-thick epilayers. The surface of epilayers grown on the CdZnTe substrate with a lattice mismatch of 0.05% shows a mirror-like flat surface. The surface of epilayers grown on the CdZnTe substrate with a lattice mismatch of 0.3% shows a finer growth pattern than that grown on the CdTe substrate. This pattern is consistent with the network pattern of misfit dislocation lines as shown in Fig. 4. We confirmed that the dislocations under the dents are misfit dislocations, and our model interpreting the growth pattern of thin epilayers is correct.

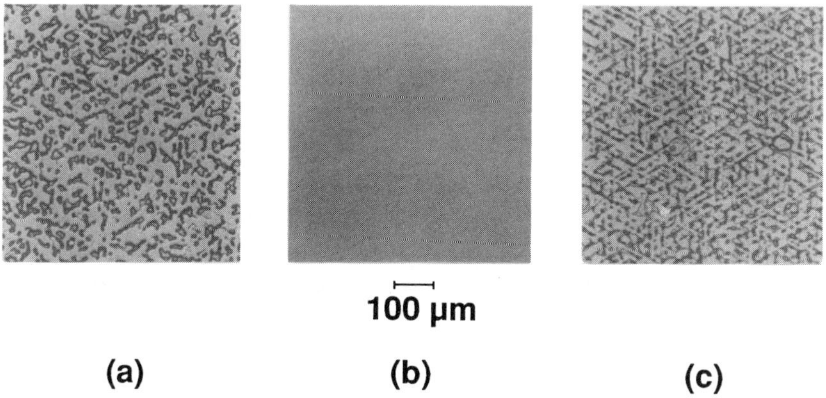


Figure 5. Photomicrographs of the surface pattern of 3-μm-thick HgCdTe epilayers grown on substrates with following lattice mismatch: (a) CdTe ($\Delta a/a$ = 0.2%), (b) $Cd_{0.97}Zn_{0.03}Te$ ($\Delta a/a$ = 0.05%), (c) $Cd_{0.91}Zn_{0.09}Te$ ($\Delta a/a$ = 0.3%).

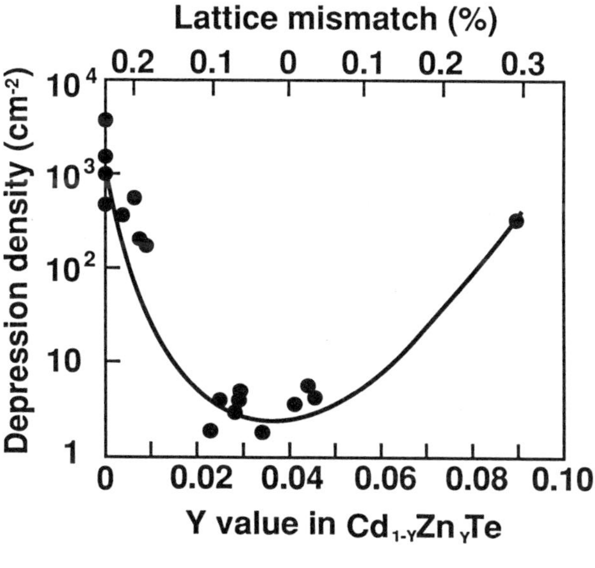


Figure 6. Dependence of the depression density on the lattice mismatch for 30-μm-thick epilayers grown on CdTe and CdZnTe substrates.

The influence of lattice mismatch on surface defects for thick epilayers was also investigated. Figure 6 shows the effect lattice mismatch has on the depression density of 30-μm-thick epilayers. The depression density of epilayers on lattice-matched CdZnTe substrates is smaller than that of epilayers on CdTe. This tendency is similar to the results reported by Tranchart et al. [5]. The depression density for thick epilayers on CdTe is about $10^3$ $cm^{-2}$, which is two orders of magnitude smaller than the dent density of thin epilayers. We therefore conclude that as an epilayer becomes thicker, most of the dents disappear. However, some dents remain as depressions.

## SUMMARY

We found that lattice-mismatch stresses strongly influence the growth pattern of thin epilayers. Epilayers grown on substrates with a lattice mismatch of more than 0.2% show island-like growth, aligned with the network of the misfit dislocation lines. Thin epilayers grown on substrates with a lattice mismatch of less than 0.05% have mirror-like flat surfaces.

## REFFERENCES

[1] T.Maekawa, T. Saito, M. Yoshikawa, and H. Takigawa, Mat. Res. Symp. Proc. 56, 109 (1986).
[2] K. Nakagawa, K. Maeda, and S. Takeuchi, Appl. Phys. Lett. 34, 574 (1979).
[3] H. Takigawa, M. Yoshikawa, M. Ito, and K. Maruyama, Mat. Res. Symp. Proc. 37, 97 (1984).
[4] W. G. Oldham and A. G. Milnes, Solid-State Electron. 7, 153 (1964).
[5] J. C. Tranchart, B. Latorre, C. Foucher, and Y. Le Gouge, J. Crystal Growth 72, 468 (1985).

# Stress in Titanium disilicide layers, during and after formation.

J.F. Jongste, F.E. Prins, G.C.A.M. Janssen and S. Radelaar,
Centre for Submicron Technology, Delft University of Technology, P.O. Box 5046, 2600 GA Delft, the Netherlands.

## INTRODUCTION

During and after formation of a thin layer of titanium disilicide ($TiSi_2$) on a silicon substrate stress is caused in several ways: Intrinsic stresses are due to the deposition process or to phase transformations and grain growth of the deposited material. Extrinsic stresses are caused by thermal effects: the difference in linear thermal expansion coefficients of the layer and the substrate respectively. Problems related to stresses can occur in semiconductor device fabrication. Stresses can deteriorate gate oxides in MOSFETs and can cause cracks in interconnect lines. Also, focusing problems in lithographic steps can occur because of wafer warpage. In this paper some examples of the different types of stress that can occur are shown and discussed. Both multilayer and self aligned Ti-Si samples have been studied: The advantage of the use of Ti-Si multilayers to produce $TiSi_2$ is that diffusion has to proceed only over a short distance i.e., the multilayer period. So the annealing time can be short. In the self aligned silicidation process, where a layer of a titanium layer on top of a silicon substrate is annealed, the diffusion length is equal to the thickness of the Ti layer. Because longer annealing times are needed, the latter type is used to monitor stress during formation.

## STRESSES

Intrinsic stress may arise from various processes taking place in the deposited layer. Thornton [1] discussed a qualitative structure zone model to describe stress development during sputter deposition. Depending on the deposition conditions (i.e. pressure, substrate temperature, amount of entrapped gas, ion bombardment during sputtering) the stress in the deposited layer can be tensile or compressive. Another mechanism is the change of volume caused by phase transformations of the deposited material. If the layer can not adapt to its new dimensions because of low atomic mobility, stress will result.

Extrinsic thermal stress is caused by elastic deformation due to a difference in linear thermal expansion coefficients of the layer and the substrate. The thermal stress in a $TiSi_2$ layer, formed on a silicon substrate at an elevated temperature, is given by Eq.(1):

$$\sigma_{//} = \frac{E_l}{1-\nu_l} (\alpha_l - \alpha_s) \Delta T \qquad (1)$$

with $E_l$ : Young's modulus of the layer,
$\nu_l$ : Poisson's ratio of the layer,
$\alpha_l, \alpha_s$: the thermal expansion coefficients of layer and substrate respectively
$\Delta T$ : temperature difference between annealing temperature and observation temperature.

## MEASUREMENT OF STRESS

In 1909 Stoney [2] derived an equation (2) describing the stress in thin films, parallel to the substrate:

$$\sigma_{//} = \frac{E_s}{6(1-\nu_s)} \frac{t_s^2}{t_l} \frac{1}{R} \qquad (2)$$

With $E_S$ : Young's modulus of the substrate,
$\nu_S$ : Poisson's ratio of the substrate,
$t_l$, $t_S$: thickness of layer and substrate
R : the radius of curvature of the substrate.

From measurements of the thickness of the layer and the radius of curvature of the substrate the stress can be calculated. Several methods exist to determine the radius: With Fizeau interferometry [3] one measures the curvature optically by observing the number of fringes resulting from the interference of monochromatic light between the sample and an optical flat. With this method anisotropy of the stress can easily be determined by analyzing the shape of the fringes.

Another method uses a dual parallel laser beam to measure wafer curvature (Teal and Murarka [4]). The laser beams are aimed at the wafer. The reflected laser beams are projected on a screen at a known distance from the wafer. From the relative positions of the spots on the screen the radius is calculated.

## EXPERIMENTAL

Titanium disilicide layers have been produced from Ti-Si multilayers as well as from titanium layers on monocrystalline silicon. The multilayers with a thickness of 200 nm have been produced by alternating sputter deposition. The Ti layers were sputtered to a thickness of 110 nm. Sputtering was done in a Leybold Heraeus Z550. Prior to sputtering the base pressure was below $4 \times 10^{-5}$ Pa for all deposition runs. Argon was used as sputtering gas at an operating pressure of 0.5 Pa. Samples were annealed in a conventional furnace or a rapid thermal processor (RTP, AG Associates Heatpulse 410). The ambient was argon or nitrogen. The curvature of samples annealed in the tube furnace was measured in situ by the laser beam method. The curvature of all samples was measured ex situ by Fizeau interferometry.

## RESULTS AND DISCUSSION

Stress after deposition.

Fig.1 shows the stress in a deposited multilayer structure as a function of the multilayer period. By varying the period the stress can be varied between 0.1 GPa (tensile) and -0.4 GPa (compressive). The data are described by a model of a layered structure in which the formation of an interfacial layer at each Ti-Si interface is assumed (Fig.2). The interfacial layer is formed by mixing the Ti-Si interface during the deposition of a new Ti or Si layer. An extensive description of this model is given by Wessels et al. [5]. The stress $\sigma_{//}$ in the multilayer structure is equal to the weighted average of the stresses $\sigma_{Ti}$, $\sigma_{Si}$ and $\sigma_{Int}$ in the titanium, silicon and interfacial layer:

$$\sigma_{//} = ( \sigma_{Ti}\, t_{Ti} + \sigma_{Si}\, t_{Si} + \sigma_{Int}\, t_{Int} ) / \delta \qquad (3)$$

with $\delta$: the multilayer period,
$t_{Ti}$, $t_{Si}$ and $t_{Int}$: the thickness of the layers.

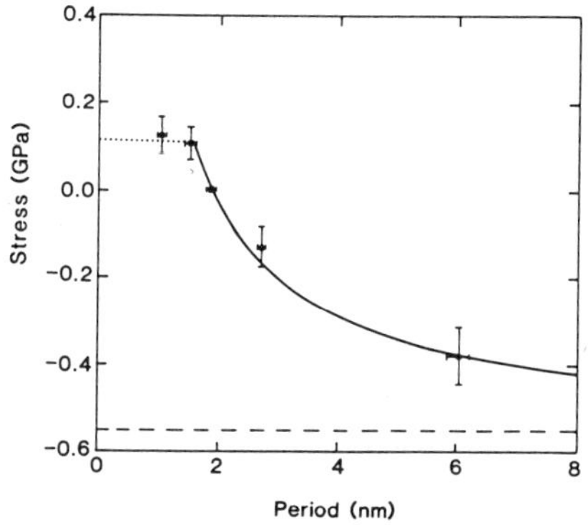


Fig.1 Stress in a Ti-Si multilayer as a function of the multlilayer period

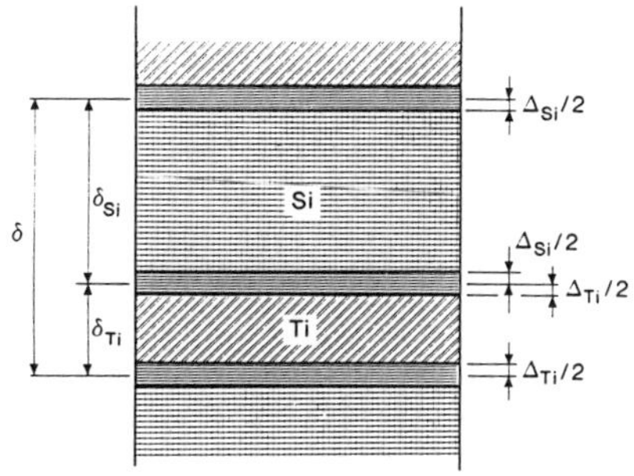


Fig.2 The layered structure model.

This model is shown in Fig.1 as the drawn curve.The limits of the model are also shown: for a multilayer period smaller than the thickness of the interfacial layer no separate Ti or Si layers exist, only the mixed layer. Accordingly the stress will remain constant for such small multilayer periods as shown by the dotted line. From the breaking down of the model for small periods the stress in the interfacial layer is estimated to be 0.11 ±0.04 GPa and its thickness 1.2 ± 0.2 nm. The other limit occurs when the interfacial layer is small compared to the multilayer period as shown by the dashed line. In this limit (-0.55 ±0.07 GPa) the observed results agree well with the weighted average of independently measured stresses in Ti and Si layers (-0.60 ± 0.03 GPa).

Determination of Poisson's ratio of titanium disilicide.

Multilayer samples were annealed at 650°C in the RTP for 10 s. The stress in the samples was measured by Fizeau interferometry. Also the strain in the $TiSi_2$ films was measured by X-ray diffraction using the $\sin^2\Psi$ method. In this method the difference in strain perpendicular and parallel to the layer ($\varepsilon_{//} - \varepsilon_{\perp}$) is determined. The relations between ($\sigma_{//}$ and $\varepsilon_{//}$) and ($\sigma_{//}$ and $\varepsilon_{\perp}$) given by Timoshenko [6] are easily combined to

$$\nu_L = \frac{E\ (\varepsilon_{//} - \varepsilon_{\perp})}{\sigma_{//}} - 1 \qquad (4)$$

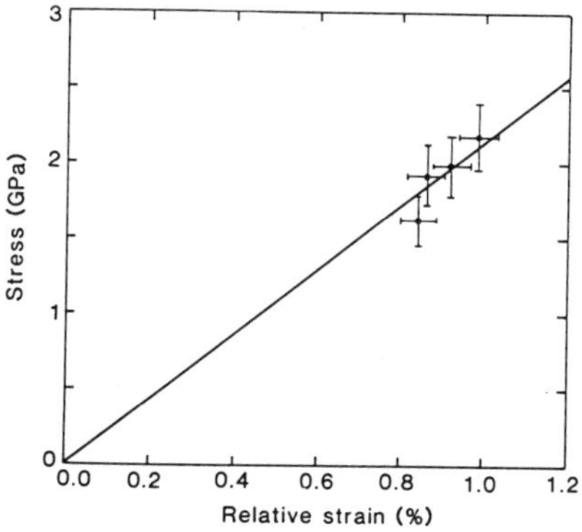


Fig.3 Plot of stress versus strain.

The results of the stress and strain measurements are shown in Fig.3. Using the value of E for bulk $TiSi_2$ (259 GPa), $\nu_L$ is determined from the plot to be 0.22 (±0.07).

Stress during formation of titanium disilicide.

A series of heat treatments of titanium on monocrystalline silicon has been performed to monitor stress during formation of the disilicide. The samples were annealed in the RTP at 690°C during 1, 4, 9, 16 and 25 seconds in an argon ambient. From X-ray diffraction experiments on these partially reacted samples it is found that $Ti_5Si_3$, $TiSi_2$ (C49) and residual Ti are present. No $TiSi_2$ (C54) is found. Using Auger measurements the concentration profiles as a function of depth of the samples are observed. The results show that even after 25 s annealing the reaction was still not completed. Fig.4 shows the Auger depth profile for a sample annealed for 16 s. The layered structure is clearly seen: the surface layer consists of Ti in which approximately 30 % oxygen is dissolved.

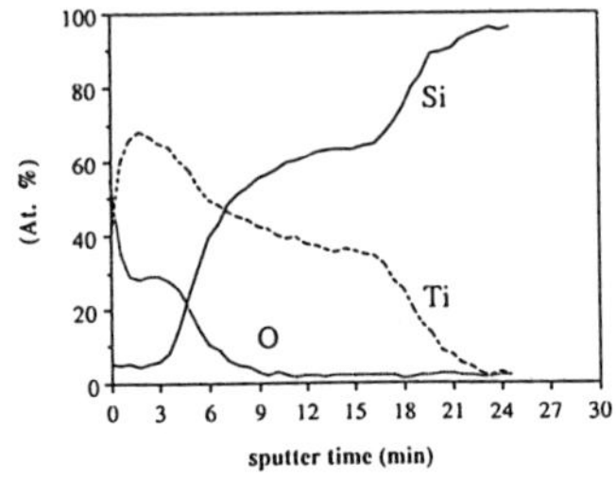


Fig.4 Auger depth profile after annealing at 690 °C for 16 s.

Below this oxygen rich titanium layer the $TiSi_2$ (C49) layer is found. Eq.(5) shows the relation between the stress and the thickness of the Ti and $TiSi_2$ layers:

$$\sigma_{//}\, t_{tot} = \sigma_{Ti}\, t_{Ti} + \sigma_{TiSi2}\, t_{TiSi2} \qquad (5)$$

with $t_{Ti}$, $t_{TiSi2}$ : the thickness of the Ti and $TiSi_2$ layers. In Fig.5 $\sigma_{//}\frac{t_{tot}}{t_{Ti}}$ versus $\frac{t_{TiSi2}}{t_{Ti}}$ is plotted. From a least squares fit of this plot the values of the stress in Ti and $TiSi_2$ (C49) were determined: $\sigma_{Ti}$ = 1.5 (±0.7) GPa and $\sigma_{TiSi2}$ = 2.8 (±0.6) GPa.

Stress during isochronal annealing.

Titanium layers (110 nm) on monocrystalline silicon were heated at a constant rate of 5K/min in a tube furnace in an argon ambient. The wafer curvature was measured in situ by the laser beam method. Because the thickness of the layer changes during heating, it is impossible to calculate the stress in the layer from the wafer curvature directly, using Eq.(2). However the curvature is inversely proportional to the stress, so plotting 1/R versus temperature will show stress development (Fig.6). Below 400°C 1/R has a negative value, meaning a compressive stress is present. Above 400°C three abrupt (reproducible) changes of the curvature are

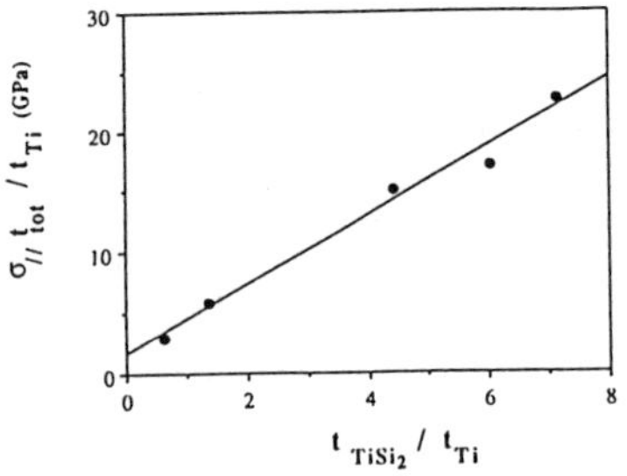


Fig.5. Fit of the stress during formation of $TiSi_2$ (C49) using Eq.5.

observed at 440°, 570° and 660°C. From X-ray diffraction experiments the phases that are present at these temperatures have been identified. At 440°C only crystalline Ti is observed. At 570°C the layer consists mainly of titanium monosilicide (TiSi), also a small amount of C49 $TiSi_2$ is observed. In the sample annealed up to 660°C only titanium disilicide in the stable C54 structure is present. It is concluded that the observed changes of curvature are caused by phase transformations. However considering the complex simultaneous behaviour of three thermally activated processes: diffusion, phase transformations and stress relaxation, it is very difficult to explain these results quantitatively in more detail. Further investigations will be necessary.

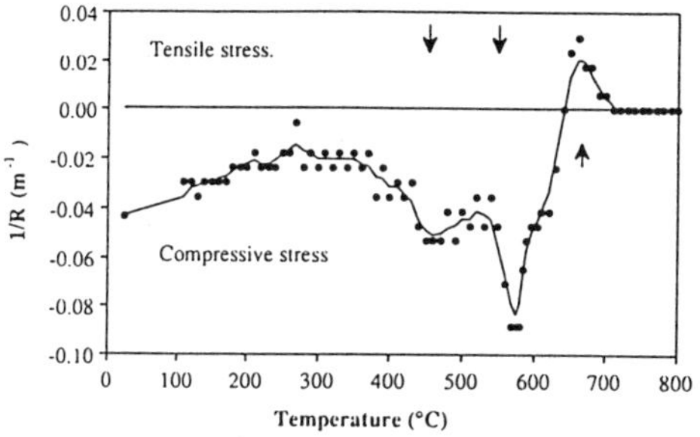


Fig.6 Wafer curvature (1/R) during isochronal annealing (Heating rate: 5°/ min)

<u>Stress after formation.</u>

Self aligned samples were annealed in a tube furnace or a RTP at 800°C. In all samples $TiSi_2$ (C54) was formed. The stress of the samples, annealed in the furnace, was measured in situ by the laser beam method at several temperatures during cooling after a heat treatment at 800°C for 30 min. The stress as a function of temperature is shown in Fig.7. It is noticed that the stress increases linearly with decreasing temperature as is predicted by Eq.(1). It is concluded that for this type of samples the stress is completely accounted for by the difference in linear thermal expansion coefficients of the layer and the substrate

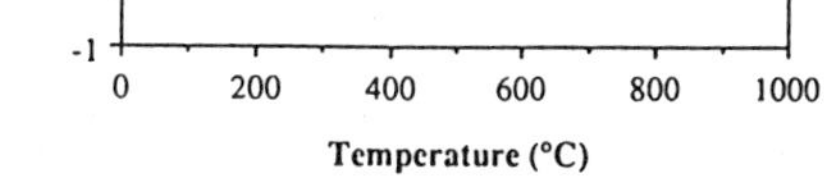


Fig.7 Stress as a function of temperature after formation of titanium disilicide at 800°C

by the difference in linear thermal expansion coefficients of the layer and the substrate (within experimental error). At room temperature $\sigma_{//} = 2.6\ (\pm 0.1)$ GPa. The least squares fit of the data yields a slope of -0.0034 GPa/K. Using the values of $E_l = 259$ GPa, $\nu_l = 0.22$, $\alpha_{Si} = 3.8 \times 10^{-6}$/K a value of $\alpha_l = 13.4 \times 10^{-6}$/K is found. According to Touloukian [7] $\alpha_l = 10.5 \times 10^{-6}$/K, measured for sintered titanium disilicide. Because of the preparation technique the cited value is believed to be in error.

## CONCLUSIONS

During and after formation stress occurs in the titanium disilicide layer. This stress is intrinsic as well as extrinsic.

In the isochronal stress measurements we have shown that the stress in the layer changes at elevated temperatures because of intrinsic causes (i.e. diffusion, phase transformations and stress relaxation).

After the self aligned formation of $TiSi_2$ (C49) at 690°C the stress in the residual Ti layer is found to be 1.5 (±0.7) GPa, and in the $TiSi_2$ (C49) layer 2.8 (±0.6) GPa.

The room temperature stress in C54 $TiSi_2$ after annealing at 800°C is 2.6 (±0.1) GPa, which can be fully accounted for by the difference in linear thermal expansion coefficients of the layer and the substrate.

## ACKNOWLEDGEMENTS

The authors would like to thank O.B. Loopstra (Laboratory of Metallurgy, University of Technology Delft) for the X-ray diffraction experiments.

This work is part of the research program of the "Stichting voor Fundamenteel Onderzoek der Materie (FOM)", which is financially supported by the "Nederlandse Organisatie voor Wetenschappelijk Onderzoek (NWO)". Financial support of the "Innovatie gericht onderzoeksprogramma voor de IC-technologie (IOP-IC)" is gratefully acknowledged.

## REFERENCES

[1]. J.A. Thornton, Deposition Technologies for Films and Coatings, R.F. Bunshah (ed.), Noyes Publications, New Jersey 1982.
[2]. G.G. Stoney, Proc. Royal Society London, A82, p172, 1909.
[3]. W.H. Steel, Interferometry, Cambridge Monographs on Physics, (Cambridge University Press, Cambridge, 1967), p211.
[4]. V.L. Teal and S.P. Murarka, J. Appl. Phys. **61**, 5038 (1987).
[5]. P.J.J. Wessels , J.F. Jongste, G.C.A.M. Janssen, A.L. Mulder, S.Radelaar and O.B. Loopstra, J. Appl. Phys. **63**, 4979 (1988).
[6]. S.P. Timoshenko and J.N. Goodier, Theory of Elasticity, 3rd ed., (McGraw-Hill, Singapore, 1982), p.8.
[7]. Y.S. Touloukian, R.K. Kirby, R.E. Taylor and T.Y.R. Lee, Thermal Expansion of Nonmetallic Solids, Thermophysical properties of matter, vol. 13, (IFI/Plenum, New York, 1977), p.1212.

# VISCOELASTIC STRESS RELIEF IN PATTERNED SILICON-ON-INSULATOR STRUCTURES

Theodore J. Letavic, Edward W. Maby and Ronald J. Gutmann, Department of Electrical, Computer and Systems Engineering and Center for Integrated Electronics, Rensselaer Polytechnic Institute, Troy, NY 12180.

## ABSTRACT

A high-temperature viscoelastic stress relief technique has been investigated as a means for reducing in-plane stress encountered during zone-melt recrystallization of patterned silicon-on-insulator structures. This technique incorporates a phosphosilicate glass layer between the silicon film and the insulating substrate to provide a viscous flow mechanism for stress relief within the composite structure. The stress relaxation can be qualitatively described by a mechanical model which couples thermal expansion and viscoelastic flow. The model predicts the time constant for stress relief at high temperatures as a function of pattern size, and the results are useful as a design aid for zone-melt recrystallization experiments.

## INTRODUCTION AND TECHNICAL BACKGROUND

The zone-melt recrystallization (ZMR) process has been extensively utilized as a means for obtaining device-quality semiconductor films on an insulating substrate. Samples which are subjected to the ZMR process are layered composites which are fabricated by the sequential deposition of thin surface films onto a thick insulating substrate. In most studies, the surface film to be recrystallized has been polycrystalline silicon, and substrates such as oxidized silicon [1], fused silica and sapphire [2], alumina [3], and various glasses [4,5] have been employed. ZMR of silicon is necessarily a high-temperature process in which the entire composite is uniformly subjected to temperatures just below the silicon melting point.

High-temperature processing of composite structures consisting of materials with different thermophysical properties can lead to excessive stress levels within the constituent films. As the temperature of a layered composite is uniformly changed, in-plane forces within the layers are generated to accommodate elastic strain due to the requirement of point-to-point coherency between adjacent materials. Mechanical failure becomes probable when the magnitude of in-plane stress approaches one tenth of the shear modulus in any one of the films. Observed failure modes during silicon ZMR over thermally mismatched substrates relate to either the substrate (plastic deformation, thermal spalling and fracture) or the overlying thin films (tensile failure and delamination).

In order to provide physical insight into the observed failure modes and indicate directions for future experiments, in-plane stresses have been evaluated using a model based upon linear elasticity theory. A cross-sectional view of a typical silicon-on-alumina composite is shown in Fig. 1, and the analysis of this configuration is made subject to two assumptions. First, it is assumed that a simple uniaxial solution will approximate a (less tractable) biaxial solution in regions sufficiently removed from a free edge if Young's modulus is scaled by the factor $1/(1 - \gamma)$, where $\gamma$ is Poisson's ratio [6]. Second, it is assumed that for the case of thin surface films on a thick mismatched substrate, the in-plane stresses in the surface films are dominated by the substrate. The amount of force which one member of the composite can exert upon another due to differential thermal expansion (DTE) is directly proportional to the ratio of the thickness of the exerting member to the thickness of the entire composite [7].

The in-plane stress within the polysilicon film of Fig. 1 was calculated for the case of a temperature ramp which proceeds from room temperature to 1200 $^oC$. At this final temperature, the in-plane stress exceeds $10^{10}$ dynes/cm$^2$, a level which is sufficient for tensile failure. Composite bending can provide some stress relaxation, but this strain relief is nearly two orders of magnitude less than the thermally induced strain.

VISCOELASTIC STRESS RELIEF

The stress levels which apply to the system of Fig. 1 can be reduced at high temperature by incorporating an additional stress relief mechanism into the layered composite. This is accomplished by adding phosphorus to the the silicon dioxide which separates the polycrystalline silicon from the alumina substrate. The viscosity of phosphosilicate glass (PSG) is concentration and temperature dependent, and phosphorus concentrations on the order of 4-10 wt % are sufficient to cause significant reduction of viscosity at a given temperature. Phosphorus outdiffusion into the upper silicon layer can be prevented by adding a thin (typically 30-nm-thick) silicon nitride diffusion barrier. This thin silicon nitride layer is not included explicitly, as the effect on the model is negligible.

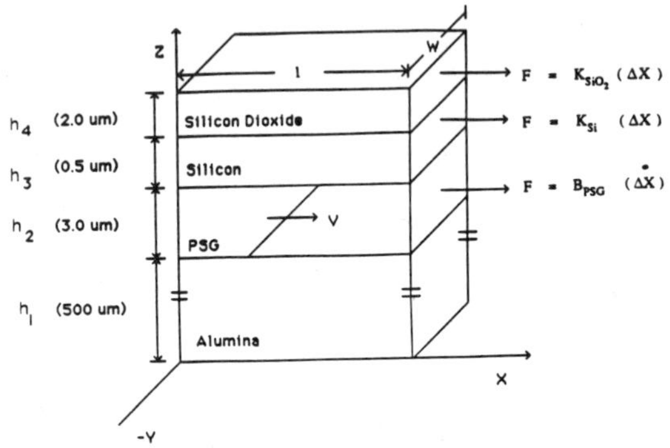


Fig. 1. A geometric model of a patterned silicon-on-insulator structure for mechanical modeling of stress relief which illustrates the relation of forces to displacements within the surface films.

To quantify the stress relief obtained by the incorporation of a PSG layer, the composite system of Fig. 1 is analyzed in terms of a simple lumped mechanical model. Figure 2a shows a spring, spring-dashpot and dashpot series representation which is used to model elastic, anelastic and viscous response, respectively, in each film. The complete model for the composite consists of four of these subnetworks connected in parallel. It can be simplified to take the form of the network shown in Fig. 2b by determining the dominant mode of temperature-induced deformation for each film and by specifying a general form of strain input into the mechanical system, as discussed below.

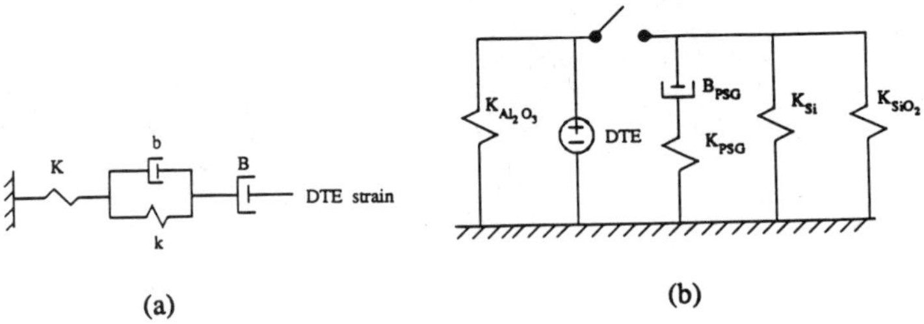


Fig. 2. (a) A general four-element model for material deformation which includes elastic, anelastic, and viscous behavior for an individual film, and (b) a simplified mechanical model for the calculation of in-plane stresses in a silicon-on-alumina composite.

Anelastic behavior such as grain-boundary sliding and diffusion effects are generally characterized by creep rates which are long compared to the duration of a typical ZMR experiment, and this behavior is ignored for all films. The uppermost (undoped) silicon dioxide film, the silicon film and the alumina substrate do not exhibit viscous behavior within the temperature range considered, so a single spring is used to represent a purely elastic thermal response. However, as a Maxwellian solid, the thermal response of PSG is dominated by elastic behavior at low temperatures and viscous behavior at high temperatures, so both a spring and a dashpot are required.

The mechanism whereby strain is introduced into the lumped-element model of Fig. 2b is dominated by the substrate, and this is represented by a displacement source with approximate value $(\alpha_{Al2O3} - \alpha_{Si})\Delta T\ l$, where $\alpha_{Al2O3}$ and $\alpha_{Si}$ are temperature-independent coefficients of thermal expansion for alumina and silicon, respectively, $\Delta T$ is the temperature change relative to room temperature and l is the film length. The coupling between this source and the thin films is temperature dependent. At low temperatures, the PSG exhibits elastic behavior, and the temperature-induced strain is transmitted across the coherent alumina/PSG and PSG/silicon interfaces. This direct coupling applies when the switch shown in the model of Fig. 2b is closed. At high temperatures, the viscosity of the PSG can be sufficiently low to result in a rate of stress relief which is much greater than the rate of stress increase which would otherwise be introduced due to changing temperature. In this case, the alumina substrate and the silicon film are decoupled, and the switch is open. The model is not valid at intermediate temperatures because it does not include lumped elements which describe partial coupling.

## LUMPED-ELEMENT PARAMETERS

In order to apply the model of Fig. 2b, the various lumped-element parameters must be related to the composite geometry and the thermophysical properties of the constituent films. In an elastic film which is modeled as a spring, an applied force is proportional to displacement, and the constant of proportionality is the spring constant given by

$$K = \frac{E\ W\ h}{l}\ , \tag{1}$$

where E is the biaxial Young's modulus [8], [9], and l, w and h are the length, width and thickness, respectively, of the film. Young's modulus for any film under consideration does not vary significantly with temperature, and thermally induced variations of film dimensions are small (so that room-temperature dimensions are applicable). Thus, it can be assumed that the spring constants in the model of Fig. 2b are temperature independent.

It is more difficult to define a dashpot coefficient which applies to the PSG film. For the case of a viscous film with the dimensional parameters defined above, one can define a shear force through the relation

$$F = \eta \left(\frac{\delta v}{\delta z}\right) l\ W\ , \tag{2}$$

where $\eta$ is the viscosity and $(\partial v/\partial z)$ is the normal velocity gradient within the film. The velocity profile is assumed to be laminar, for film thicknesses much less than either lateral dimension, and linear, with a maximum relative velocity ($v_{max}$) at the PSG/silicon interface for the case of tensile stress in the silicon film. Thus,

$$\frac{\delta v}{\delta z} = \frac{v_{max}}{h}\ . \tag{3}$$

By convention, the force applied to a dashpot is proportional to the displacement velocity, and the constant of proportionality is the dashpot coefficient (B). An average viscous-flow velocity ($v_{max}/2$) is used to obtain

$$B = \frac{2\ \eta\ l\ W}{h}\ . \tag{4}$$

The viscosity used in this expression is strongly temperature dependent. For the case of simple glass-forming oxides, the viscosity in the sub-transition region fits an Arrhenius relation of the form

$$\eta = \eta_o \exp\left(\frac{Q}{RT}\right) \qquad (5)$$

where Q is an activation energy and $\eta_o$ is a pre-exponential factor. For the case of PSG with 8.7 wt-% P, Q and $\eta_o$ are 110 kcal/mole and 5.75 x $10^{-7}$ poise, respectively, and these values are based on the calculation of the fractional content of Si-O and P-O bonds in the PSG [10]. The accuracy of viscosity values which have been calculated in this manner have been verified by fusion-temperature experiments [11].

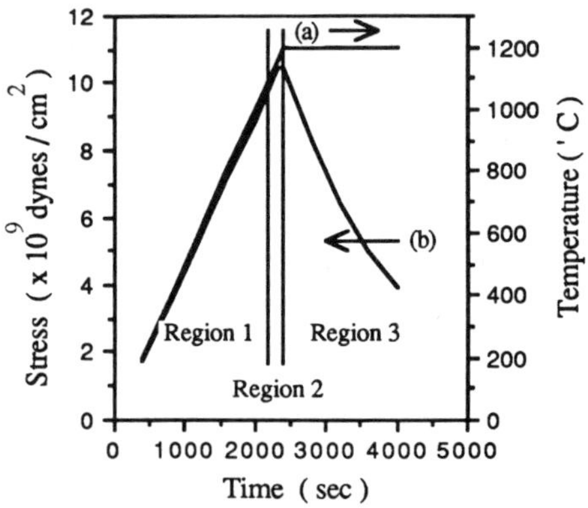


Fig. 3. (a) A plot of temperature versus time which is typical for ZMR of polysilicon on a thermally mismatched substrate, and (b) a plot of the in-plane stress within the polysilicon film versus ZMR process time.

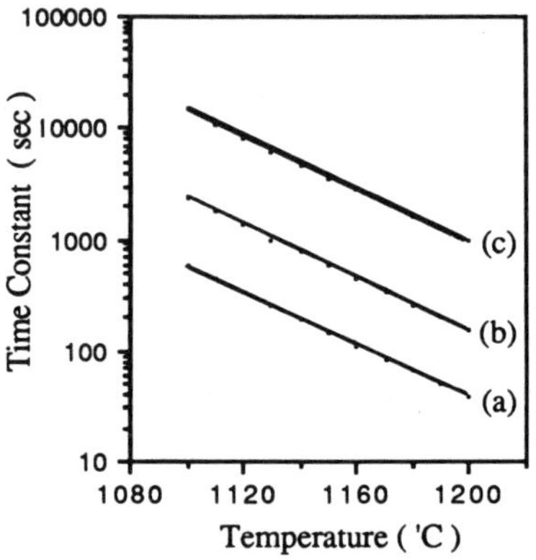


Fig. 4. A parametric plot of time constant for stress relief versus temperature for pattern dimensions ( l = W) of (a) 100, b) 200, and (c) 500 um.

DYNAMIC RESPONSE

Fig. 3 shows a hypothetical stress response under typical ZMR process conditions for the case of a particular composite system (l = w = 500 μm) which is governed by the model of Fig. 2b. This example features a temperature increase from 25 $^oC$ to 1200 $^oC$, with a "slow" ramp rate of 30 $^oC$ per minute, and a prolonged 1200 $^oC$- "soak" period. Three distinct solution regimes apply to Fig. 3. Region 1 applies to low temperatures when the PSG viscosity is very large, and the alumina substrate and the silicon film are directly coupled (switch closed). In this region, the stress in the silicon film follows the linear expansion of the substrate. Region 3 applies to the high- and constant-temperature soak period when the PSG viscosity is very small, and the alumina substrate and the silicon film are not coupled (switch open). In this region, the stress in the silicon film is governed by

$$\frac{dF}{dt} + \left(\frac{K_{eff}}{B_{PSG}}\right) F = 0 \quad , \qquad (6)$$

where

$$K_{eff} = \frac{K_{PSG}\,(\,K_{Si} + K_{SiO2}\,)}{K_{Si} + K_{SiO2} + K_{PSG}} \tag{7}$$

The solution to Eq. 6 is an exponential decay with time constant

$$\tau = \frac{2\,\eta\,l^2}{K_{eff}\,h_2} \tag{8}$$

Figure 4 shows this time constant as a function of temperature for several (square) island sizes with w = l.

Unfortunately, an accurate stress solution cannot be obtained in Region 2 since the model of Fig. 2b does not permit partial coupling between the alumina substrate and the silicon film. Thus, it is not possible to specify the "initial" stress condition at the onset of region 3. Indeed, if the stress relaxation is substantial in region 2, as expected if the ramp rate to a high soak temperature is very slow, then the levels of stress in region 3 will be insignificantly small.

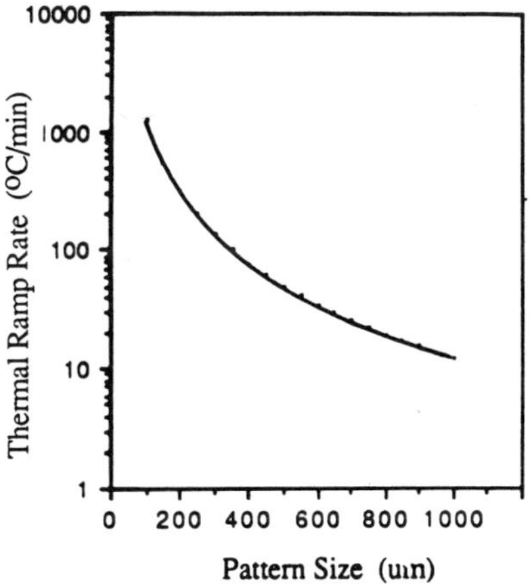


Fig. 5. Thermal ramp rate versus pattern size (l = w) for a maximum stress level of 1.5 x $10^{10}$ dynes/cm$^2$.

DISCUSSION

Notwithstanding these difficulties with region 2, one can estimate the maximum stress which is achieved therein by assuming that the maximum-stress condition applies when the rate of stress increase with direct substrate coupling is comparable to the rate of stress relaxation with an uncoupled substrate. If $\sigma_{max}$ represents the maximum stress, then over the time interval [t, t + Δt],

$$\sigma_{max}\left[\,1 - \exp\left(\frac{\Delta t}{\tau\,(\sigma_{max})}\right)\right] = \dot{\sigma}\,\Delta t \quad , \tag{9}$$

where σ is the rate of stress throughtout the temperature ramp period. If $\Delta t << \tau\,(\sigma_{max})$, the left-hand side of Eq. 8 can be expanded in a Taylor series, and, by dropping all but the lowest-order term, one obtains

$$\frac{\sigma_{max}}{\tau\,(\sigma_{max})} = \dot{\sigma} \tag{10}$$

Assuming a linear stress increase up to $\sigma_{max}$, Eqs. 7 - 10 yield the the result

$$m\, l^2 \cong \frac{T^* K_{eff}\, h^2}{2\eta_o} \exp\left[\frac{-Q}{R(T_o + T^*)}\right] , \qquad (11)$$

where m is the thermal ramp rate and

$$T^* = \frac{\sigma_{max}(1-\gamma_{Si})}{E_{Si}\,(\alpha_{Al2O3} - \alpha_{Si})} . \qquad (12)$$

The maximum stress which one would expect to observe will be smaller than the value used in Eq. 12 (due to the uncertainty in region 2 noted earlier), but the difference should be small since the viscosity of the PSG is changing rapidly in the vicinity of $\sigma_{max}$. Figure 5 shows thermal ramp rate versus (square) island pattern size for the case when $\sigma_{max}$ = 1.5 x $10^{10}$ dynes/cm$^2$.

The allowable island (or pattern) size is strongly dependent upon the activation energy Q and the allowable maximum stress level $\sigma_{max}$. The activation energy of 110 kcal/mole needs to be experimentally confirmed in the temperature regime of interest, and the maximum allowable stress level of 1.5 x $10^{10}$ dynes/cm$^2$ (approximately 0.01 times Young's modulus of silicon) used in Fig. 5 needs further verification. If the $\sigma_{max}$ value used in Fig. 5 is sufficiently conservative, then the 500- to 1000-μm island sizes depicted in the figure (for 40- to 10-$^o$C/min thermal ramp rates, respectively) appear promising for future ZMR experiments.

## ACKNOWLEDGEMENTS

This work was supported, in part, by the Rome Air Development Center, Air Force Systems Command (Hanscom), Contract #F19628-85-K-0038. One of the authors (TJL) has received both fellowship and summer support from the North American Philips Laboratory. The authors gratefully acknowledge their interactions with E. Arnold and others at Philips.

## REFERENCES

1. E. W. Maby, M. W. Geis, Y. L. Le Coz, D. J. Silversmith, R. W. Mountain and D. A. Antoniadis, Electron Device Lett. EDL-2, 241 (1981).
2. B-Y Tsaur, J. C. C. Fan and M. W. Geis, Appl. Phys. Lett. 40, 323 (1982).
3. T. J. Letavic, E. W. Maby and R. J. Gutmann, Silicon-on-Insulator and Buried Metals in Semiconductors, edited by J. C. Sturm, C. K. Chen, L. Pfeiffer and P. L. F. Hemment, Mater. Res. Soc. Symp. Proc. 107, 183 (1988).
4. R. A. Lemons, M. A. Bosch, A. H. Dayem, J. K. Grogan and P. M. Mankiewich, Appl. Phys. Lett. 40, 469 (1982).
5. A. Martinez, R. Pandya and E. Arnold, Silicon-on-Insulator and Buried Metals in Semiconductors, edited by J. C. Sturm, C. K. Chen, L. Pfeiffer and P. L. F. Hemment, Mater. Res. Soc. Symp. Proc. 107, 415 (1988).
6. R. Zeyfang, Solid-State Electron. 14, 1035 (1971).
7. G. H. Olsen and M. Ettenberg, Crystal Growth II, edited by C. H. L. Goodman (Plenum Press, New York, 1978) p. 49.
8. I. A. Blech and E. S. Meieran, J. Appl. Phys. 38, 2913 (1967).
9. R. J. Jaccodine and W. A. Schlegel, J. Appl. Phys. 37, 2429 (1966).
10. R. H. Doremus, Glass Science (Wiley, London, 1973), pp. 101-114.
11. C. R. Hammond, Phys. and Chem. of Glasses 19, 41 (1978).

# THERMALLY INDUCED STRESSES AND STRAINS IN LASER PROCESSING OF THIN FILMS

Judah A. Tuchman, Lisa P. Welsh and Irving P. Herman
Department of Applied Physics, Columbia University, New York, NY 10027

## ABSTRACT

The stresses and strains induced in thermal laser processing of substrates and thin films on substrates, are obtained in terms of single integrals by solving the thermoelastic equations using a Gaussian profile laser as the heating source. This analysis is applied to silicon thin films on fused silica and sapphire substrates. In part, this study shows that defects can form in the films because the stresses induced during high temperature laser processing of silicon and similar materials can exceed the yield stress under certain experimental conditions.

## INTRODUCTION

During laser processing of surfaces, thermal heating will cause the formation of stresses and strains throughout the heated material, which must be analyzed because of their potential long term effects. In the extreme case, damage to the material system may result by the formation of defects. However, even the creation of built-in-stresses and strains without defects can alter properties. For example, laser-induced elastic strains will perturb phonon frequencies, thereby complicating in-situ temperature measurements by Raman microprobe analysis [1].

In this paper, solutions of the thermoelastic equations for local heating of isotropic substrates by a $TEM_{00}$ Gaussian laser beam are presented. These solutions are extended to the case of thin films on substrates, as in Reference [1], to assess potential contributions to material damage by defect formation.

## SOLUTION – THERMOELASTIC EQUATIONS

The temperature rise due to a focused laser beam with cylindrical symmetry, has been obtained by Lax [2] by solving the heat flow equation under steady state conditions. In the particular case of a $TEM_{00}$ Gaussian beam with power $P_0$, beam waist $\omega$, and intensity

$$I(r, z = 0) = \frac{P_0}{\pi\omega^2} e^{-r^2/\omega^2} \qquad (1)$$

 

Table I Total strains and residual stresses in a substrate heated by a gaussian laser beam.

$$\varepsilon^s_{rr} = -\varepsilon_0 \int_0^\infty \left(J_0(\lambda R) - \frac{J_1(\lambda R)}{\lambda R}\right) F(\lambda) \left[A(\lambda)\exp(-\lambda Z) - B(\lambda)\exp(-WZ) - C(\lambda) Z \exp(-\lambda Z)\right] d\lambda$$

$$\varepsilon^s_{\theta\theta} = -\varepsilon_0 \int_0^\infty \frac{J_1(\lambda R)}{\lambda R} F(\lambda) \left[A(\lambda)\exp(-\lambda Z) - B(\lambda)\exp(-WZ) - C(\lambda) Z \exp(-\lambda Z)\right] d\lambda$$

$$\varepsilon^s_{zz} = \varepsilon_0 \int_0^\infty J_0(\lambda R) F(\lambda) \left[\left(\frac{2W}{\lambda} B(\lambda) - A(\lambda)\right)\exp(-\lambda Z) - \frac{W^2}{\lambda^2} B(\lambda)\exp(-WZ) - C(\lambda) Z \exp(-\lambda Z)\right] d\lambda$$

$$\sigma^s_{rr} = -\sigma_0 \int_0^\infty J_0(\lambda R) F(\lambda) \left[\left(\frac{2W}{\lambda} - 1\right) B(\lambda)\exp(-\lambda Z) - \frac{W^2}{\lambda^2} B(\lambda)\exp(-WZ) - C(\lambda) Z \exp(-\lambda Z)\right] d\lambda$$

$$+ \sigma_0 \int_0^\infty \frac{J_1(\lambda R)}{\lambda R} F(\lambda) \left[A(\lambda)\exp(-\lambda Z) - B(\lambda)\exp(-WZ) - C(\lambda) Z \exp(-\lambda Z)\right] d\lambda$$

$$\sigma^s_{\theta\theta} = \sigma_0 \int_0^\infty J_0(\lambda R) F(\lambda) \left[\left(A(\lambda) + B(\lambda)\left(1 - \frac{2W}{\lambda}\right)\right)\exp(-\lambda Z) + \frac{W^2 - \lambda^2}{\lambda^2} B(\lambda)\exp(-WZ)\right] d\lambda$$

$$- \sigma_0 \int_0^\infty \frac{J_1(\lambda R)}{\lambda R} F(\lambda) \left[A(\lambda)\exp(-\lambda Z) - B(\lambda)\exp(-WZ) - C(\lambda) Z \exp(-\lambda Z)\right] d\lambda$$

$$\sigma^s_{zz} = \sigma_0 \int_0^\infty J_0(\lambda R) F(\lambda) \left[B(\lambda)\left(\exp(-\lambda Z) - \exp(-WZ)\right) - C(\lambda) Z \exp(-\lambda Z)\right] d\lambda$$

$$\text{where } \varepsilon_0 = \frac{1+\nu}{1-\nu} \frac{2\alpha' W}{\sqrt{\pi}} T_{max}, \quad \sigma_0 = \frac{E}{1+\nu} \varepsilon_0$$

$$\text{and } A(\lambda) = \frac{\lambda^3 - (W+2\lambda)(1-\nu)(W-\lambda)^2}{(W^2-\lambda^2)^2}, \quad B(\lambda) = \frac{\lambda^3}{(W^2-\lambda^2)^2} \text{ and } C(\lambda) = \frac{\lambda^3}{(W+\lambda)(W^2-\lambda^2)}$$

Table II Total strains and residual stresses in a laser heated thin film on a substrate.

$$\varepsilon^f_{rr}(r) = \varepsilon^s_{rr}(r, z=0)$$

$$\varepsilon^f_{\theta\theta}(r) = \varepsilon^s_{\theta\theta}(r, z=0)$$

$$\varepsilon^f_{zz}(r) = \frac{\nu^{(f)}}{\nu^{(f)}-1} \left[\varepsilon^s_{rr}(r, z=0) + \varepsilon^s_{\theta\theta}(r, z=0)\right] + \frac{1+\nu^{(f)}}{1-\nu^{(f)}} \alpha'^{(f)} T(r, z=0)$$

$$\sigma^f_{rr}(r) = \frac{E}{(1-\nu^{(f)2})} \left(\varepsilon^s_{rr}(r, z=0) + \nu^{(f)} \varepsilon^s_{\theta\theta}(r, z=0) - (1+\nu^{(f)}) \alpha'^{(f)} T(r, z=0)\right)$$

$$\sigma^f_{\theta\theta}(r) = \frac{E}{(1-\nu^{(f)2})} \left(\nu^{(f)} \varepsilon^s_{rr}(r, z=0) + \varepsilon^s_{\theta\theta}(r, z=0) - (1+\nu^{(f)}) \alpha'^{(f)} T(r, z=0)\right)$$

$$\sigma^f_{zz}(r) = 0$$

incident on a semi-infinite elastic substrate, the resultant temperature rise takes the form

$$T(R, Z, W) = \frac{\alpha P_0(1-R_0)}{\pi K} \int_0^{\infty} J_0(\lambda R)\, F(\lambda)\, \frac{We^{-\lambda Z} - \lambda e^{-WZ}}{W^2-\lambda^2}\, d\lambda \tag{2}$$

with $\alpha$ the absorption coefficient, K the thermal conductivity, and $R_0$ the surface reflectivity, each assumed to be independent of temperature for a given substrate [2]. Here, $F(\lambda)$ is the Bessel transform of the laser lineshape f(R), so that

$$F(\lambda) = \int_0^{\infty} J_0(\lambda R)\, R\, e^{-R^2}\, dR = \frac{1}{2} e^{-(1/4)\lambda^2} \tag{3}$$

Use has also been made of the normalized coordinates $R = r/\omega$, $Z = z/\omega$, and $W = \alpha\omega$ . This temperature rise may now be included as the driving term in the thermoelastic equations, to yield the thermally induced stresses and strains.

The thermoelastic equations themselves are obtained from the stress-strain relations and from the equations of equilibrium for an isotropic system. Using stress functions $\Omega$ and $\Psi$ defined in terms of the stresses $\sigma$ and total strains $\varepsilon$ by Youngdahl [3], they take the form

$$\nabla^2\Omega = \frac{1}{2(1-\nu)} \left( -\frac{\partial^2\Psi}{\partial Z^2} + 2(1+\nu)\alpha'\omega^2 T \right) \tag{4}$$

$$\nabla^2\Psi = 0 \tag{5}$$

where $\alpha'$ is the thermal expansion coefficient, and $\nu$ is Poisson's ratio. Equations 4 and 5 are easily solved for the substrate [1] and expressed directly in terms of the residual stresses and total strains, where the total strain is the sum of the elastic strain and the thermal strain, yielding the relations in Table I. Note that the off-diagonal elements of the stress and strain tensors are identically zero for all but the rz components, which will also be ignored since they are very small [1].

These solutions may now be extended to the case of a thin film on a substrate. It is assumed that the thin film is sufficiently thin so that it does not perturb the stresses and strains in the substrate and that radial thermal conduction in the film is negligible relative to that in the substrate. Then, the resultant stresses and strains in the substrate are determined by linear superposition of two separate thermal driving terms, corresponding to laser absorption at the surface due to the thin film (of thickness d) and absorption within the substrate. For heating at the surface, the temperature rise of Equation (2) is used in the $W \rightarrow \infty$ limit with diminished effective incident power, $P_0^{eff} = P_0(1-e^{-\alpha_f d})$ . This expression ignores multiple reflections in the film, which could be included when necessary. For heating due to absorption in the substrate, the temperature rise of Equation (2) is used with $P_0^{eff} = P_0 (1 - R_1) e^{-\alpha_f d}$, where $R_1$ is the reflectivity at the film/substrate interface, again ignoring multiple reflections. From the solutions for the stresses and strains in the substrate obtained using both thermal heating terms, the stresses and strains in the thin film are now determined by matching boundary conditions to the substrate.

The planar strains in the film may be obtained from the continuity of planar total strains at the film/substrate interface. Film strain in the z direction is derived from the stress-free condition ($\sigma_{zz}^f(r) = 0$) for the film/air interface using the corresponding stress-strain relation. The remaining two diagonal film stresses are found by direct substitution of the other known quantities in the stress-strain relations.

The six solutions for stress and total strain within the film are given in Table II in terms of the quantities of Table I and Equation 2, where E is Young's modulus and where the s and f superscripts refer to the substrate and the film, respectively. The elastic strains $\bar{\varepsilon}$ may be obtained directly from the total strains by subtracting the thermal strain $\alpha'$ T(r, z=0) from each component. Note that since the film is assumed to be very thin, all physical quantities within it are independent of z .

The expressions in Table II for stress and strain within the film may be numerically integrated, with the results displayed in Figures 1-3 for a silicon film on sapphire, a silicon film on fused silica, and for a silicon substrate (or, equivalently, a silicon film on a silicon substrate). Only the $W \rightarrow \infty$ (W = 15) case has been shown, corresponding to total laser absorption within the thin film.

The three cases are plotted for different effective laser powers, chosen so that the ratio $[P_0^{eff}(1-R_0)]/\omega K_{substrate}$ was held constant. In particular, the ratio was chosen so that the maximum temperature rise, $T_{max}$ = T(R=0, Z=0, W=∞) = $[P_0^{eff}(1-R_0)]/(2\sqrt{\pi}\ \omega K_{substrate})$ determined from Equation (2), would correspond to heating the silicon thin film to the melting point temperature of 1420°C ($T_{total} = T + T_{ambient}$ = 1420°C). The other physical quantities which have been used are: $\nu = 0.42$, $E = 1.13 \times 10^{12}$ dyn/cm$^2$, $\alpha' = 4.0 \times 10^{-6}$/°C for silicon; $\nu = 0.17$, $E = 7.58 \times 10^{11}$ dyn/cm$^2$, $\alpha' = 5.5 \times 10^{-7}$/°C for fused silica; and $\nu = -0.02$, $E = 3.65 \times 10^{12}$ dyn/cm$^2$, $\alpha' = 8.6 \times 10^{-6}$/°C for sapphire, with all of these constants assumed to be independent of temperature.

As seen from the figures, the qualitative trends are similar for silicon thin films on either a silicon or fused silica substrate. However, results for silicon thin films on sapphire are quite different, due to the relatively large difference in the physical quantities for sapphire compared to the other two materials (particularly for Poisson's ratio).

Of course, since the laser increases the temperature everywhere within the film, the film expands in all directions. This is demonstrated by the positive values for the total strains everywhere. However, since the thin films are primarily influenced by the substrate, the relative magnitudes of the individual stress and strain components depend entirely on the particular

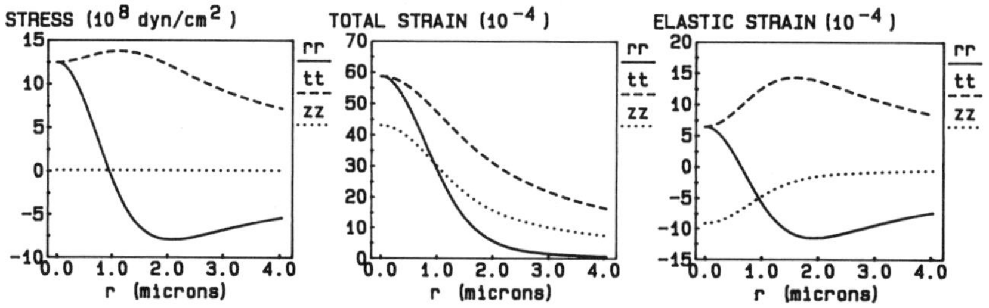


Figure 1. The residual stresses, total strains, and elastic strains are shown as a function of r for a silicon thin film on sapphire, using $\omega = 1$ μm, $T_{max} = 1400$ K, and W = 15. Note that rr, tt, and zz in the figure correspond to the rr, θθ, and zz components, respectively.

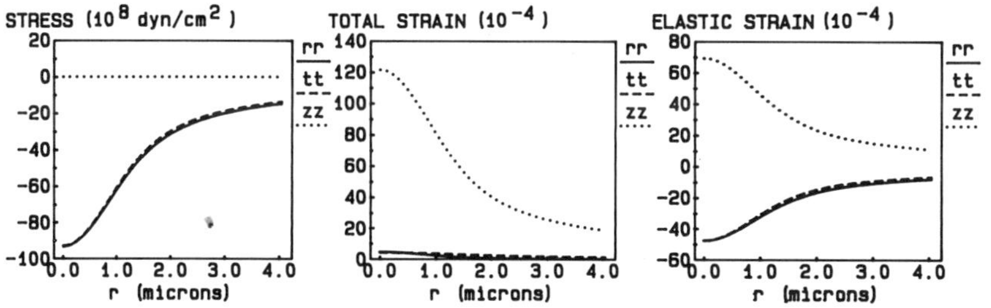


Figure 2. Same as Fig. 1 but for a silicon thin film on fused silica.

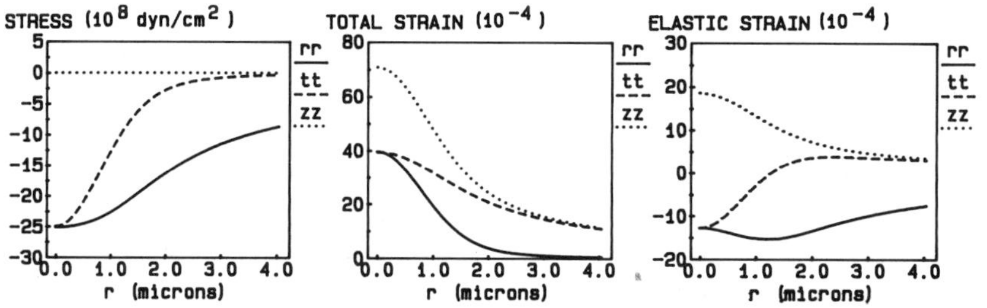


Figure 3. Same as Fig. 1 but for a silicon thin film on silicon. These results are identical to those obtained in Ref. [1] for a silicon substrate.

film/substrate system. As a result, a silicon film on sapphire is under tensile stress near beam center (positive residual stress) and has relatively less bowing (or displacement) in the vertical direction ($\varepsilon_{zz} < \varepsilon_{rr}, \varepsilon_{\theta\theta}$), since the thermal expansion coefficient of the substrate is greater than that of the film. For silicon on fused silica, on the other hand, the planar stress is compressive, and the strain is greater vertically than in the planar direction ($\varepsilon_{zz} > \varepsilon_{rr}, \varepsilon_{\theta\theta}$), since the thermal expansion coefficient of the substrate is here less than that of the film. For silicon on silicon, again the stress is compressive, and $\varepsilon_{zz} > \varepsilon_{rr}, \varepsilon_{\theta\theta}$ since expansion in the planar direction is inhibited by decreasing temperature in the substrate away from the beam center.

This analysis has assumed that the silicon films have no built-in stresses when under ambient conditions, with no laser incident. However, because silicon films are often deposited at elevated temperatures ($T_{dep} \sim 640\ ^oC$), adhering silicon films on sapphire will actually be under compressive stress at ambient temperature, while silicon films on fused silica will be under tensile stress. Qualitatively, this means that laser heating will counterbalance the built-in-stress of the system (which is compressive for silicon on sapphire) at low laser powers, and will generate stress in the opposite sense (tensile for silicon on sapphire as in Figure 1) only at higher laser powers. However, since the laser does not heat the surface uniformly, the cross-over from compressive to tensile stress (for silicon on sapphire) or tensile to compressive stress (for silicon on fused silica), will actually take place when the laser heats the substrate at r=0 to a temperature slightly higher than $T_{dep}$. A more refined treatment would require inclusion of the built-in stresses in solving the thermoelastic equations.

## DEFECT FORMATION

Defects will occur when laser heating leads to either tensile or compressive conditions if the magnitude of the stress exceeds certain yield conditions. Two types of damage are possible: that within the bulk of the film or substrate and that at the film/substrate interface. Only damage within the silicon thin film will be considered here.

Material damage will occur when the stress in the material exceeds a certain yield stress. For many materials, the yield stress is a function of the rate of strain [4], with

$$\sigma_E = C_0 \left(\frac{\partial |\bar{\varepsilon}|}{\partial t}\right)^{1/n} \exp\left(\frac{U}{nk_B T_{total}}\right) \tag{6}$$

where U is the activation energy for glide movement, and $C_0$ and n are constants. For the silicon thin films under consideration, U = 2.3 eV, n = 2.1 and $C_0 = 1.7 \times 10^5\ (\mathrm{dyn/cm^2})\mathrm{s}^{1/n}$, and for laser heating by a scanning Gaussian beam, with scan velocity $|v|$, then the dwell time of the laser over a distance of the spot size $\omega$ is $\frac{\omega}{v}$, so that $\left|\frac{\partial|\bar{\varepsilon}|}{\partial t}\right| \approx \frac{v}{\omega}|\bar{\varepsilon}|$.

In silicon, stress in excess of the yield stress will result in dislocations on the {111} slip planes in the <110> slip directions. The stresses within the silicon thin films may therefore be projected along the twelve combinations of slip directions on slip planes, for comparison with the yield stress. This analysis was performed in Reference [1] for the case of a silicon substrate (or equivalently a silicon film on a silicon substrate). It was demonstrated that for laser powers such as those considered here (where $T_{max} \sim 1400°C$) defects will form when the laser scan speed is less than ~ 10 mm/sec, corresponding to the fastest demonstrated speeds for direct laser writing.

Similar limiting conditions may be obtained for silicon thin films on sapphire and fused silica. In particular, it may be shown that for silicon on sapphire, the minimum possible scan speed with no defect formation is decreased by approximately a factor of two vis-a-vis that for Si on Si, whereas for silicon on fused silica this minimum is increased by more than a factor of four. Therefore damage is more likely during laser heating of Si films on fused silica and less likely for Si films on sapphire than for laser heating of Si substrates.

Of course, the scan velocities may be decreased without producing damage if the effective laser power is decreased, thereby decreasing $T_{max}$. Moreover, inclusion of built-in stresses will allow for even greater temperatures and even slower scan speeds with no defect formation, since built-in stresses and stresses due to laser heating have opposite signs, as stated above.

## CONCLUSIONS

The stresses and strains induced during laser processing of thin films on substrates have been evaluated in this study. Results indicate that in certain experimental regimes thermoelastic stresses may be large enough to induce material damage. More refined treatment of this problem could include built-in stresses, deviations from isotropic structure and perfect elasticity, temperature-dependent optical and thermal parameters, the relaxation of the ideal thin film assumptions, and a more detailed model of defect formation.

## ACKNOWLEDGMENT

This work was supported by the Office of Naval Research.

## REFERENCES

1. L. P. Welsh, J. A. Tuchman and I. P. Herman, J. Appl. Phys 64, 6274 (1988).

2. M. Lax, J. Appl. Phys. 48, 3919 (1977); Appl. Phys. Lett. 33, 786 (1978).

3. C. K. Youngdahl, Int. J. Engin. Sci. 7, 61 (1969).

4. H. Siethoff and P. Haasen, in Lattice Defects in Semiconductors, edited by P. R. Hasiguti (University Press, Tokyo, 1968), p. 491.

# THIN FILM STRESSES IN TiW/AlCuSi/TiW SANDWICH STRUCTURES

ITON WANG
Circuit Technology Group, Hewlett Packard Company, 5301 Stevens Creek Blvd., Santa Clara, CA 95051

## ABSTRACT

A highly reliable metallization for high current bipolar integrated circuits was developed with the aid of thin film stress analysis to optimize the process. The final structure was a sandwich layer of TiW and AlCuSi alloy. Film stresses were measured for individual TiW layers as well as for the sandwich layers of TiW/AlCuSi/TiW. In this paper, it will be shown how sputter deposition pressure can drastically affect film stress. For individual TiW layers, the film stress can be varied between compressive and highly tensile by adjusting the Argon pressure. Meanwhile, the AlCuSi layers were always tensile under the normal range of sputtering pressures. By controlling the individual film stresses, a sandwich layer can be designed so that the overall structure has a very low average stress. This work will also show how a TiW capping layer has a major effect on the final average stress of the sandwich structure.

## INTRODUCTION

In order to improve performance as well as to increase packing density, it has become necessary to use multilevel metallizations. The universal trend toward using diffusion barriers [1,2] and planarization techniques [3] has created a nightmare for the thin film engineer. Some very complex structures are being used to improve device performance and yield. Processes such as CVD tungsten [4] and metal pillars [5] are being developed for contact and via fills. Planarization schemes include oxide etch-back, polyimide, SOG, BPSG reflow, BSQ and hot substrate metal depositions. All these processes go through some temperature cycling and create an interfacial layer which can and usually does generate stress. If these stresses are not controlled, device degradation or failure can occur. It is therefore important to understand the potential problems that can be caused by thin film stresses.

The impact of the substrate on stress measurements [6] should not be overlooked. Ionic Systems did a study comparing prime silicon wafers with test or monitor grade wafers and found the measured values of stress to vary by an order of magnitude. Based on test results using a single point measurement at the center of the wafer, the following recommendations were made:

1. For the greatest accuracy and repeatability, only prime silicon wafers should be used for stress measurements.
2. Prime wafers that do not have initial readings in the main grouping should be discarded.

Other tests have shown that the crystal orientation of silicon wafers can have a major effect on the magnitude and polarity of stresses measured. This is not to imply that the quality of the wafer or its orientation affects film stress, but the accuracy of measurement depends on measurement techniques.

In systems such as the Canon LSF-500 or Ionic Systems' Stressgauge, the property being measured is the bow of the wafer or substrate. The deflection in the wafer is primarily due to three different phenomena [7]:

1. COEFFICIENT OF THERMAL EXPANSION MISMATCH: Most thin films are formed with the assistance of heat or produce heat as a side effect. As the substrate heats or cools, the mismatch in the coefficient of thermal expansion will provide a bow in the substrate.
2. INTERNAL FILM EFFECT: This effect is due to deposition conditions. Deposition parameters such as gas pressure and power density can affect the stoichoimetry of the deposited film, resulting in a net stress in the material.
3. EXTERNAL FILM EFFECT: This effect is due to a structural mismatch at the interface between the deposited film and substrate. Only epitaxial layers are able to have identical repeating structures to the substrate. This effect is primarily responsible for stress variations with film thickness. Films with a thickness less than about 1,000 Angstroms will have stresses due mostly to this effect. Thicker films will be affected more by the first two phenomena.

Studies have shown that deposition parameters have to be carefully considered in order to obtain meaningful test data. Even a thin oxide layer on the substrate can drastically affect the apparent stress of the deposited film. If at all possible, duplicate the actual structure being studied, including any pre-cleaning steps such as a chemical etch.

## EXPERIMENTAL PROCEDURE

The substrates used were 75mm silicon wafers lapped down to 10 mils. The initial curvature of the wafers was measured using a Canon LSF-500 surface flatness measurement system with an H-P 182A scope for determining the direction of bow. Thin films were then deposited in a Perkin-Elmer 4410 sputter deposition system using conditions shown in Table I.

TABLE I

SPUTTER DEPOSITION CONDITIONS FOR P-E 4410

| | |
|---|---|
| <u>BASIC</u> <u>CONDITIONS</u>: | |
| Base pressure | 5 x 10E-7 Torr |
| Table spacing | 2.5 " |
| Table rotation | 6 rpm |
| Process gas | UHP Argon |
| <u>TiW</u> <u>DEPOSITION</u>: | |
| DC Magnetron, Delta configuration | |
| Sputtering power | 1.50 KW |
| Sputter clean time | 5 min. |
| Deposition time | 2-5 min. |
| Deposition pressure | 6-10 mT |
| <u>Al/2%Cu/1%Si</u> <u>DEPOSITION</u>: | |
| DC Magnetron, Delta configuration | |
| Sputtering power | 9.00 KW |
| Sputter clean time | 5 min. |
| Deposition time | 4 min. |
| Deposition pressure | 4-10 mT |

Using conditions specified above, the various films deposited are shown in Table II:

TABLE II

FILMS DEPOSITED

1300 A TiW, pressure range 6-10 mTorr
7500 A AlCuSi, pressure range 4-10 mTorr
1300 A TiW/7500 A AlCuSi, 6 mTorr
1300 A TiW/7500 A AlCuSi/1300 A TiW, 6 mTorr
1300 A TiW/7500 A AlCuSi/667 A TiW, 6 mTorr

After deposition, the wafer curvature was measured and compared against the initial readings. The average deflection from the zero degree and 90 degree readings were calculated. Film stress was calculated using the method by Glang, Holmwood and Rosenfeld [8].

## RESULTS AND DISCUSSION

The wafer deflection as a function of Argon sputtering pressure is shown below. The deflection is measured at 1.27 cm from the center. It is the difference between the initial and final wafer curvature. Fig. 1 shows the results for a single layer of 1300 Angstrom TiW, while Fig. 2 is for a 7500 Angstrom layer of AlCuSi. Negative values of wafer deflection indicate films are in compression; positive values are for films in tension. For the range of sputtering pressures studied, the TiW films varied from compressive stress to tensile stress as the Argon pressure was increased. This corresponds to a compressive stress of 2.66E10 dyne/cm2 (at 6 mT) to a high tensile stress of 4.51E10 dyne/cm2 (at 10 mT). Note the extremely sharp transition from compressive to tensile stress at approximately 9 mT sputtering pressure. Process control would be very difficult if this were the operating pressure selected.

On the other hand, the 7500 Angstrom AlCuSi films were always under tensile stress over the Argon pressure range studied. The highest stress was at

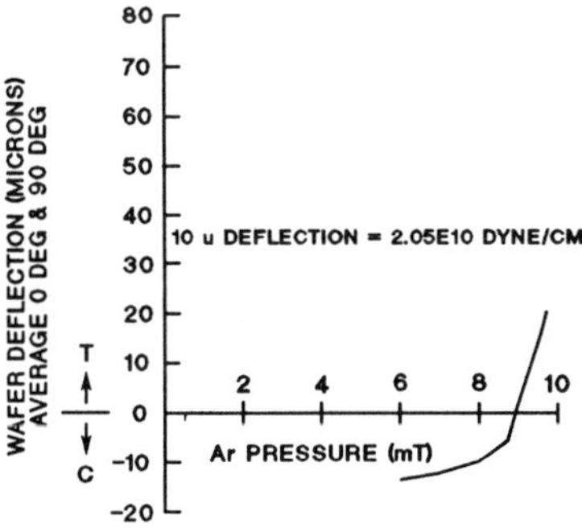


Fig. 1 Wafer deflection vs Ar pressure for 1300 A TiW layer.

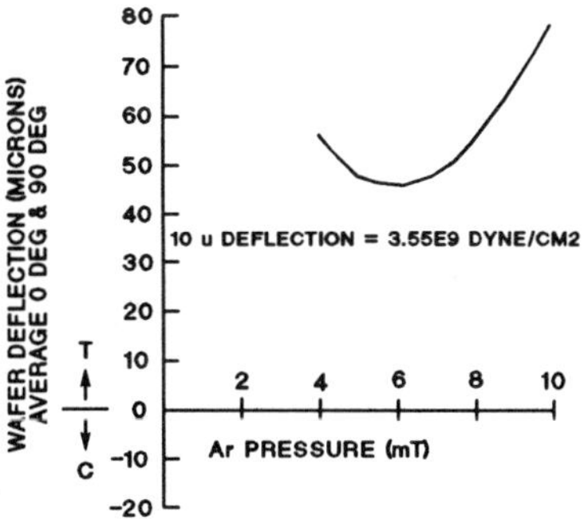


Fig. 2 Wafer deflection vs Ar pressure for 7500 A AlCuSi layer.

10 mTorr and corresponded to a value of 2.75E10 dyne/cm2. From Fig. 2, it can be seen that even at the optimum Argon pressure of 6 mTorr, the wafer deflection was 45 microns (1.60E10 dyne/cm2). The data from Fig. 1 and Fig. 2 indicate that the curves are relatively flat and process control should be easy. A test was run at 6 mTorr to see if the compressive stress of the TiW layer would cancel the tensile stress of the AlCuSi layer. A 2-layer structure with the bottom layer of 1300 Angstroms of TiW and top layer of 7500 Angstrom of AlCuSi was deposited. Fig. 3 shows the wafer deflection of this structure which represents an overall film stress of 1.56E10 dyne/cm2 tensile.

In another study, a sandwich layer of TiW/AlCuSi/TiW was deposited. The top layer of TiW was chosen to be 1300 Angstroms and 667 Angstroms, which corresponds to 1/2 and 1/4 wavelength of our uv exposure system. The bottom TiW layer was again 1300 Angstroms thick;the AlCuSi was 7500 Angstroms thick. The wafer deflection is shown in Fig. 4. When a TiW capping layer of 1300 Angstroms is used, the stress of the sandwich structure was 8.00E8 dyne/cm2 The structure has a tensile stress of 5.32E9 dyne/cm2 if the TiW cap thickness is reduced to 667 Angstroms. These results point to a major effect of the TiW capping layer on the average film stress of the sandwich structure. Even a thin 667 Angstrom cap reduced the stress of the composite structure by by more than a factor of three to 5.32E9 dyne/cm2 tensile. If the capping thickness is doubled (increasing the compressive stress of the cap), the sandwich stress changes signs and becomes compressive at 8.00E8 dyne/cm2. Even though the individual layers may have relatively high stresses, the average stress of the tri-layer film may be very low due to opposite and compensating stresses of the AlCuSi layer and the TiW layers.

Electromigration tests have been done on the sandwich structures and results indicate reliability which exceeds our GSS specifications. These tests were done on PECVD oxide passivated patterns with a 2 micron line width and a length of 1000 microns. The devices were annealed at 450 degrees C for 30 minutes. Test conditions were as follows: J = 2E6 Amps/cm2, T = 200 deg C and 220 deg C.

We have also looked at the reflectivity of the sandwich films and have found a reduction of 30-50 % compared to AlCuSi layers with no TiW caps. This lower reflectance aids photolithography considerably and is almost mandatory if one is working on sub-micron geometries. Tests have shown that hillock density is reduced considerably when a TiW cap is used.

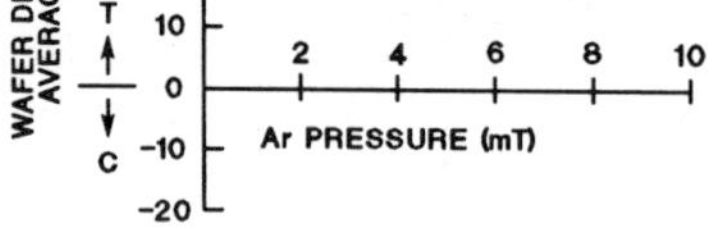


Fig. 3 Wafer deflection of a 2-layer structure of TiW/AlCuSi deposited at 6 mT Ar pressure.

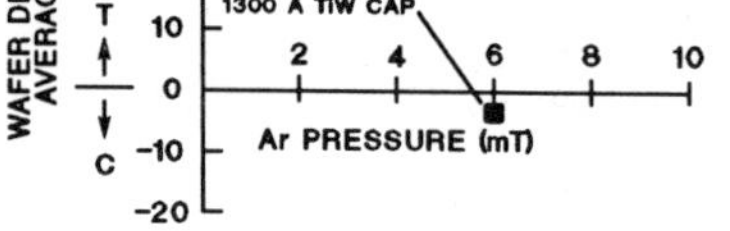


Fig. 4 Wafer deflection of a sandwich layer of TiW/AlCuSi/TiW with TiW capping layers of 667 A and 1300 A.

## CONCLUSIONS

A highly reliable TiW/AlCuSi/TiW sandwich structure has been developed with the aid of thin film stress analysis. It has been shown that film stress is a controllable process parameter and by optimizing each individual layer the average stress of a sandwich structure can be reduced significantly. This in turn reduces the wafer curvature and aids in photolithography where depth of field has to be considered. Hillock reduction and lower reflectance are also important side benefits of the metallization system. Although film stress is not the only important consideration in selecting an interconnect metal scheme, it should not be overlooked. If thin film stresses are controlled, reliability problems such as film cracking or stress-induced voids [9,10] can be avoided of greatly reduced.

## ACKNOWLEDGEMENTS

The author wishes to thank Falke Hennig for his support; Dick Coen and Dave Lopes for assistance with the deposition equipment.

## REFERENCES

[1] M.E. Thomas, M.P. Hartnett, J.E. McKay, A.K. Kapoor and J.D. Chinn, IEEE Proc. VLSI Multilevel Interconnection Conference, Santa Clara, CA, 1988, pp. 183-190.

[2] R.S. Nowicki, J.M. Harris, M.A. Nicolet and I.V. Mitchell, Thin Solid Films, 53, 195 (1978).

[3] C.H. Ting and H.Y. Lin, IEEE Proc. VLSI Multilevel Interconnection Conference, Santa Clara, CA, 1987; pp. 61-77; B. Lee and E.C. Douglas, ibid pp. 85-92; P. Pai, W.G. Oldham and C.H. Ting, ibid., pp. 364-370.

[4] E.K. Broadbent, J. Vac. Sci. Technol. B 5 (6), 1661 (1987).

[5] R.E. Oakley, S.J. Rhodes, E. Armstrong and A. Marsh, IEEE Proc. VLSI Multilevel Interconnection Conference, New Orleans, LA, 1984, pp. 23-29.

[6] Ionic Systems Note, " Substrate Impact on Stress Measurement ", June 10, 1988.

[7] Ionic Systems Note, " Tech Topic on Film Stress ", June 10, 1988.

[8] R. Glang, R.A. Holmwood, and R.L. Rosenfeld, The Review of Scientific Instruments, 36 (1), Jan. 1965.

[9] H. Koyama, Y. Mashiko, T. Nishioka, " Suppression of Stress Induced Aluminum Void Formation ", 24th Annual Proc. in IEEE Int. Reliability Phys. Symp., 24 (1986).

[10] J. Yue, W. Funsten, R. Taylor, " Stress Induced Voids in Aluminum Interconnects During IC Processing ", 23rd Annual Proc. in IEEE Int. Reliability Phys. Symp., 126 (1985)

# THE RELATIONSHIP BETWEEN INTRINSIC STRESS OF SILICON NITRIDE FILMS AND ION GENERATION IN A 50 kHz RF DISCHARGE

K. AITE*, J. HOLLEMAN*, J. MIDDELHOEK* AND R. KOEKOEK**
*University of Twente, ICE Lab., P.O.B. 217, 7500-AE Enschede, The Netherlands. **Tempress BV., Marconistraat 14, 7903-AG Hoogeveen, The Netherlands.

## ABSTRACT

Silicon nitride films prepared by the Plasma Enhanced Chemical Vapor Deposition technique (PECVD) are widely used in microelectronics. The intrinsic stress value of the silicon nitride films is a key factor which determines their reliability. A low frequency (50 kHz) RF discharge and a $N_2/NH_3/SiH_4$ gas mixture were used to deposit silicon nitride films on 4-inch silicon wafers in a horizontal hot-wall reactor at a temperature of 350 °C.
The compressive stresses in the deposited films were found to increase linearly with the nitrogen ionization created in the glow discharge. The nitrogen ion bombardment at the film surface induces ion implantation into the deposited film which results in film expansion. The stresses were deduced by the interference fringes technique and from surface profiler measurements. Optical Emission Spectroscopy of the $N_2/NH_3/SiH_4$ discharge was used for measuring the $N_2^+$ peak intensity variation with the partial pressure of the nitrogen gas.
When the nitrogen ionization was increased, a shift of the main infrared transmission peak corresponding to the Si-N bond in the silicon nitride films was observed using Fourier Transform Infrared Spectroscopy (FTIR). The Si-H and N-H infrared peaks in the films were not affected by the ion bombardment. A linear relationship was found between the wavenumber of the main infrared Si-N peak $\nu_M$ ( 835-875 $cm^{-1}$) and the intrinsic stress $\sigma_i$ of the deposited nitride films with the slope $d\nu_M/d\sigma_i$ $-4.85\times10^{-8} cm^{-1}/Pa$.

Mat. Res. Soc. Symp. Proc. Vol. 130. 

## INTRODUCTION

PECVD silicon nitride is a hard material, impermeable to mobile ions, mainly used for passivation and scratch protection in the field of integrated circuits fabrication [1]. Therefore, mechanical stability of PECVD silicon nitride films is an essential property. Large internal stresses are the cause of irreversible degradation of the film which can occur during deposition or after subsequent annealing steps. High stress can cause cracking of the nitride film. A strong compressive stress results in voids and cracks in the aluminum lines.

It is well known that the internal stresses of PECVD silicon nitride films are related to the discharge frequency [2-3], and are generally compressive or tensile when deposited at low or high frequencies, respectively. The mechanical properties of thin films can be altered by the implantation or incorporation of impurities during or after growth [4,5].

This work shows the influence of nitrogen incorporation to the $SiH_4$-$NH_3$ gas mixture in a low frequency (50 kHz) rf discharge on the intrinsic stress of the resulting silicon nitride films. The increase of the intrinsic stress due to $N_2^+$ implantation into the film is correlated with a shift of position of the main infrared Si-N peak of the FTIR (Fourier Transform Infrared) spectrum.

## EXPERIMENTAL

The deposition experiments were performed in a horizontal TEMPRESS hot-wall reactor with a multielectrode system. The deposition system consists of a reactor chamber, a removable wafer boat, a furnace , an rf generator, gas controllers, and a pumping sub-system . The whole system was controlled by a programmable digital process controller. The reaction chamber is made of quartz tube and heated by an external three-zones furnace. The nitrogen concentration was changed by varying the ratio of the input gases $N_2$ and $NH_3$ while the ( $N_2$+$NH_3$ ) and $SiH_4$ flows were kept constant, at 550 and 50 sccm, respectively. The substrate temperature and total pressure were set to 350 °C and 100 Pa, respectively. The rf power was 50 watts and the rf frequency 50 kHz. P-type <100> oriented Si wafers 100 mm in diameter and 25-60 ohm.cm resistivity were used. The optical emission spectra of the discharge were measured during deposition through a quartz window in the front door of the reactor. FTIR studies were performed with silicon nitride films having a minimum thickness of 500 nm, deposited on bare silicon substrates. Background absorption of the substrate was substracted from the spectra to obtain the bulk film spectra. High resistivity silicon substrates polished on both sides were used in order to prevent dopant absorption and to reduce scattering of IR radiation, respectively.

A silicon nitride film deposited on a silicon substrate is in a state of stress that can be tensile or compressive . Such stresses $\sigma_{SiN}$ are generally consisting of two components, the thermal stress $\sigma_{th}$ and the intrinsic stress $\sigma_i$ :

$$\sigma_{SiN} = \sigma_{th} + \sigma_i \tag{1}$$

The thermal stress $\sigma_{th}$ results from a difference between the thermal expansion coefficient of the film, $\alpha_{SiN}$, and that of the substrate, $\alpha_{Si}$. Quantitatively, $\sigma_{th}$ is usually given by :

$$\sigma_{th} = \frac{E_{SiN}}{1 - \nu_{SiN}} (\alpha_{Si} - \alpha_{SiN})(T_{dep} - T_{meas}) \tag{2}$$

where $E_{SiN}$ and $\nu_{SiN}$ are the Young's modulus and the Poisson's ratio of the film, respectively; $\alpha_{Si}$ and $\alpha_{SiN}$ are expansion coefficients for the Si substrate and the silicon nitride film, respectively. $T_{dep}$ is the temperature at which the deposition takes place and the stress measurement is done at $T_{meas}$. For the film constants we used the values reported by Retajczyk and Sinha [6]: $\alpha_{SiN}=1.5\times10^{-6}$/deg.C, and $E_{SiN}/(1-\nu_{SiN}) = 1.1\times10^{11}$ Pa. For Si we have taken $\alpha_{Si}= 2.6\times10^{-6}$/deg.C [7]. The total stress in the layers was determined at room temperature by measuring the radius of curvature of the wafers before ($R_{bef}$) and after ($R_{aft}$) film deposition with the Newton 's rings technique and profilometry (Sloan "Dektak"). Both techniques lead to comparable results within the experimental error. The following equation allows calculation of the average film stress [8] :

$$\sigma_{SiN} = \pm \frac{E_{Si}}{6(1-\nu_{Si})} \times \frac{t_{Si}^2}{t_{SiN}} \times \left[ \frac{1}{R_{aft}} - \frac{1}{R_{bef}} \right] \tag{3}$$

The sign + and - stands for tensile and compressive stresses, respectively. Where $t_{Si}$ and $t_{SiN}$ are the thicknesses of the substrate and the film, respectively. $E_{Si}$ and $\nu_{Si}$ are the Young 's modulus and the Poisson 's ratio of the substrate, respectively. For Si <100>, the value of $E_{Si}/(1-\nu_{Si})$ is $1.8\times10^{11}$ Pa. Finally the intrinsic stress component is deduced from eqs.(1), (2) and (3). Films with a minimum thickness of 500 nm were used for the measurements. The stress in the deposited films was independent of the thickness. The flatness of the wafers was better than 2 $\mu$m.

## RESULTS AND DISCUSSION

The plasma deposited silicon nitride films were found to be all in compression. Figure 1 shows that a linear realtionship exists between the intrinsic stress of the films and the partial pressure of nitrogen. The compressive stress in the silicon nitride films increases linearly with the addition of nitrogen to the gas mixture. Figure 2 is a plot of the optical emission spectrum of a nitrogen plasma. An intense optical emission line corresponding to the emission of ionized nitrogen molecules is visible at the wavelenght of 392 nm. The intensity of this line was measured during all depositions. The nitrogen partial pressure was varied in the $N_2/NH_3/SiH_4$ gas mixture. Figure 3 shows that the optical emission intensity of the $N_2^+$ line increases linearly with the nitrogen partial pressure in the discharge. The partial pressure of $NH_3$ has almost no influence on the intrinsic stress of our plasma deposited nitride films deposited at a total pressure of 100 Pa [9]. We conclude that a correlation may exist between the nitrogen content in the discharge and the enhancement of the compressive stress in the films.

Electrons are accelerated by the alternating electric field in a rf glow discharge. Excitation, dissociation and ionization result from inelastic collisions between highly energetic electrons and neutral gas molecules. The mobility of electrons is higher than that of the ions, therefore the electrode surface charges up to a negative value $V_s$ with respect to the plasma potential [10,11]. The potential distribution in a rf glow discharge is schematically represented in Figure 4. A so-called sheath region or dark space exists between the glow region where the potential $V_p$ is almost uniform and the electrode surface. The ions can follow the low frequency alternating electric field [12], thus $N_2^+$ ions diffusing to the sheath edge are accelerated to the electrode surface with a maximum energy of $(V_p\text{-}V_s)$eV. The number of $N_2^+$ ions increases with the nitrogen partial pressure. Therefore the ion bombardment of the growing film surface is enhanced. Few ions will be lost through recombination. The film wants to expand parallel to the surface as a consequence of the ion implantation, thus the stress in the film becomes more compressive. The density of the deposited films also increased from 2.5 to 2.9 g/cm$^3$ for values of $N_2/(N_2+NH_3)$ of 0.10 and 0.75, respectively.

The hydrogen content of the film 21 a/0 H, deduced from FTIR spectra was not affected by the variation of the nitrogen partial pressure. A decrease of the hydrogen content is not observed after lowering the amount of ammonia in the discharge. The hydrogen content decrease may be compensated by the $H^+$ implantation into the film during deposition. In fact, it was reported that the hydrogen content in a-Si films deposited in a triode configuration

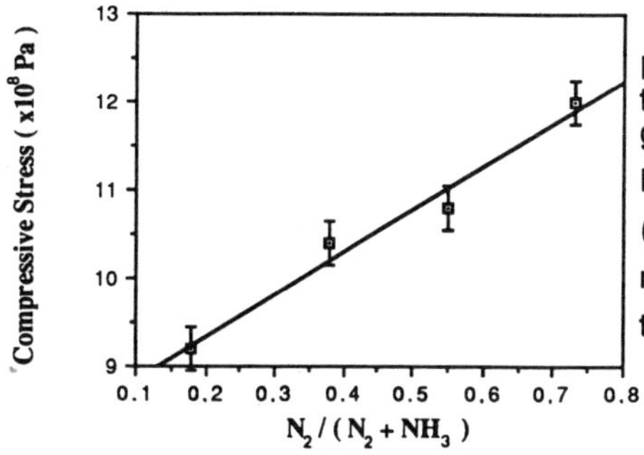


Figure 1. Intrinsic stresses of the PECVD silicon nitride layer vs. gas flow ratio $N_2/(N_2 + NH_3)$.
Deposition Conditions :
$(N_2+NH_3)$ = 550sccm, $SiH_4$ = 50sccm, rf power = 50W, rf frequency = 50kHz, total pressure = 100Pa, T = 350 °C.

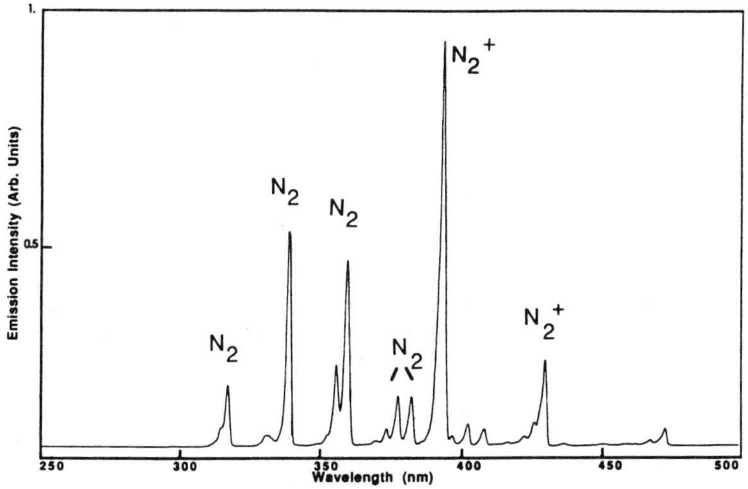


Figure 2. Optical emission spectrum of a nitrogen glow discharge. Pressure = 133 Pa, $N_2$ flow = 500 sccm, rf frequency = 50 kHz, rf power = 50 W, T = 350 °C.

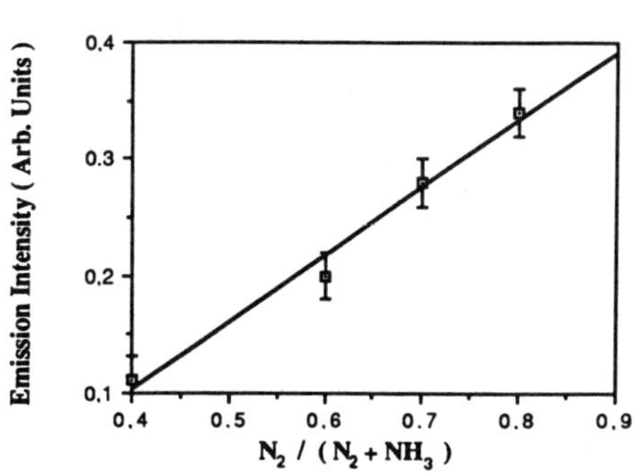


Figure 3. Relationship between the optical emission intensity of the $N_2^+$ emission line and the nitrogen partial pressure expressed by the gas flow ratio $N_2/(N_2 + NH_3)$.

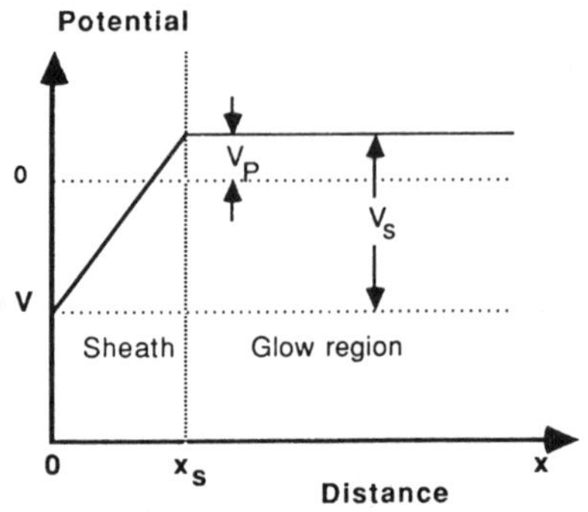


Figure 4. Schematic representation of the potentials established in a rf discharge vs. distance, where the powered electrode is at x = 0.

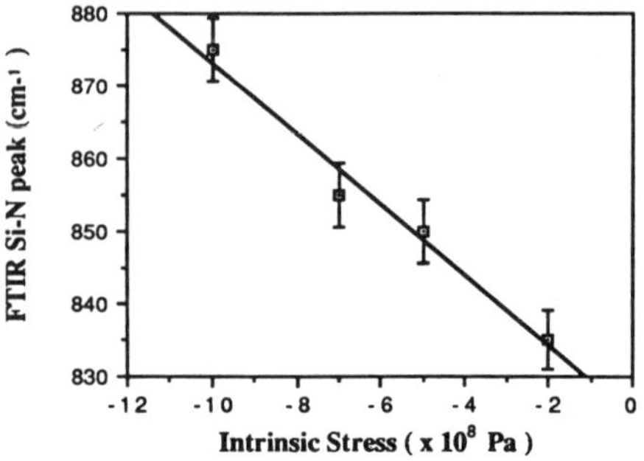


Figure 5. Relationship between the infrared transmission peak position of the Si-N bond and the intrinsic stress in the PECVD silicon nitride films.

plasma increased with the $H^+$ ion flux striking the growing film surface[13]. The main infrared transmission peak corresponding to the Si-N bond in the PECVD silicon nitride film is shifted to higher wavenumbers when the nitrogen partial pressure is increased. A linear relationship exists between the position of the main Si-N infrared peak $\nu_M$ and the intrinsic stress $\sigma_i$ in the films (Fig.5) with the slope $d\nu_M/d\sigma_i$ -4.85 x $10^{-8}$/Pa. The positions of Si-H and N-H infrared peaks as well as their respective areas were not affected by the $N_2^+$ bombardment. Although nitrogen has a higher dissociation energy compared to ammonia, the Si/N ratio in the films was changing only slightly between 0.82 and 0.93 when increasing the value of $N_2/(N_2+NH_3)$ from 0.10 to 0.75. This may be explained by the contribution of dissociative adsorption of $N_2^+$ ions ( $N_2^+ + e^- \longrightarrow 2N$ ) [14].

## CONCLUSION

Intrinsic stresses of PECVD silicon nitride films deposited at low frequency (50 kHz) using $N_2/NH_3/SiH_4$ gas mixture increase linearly with the nitrogen partial pressure. This is attributed to the increase of $N_2^+$ ion implantation of the growing surface. A linear relationship exists between the position of the main Si-N infrared peak $\nu_M$ and the intrinsic stresses $\sigma_i$ of the deposited nitride films with the slope $d\nu_M/d\sigma_i$ - 4.85x$10^{-8}$cm$^{-1}$/Pa.

## REFERENCES

[1] J.R. Hollahan and R.S. Rosler, in "Thin Film Processes", edited by J.L. Vossen and W. Kern, Academic Press, N.Y., p. 335, (1978).

[2] W.A.P. Claassen, W.G.J.N. Valkenburg, M.F.C. Willemsen, and W.M.v.d. Wijgert, J. Electrochem. Soc., **132**, p. 893, April 1985.

[3] A.K. Sinha, H.J. Levinstein, T.E. Smith, G. Quintana, S.E. Haszako, J. Electrochem. Soc., **125**, pp. 601-8, (1978).

[4] Thornton J.A., In Deposition Technologies for Films and Coatings, ed. R.F. Bunshah, pp. 170-243. Park Ridge : Noyes publications, (1982).

[5] Eernisse E.P., J. Appl. Phys., **48**, 3337-41, (1977).

[6] T.F. Retajczyk, Jr., and A.K. Sinha, Thin Sol. Films, **70**, 241 (1980).

[7] P.J. Burkhardt and R.F. Marvel, J. Electrochem. Soc., **116**, 864 (1969).

[8] "Physics of Thin Films", Vol.2 , pp.211-273, edited by G.Haas and R.F. Thun, Academic Press, New York (1966).

[9] K. Aite, J. Holleman, J. Middelhoek and R. Koekoek, to be published in EURO CVD 7, Perpignan, France, 19-23 June 1989.

[10] J.W. Coburn and E. Kay, J. Appl. Phys., **43**, 4365 (1972).

[11] J.L. Vossen, J. Electrochem. Soc., **126**, 319 (1979).

[12] Hirose M., Semiconductors and Semimetals, Vol. 21 A, Pankove J.I., Ed., Academic Press, New York, p.12, (1984).

[13] K. Ando, M. Aozasa, R.S. Pyon, Appl. Phys. Lett., **44**, pp. 413-15, 1984.

[14] H.F. Winters, in "Topics in current Chemistry,"S. Veprek and M. Venugopolan, Editors, Chap.3, p. 120, Plasma Chemistry III, Springer-Verlag, Berlin (1980).

# INTRINSIC STRESS IN a-GERMANIUM FILMS DEPOSITED BY RF-MAGNETRON SPUTTERING

D. Fahnline, B. Yang, K. Vedam, R. Messier, and L. Pilione
Materials Research Laboratory, The Pennsylvania State University, University Park, PA 16802

## ABSTRACT

Thin a-Ge films deposited by rf-magnetron sputtering undergo a change from compressive to tensile stress as the argon gas pressure is increased. Spectroscopic ellipsometry, neutron activation analysis, and a dual laser beam reflection technique were used to determine the film's void fraction content, Ar/Ge atomic percent, and stress, respectively. The results of these three measurements indicate that as the film's Ar/Ge content increases, the void fraction decreases and the intrinsic stress becomes compressive. These findings are consistent with currently accepted bombardment mechanisms.

## INTRODUCTION

The intrinsic stress of thin films prepared by physical vapor deposition techniques is known to vary with preparation conditions. In particular, numerous reports have indicated that sputtering at low gas pressures results in compressively stressed metallic, semiconducting, and insulating films, while high gas pressures produce tensile films [1-6]. It is generally believed that changes in the arrival energies and numeric ratio of gas atoms reflected by the target and atoms sputtered from the target play a dominant role in the evolution of film structures accounting for these stresses. At low gas pressures the arrival energies of atoms of both groups are only slightly modified by collisions as they move from the target to the substrate. As a result energetic atoms from both groups bombard the uppermost layers of the film causing densification through atomic rearrangement as well as compressive film stress [7]. At high gas pressures collisions reduce such bombardment and increase the oblique angle of incidence of sputtered atoms arriving at the surface of the growing film causing atomic shadowing. This combination of effects results in films with a microstructure that is columnar and has voided growth boundaries. Such films are associated with in-plane tensile stresses resulting from attractive forces between columns [4]. The total void volume, distribution, and connectivity are also thought to affect a number of other film properties [2, 4, 6, 8, 9].

To study these relations experimentally we rf-magnetron sputtered a series of a-Ge films at various argon gas pressures and a fixed target to substrate distance. Spectroscopic ellipsometric measurements were conducted and subsequent data analysis yielded the void fraction of each film. Neutron activation analysis was performed to determine the argon content, and a laser beam technique was utilized to determine the intrinsic film stress.

## DEPOSITION CONDITIONS

Thin a-Ge films were rf-magnetron sputtered simultaneously onto paired c-Si and polycarbonate substrates at forward and reflected powers of 100 W and 3 W, respectively. The c-Si substrates were strips broken from 14-mil thick wafers; they were partially covered by broken glass cover slides to provide steps for film thickness measurements. The polycarbonate substrates were 1.1-cm diameter discs cut from 10-mil thick Mobay Makrofol DE film. Gas pressures were regulated by a mass flow controller and monitored by a capacitance manometer. A 5 cm diameter polycrystalline Ge target (99.999%) was fixed 80 mm above a rotatable platform loaded with multiple pairs of substrates. The target was typically pre-sputtered for 30 minutes at a 15 mTorr argon pressure after the deposition chamber had been pumped to a base pressure of approximately $2 \times 10^{-7}$ Torr. The throttle pressure was approximately $1 \times 10^{-6}$ Torr. A mask over the platform exposed one pair of substrates at a time to 15 minute depositions at fixed argon gas pressures ranging successively from 5 to 25 mTorr. A chromel-alumel thermocouple embedded in the mask indicated a temperature rise from room temperature to a maximum value of 70°C after 15 minutes of sputtering.

## FILM CHARACTERIZATION AND INTERPRETATION

Spectroscopic ellipsometric measurements were performed on the c-Si substrate samples over an energy interval from 1.8 to 4.5 eV, with an average probe depth below the film surface of 500 Å. The imaginary part $\varepsilon_2$ of the pseudodielectric function was calculated from the $\Delta$ and $\Psi$ measurements. The Bruggeman effective medium theory was then used in conjunction with reference data [10] on dielectric functions for a-Ge and $GeO_2$ to produce the void fraction versus argon gas pressure graph shown in Fig. 1. (The three negative values simply indicate densities greater than that of the a-Ge reference.) The modeling techniques used to account for the void fraction, surface oxide, and microscopic roughness of the film are described elsewhere [11].

Neutron activation studies were conducted on the c-Si substrate samples at the National Bureau of Standards. Briefly, the technique involves activating Ar and Ge atoms with thermal neutrons and measuring the energy and intensity of the ensuing gamma radiation. This information then yields the Ar/Ge atomic ratio, which is the ratio of the number of Ar atoms to the number of Ge atoms deposited in the film. Figure 2 shows the resulting Ar/Ge atomic ratio for the samples under discussion as a function of argon gas pressure. Details of these experiments will be reported elsewhere [12].

An optical lever technique which will also be described elsewhere was used to measure the change $\Delta(1/R)$ in the radius of curvature of the polycarbonate discs from before to after deposition. The thickness t of each film was found by measuring the step in the film on the corresponding c-Si substrate using a Taylor-Hobson surface profilometer; the thicknesses varied from 3700 Å at 5 mTorr to 2000 Å at 25 mTorr. (However, the stress measurements were not conducted on the same set of samples as the other measurements.) Film stresses were obtained through the equation

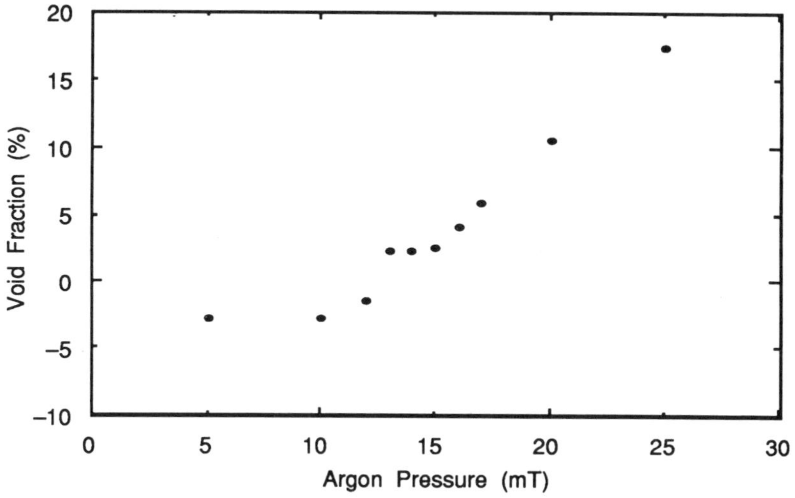


Fig. 1. Void fraction versus argon pressure.

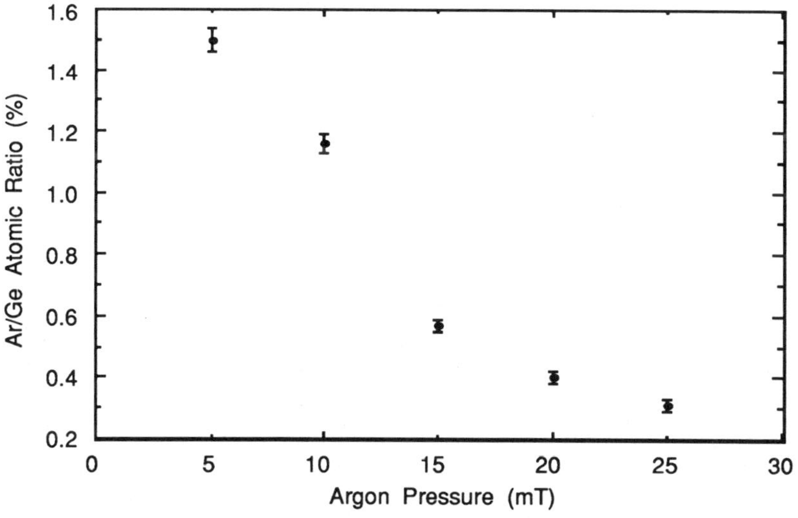


Fig. 2. Ar/Ge atomic ratio versus argon gas pressure.

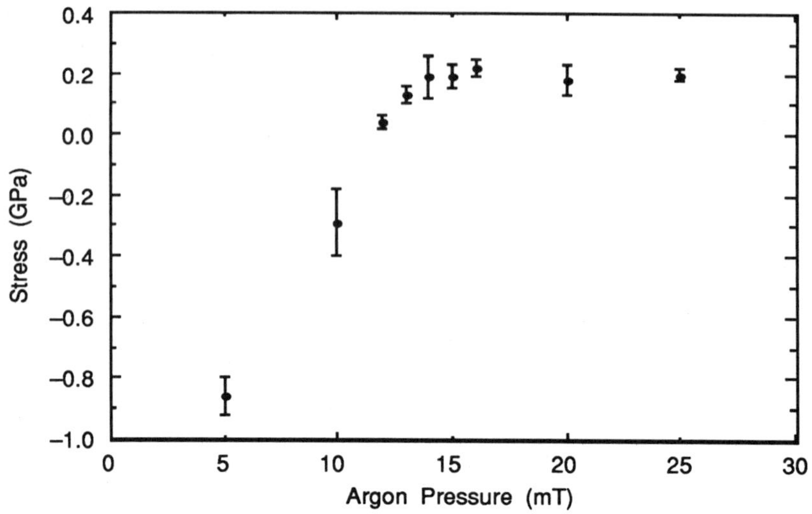


Fig. 3. Intrinsic stress versus argon pressure.

$\sigma = E\, d^2\, \Delta(1/R) / 6\,(1-\nu)\,t$, where E, d, and $\nu$ are the polycarbonate substrate's Young's modulus, thickness, and Poisson's ratio, respectively [13, 14]. The results are reported in Fig. 3 as intrinsic stress values because the temperature change during deposition was small.

## DISCUSSION

The incorporation of argon in the evolving film is a function of its arrival energy at the film surface, which is in turn a function of its energy loss as it travels from the target to the substrate after reflection from the target surface. Slowing down mechanisms have been modeled by a number of investigators. Recent reports have incorporated energy dependent collision cross-sections [15] and Monte Carlo calculations [15,16] to characterize the trajectories of individual atoms. The results of such modeling indicate the arrival energy [15] and angle of incidence [16] at the film are strong functions of the product of the gas pressure and target to substrate distance. Since the latter is fixed at 80 mm in the present experiment, they should here depend strongly on the argon gas pressure.

The data on argon gas incorporation in Fig. 2 quantitatively confirm the expected effects of this dependence. Even though there is five times more argon in the plasma at 25 mTorr, there is approximately one-fifth as much argon in the film prepared at 25 mTorr as in the one prepared at 5 mTorr.

The predicted strong dependence of arrival energy and angle of arrival on argon gas pressure should also affect void formation. The void fraction values shown in Fig. 1 show a large change between 5 and 25 mTorr quantitatively verifying this expectation. Finally, Fig. 3 exhibits compressive and tensile stresses in the argon gas pressure ranges corresponding to dense and voided films, respectively.

## SUMMARY

Rf-magnetron sputtered a-Ge films can have intrinsic stress ranging from compressive to tensile. The magnitude and type of stress is related to film bombardment due to reflected argon neutrals and sputtered germanium atoms, which can be controlled by the total argon pressure in the chamber. Increased bombardment by reflected argon neutrals can be evidenced by an increase in argon incorporation in the film and an accompanying decrease in void content. These two interrelated effects result in increased compressive stress in the film.

## ACKNOWLEDGEMENTS

The authors are grateful for the assistance of R. M. Lindstrom of the National Bureau of Standards and for the financial support of E. I. duPont de Nemours Inc. and The Commonwealth Educational System of The Pennsylvania State University.

## REFERENCES

1. J.A. Thornton and D.W. Hoffman, J. Vac. Sci. Technol. 14, 164 (1977).

2. J.E. Harper, J.J. Cuomo, R.J. Gambino, and H.R. Kaufman, Nucl. Instr. and Meth. in Phys. Res. B 7/8, 886 (1985).

3. H. Windischmann, J. Appl. Phys. 62, 1800 (1987).

4. B. Window, F. Sharples, and N. Savvides, J. Vac. Sci. Technol. A 6, 2333 (1988).

5. G. Este and W.D. Westwood, J. Vac. Sci. Technol. A 5, 1892 (1987).

6. G.L. Huffman, D.E. Fahnline, R. Messier, and L.J. Pilione, submitted to J. Vac. Sci. Technol.

7. K.H. Muller, J. Appl. Phys. 62, 1796 (1987).

8. L.J. Pilione, K. Vedam, J.E. Yehoda, and R. Messier, Phys. Rev. B 35, 9368 (1987).

9. B. Yang, L.J. Pilione, J.E. Yehoda, K. Vedam, and R. Messier, Phys. Rev. B 36, 6206 (1987).

10. G.A.N. Connell, in Amorphous Semiconductors, Topics in Applied Physics, vol 36, edited by M.H. Bronsky (Springer-Verlag, Berlin, 1985), p 73.

11. P.J. McMarr, J.R. Blanco, K. Vedam, R. Messier, and L.J. Pilione, Appl. Phys. Lett. 49, 328 (1986).

12. J.E. Yehoda, R.M. Lindstrom, and L.J. Pilione, to be published.

13. S.M. Rossnagel, P. Gilstrap, and R. Rujkorakarn, J. Vac. Sci. Technol. 21, 1045 (1982).

14. P.A. Flinn, D.S. Gardner, and W.D. Nix, IEEE Trans. on Electron Devices, ED-34, 689 (1987).

15. R.E. Somekh, Vacuum 34, 987 (1984).

16. T. Motohiro and Y. Taga, Thin Solid Films 112, 161 (1984).

# CONTROL AND VARIATION OF STRESS IN PECVD $SiN_x$ FILMS ON InP

J. LOPATA, W. C. DAUTREMONT-SMITH AND J. W. LEE
AT&T Bell Laboratories, Murray Hill, NJ 07974

## ABSTRACT

Stress in plasma enhanced chemical vapor deposited (PECVD) $SiN_x$ films on InP has been evaluated as a function of source gases ($NH_3/SiH_4$ or $N_2/SiH_4$) and plasma operating frequency (high, >> 1 MHz or low, << 1 MHz). All films were deposited at 300°C in the same parallel-plate, radial flow plasma reactor. Levels of stress in PECVD $SiN_x$ on InP within a continuous range from moderately high tensile ($\sim 5 \times 10^9$ dyne $cm^{-2}$) to very high compressive ($2 \times 10^{10}$ dyne $cm^{-2}$) were obtained from appropriate choices of deposition parameters. Deposition from $NH_3/SiH_4$ at high frequency produces tensile stress, of magnitude increasing with $NH_3/SiH_4$ flow ratio. Deposition from $N_2/SiH_4$ at high frequency produces zero to low compressive stress. At low frequency compressive stress is always produced; for $N_2/SiH_4$ increasing the gas flow ratio from 25:1 to 500:1 reduces the compressive stress from $1.8 \times 10^{10}$ to $7 \times 10^8$ dyne $cm^{-2}$. The ability to vary the stress in a dielectric film of approximately constant chemical composition over such a broad range is beneficial for assessing the effects of stress on device performance.

## INTRODUCTION

Silicon Nitride ($SiN_x$) thin films are widely used in semiconductor device fabrication as passivating coatings and insulating layers. One property of $SiN_x$ films which can affect semiconductor device performance is the level of stress in the film. At a high enough level stress can promote defect migration in devices and lead to premature failures. Thus it is imperative to control the level of stress in the dielectric films used in device fabrication.

In the plasma enhanced chemical vapor deposition (PECVD) of $SiN_x$ films the level of stress has been shown to be strongly dependent on deposition parameters.[1-3] Variation of RF power, plasma operating frequency, substrate temperature, gas pressure, and gas composition have resulted in $SiN_x$ films with stress ranging from highly compressive to tensile. One deposition parameter, RF plasma operating frequency, not only can affect the chemical properties of the discharge, but also the amount of ion bombardment of the film as it is deposited. Previous investigations on the $SiN_x$ stress dependence on RF plasma operating frequency have found an abrupt change in stress from compressive to tensile for increasing frequency. In these studies $SiN_x$ films were deposited from $NH_3/SiH_4$,[4] $NH_3/N_2/SiH_4$,[5] or $NH_3/N_2/SiH_4/Ar$[6] plasmas with compressive stress obtained at low frequencies (<< 1 MHz) and tensile stress at high frequencies (>> 1 MHz).

In a detailed study of mechanical stresses in $SiN_x$ thin films Claassen[5] has shown the $SiN_x$ film stress to abruptly cross over from compressive for low frequency deposition to tensile for high frequency deposition from an $NH_3/N_2/SiH_4$ source gas mixture. This abrupt stress change occurs at $\sim 4$ MHz and by comparison only gradual increases in the level of stress are observed as the frequency is increased or decreased from 4 MHz. A similar abrupt crossover from compressive to tensile stress for increasing plasma operating frequency was observed by Koyama et al.[4] at $\sim 2$ MHz for $SiN_x$ depositions from

 

$NH_3/SiH_4$. The level of compressive film stress increased gradually with decreasing frequency from the crossover at ~2 MHz, and similarly tensile film stress increased gradually for frequencies increasing from ~2 MHz. It would be difficult to obtain $SiN_x$ films of very low or zero stress from depositions using similar conditions solely by controlling the plasma operating frequency, because of the abrupt change in the sign of the stress. The dependence of the gas composition on film stress must also be considered.

A comparison of literature values of tensile stress in $SiN_x$ films deposited at high frequency shows a decrease in the magnitude of tensile stress for increasing $N_2$ content in the source gas. Deposition from $NH_3/SiH_4$ produced films with $\sim 8 \times 10^9$ dyne/cm$^2$ tensile stress[4] while films from $NH_3/N_2/SiH_4$ were $\sim 4 \times 10^9$ dyne/cm$^2$.[5] Similar deposition from $NH_3/N_2/SiH_4/Ar$ produced films with $2 \times 10^9$ dyne/cm$^2$ tensile stress.[6] The replacement of $NH_3$ by $N_2$ as the source of nitrogen for $SiN_x$ films deposited at high frequency tended to decrease the level of tensile stress. Compressive stress in $SiN_x$ films for depositions from the same gas mixtures at low frequency was larger in magnitude for $NH_3/SiH_4$ depositions than for those which contained $N_2$ and $NH_3$. The addition of $N_2$ also tended to decrease the level of compressive stress in $SiN_x$ films for low frequency deposition. By varying the amount of $N_2$ in the source gas it may be possible to controllably obtain low stressed $SiN_x$ films at either low or high frequency and avoid the abrupt change in stress observed at ~4 MHz.

In this paper we investigate the level of stress in thin PECVD $SiN_x$ films as a function of RF plasma operating frequency and source gas composition. With the appropriate choice of either $NH_3/SiH_4/Ar$ or $N_2/SiH_4/Ar$ gas mixtures we obtained $SiN_x$ films with stress ranging from high compressive to moderately high tensile. Films with very low compressive or essentially zero stress were obtained from depositions at high frequency (13.56 MHz) and at low frequency (50 kHz). The effect on $SiN_x$ wet etch rate for varied source gas composition and plasma operating frequency is also reported.

## EXPERIMENTAL

The substrate material used was n-type InP(100) 270 μm thick. Prior to deposition all substrates were cleaned with organic solvents and dilute HF to remove the native oxide and subsequently loaded onto a heatable table of a parallel-plate radial flow PECVD reactor. The deposition conditions used to obtain $SiN_x$ films for this stress study are shown in Table 1. For the series of depositions from either $NH_3/SiH_4/Ar$ or $N_2/SiH_4/Ar$ gas mixtures, the flow of 2% $SiH_4$ diluted with Ar was held constant and either $NH_3$ or $N_2$ flows were changed. The gas ratios were varied from 2:1 to 9:1 for $NH_3/SiH_4$ depositions and from 25:1 to 500:1 for $N_2/SiH_4$ deposition at either 50 kHz or 13.56 MHz. The plasma deposition reactor was standardly equipped with a high frequency (13.56 MHz) RF generator. Depositions at 50 kHz took place in the same reactor by replacing the RF power supply would be operated at varying frequency. Thicknesses and refractive indeces of the films were measured by ellipsometry and profilimetry when steps were etched into the films. Wet etch rates were determined using 6:1 buffered oxide etchant at room temperature.

The radii of curvature of the InP substrates with $SiN_x$ films were measured by X-ray diffraction using the Automatic Bragg Angle Control (ABAC) technique.[7,8] Film stress was calculated from the formula[9]

$$\sigma_f = \frac{1}{6}\left(\frac{E_s}{1-\nu}\right)\frac{t_s^2 R}{t_f} \qquad (1)$$

where $\sigma_f$ is the film stress, $E_s$, and $\nu$ are Young's modulus and Poisson's ratio for the substrate, $t_s$ and $t_f$ are the substrate and film thicknesses, and R is the radius of curvature of the substrate. For InP substrates the elastic stiffness parameter, $\frac{E_s}{(1-\nu)}$, used was $9.5 \times 10^{11}$ dyne/cm$^2$ for the film stress calculations. All radii of curvature measurements were made with the samples at room temperature. $SiN_x$ film thicknesses ranged from 400 to 4200Å on the nominally 270 $\mu$m thick InP substrates, and resulted in radii of curvature measurements varying from 7 to 200m.

**Table 1: DEPOSITION CONDITIONS USED TO OBTAIN $SIN_x$ FILMS IN A PARALLEL-PLATE, RADIAL FLOW REACTOR**

| Deposition Parameter | $SiH_4/NH_3/Ar$ Plasma | $SiH_4/N_2/Ar$ Plasma |
|---|---|---|
| Gas Flows | | |
| 2% $SiH_4$:Ar | 1200 sccm | 100 sccm |
| $NH_3$ | 50 - 220 sccm | — |
| $N_2$ | — | 50 - 1000 sccm |
| Deposition Rate | 70 - 220 Å/min | 25 - 105 Å/min |
| Pressure | 1 Torr | |
| Temperature | 300°C | |
| RF Power Density | 0.08 - 0.18 w/cm$^2$ | |
| Electrode Spacing | 2.54 cm | |
| RF Frequency | 50 kHz - 13.56 MHz | |
| Film Refractive Index | 1.85 - 2.15 | |

## RESULTS AND DISCUSSION

As shown in Table 1 the $SiN_x$ films used in this stress study were those with refractive indeces within the range 1.85 to 2.15 in order to maintain approximately constant chemical composition in the films. Gas flow ratios were consequently restricted to produce films of refractive index within this range. Above 2.15 the $SiN_x$ films were Si rich and below 1.85 they were N rich. The RF power was increased for some depositions in order to obtain deposition rates close to 100 Å/min. Sinha showed that for films deposited for $NH_3/SiH_4/Ar$ gas mixtures at 13.56 MHz, the tensile stress produced was relatively invariant with increasing power up to $\sim 0.20$ W/cm$^2$.(1) Above this RF power level the film stress changed from tensile to compressive and became increasingly compressive as the RF power increased. Claassen noted that for low frequency deposition the variation RF power had almost no influence on stress.(10) Because of the low power densities used to deposit $SiN_x$ films for this study, the variation in power had little or no effect on the level of stress.

The $SiN_x$ film properties shown in Table 2 resulted from depositions from varied ratio $NH_3/SiH_4$ source gases and varied plasma operating frequency. The deposition from a 2:1 $NH_3/SiH_4$ gas mixture at 50 kHz produced films with compressive stress while tensile stress was produced at 13.56 MHz. These data are consistent with those of Koyama and Claassen and a similar abrupt change in the sign of the stress would be expected to occur at $\sim 4$ MHz. The change from compressive to tensile stress for $SiN_x$ films deposited with increasing plasma operating frequency resulted from a decrease in the amount of ion bombardment the film receives as it is deposited. $SiN_x$ films deposited with a higher level of ion bombardment have less trapped gas and as a result exhibit less film shrinkage after deposition is completed. This shrinkage of the film results in the formation of tensile stress.(2) A comparison of the tensile stress in $SiN_x$ films deposited at 13.56 MHz from increasing gas ratio $NH_3/SiH_4$ is shown in Fig. 1. The range of $NH_3/SiH_4$ gas ratios was limited from 2:1 to 9:1 because only films in the refractive

**Table 2: $SiN_x$ FILM PROPERTIES FOR VARIED RATIO $NH_3/SiH_4$ SOURCE GASES AND RF PLASMA OPERATING FREQUENCY**

| Flow Ratio $NH_3/SiH_4$ | Frequency | Refractive Index | Wet Etch Rate (Å/min) | Film Stress ($\times 10^9$ dyne/cm$^2$) |
|---|---|---|---|---|
| 2:1 | 50 kHz | 2.01 | ~70 | 11C |
| 2:1 | 13.56 MHz | 2.15 | ~2000 | 2.3T |
| 9:1 | 13.56 MHz | 1.85 | ~3000 | 4.5T |

index range of 1.85 to 2.15 were used for this study. Below a 2:1 $NH_3/SiH_4$ gas ratio the $SiN_x$ films tended to be Si rich with refractive indeces >2.15, above 9:1 the films were N rich. The increase in tensile stress for increasing $NH_3/SiH_4$ gas ratio was caused by more trapped gas in the film as is was deposited. This resulted in more gas desorption and film shrinkage, and a higher level of tensile stress. The wet etch rate data shown in Table 2 are consistent with literature values in that films with higher tensile stress have higher etch rates. The level of porosity of the film which results from the amount of gas desorption and the formation of tensile stress, determines the wet etch rate.

The film properties for $SiN_x$ depositions from varied gas ratio $N_2/SiH_4$ source gases and varied RF plasma operating frequency are shown in Table 3. For low frequency depositions at 50 kHz and varied $N_2/SiH_4$ gas ratio the $SiN_x$ film stress was always compressive, very high $1.8 \times 10^{10}$ dyne cm$^{-2}$ at a 25:1 ratio and very low ($7 \times 10^8$ dyne cm$^{-2}$) at 500:1 as shown in Fig. 2. The gas ratio range was again limited to produce $SiN_x$ films within the refractive index range of 1.85 to 2.15. The level of compressive stress can be varied over a wide range simply by varying the $N_2/SiH_4$ gas ratio for $SiN_x$ films deposited at 50 kHz. The high frequency depositions at 13.56 MHz from 500:1 $N_2/SiH_4$ resulted in very low compressive to essentially zero, stress i.e. below the limit of detectability by the ABAC technique ($\sim 2 \times 10^8$ dyne cm$^{-2}$). Both the 50 kHz and 13.56 MHz depositions resulted in $SiN_x$ films with very low compressive stress for depositions from a 500:1 $N_2/SiH_4$ gas ratio unlike depositions from $NH_3/SiH_4$ which produced films with tensile stress at high frequency. The increasing wet etch rate (shown in Table 3) for $SiN_x$ films deposited at 50 kHz for increasing $N_2/SiH_4$ gas ratio is caused by a greater amount of $N_2$ gas incorporation in the film. Higher etch rates result from the desorption of the trapped gas and subsequent increase in film porosity. An increase in wet etch rate is also observed for increasing plasma operating frequency. The higher level of ion bombardment the film receives during low frequency deposition would cause less gas to be incorporated.

**Table 3: $SiN_x$ FILM PROPERTIES FOR $N_2/SiH_4$ SOURCE GASES AT VARIED RATIOS AND RF PLASMA OPERATING FREQUENCIES**

| Flow Ratio | Frequency | Refractive Index | Wet Etch Rate (Å/min) | Film Stress ($\times 10^9$ dyne/cm$^2$) |
|---|---|---|---|---|
| 25:1 | 50 kHz | 2.16 | ~45 | 18C |
| 500:1 | 50 kHz | 1.85 | ~300 | 0.7C |
| 500:1 | 13.56 MHz | 2.15 | ~800 | 0.2 - 0.7C |

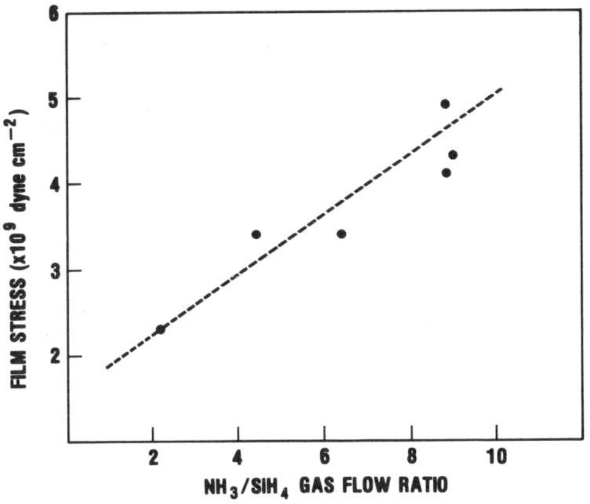


FIGURE 1. RANGE OF TENSILE STRESS IN $SiN_x$ FILMS DEPOSITED FROM $NH_3/SiH_4$ AT 13.56 MHz.

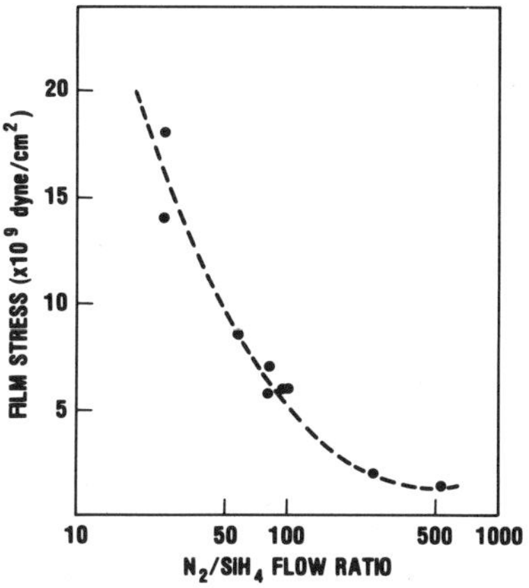


FIGURE 2. COMPRESSIVE STRESS IN $SiN_x$ FILMS DEPOSITED FROM $N_2/SiH_4$ SOURCE GASES AT 50 kHz.

The accessible range of $SiN_x$ film stress for varied gas ratio and RF plasma operating frequency is summarized in Fig. 3. Tensile stressed films resulted from $NH_3/SiH_4$ depositions at 13.56 MHz with stress increasing for increasing gas ratio. $SiN_x$ films deposited from $N_2/SiH_4$ resulted in a very wide range of compressive stress. For 50 kHz depositions at a 25:1 $N_2/SiH_4$ gas ratio very high compressive stress resulted. This level of compressive stress could be varied continuously from very high to very low by increasing the $N_2/SiH_4$ gas ratio to 500:1. Even lower values of compressive stress were

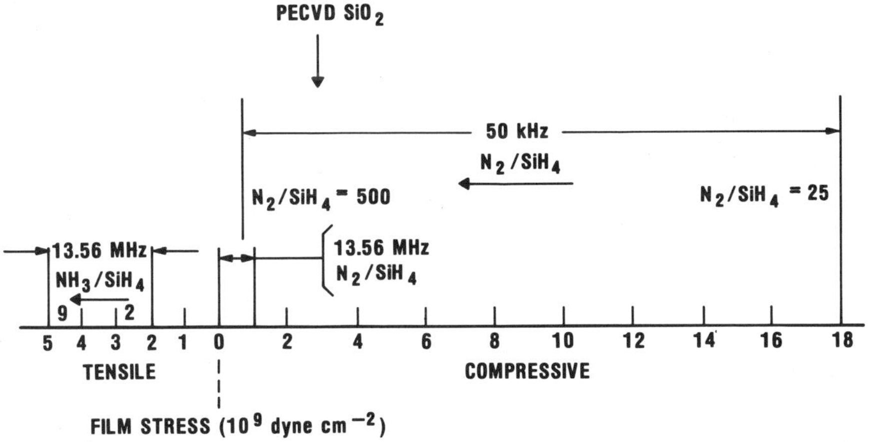


FIGURE 3. PECVD $SiN_x$ ON InP — ACCESSIBLE RANGE OF STRESS VALUES FOR FILMS DEPOSITED AT 300°C.

obtained from the high frequency deposition at 500:1 $N_2/SiH_4$. The level of stress in PECVD $SiO_2$, which is relatively invariant with deposition parameters,[3] is indicated for comparison. This wide range of stress accessible for $SiN_x$ films deposited in the same parallel-plate, radial flow reactor allows for the determination of the effects of stress on semiconductor device performance and reliability simply by the variation of gas flow ratios and plasma frequency.

## SUMMARY

The stress in PECVD $SiN_x$ films on InP substrates has been measured as a function of source gas composition and RF plasma operating frequency. The films were deposited in the same parallel plate, radial flow reactor under identical conditions except for varying $NH_3/SiH_4$ or $N_2/SiH_4$ gas ratios and either 13.56 MHz or 50 kHz plasma frequencies. Tensile stressed $SiN_x$ films were deposited from $NH_3/SiH_4$ source gases at high frequency (13.56 MHz). The magnitude of the tensile stress decreased from $5 \times 10^9$ dyne cm$^{-2}$ to $2 \times 10^9$ dyne cm$^{-2}$ by decreasing the $NH_3/SiH_4$ gas ratio from 9:1 to 2:1. Low frequency deposition from $NH_3/SiH_4$ also at 2:1 gas ratio produced films with high compressive stress at $1.1 \times 10^{10}$ dyne cm$^{-2}$. Deposition from $N_2/SiH_4$ source gases at high frequency at a gas ratio of 500:1 resulted in $SiN_x$ films with low compressive stress ($7 \times 10^8$ dyne cm$^{-2}$) to zero stress (unmeasureable by the ABAC technique). At low frequency $SiN_x$ films deposited from $N_2/SiH_4$ gas mixtures had compressive stress ranging in magnitude from $1.8 \times 10^{10}$ dyne cm$^{-2}$ for 25:1 $N_2/SiH_4$ gas flow ratio to $7 \times 10^8$ dyne cm$^{-2}$ for 500:1 gas mixtures. The ability to vary the stress in $SiN_x$ films from high compressive values to moderately high tensile may be useful in assessing the effects of stress on semiconductor device performance.

References

1. A. K. Sinha, H. J. Levinstein, T. E. Smith, G. Quintana, and S. E. Haszko, J. Electrochem. Soc., *125*, 601 (1978).

2. W. A. P. Claassen, W. G. J. Valkenburg, W. M. v. d. Wijgert, and M. F. C. Willemsen, Thin Solid Films, *129*, 239 (1985).

3. W. C. Dautremont-Smith, R. A. Gottscho, R. J. Schutz, in *Semiconductor Materials and Process Technology Handbook*, G. E. McGuire (ed.), Microelectronics Center of North Carolina, (1988).

4. K. Koyama, K. Takasaki, M. Maeda, and M. Takagi, in Proc. 3rd Symp. on Plasma Processing, J. Dieleman, R. G. Frieser, and G. S. Mathad (eds.), Tne Electrochemical Society, Vol. 82-6, 478 (1982).

5. W. A. P. Claassen, W. G. J. N. Valkenburg, M. F. C. Willemsen, and W. M. v. d. Wijgert, J. Electrochem. Soc., *132*, 898 (1985).

6. F. Martinet, G. Geugan, J. C. Jesionka, in Proc. Symp. on Silicon Nitride Thin Insulating Films, The Electrochemical Society, Vol. 83-8, 190 (1983).

7. G. A. Rozgonyi and D. C. Miller, Thin Solid Films 41, 57 (1977).

8. G. A. Rozgonyi and D. C. Miller in Crystal Growth: A Tuturial Approach, W. Bardsley, D. J. J. Hurle, J. B. Mullins (eds.), North-Holland, 307 (1979).

9. A. Brenner and S. Senderoff, J. Res. Nat. Bur. Stand., 42, 105 (1949).

10. W. A. P. Claassen in Plasma Chemistry and Plasma Processing, Plenum Vol. 7, No. 1, 109 (1987).

# EFFECTS OF STRESS ON THE ADHESION OF METALS TO POLYIMIDES

K. SESHAN AND R. H. LACOMBE
International Business Machines Corporation
East Fishkill Facility, Route 52
Hopewell Junction, New York 12533

J. B. WAGNER, JR.
Center for Solid State Science
Arizona State University
Tempe, Arizona 85251

## ABSTRACT

This paper describes the effects of ambient, thermal and mechanical stress on the adhesion of titanium (Ti) to polyimide (PI). Pull testing on the Ti/PI system shows that metal/polyimide bonding degrades when the composite is thermally cycled. A thermochemical mechanism is proposed that accounts for the interface degradation. We do not treat the adhesive interface as a discrete layer, but rather as a gradual "transition zone" between metal and polymer -- a zone that may grow, and/or change in composition and stress state, thus altering the adhesive properties. The mechanism predicts discontinuities that may develop in the transition zone. Through the use of finite element techniques, it is demonstrated that when loads and displacements are imposed on the composite containing interface discontinuities, large localized stresses develop. Such stresses could explain the experimentally observed low strength failures.

## INTRODUCTION

It is critical in various applications in the electronics industry that there is maximal adherence of metals to polyimides. To ensure good bonding, an adhesion layer is included in a multilayered thin film structure. Systems of primary importance, listed by Mattox [1], are:

TABLE I

MULILAYER SYSTEMS

| 2 LAYER | 3 LAYER | 4 LAYER |
|---|---|---|
| Au/Ti | Au/Pd/Ti | Au/Rh/Pt/Ti |
| Cu/Cr | Au/Pt/Ti | Au/Ni/Cu/Ti |
| Au/Mo | Pd/Cu/Ti | Etc. |
| Au/Nb | Au/Mo/Ti | |
| | Au/W/Ti | |

These films follow a pattern: an adhesion layer -- usually a transition metal -- followed by functional layers, and capped by passivating layers.
(Fig. 1).

3 - PASSIVATION

2 - FUNCTION

1 - ADHESION

PI

SUBSTRATE

Fig.1. Typical metal multilayer structure, deposited on polyimide (PI) on substrate, usually silicon or ceramic.

Polymides used in this study are of the type PMDA-ODA made by E. I. DuPont de Nemours Inc., and by Hitachi Corp. under various trade names [2]. (PMDA =pyromellitic di anhydride; ODA = oxydianiline). Bond energy calculations by Ho et al. [3] and Johnson et al. [4] show that bonding occurs chiefly via the oxygen atoms of the polyimide and the metal atoms. The adhesion of the metal layers are typically in the 30-50 gm/mm range as deposited [5,6]. It is evident that the adhesion will change upon heating and exposure to ambient gases. Such changes are discussed in the next section.

## EXPERIMENTAL RESULTS

Results of thermal cycling on the failure modes of titanium and chromium adhesion layers deposited on PI [7,8] are summarized in Figs. 2-5. These findings provide the experimental basis for the proposed thermo-chemical mechanism. These references reflect a change from a cohesive failure mode to an adhesive failure mode upon thermal cycling. Cohesive failure involves polyimide removal, whereas adhesive failure is characterized by low pull strengths and a clear separation between metal and polymer.

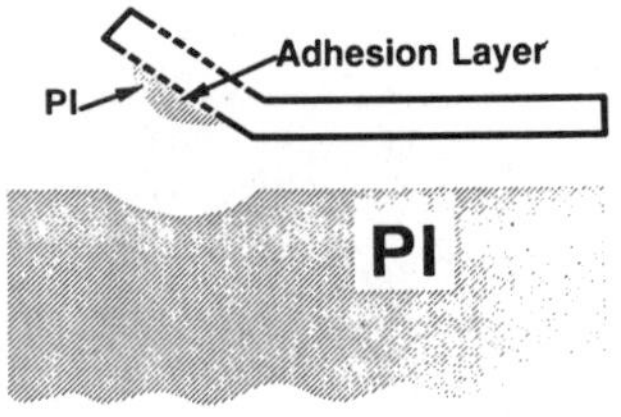


Fig. 2. Cohesive failure in the polyimide.

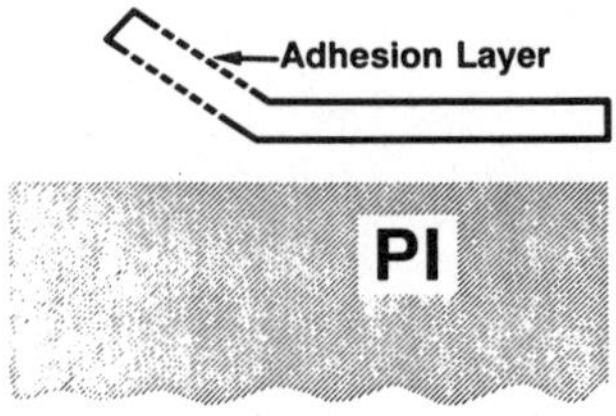


Fig. 3. Adhesive failure in the polyimide.

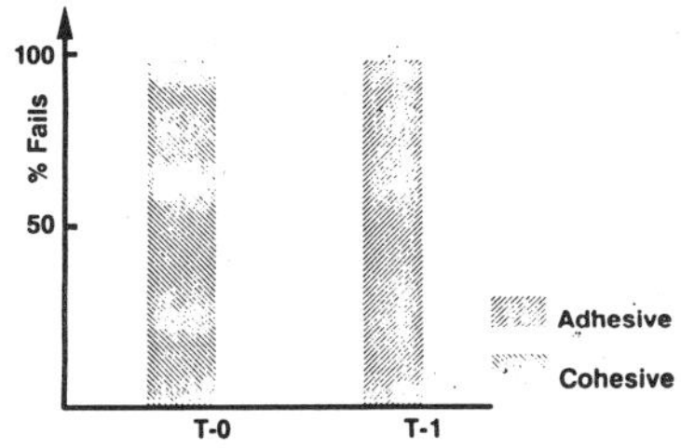


Fig. 4. Results of cycling Ti/PI.

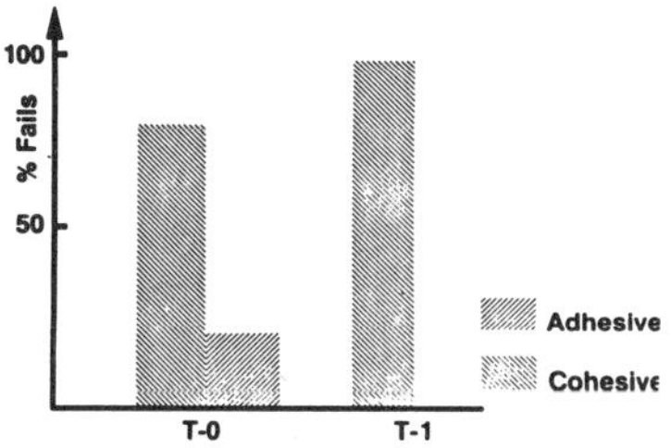


Fig. 5. Results of cycling Cr/PI.

Figures 2,3 delineate the difference between cohesive and adhesive failures. The former involves removal of polyimide, indicating a good bond; the latter indicates degradation of the interface.

Figures 4,5 show the difference between the behavior of Cr and Ti as adhesion layers. The failure mechanisms for Cr does not change; the ones for Ti do. T-0 indicates the as-deposited state; T-1 indicates testing after thermal cycling to 400°C for 1 hour.

The paramount result of our investigation is that moisture in the polyimide plays a part in the deterioration of adhesion. This was confirmed by the use of $H_2{}^{18}O$; $^{18}O$ segregates to the interface; the Ti reacts with the newly introduced oxygen species [7]. Auger and SIMS depth profiling showed that the adhesion layer in the as-deposited and the post-heat cycled condition can be depicted as in Fig. 6. At first the bonds are mainly between the metal and the oxygen of the polyimide. After cycling a discontinuous distribution of oxide and hydride, species were found in a zone close to the original interface as shown in Fig 6b [7].

The chrome-polyimide interface does not degrade as fast as the titanium interface (Fig. 5.) It has been experimentally established that prolonged heating causes that interface to fail as well. Therefore, the rest of this paper is confined to the titanium interface.

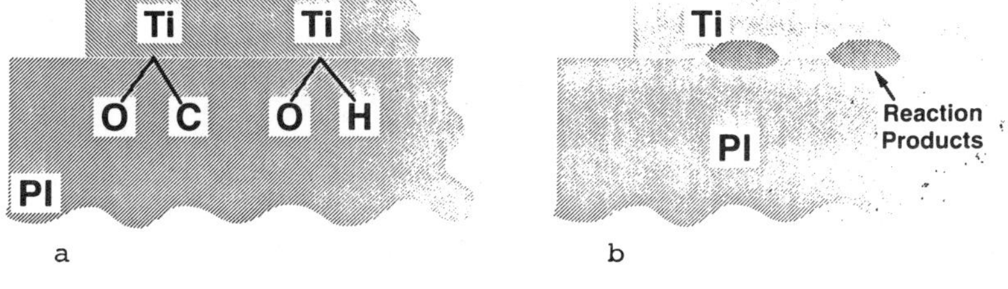


Fig.6. (a) Interface soon after the Ti is deposited; the bonds between the Ti metal and the O- and the C- of the polyimide are depicted. (b) Situation after thermal cycling; now the original Ti/PI bonds are replaced by a reaction layer, which is discontinuous, with reaction products -- like the oxides and hydrides of titanium.

A THERMOCHEMICAL MECHANISM

In order to model the previously reported changes, it is presumed that the transition from the metal to the polyimide is not abrupt; rather, that there is a transition zone with a finite width. It is assumed that in this region gases like oxygen and hydrogen, denoted with ( ) brackets, coexist with solid state bonds between the metal and the oxygen and hydrogen of the polyimide, denoted with < > brackets. The changes that occur within this zone as temperature or ambient is changed, can be examined using the standard methods of thermochemical stability as outlined by Kubaschewski and Alcock [9]. Their notation system is also used here.

The question of how the interface interacts to changes in temperature and gaseous partial pressure can now be answered by examining the free-energy diagrams shown in Fig. 7. These have been derived and plotted for 400° C using standard thermochemical data [9]. Figure 7 is a plot of the stabilities of the oxides, carbides and hydrides of titanium. The effect of increasing the oxygen partial pressure is to drive the system in the direction of the oxides (indicated by arrows). Similarly, the effect of hydrogen is diagrammed.

Although not shown here, a chromium diagram can be derived that would be similar to the titanium system. Further, the arguments presented for titanium also generally apply to chromium.

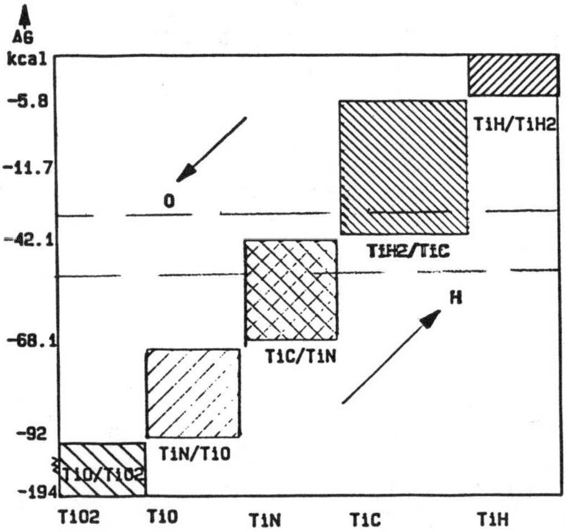


Fig. 7. Stability of various compounds of Ti. Phases that result by increasing oxygen or hydrogen partial pressures are indicated by arrows.

The stability of the newly-formed metal-ambient gas species, with respect to the original metal-polymer bonds, can now be examined. Knowing that the bond energies of the metal and polyimide is in the range of 40-50 kcal per mole [3,4] it can be seen that the presence of oxygen will cause new oxide bonds at the expense of the existing bonds. Such a reaction causes debonding to occur. The generic debonding reaction of the metal-bound polyimide reacting with newly introduced gaseous species can then be written as:

$$<M - O_{PI}> + ({}^{18}O) = <M - {}^{18}O> + <O_{PI}>$$

Specific reaction with respect to Ti may be written thus:

$$<Ti-O(PI)> + ({}^{18}O) = <Ti-{}^{18}O> + <O(PI)>$$

Similar replacement reactions can be written for the C, N, and H of the polyimide. Such replacement reactions are driven by the extreme stability of the oxides. These new phases form within the transition zone.

In this description of the interface, replacement reactions proceed within the transition zone. Metal-oxide reactant species form in the presence of oxygen, causing a discontinuous distribution of reactants in the transition zone. Such species have been experimentally observed at the interface [7]. Now if the loads and displacements are imposed on the composite, with discrete particles at the interface, large localized stresses could develop.

THE STRESS MODEL

The ANSYS [10] finite element program was used to analyze the effects of loads and displacements on the composites. The load system used is depicted in Fig. 8. A combined normal load of 40-60 gms and displacements of 0.1-0.5 micrometers were imposed on a 100 x 38 micrometer area.

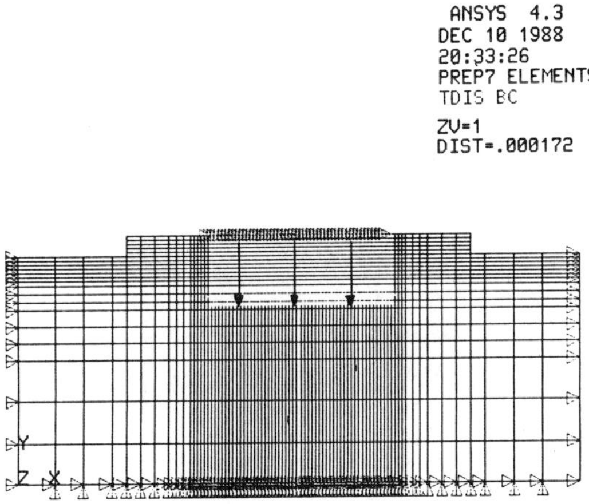


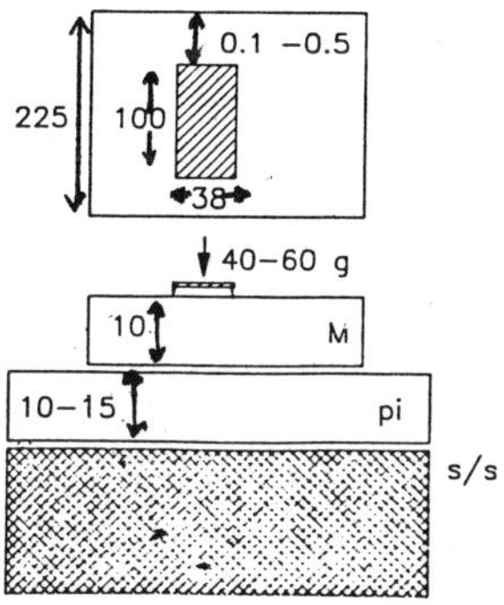


Fig. 8. Pad under compressive stress.
(a) Boundary conditions, the loads and displacements.
(b) Top and side views of same system (all dimensions in micrometers).

A defect-free interface was modelled first. The stresses that developed with and without the polyimide were compared. Results shown in Fig.9 are the stress components sx and sy (a) without and (b) with the polyimide. The first set of results examines the ideal interface, and shows the benefit of interposing a compliant layer, like polyimide, to reduce stress.

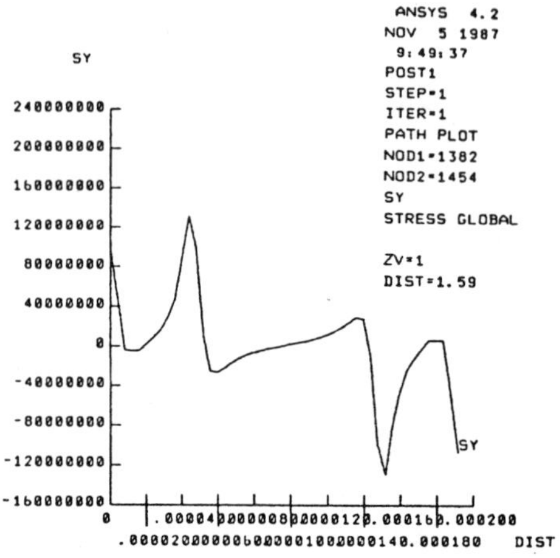


a) Normal stress in copper pad just above ceramic - no polyimide.

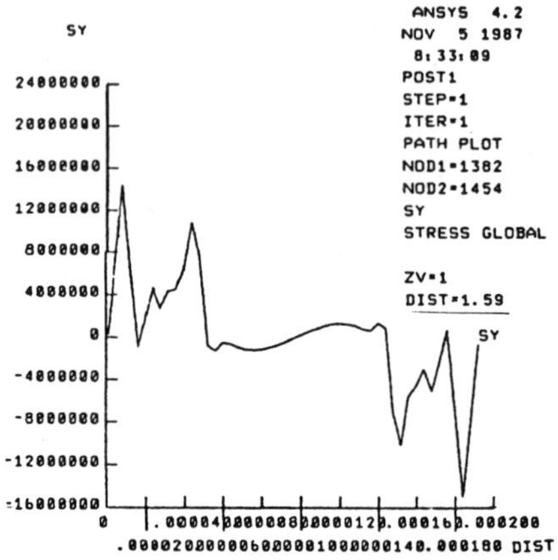


b) Normal stress in copper pad just above polyimide.

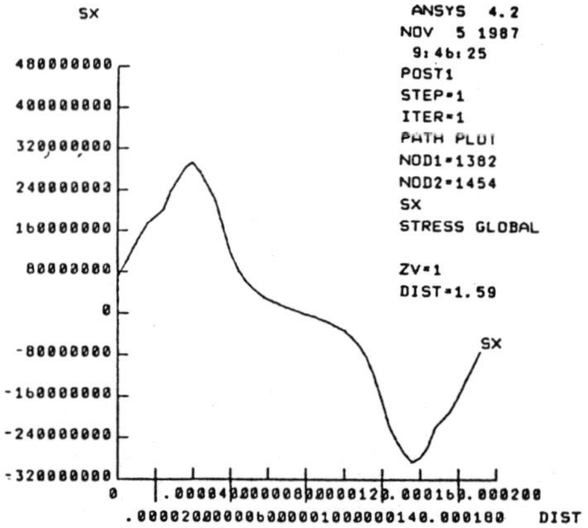


c) Axial stress in copper pad just above ceramic - no polyimide.

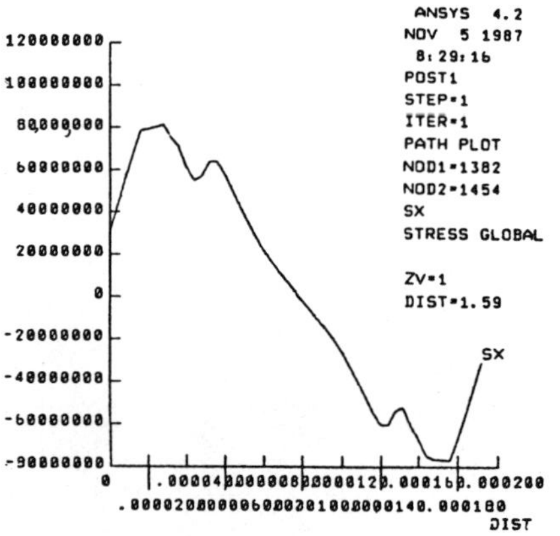


d) Axial stress just above polyimide in copper pad.

Fig. 9. Normal stress sy (a) without and (b) with polyimide. Axial stress, sx, is in the plane of the pad (c) without and (d) with polyimide. Note reduction of stress with the polyimide interposition.

TABLE II

SUMMARY OF STRESS DATA IN (MPa)

| | METAL ON CERAMIC | METAL ON POLYIMIDE |
|---|---|---|
| sx(max) | 320 | 80 |
| sy(max) | 120 | 12 |

The reduction in stress upon interposing polyimide is shown in Table II. However, the beneficial effects of stress reduction at the interface with the polyimide interposed will be compromised if discontinuities are present. A mechanism of exactly how such discontinuities develop has been presented. Introducing isolated particles in the reaction layer causes high, localized stresses. In order to analyze this, the interface was modelled with a discontinuity. Results are shown in Figs. 10, 11.

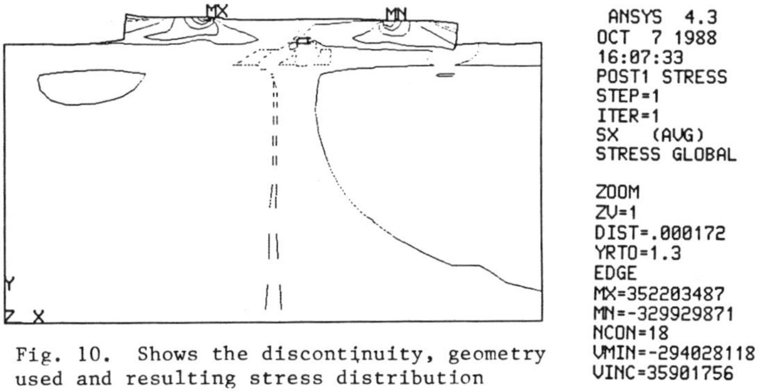


Fig. 10. Shows the discontinuity, geometry used and resulting stress distribution for sx.

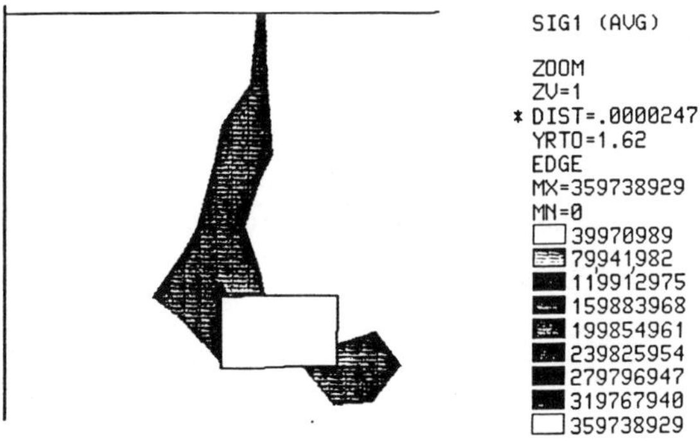


Fig. 11. Shows distribution of principal stress SIG1(AVG). Assymmetry results from the loading condition. All regions within shaded area exceed yield point of the polyimide.

Figure 8 shows the geometry used; the discontinuity is placed at the metal polyimide interface. The stress component, sx, is shown. The loading conditions cause the asymmetry in the contours. While all the components of stress and displacements were calculated, only the most informative is addressed. This is the principal stress component, sig1(avg), shown in Fig. 11. The disruptive nature of interface discontinuities is evident. Stresses within the shaded region exceed 80 MPa; this far exceeds the yield point of the polyimide. It is believed that such stresses play a part in the nucleation and propagation of fractures at the interface.

It must be cautioned that the stresses show in Figs. 10,11 are overestimates because viscoelastic and other non-linear effects, which play a role but are difficult to model, were not modelled in this study. Nevertheless, several useful conclusions can be drawn:

- As stress concentrations increase, the disparity in modulus between host and discontinuity becomes greater.
- For each value of applied stress there is a critical particle size that will cause failure.
- There most certainly exists a range of particle sizes at the interface.

Therefore, interface failures at low loads can be expected.

## CONCLUSIONS

A detailed analysis of metal/polymer interface has been presented. The interface, in this view, is not abrupt. Rather, a transition zone between the metal and polymer is postulated. This zone may change in size, stress state, and composition as the interface is heated in oxidizing or reducing ambients. The heating process causes new reaction products to form at the interface. These new products act as discontinuities and when loads and displacements are applied, they cause large stress concentrations. The stress concentrations initiate and propagate adhesive failures at low applied loads, consistent with experimental observations [7,8].

## ACKNOWLEDGMENT

Thanks to L. A. Plaushin for editorial assistance.

## REFERENCES

1. D. M. Mattox, J. Electrochem. Soc. 10, (1) 20 (1983).

2. Literature - Hitachi Corp., and E. I. DuPont de Nemours, Inc., Wilmington, Del.

3. P. S. Ho et al., IBM J. Res. Develop. 32 (5) (1988).

4. K. H. Johnson, F. Herman and R. Kjellander in Electronic Structure of Polymers and Molecular Crystals, edited by J. M. Andre and J. Ladik, (Plenum, NY, 1974), p.601.

5. Adhesion Measurements in Thin Films, Thick Films and Bulk Coatings, edited by K. L. Mittal (American Society for Testing Materials STP 640 1978).

6. R. H. Lacombe in Surface and Colloid Science in Computer Technology, edited by K. L. Mittal (Plenum, NY, 1987), p.179.

7. B. K. Furman, S. Purushottam et al., Proc. Fall Meeting. Electrochem. Soc., San Diego, CA, Oct. 1986.

8. K. Seshan, S.N.S. Reddy and S.K. Ray, Proc. 604, 87th IEEE Electronic Components Conf., Boston, MA, (1987).

9. O. Kubaschewski and C. B. Alcock in Metallurgical Thermochemistry, (Pergamon, N.Y., 1979).

10. ANSYS - registered trademark of Swanson Analysis Systems, Inc., Houston, PA 15342.

# BEHAVIOR OF DEFECTS RELATED TO INTERFACE-STRESSES IN MODEL SUBMICRON SOI STRUCTURES

N.D. Theodore and C.B. Carter
Department of Materials Science and Engineering, Cornell University, Ithaca, NY 14853
S.C. Arney and N.C. MacDonald
School of Electrical Engineering, Cornell University, Ithaca, NY 14853

## ABSTRACT

The origin and propagation of defects in novel submicron SOI structures have been investigated by transmission electron microscopy. The majority of the defect configurations observed could be explained in terms of dislocations generated as a result of stresses induced by the oxidation process. Dislocations were found to propagate into the silicon along particular {111} planes in conformity with stress-modelling. The examination of particular dislocation configurations allowed an estimation of the stresses at the silicon/ silicon dioxide interface. Defect densities were observed to depend on the geometry of the structures.

## INTRODUCTION

The need for increasing device packing densities has led to the scaling down of silicon devices into the submicron regime. In this regime, parasitic circuit elements due to wiring and junction capacitance become significant. Isolation schemes are therefore required that can reduce these elements. A variety of Silicon-on-Insulator (SOI) isolation schemes are to be seen in literature [1-3]. Selective oxidation isolation techniques are quite well established for both bipolar and metal-oxide-silicon large-scale integration applications. Conventional local oxidation (LOCOS) [4] for instance involves the masking of active regions with a nitride film during the isolation oxidation step. A problem typical of this and similar technologies is the generation of dislocations at film edges due to high-stresses in the nitride films [5]. Several investigators have looked at the origin and behavior of dislocations arising in the course of selective oxidation of silicon films or structures [5-10]. Transmission electron microscopy (TEM) has been used occasionally in these investigations; a variety of TEM techniques are in fact available for these and similar characterization of device structures [11].

One possible isolation scheme involves the fabrication of silicon device-structures on an oxide layer. A recently developed technology uses selective lateral oxidation of electron beam-patterned silicon islands to produce submicron-width SOI structures [12]. The quality of the resulting silicon islands is important because of the influence of structural defects on the operational viability of possible devices. In this study, TEM specimens were prepared to evaluate the defect configurations present in these structures with a view to understanding the origin and propagation of the defects. This could possibly lead to methods of defect-elimination.

## EXPERIMENTAL

The novel submicron SOI structures used for this study were prepared in the following manner. First. silicon islands 0.25 µm wide and 1-2 mm long were defined by electron-beam lithography. The top and side walls were next capped with silicon nitride, which acts as an oxidation-barrier during subsequent processing steps. Selective oxidation was then performed; the oxide encroaches laterally beneath the silicon islands. Continued oxidation causes oxide regions to merge, isolating the silicon island from the substrate. Following an anneal, the structures were coated with a protective overlayer of polysilicon or CVD oxide. TEM cross-section specimens were prepared; [011] geometry. Differential milling of the various component materials necessitated termination of ion-thinning before portions of the structure were lost. This resulted in the silicon in the vicinity of the nitride being 1 to 2 microns thick; high accelerating voltages were therefore required for microscopy. A JEOL 4000EX microscope was used, operating at 400 kV.

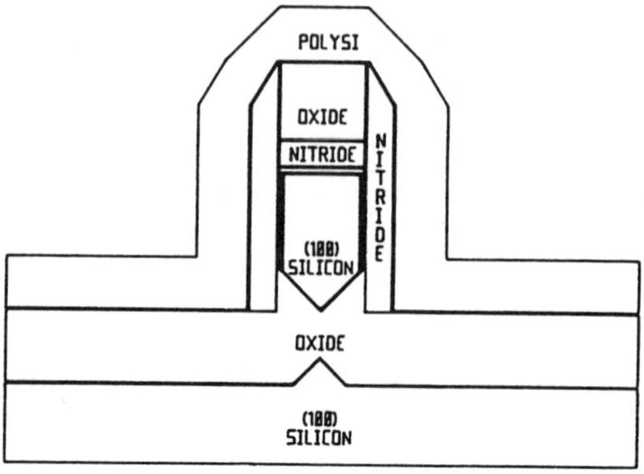


Fig. 1. Schematic cross-section of structure; fully isolated island of (100) silicon over thermally grown silicon dioxide.

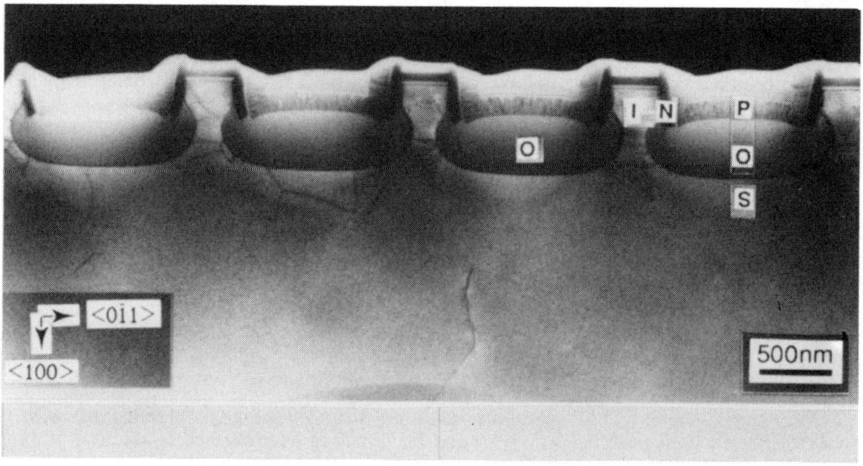


Fig. 2. TEM micrograph showing array of fabricated structures; 200 reflection. Note increased defect densities in silicon between oxide regions; dislocations do propagate into silicon islands and into substrate. Legends 'O' for oxide, 'I' for island, 'N' for nitride, 'P' for polysilicon and 'S' for substrate.

## OBSERVATIONS AND DISCUSSION

Two kinds of structures were investigated: partially-isolated silicon islands, where oxidation was terminated prior to isolation of the island by encroaching oxide, and completely isolated-islands. Figure 1 for instance shows a schematic cross-section of the second kind of structure; a fully isolated island of (100) silicon over a thermally grown silicon dioxide. The large stresses associated with the two-dimensional oxidation are found to give rise to the generation and propagation of dislocations in the silicon island during the processing sequence, oxidation followed by cool-down to room-temperature. Figure 2 shows a TEM micrograph of an array of the fabricated structures. The figure shows increased defect densities in the silicon between oxide regions. The dislocations can be seen to propagate into the silicon islands and also into the substrate. Oxidation-induced stresses in the partially isolated islands can be modelled by <011> and <111> deformations. The silicon 'neck' in figure 3 shows signs of activation of the $(\bar{1}1\bar{1})$ and the $(1\bar{1}\bar{1})$ glide planes. Dislocations propagate out on these planes in conformity with calculated Schmid factors. The other {111} planes are not activated; calculated Schmid factors for these planes turn out to be zero. Stresses in the region of the silicon 'shoulder' can be modelled by a <111> deformation; calculated Schmid factors show that the $(1\bar{1}1)$ and the $(\bar{1}\bar{1}1)$ planes are activated in addition to either the $(\bar{1}1\bar{1})$ or the $(1\bar{1}\bar{1})$ glide

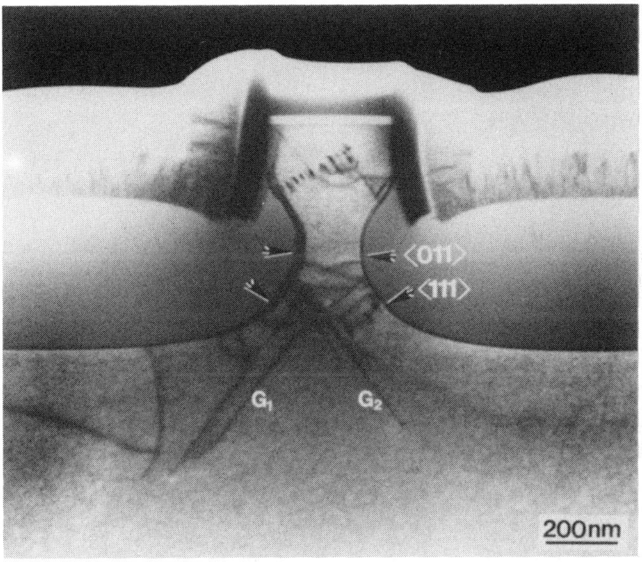


Fig. 3. TEM micrograph showing defect configuration; 200 reflection. Stresses in regions marked by arrows can be modelled in terms of <011> and <111> deformations, as indicated. G1 and G2 indicate dislocations propagating out on $(\bar{1}1\bar{1})$ and $(1\bar{1}\bar{1})$ glide planes respectively in conformity with stress-modelling.

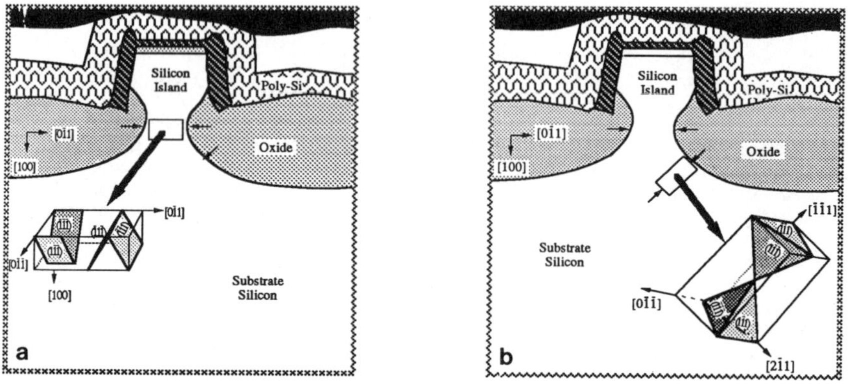


Fig. 4. (a) Schematic of specimen-area shown in figure 3 indicating orientation of {111} planes with respect to direction of application of stress $[0\bar{1}1]$; only the $(\bar{1}1\bar{1})$ and $(\bar{1}\bar{1}1)$ planes are activated. (b) Schematic of specimen-area shown in figure 3 indicating orientation of {111} planes with respect to direction of application of stress $[1\bar{1}1]$; only the $(\bar{1}11)$, $(1\bar{1}1)$ and $(\bar{1}\bar{1}\bar{1})$ planes are activated.

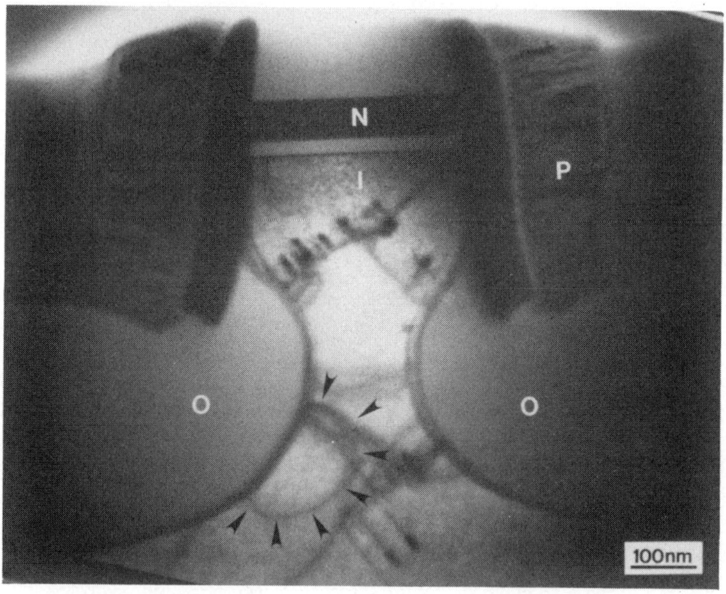


Fig. 5. TEM micrograph showing a dislocation half-loop propagating out from the silicon/ silicon dioxide interface, on the $(\bar{1}\bar{1}\bar{1})$ or the $(1\bar{1}\bar{1})$ glide planes. This is in conformity with a <111> deformation in that region; defect-loop is indicated by the arrows.

plane, depending on the shoulder. The orientations of these planes for the direction of stress corresponding to the two deformation geometries is presented in figures 4a and 4b. Several TEM micrographs (e.g., Fig. 5) show a half-loop propagating from the silicon/ silicon dioxide interface at the shoulder, on the $(\bar{1}\bar{1}1)$ or the $(1\bar{1}\bar{1})$ glide plane as expected from the model. Figure 6 shows a schematic of the area seen in the previous figure. The diameter of the half-loop seen in the figures can be measured. A simple balancing of the line tension of the curved dislocation by the stress present at the interface enables us to calculate this stress. Given a radius of 95 nm, the stress at the silicon / silicon dioxide interface is calculated to be about 20MPa.

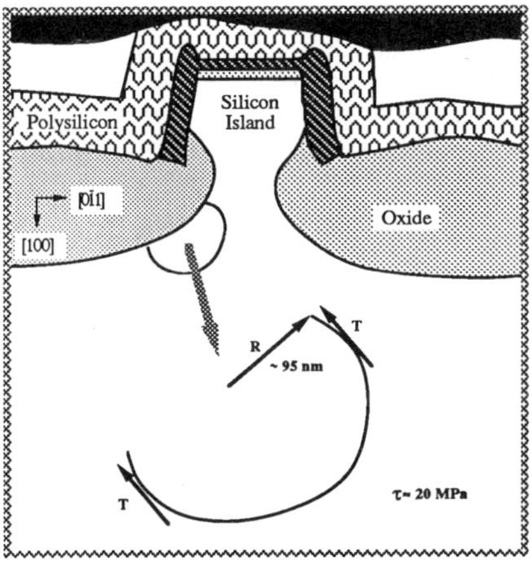


Fig. 6. Schematic of specimen-area shown in figure 5 indicating presence of half-loop propagating out from the silicon/ silicon dioxide interface, on the $(\bar{1}\bar{1}1)$ or the $(1\bar{1}\bar{1})$ glide planes. Diameter of half-loop permits estimation of stress-levels at interface.

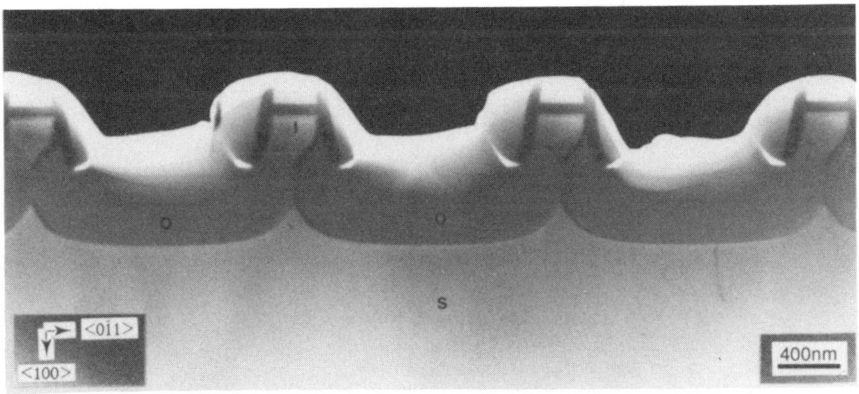


Fig. 7. TEM micrograph showing array of fabricated structures that have been oxidized to achieve full isolation of the silicon island. Note dramatic reduction in dislocation density.

TEM micrographs reveal dislocation interactions. Figure 7 shows structures that have been oxidized to attain complete isolation of the islands. Note the dramatic reduction in dislocation density. This lowered dislocation density is possible because of decreased stress-levels caused by the silicon islands floating free on the oxide. Similar structures are seen upon annealing to possess virtually defect-free silicon islands. Work is in progress to determine if the dramatic reduction in defect-density is caused by stress-reduction as a result of the isolation or as a result of the anneal.

## SUMMARY

The origin and propagation of defects in novel submicron SOI structures have been investigated by transmission electron microscopy. The majority of the defect configurations observed could be explained in terms of dislocations generated as a result of stresses induced by the oxidation process. Dislocations were found to propagate into the silicon along particular {111} planes. Particular dislocation configurations facilitated estimation of stresses at the silicon/ silicon dioxide interface. Defect densities were observed to depend on the geometry of the structures.

## ACKNOWLEDGEMENTS

The authors thank Mr. Ray Coles for maintaining the microscopes and Ms. M. Fabrizio for photographic work. The microscope facility is supported, in part, by the NSF through the Material Science Center at Cornell. The authors acknowledge support by the Semiconductor Research Corporation (Grant 88-DC-069).

## REFERENCES

1. K. Imai and H. Unno, IEEE Trans. Elect. Dev. ED-31 (3), 297 (1984)
2. M.W. Geis, H.I. Smith, B-Y. Tsaur, J.C.C. Fan, D.J. Silversmith, and R.W. Mountain, J. Electrochem. Soc. 129 (12), 2812 (1983)
3. P.L.F. Hemment, E. Maydell-Ondrusz, K.G. Stephens, J.B. Butcher, D. Ioannou, and J.C. Alderman, Nucl. Inst. and Methods, 157 (1983). The Netherlands: North Holland.
4. J.A. Appels, E. Kooi, M.M. Paffen, J.J.H. Schatorje, and W.H.G.G. Verkuylen, Phillips Res. Rep. 25, 118 (1970)
5. J. Vanhellemont, S. Amelinckx and C. Claeys, J. Appl. Phys. 61 (6), 2176 (1987)
6. K. Shibata and K. Taniguchi, J. Electrochem. Soc. 127 (6), 1383 (1980)
7. Y. Tamaki, S. Isomae, S. Mizuo and H. Higuchi, J. Electrochem. Soc. 128 (3), 644 (1981)
8. Y. Tamaki, S. Isomae, S. Mizuo and H. Higuchi, J. Electrochem. Soc. 130 (11), 2267 (1983)
9. A. Bohg and A.K. Gaind, Appl. Phys. Lett. 33 (10), 895 (1978)
10. K. Sagara, Y. Tamaki, and M. Kawamura, J. Electrochem. Soc. 134 (2), 500 (1987)
11. T.T. Sheng and R.B. Marcus, J. Electrochem. Soc.127 (3), 737 (1980)
12. S.C. Arney and N.C. MacDonald, J. Vac. Sci. and Tech. B6 (1), 341 (1988)

# STRESSES IN FILMS DEPOSITED BY PLANAR MAGNETRON SPUTTERING ON GLASS SUBSTRATES OF DIFFERENT THICKNESS

Zhi-Feng ZHOU AND Yu-Dian FAN
Department of Materials Science and Engineering, Tsinghua University, Beijing, CHINA

## ABSTRACT

Stresses affect the structure and properties of the films significantly. In this paper, Co-Cr films were deposited by D.C. planar magnetron sputtering on rectangular glass plates of thickness varied from 0.18 to 2.00mm. The measurement of stress was performed by the bending plate technique with the detection of capacitance change. The temperature of the bending plate was measured by an Fe-Ni thin film thermocouple on the back side of the substrate. The experimental results showed that the stresses were all tensile, but in the films on the thicker substrates were smaller than that on the thinner ones. And apparently the temperature rise of the thicker substrates during deposition was lower than that of the thinner ones. The magnetic characteristics of the Co-Cr films related to the substrate thickness may be attributed to both stress and temperature difference in the films deposited on the substrates of different thickness.

## INTRODUCTION

In recent years, as an applicable perpendicular magnetic recording medium, Co-Cr films are being studied widely [1], but few attention has been paid to their mechanical properties such as internal stress. Desserre *et al.*[2] for the first time found an interesting phenomenon that the perpendicular coercivity and the C-axis orientation of Co-Cr films had some relation with the thickness of substrates on which the films were deposited, but the reason was not made certain. In this paper, the dependence of the structure and the magnetic properties of Co-Cr films on the thickness of substrate was also found. For the purpose of investigating this phenomenon,the surface temperature and the stress of the films were measured respectively.

## EXPERIMENTAL PROCEDURE

Among many techniques for the measurement of stress in thin films, the cantilevered beam technique is widely used, especially in the case for evaporated films [3]. In general, the small deflection of the free end of the cantilevered beam is detected optically or capacitively, and the latter is more sensitive. Unfortunately, the capacitive detection can not be applied as an *in situ* stress measurement method to sputtered films because of the serious interruption of glow discharge, but the stress can be measured after sputtering.

By the bending beam technique, the stress in a film, S, can be calculated by the following well known formula[4] :

$$S = E_s H^2 \delta / 3(1-\nu) L^2 f \qquad (1)$$

where $\delta$ is the deflection of the free end of the cantilevered beam (or plate), f is the film thickness, L and H are the length and the thickness of the substrate respectively, and Es and $\nu$ are the Young's modulus and the Poisson's ratio of the substrate material respectively ( here $\nu$ is taken to be 0.22 ). The deflection of the free end, $\delta$, can be detected by the change of capacity C between the metallized side of the substrate and a fixed metal plate. The relation between C and $\delta$ has been given by Wilcock *et al.*[5].

The stress measuring apparatus is schematically illustrated in Fig.1. The glass substrate used as a cantilevered plate had been annealed up to 300℃ for about one hour in order to increase its structure stability. The backside of the substrate was metallized by depositing a Ti film of 1000Å thick as the difference in thermal expansion coefficient between the glass and Ti is small. Then one end of the substrate was clamped rigidly to the jig and a kind of epoxyn was also used in cementing to avoid the possible small sliding of the substrate [6], so a stable parallel-plate capacitor was formed. The small changes in electric capacity were measured by a CY2611A type capacitance measuring apparatus in which the five terminal method was used. The nominal accuracy of this apparatus is 0.1% and the resolution is 0.01pF. All the connecting wires were coaxial. The metal vacuum chamber was earthed to ensure an electrostatic screen. Under the above arrangement, the initial capacitance was nearly unchanged before sputtering for a long time.

To calibrate the stress measuring apparatus, a series of known weights were hung to the free end of the cantilevered plate and the corresponding deflections were measured. With the aid of the least square curve fitting, the Young's modulus of the glass substrate was calculated on average to be $5.7\times10^{10}N/m^2$, which was similar to the value given by Wilcock and Hoffman [5,6]. The reproducibility of this measurement was within ± 5%. When the load was put away from the free end, the substrate recovered immediately to its original position, this meant that the curvature of the substrate was elastic and no sliding occurred under the action of the applied load. Many experiments showed that the stress measuring reproducibility was within ± 10%.

Co-Cr films were deposited by D.C. planar magnetron sputtering from a alloy target (19.3 wt% Cr) onto rectangular glass plates of different thickness varied from 0.18 to 2.00mm. The size of the target was 240mm×110mm with a cathode power of 2.0kW, and the target to substrate distance was kept to be 75mm. The background pressure and Ar working pressure were $4.0\times10^{-4}$Pa and 0.4Pa respectively. The deposition rate was 0.34μm/min.

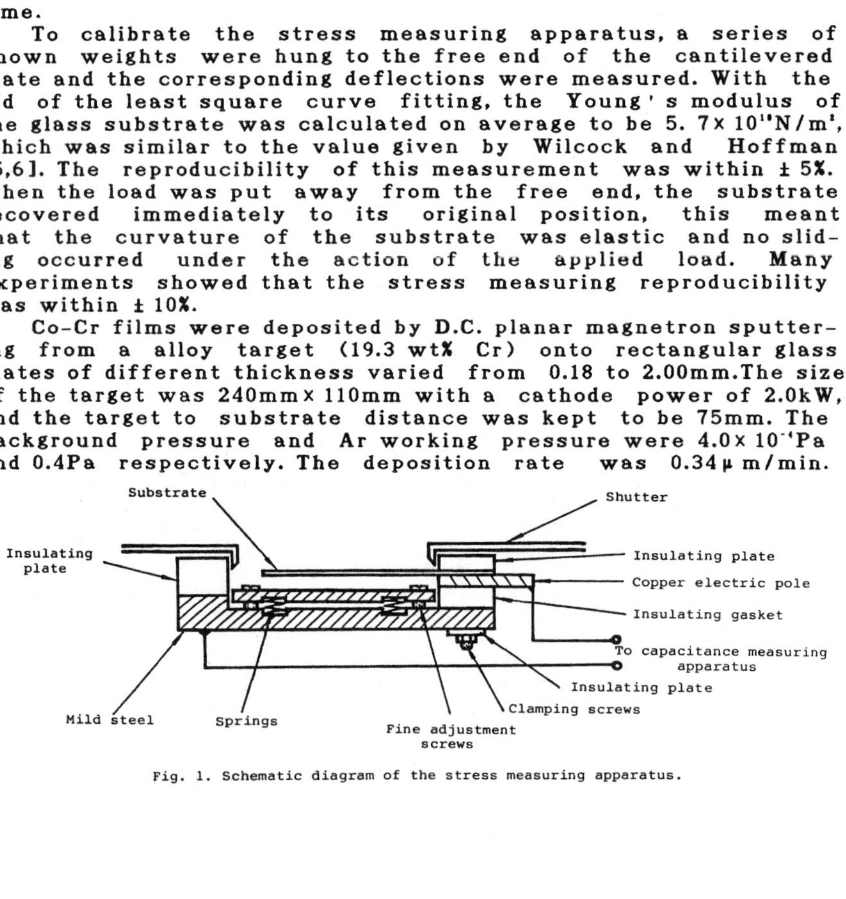


Fig. 1. Schematic diagram of the stress measuring apparatus.

The final film thickness was controlled to be 0.50μm. Measuring the film thickness by optical interferometry indicated that the thickness uniformity was within ± 2%.

Before sputtering, the substrate was heated for outgasing and cleaned by plasma bombardment successively. The impurities existing on the surface of the substrate may dramatically lower the film stress. In order to further avoid the influence of contamination and to modify the magnetic properties of the Co-Cr film a Ti film of 0.1μm thickness was pre-deposited[7]. The substrate temperature was measured by an Fe-Ni thin film thermocouple on the back side of the substrate in a separated experiment [ 8]. Since the temperature change caused by deposition could be simulated approximately by heating the substrate with an iodized tungsten lamp placed in the vacuum chamber, the thermal process simulation was performed after the deposition. It was shown that the capacitance change caused by the thermal process was reversible within the highest deposition temperature range. With this method, the total stress could be corrected for the thermal effect approximately.

In addition, the C-axis dispersion of the Co-Cr films with HCP structure was evaluated by the X-ray rocking curve technique, and the magnetic properties were measured by a vibrating sample magnetometer(VSM).

## RESULTS

The temperature changes after sputtering for the substrates of different thickness are shown in Fig.2. It is clear that the temperature at the end of deposition is higher for the thinner substrate with a smaller thermal capacity,but the cooling rate is faster. As the deposition time was constant, the thinner the substrate, the higher the average substrate temperature Ts during sputtering.

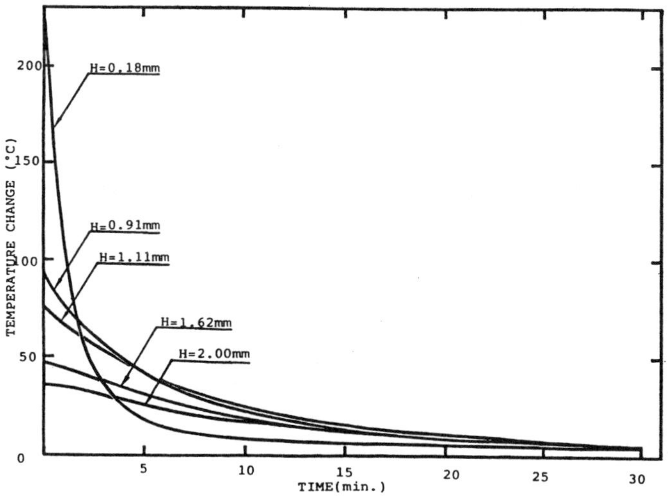


Fig. 2. Temperature change after sputtering for substrates of different thickness.

Fig. 3 demonstrates the dependence of the C-axis dispersion $\Delta\theta_{50}$, the saturation magnetization Ms, the perpendicular coercivity Hc(⊥) and the in-plane coercivity Hc(//) on the thickness of substrate. Since the substrates were not heated externally during sputtering, the values of Hc(⊥) were relatively low [9]. All these experimental curves take a parabolic shape with the minimum values at about H=1mm.

After deposition, the capacitance change was recorded for 30 min., and the corresponding stress in this cooling period was calculated. The thermal cycling experiment showed that the thermal stress only took a small part of the total stress. After the correction for thermal stress, the intrinsic stress was nearly constant throughout the recording time (30 min.). This means that no relaxation of intrinsic stress took place during cooling. Fig.4 shows the relation between the equilibrium intrinsic stresses and the thickness of subatrate. The stresses were all tensile and decreased monotonically with the thickness of substrate. When air was admitted into the vacuum chamber, the film stress was nearly unchanged.

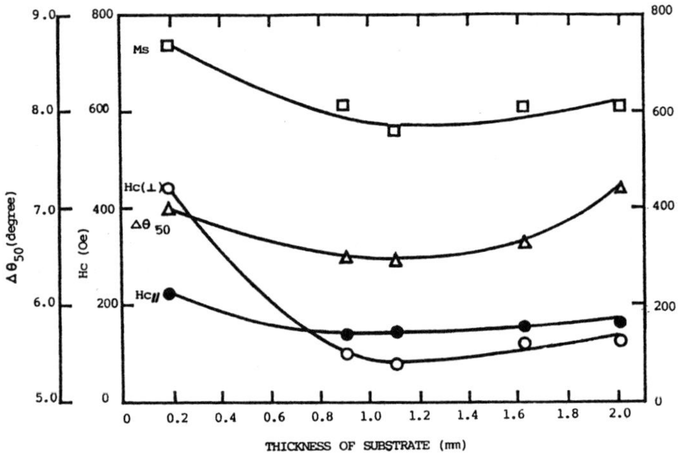


Fig. 3. Dependence of Ms,$\Delta\theta_{50}$, Hc(⊥) and Hc(//) on the thickness of substrate.

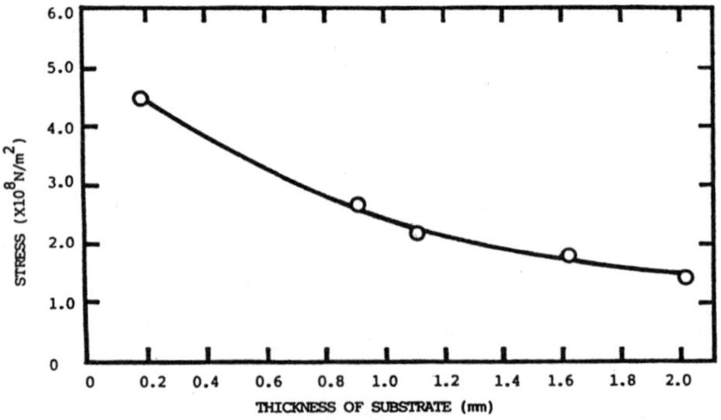


Fig. 4. Dependence of the equilibrium intrinsic stress on the thickness of substrate.

## DISCUSSION

The origin of stress can be traced to the structure of films and then the corresponding deposition conditions [3]. It was pointed out previously by Hoffman *et al.*[10] that in the case of sputtering the stress transition phenomenon existed and the model called "atomic peening" of deposited films by fast neutral gas atoms was proposed. In general, the tensile stress will transit to the compressive stress with the decreasing of working gas pressure. When the atomic mass of the target materials is small, such as in the case of pure Cr or Co target, the transition pressure becomes very low, sometimes even below the minimum working pressure of discharge. So the tensile stress existed in Co-Cr films seems to be reasonable, since the 'atomic peening' effect is not predominant.

While the film is deposited on a thinner substrate, the condensation heat delivered will make the substrate to a higher temperature, thus the mobility of atoms increases and more defects (e.g., vacancies and grain boundaries) will be eliminated to form a more denser structure. In this case, a larger tensile stress is produced.

It was observed that after annealing the deposited Co-Cr films at a temperature of 200℃ (nearly the highest substrate temperature on deposition), the tensile stress in the film on the thickest substrate increased, but in the film deposited on the thinnest substrate unchanged . This result confirms the above theoretical prediction,and is similar to the annealing effect of evaporated films deposited at very low temperature [11].

In addition, it may be considered that the substrate bending and the corresponding stress relaxation will take place for the films on the thinner substrates [12]. But the ratio of the film thickness, f, to the substrate thickness, H, is quite small in our experiment ($f=0.5\mu m$, $H=0.18-2.00mm$). A calculation [13] indicated that even for the thinnest substrate, the relief of stress in the film was only 4%(here the Young's modulus of the film Ef was taken to be $21\times10^{10}$ $N/m^2$[14]). So all the substrates used can be thought to be rigid, and this stress relief mechanism should not be considered.

The dependence of the magnetic properties on the thickness of substrate is not clear, perhaps it may be concerned with the coaction of temperature and stress effects. The segregation of Cr atoms at the grain boundaries seems to be instructive to this problem [15].A further study is needed.

## CONCLUSIONS

The cantilevered plate technique with capacitance change can be applied to the measurement of stress in films deposited by sputtering. The structure and magnetic properties of Co-Cr films have some relation with the thickness of substrate. The temperature rise of the thinner substrates is higher than that of the thicker ones. The tensile stresses in the films deposited on the thinner substrates are larger than that on the thicker ones. The difference of the tensile stress in the films deposited on substrates of various thickness can be explained by the defect elimination model. The effect of substrate thickness on the structure and the magnetic properties of the Co-Cr films may be attributed to the coaction of both temperature and stress.

ACKNOWLEDGEMENTS

This work was funded by the Natural Science Foundation of China (NSFC Contract No. 9587007-03). The VSM measurement was performed by the Physics Department, Beijing University. The authors would like to thank Ms. Y.J. You and Ms. P.F. Sun for their assistance in the experiments.

REFERENCES

1. K. Ouchi and S. Iwasaki, IEEE Trans. Magn. MAG-23,2443 (1987).

2. J. Desserre, J. C. Bouchand, and D. Jeanniot, J. Magn.and Magn. Mat. 54-57, 1683 (1986).

3. D. S. Campbell, in Handbook of Thin Film Technology, edited by L. I. Maissel and R. Glang (McGraw-Hill, New York, 1970), p. 12-21.

4. P. R. Stuart, Vacuum 19, 507 (1969).

5. J. D. Wilcock and D. S. Campbell, Thin Solid Films 3, 3 (1969).

6. R. E. Rottmayer and R. W. Hoffman, J. Vac. Sci. Tech. 7, 461 (1970).

7. O. Kitakami, K. Ojima, Y. Ogawa, T. Maro, and H. Fujiwara, IEEE Trans. Magn. MAG-23, 2797 (1987).

8. M. Laugier, Thin Solid Films 67, 163 (1980).

9. R.Sugita and F. Kobayashi, IEEE Trans. Magn. MAG-18, 1818 (1982).

10. D. W. Hoffman and J. A. Thornton, J. Vac. Sci. Tech. 20, 355 (1982).

11. R. W. Hoffman, R. D. Daniels, and E. C. Crittenden, Jr.,Proc. Phys. Soc. london B67, 497 (1954).

12. A.Brenner and S. Senderoff, J. Res. Nat. Bur. Stand. 42 , 105 (1949).

13. R. W. Hoffman, in Physics of Nonmetallic Thin Films, edited by C. H. S. Dupuy and A. Cachard (NATO Adv. Study Inst. Ser., 1976), B14, p. 273.

14. Y. Niimura and M. Naoe, IEEE Trans. Magn. MAG-21, 1447 (1985).

15. H.Masuya and H. Awano, IEEE Trans. Magn. MAG-23 ,206 (1987).

# THE BEHAVIOR OF POLYSILICON THIN FILM STRESS AND STRUCTURE UNDER RAPID THERMAL PROCESSING CONDITIONS

WALTER HUBER*, G. BORIONETTI AND C. VILLANI
DNS Electronic Materials, Viale Gherzi, 31, 28100 Novara, Italy

## ABSTRACT

Undoped polysilicon layers deposited at 620 °C onto a silicon wafer induce a compressive stress of approximately 1 x $10^9$ dynes/cm$^2$ resulting in a bow of the wafer. This stress can be relieved by rapid annealing at temperatures above 1000 °C. A comparison with conventional furnace annealing reveals that the stress relaxation is a weak function of time and strongly depends on temperature. Rapid thermal processing also causes immediate rearrangement of the film structure, as observed by X-ray diffraction. Although both stress and film texture change with annealing, no conclusive relationship is observed.

## INTRODUCTION

Polysilicon layers have a wide range of applications in the IC manufacturing technology. For example, polysilicon layers are used for gettering deleterious impurities, as well as a MOS gate contact material. For the purpose of gettering a continuous layer of 1-2 micron in thickness is deposited onto the backside of a silicon wafer [1]. The deposition temperature usually is below 700 °C which results in a small grained film with a high degree of lattice disorder. Such a polysilicon layer creates a strain to the wafer which gives rise to a wafer bow. On one hand it is well established that the degree of induced strain is strongly influenced by the deposition conditions [2]. On the other hand, any subsequent high temperature anneal changes significantly the properties of the polysilicon [3]. After an extended high temperature treatment, a significant stress relief and modification of the film structure is measured.

It is the purpose of this work to demonstrate that the structure and stress of the polysilicon layers undergo instantaneous changes at high temperatures. The results establish the temperature as the most significant anneal parameter, while time and ambient play only a minor role.

## EXPERIMENTAL PROCEDURES

Undoped polysilicon layers were deposited in a commercially available LPCVD reactor onto a 100 mm P type silicon wafer having an orientation slightly off the <111> direction. During the deposition a flat temperature profile of 620 °C and a pressure of 175 milli Torr was maintained. The polysilicon film was deposited onto a chemically etched surface with a deposition rate of approximately 100 Angstroms/min. The thickness was determined by interferometry of polysilicon deposited onto oxidized monitor wafers that were processed simultaneously. The measured thickness are in the order of 1.5 microns.

After the polysilicon deposition both surfaces of the wafer are covered with a continuous polysilicon layer. One group of wafers were rapidly annealed right after the polysilicon deposition . Another set of samples was prepared by first polishing one side of the wafer to a mirror-like finish. In this case, only the backside of the wafer was covered with polysilicon during the high temperature anneals.

The wafers were either heat-treated in a commercially available rapid thermal processor [4] or in a conventional tube furnace. The temperature range between 1000 $^oC$ and 1200 $^oC$ was investigated and the anneal time was varied between 10 seconds and 12 hours. The anneals were either carried out in an inert or oxidizing ambient. Any oxide which has grown during annealing was removed with an HF mixture before characterization.

The strain of the polysilicon layer to the silicon wafer was determined by measuring the absolute value and the direction of the wafer curvature with a capacitance gauge. X-ray diffraction was used to examine the texture and crystallite size of as-grown and annealed films. For the determination of the film texture the <111>, <220>, <311> and <331> were recorded. The grain size of the polysilicon film was inferred from the the half widths of the <111>, <220> and <331> peaks applying the Scherrer formula which takes into account the line splitting of the X-ray spectrum and instrument broadening [5]. It was assumed that the half width of the peak is only a function of the crystallite size, thus neglecting any strain effects. The polysilicon/silicon interface was also characterized by delineating defects with a Schimmel etch solution.

## RESULTS AND DISCUSSION

An undoped polysilicon film deposited onto a silicon wafer always exerts a compressive stress to the wafer resulting in an increase of the wafer bow of 27 microns. For a bow of this magnitude a stress constant of $1.1 \times 10^9$ dynes/cm2 is calculated and is in good agreement with published data [6].

The change in wafer curvature that were rapidly annealed after polysilicon deposition are summarized in Table 1.

**TABLE 1**

WAFER BOW UNDER VARIOUS RAPID THERMAL ANNEALING CONDITIONS AFTER POLYSILICON DEPOSITION

| RAPID THERMAL PROCESSING CONDITIONS | INCREASE IN WAFER BOW (Microns) (MEASUREMENT (B) - MEASUREMENT (A)) |
|---|---|
| CONTROL | 27.5 +/- 0.5 |
| 1000 °C, 120 SEC | 14.3 +/- 0.9 |
| 1100 °C, 30 SEC | 4.0 +/- 0.5 |
| 1200 °C, 30 SEC | 0.4 +/- 0.5 |

The listed values are the difference between the wafer bow measured before the polysilicon deposition (Measurement (A)) and after polishing and rapid thermal annealing (Measurement (B)). It can be seen that after a 120 seconds anneal at 1000 °C the increase in curvature is only half the value measured in the control group. Only a marginal bow increase is measured in wafers that were treated at 1100 °C for 30 seconds. Within the experimental error, a complete relaxation is observed after 30 seconds at 1200 °C.

The effects of anneal temperature, time and ambient were closely investigated on wafers that were annealed after polishing. The decrease in wafer bow under various anneal conditions is shown in Table 2.

TABLE 2 - WAFER BOW FOR VARIOUS ANNEAL CONDITIONS AFTER POLISHING

| TEMPERATURE, °C | ANNEAL TIME, sec RAPID | | ANNEAL TIME, sec CONVENTIONAL | AMBIENT | DECREASE IN WAFER BOW microns (Meas. (B) - Meas. (C)) |
|---|---|---|---|---|---|
| 1000 | 30 | | | Inert | 9.7 +/- 0.6 |
| 1000 | 120 | | | Inert | 10.9 +/- 0.3 |
| 1000 | | | 3600 | Inert | 6.4 +/- 0.7 |
| 1100 | 30 | | | Inert | 17.9 +/- 0.9 |
| 1100 | | | 3600 | Inert | 16.4 +/- 0.9 |
| 1100 | | | 3600 | Oxygen | 16.6 +/- 1.3 |
| 1100 | | | 43200 | Inert | 21.3 +/- 2.7 |
| 1200 | 10 | | | Inert | 27.5 +/- 4.5 |
| 1200 | 30 | | | Inert | 23.8 +/- 2.5 |
| 1200 | 60 | | | Inert | 25.5 +/- 2.6 |
| 1000 | 30 | and | 3600 (1100°C) | Inert | 19.0 +/- 2.0 |
| 1200 | 30 | and | 3600 (1000°C) | Inert | 20.4 +/- 3.9 |

The difference in bow was obtained by comparing the bow after polishing (Measurement (B)) and after annealing (Measurement (C)). The data in Table 2 clearly establish the anneal temperature as the most important parameter for reducing the strain induced by the polysilicon film, while time and ambient play only a minor role. For instance, a rapid anneal at 1200 °C for 10 seconds is more effective than a conventional 1 hour heat treatment at 1100 °C.

Double anneals, where the rapid thermal treatment is followed by a conventional cycle at a different temperature, indicate that the highest temperature determine the final bow of the silicon wafer. A 1100 °C treatment further reduces the wafer bow when it's carried out after a rapid anneal at 1000 °C. On the other hand, a conventional treatment does not significantly degrade the effects of a 30 seconds anneal at 1200 °C.

From the X-ray diffraction results a preferred <110> orientation is determined, as indicated at the bottom of Table 3. The as-grown intensities are normalized for the intensities expected from a silicon powder and corrected for film thickness. To show the effects of high temperature treatments on the X-ray peak intensities, each peak is reported relative to the as-grown texture.

TABLE 3 - INTENSITY OF X-RAY DIFFRACTION PEAKS FOR VARIOUS ANNEALS IN AN INERT AMBIENT

| ANNEAL CONDITIONS | PEAK INTENSITY RELATIVE TO AS-GROWN TEXTURE, ARBITRARY UNITS | | | |
|---|---|---|---|---|
| | <111> | <220> | <311> | <331> |
| 1000 °C, 30 sec | 1.5 | 1.0 | 1.3 | 1.3 |
| 1000 °C, 120 sec | 1.5 | 0.9 | 1.3 | 0.7 |
| 1000 °C, 3600 sec | 2.0 | 1.0 | 0.8 | 1.9 |
| 1100 °C, 30 sec | 2.0 | 0.8 | 2.4 | 1.1 |
| 1100 °C, 3600 sec | 10.0 | 0.5 | 1.7 | 3.5 |
| 1200 °C, 10 sec | 4.0 | 0.9 | 2.1 | 2.7 |
| 1200 °C, 30 sec | 3.0 | 0.7 | 1.2 | 1.5 |
| 1200 °C, 60 sec | 6.0 | 0.8 | 2.4 | 3.7 |
| as grown, 620 °C [a)] | 0.1 | 11.7 | 0.25 | 1.7 |

a) NORMALIZED FOR THE PEAK INTENSITY OF SILICON POWDER AND FILM THICKNESS.

The crystallite size as determined from the half-width of the X-ray peaks is summarized in table 4

TABLE 4 - GRAIN SIZE UNDER VARIOUS ANNEAL CONDITIONS IN AN INERT AMBIENT

| ANNEAL CONDITIONS | GRAIN SIZE, Angstroms | | | DECREASE IN WAFER BOW microns (DATA FROM TABLE 2) |
|---|---|---|---|---|
| | <111> | <220> | <331> | |
| as grown, 620 °C | 190 | 1700 | 180 | 27.0 +/- 5.0 |
| 1000 °C, 30 sec | 170 | 1500 | 250 | 9.7 +/- 0.6 |
| 1000 °C, 120 sec | 190 | 2000 | 185 | 10.9 +/- 0.3 |
| 1000 °C, 3600 sec | 350 | 2200 | 400 | 6.4 +/- 0.7 |
| 1100 °C, 30 sec | 320 | 2500 | 330 | 17.9 +/- 0.9 |
| 1100 °C, 3600 sec | > 3500 | 3100 | > 3500 | 16.4 +/- 0.9 |
| 1200 °C, 10 sec | 900 | 2500 | 900 | 27.5 +/- 4.5 |
| 1200 °C, 30 sec | 2000 | 2800 | 730 | 23.8 +/- 2.5 |
| 1200 °C, 60 sec | 1700 | > 5000 | 1800 | 25.5 +/- 2.4 |

The texture and grain size are typical for polysilicon films deposited under these conditions [7]. The structure of the polysilicon layer are almost preserved during heating the samples upon 1000 °C. At higher temperatures, the dominant <110> texture starts to diminish and is accompanied by rapid grain growth in the <111> and <331> directions. The results of Table 4 indicate that grain growth occurs instantaneously at 1200 °C.

A comparison between the wafer bow and polysilicon structure after rapid thermal annealing does not reveal a relationship. The absence of a clear correlation between film structure and wafer curvature suggests that the overall morphology does not necessarily determine the strain to the wafer. It can be inferred that the wafer bow is decreased by epitaxial realignment of the polysilicon layer upon heating. This is a feasible mechanism because rapid realignment between the polysilicon layer and a crystalline silicon surface has been reported [8]. Defect etching of an angle lapped wafer also did not delineate any defects extending into the silicon wafer, consistent with the conjecture of epitaxial realignment.

## CONCLUSIONS

Polysilicon layers deposited at 620 °C exert a compressive stress to the silicon wafer that can effectively relieved by rapid thermal processing at temperatures above 1000 °C. A comparison with conventional furnace treatments establish that the relaxation strongly depends only on temperature. The other anneal parameters, time and ambient, do not have any significant influence.

During high temperature processing, an instantaneous growth of the crystalline size takes place when the temperature reaches 1200 °C. The predominant <110> texture diminishes with increasing temperature and time. Although both the structure and compressive stress of the polysilicon film change with rapid thermal processing, their clear correlation could not be established.

## REFERENCES


1. M.C. Arst and J.G. De Groot, J. Electronic Materials 13, 763 (1984).
2. G. Borionetti, M. Domenici and G. Ferrero, presented at the 5th International Symposium on Semiconductor Processing, Santa Clara, CA, 1988 (to be published).
3. J. Kamins, J. Electrochem. Soc. 125, 927 (1978).
4. Heatpulse 210, Rapid Thermal Processor, AG Associates , Sunnyvale, CA.
5. B.D. Cullity, Elements of X-Ray Diffraction, (Addison-Wesley Reading, MA, 1967), p. 99.
6. E.W. Hearn, D.J. Werner and D.A. Doney. J. Electrochem. Soc. 133, 1749 (1986).
7. C. Harbike, L. Krausbauer, E.F. Steigmeier, E.A. Widner, H.F. Kappert and G. Neugebauer, J. Electrochem. Soc. 131, 675 (1984).
8. H.B. Harrison, S.T. Johnson, Y. Komen, C. Wang and S. Cohen in Materials Issues in Silicon Integrated Circuit Processing, edited by M. Wittner, J. Stimmell and M. Strathmann (Mater. Res. Soc. Proc. 71 Pittsburgh, PA, 1986) p. 455.

PART V

# Abstracts of Unpublished Papers

# INTERFACIAL FRACTURE MECHANICS

JOHN W. HUTCHINSON
Division of Applied Sciences, Harvard University, 29 Oxford Street, Cambridge, MA 02138

## EXTENDED ABSTRACT

The mechanics of interface cracks has recently been clarified (e.g., [1]) and a number of efforts are underway to develop a mechanics of interface fracture with applications to composite materials, thin film/substrate systems, and coatings. An intrinsic feature of interfacial fracture is its mixed mode character wherein both shear and normal stresses act on the interface directly ahead of the crack. Depending on geometry and loading, the mixture of modes can range from purely normal stresses (mode I) to purely shear stresses (mode II). Toughness of an interface is characterized by critical combinations of mode I and mode II stress intensity factors (i.e., a locus of critical combinations) rather than just the single critical mode I stress intensity factor in the fracture of homogeneous materials. Equivalently, the critical energy release rate depends on the mode combination for interfacial fracture. Solutions are now available to the following problems: an interface crack in a layered structure where a very thin layer is sandwiched between two thick layers of different material [2], an interface crack between two layers of arbitrary thickness subject to arbitrary combinations of bending and stretching loads [3], and a crack in a substrate paralleling an interface between a thin film and the substrate driven by a variety of loadings including residual tension in the film [4]. These solutions can be used to analyze specimens for determining interfacial toughness and for predicting cracking in thin film or layered structures.

## References

[1] J. R. Rice, J. Appl. Mech. 55, 98 (1988).
[2] Z. Suo and J. W. Hutchinson, "On sandwich test specimens for measuring interface crack toughness", to be published in J. Materials Engng.
[3] Z. Suo and J. W. Hutchinson, "Interface crack between two elastic layers", to be published in Int. J. Fracture.
[4] Z. Suo and J. W. Hutchinson, "Steady-state cracking in brittle substrates beneath adherent films" submitted to Int. J. Solids Struct.

# Author Index

# Subject Index